Sensory Systems

ANATOMY AND PHYSIOLOGY

Sensory Systems

ANATOMY AND PHYSIOLOGY

Aage R. Møller

School of Human Development
Callier Center for Communication Disorders
The University of Texas at Dallas
Dallas, Texas

Amsterdam Boston London New York Oxford Paris
San Diego San Francisco Singapore Sydney Tokyo

This book is printed on acid-free paper. ♾

Academic Press
An imprint of Elsevier Science.
525 B Street, Suite 1900, San Diego, California 92101-4495, USA
http://www.academicpress.com

Academic Press
84 Theobald's Road, London WC1X 8RR, UK
http://www.academicpress.com

Library of Congress Catalog Card Number: 2002107709

International Standard Book Number: 0-12-504257-4

PRINTED IN THE UNITED STATES OF AMERICA
02 03 04 05 06 07 MM 9 8 7 6 5 4 3 2 1

CONTENTS

General Introduction xi

Acknowledgments xv

Preface xvii

1 Basic Psychophysics

Abbreviations 1
Abstract 2
I. Introduction 2
II. Threshold of Detection 3
A. Factors That Affect Detection of Faint Stimuli 4
III. Discrimination Between Different Stimuli 18
A. Perception of Stimulus Strength 19
B. Discrimination of Small Differences Between Stimuli 19
C. Temporal Resolution 20
D. Spatial Resolution 24
References 30

2 Anatomy and Physiology of Sensory Organs

Abbreviations 33
Abstract 34

I. Introduction 35
II. Anatomy of Sensory Organs 37
A. The Ear 37
B. The Eye 40
C. Skin 42
D. The Chemical Senses 42
III. Sensory Receptors 43
A. Anatomy of Sensory Receptors 44
B. Innervation of Sensory Cells 51
IV. Conduction of the Physical Stimulus to the Receptors 56
A. Improvement of Transmission to the Receptors 56
B. Modification of the Physical Stimulus 57
V. Physiology of Sensory Receptors 63
A. General Principles for Sensory Transduction 64
B. Information Processing at the Receptor Level 68
C. Amplitude Compression 69
D. Selectivity of Receptors 70
E. Receptive Fields 71
References 72

3 Sensory Nervous Systems
Abbreviations 75
Abstract 76
Anatomy 76
Information Processing in Ascending Sensory Pathways 78
I. Introduction 80
II. Anatomy of Sensory Nervous Systems 81
A. Ascending Sensory Pathways 81
B. Descending Systems 97
C. The Thalamus is the Gateway to the Cerebral Cortex 99
D. Anatomy of Sensory Cerebral Cortices 102
E. Corpus Callosum 107
III. Information Processing in the Sensory Nervous System 108
A. Information Processing in the Classical Ascending System 109
B. Processing of Object and Spatial Properties of Sensory Stimuli 116
C. Processing of Spatial Information 127
IV. Neural Control of Sensory Processing in Ascending Sensory Pathways 140
A. Arousal 141
B. Attention 142

C. Control of Ascending Neural Activity by Descending Pathways 143
V. Processing of Information in the Nonclassical Pathways 145
A. Developmental Aspects of Classical and Nonclassical Pathways 146
B. Processing of Information in Descending Nonclassical Systems 147
VI. How is the Neural Code of Sensory Information Interpreted? 148
A. Coding in Single Nerve Cells 148
B. Coherent Firing of Groups of Neurons 149
C. Maps 149
D. Where is the Anatomical Location for Interpretation of Sensory Information? 152
VII. Sensory Information Can Reach Nonsensory Regions of the CNS 153
A. Motor Systems 153
B. Autonomic Reactions to Sensory Stimulation 154
C. Emotional Reactions to Sensory Stimuli 155
VIII. Processing of Information in the Sensory Nervous System is Dynamic 164
A. Neural Plasticity 165
B. Plastic Changes in the Nervous System May Cause Symptoms and Signs of Disease 170
References 172

4 Somatosensory System

Abbreviations 185
Abstract 186
Classical Somatosensory System 186
Nonclassical Somatosensory System 188
Pain 189
I. Introduction 189
II. Anatomy of the Somatosensory System 190
A. Classical Somatosensory System 190
B. Nonclassical System 205
C. Descending Somatosensory Pathways 217
III. Physiology of the Somatosensory System 220
A. The Classical Somatosensory System 221
B. Physiology of the Nonclassical System 238
C. Function of Descending Systems 241
D. Pain 241
E. Itch 262
References 263

5 Hearing

Abbreviations 271
Abstract 272
Conduction of Sounds to the Receptors 272
Receptors 273
Classical Auditory Pathways 274
Physiology of the Classical Auditory System 274
Nonclassical Pathways 276
Physiology of the Nonclassical Pathways 276
I. Introduction 276
II. The Ear 277
A. Anatomy 278
B. Physiology of the Ear 284
III. The Auditory Nervous System 305
A. Anatomy of the Classical Ascending Auditory Pathways 305
B. Physiology of the Classical Auditory Nervous System 316
C. Descending Systems 346
IV. The Nonclassical Ascending Pathways 347
A. Anatomy of the Nonclassical Ascending Pathway 347
B. Physiology of the Nonclassical System 353
V. Neural Plasticity 357
A. Neural Plasticity Can Alter the Threshold and Perception of Sounds 358
B. Neural Plasticity from Overstimulation 359
References 359

6 Vision

Abbreviations 373
Abstract 374
The Eye 374
Visual Nervous System 374
I. Introduction 376
II. The Eye 376
A. Anatomy 377
B. Physiology of the Eye 380
III. The Visual Nervous System 387
A. Anatomy of the Classical Ascending Visual Nervous System 388
B. Physiology of the Visual Nervous System 400
IV. Nonclassical Visual Pathways 416
A. Anatomy of Nonclassical Visual Pathways 416
B. Physiology of the Nonclassical System 419
References 421

7 Chemical Senses: Olfaction and Gustation

Abbreviations 425
Abstract 425
Taste 426
Olfaction 426
I. Introduction 427
II. Anatomy 428
A. Receptors 429
B. The Media that Conduct the Stimulus to the Receptors 431
C. Innervation of Receptors 432
D. Gustatory and Olfactory Neural Pathways 435
III. Physiology of the Chemical Senses 443
A. Receptors 443
B. Coding of Information in the Gustatory and Olfactory Nervous Systems 448
References 449

Index 451

GENERAL INTRODUCTION

This book is written for students of physiology, anatomy, and behavioral sciences, and for all individuals who want to understand some of the most fascinating wonders in biology. René Descartes, (1596–1650) summarizes his line of reasoning in the famous phrase, 'I think, therefore I am.' This is true, but without sensory systems there would not be an intellect. While many people are fascinated by our technological achievements in such areas as computers and communication systems, the efficiency and complexity of sensory systems far exceeds even the most sophisticated man-made systems. Sensory systems not only interpret physical stimuli such as those carried by sound, light, and odors but they also provide input to our emotional brain either consciously or unconsciously. Understanding the function of sensory systems is important for many professions. This book discusses many aspects of sensory systems and their interaction with many other parts of the brain. *Sensory Systems: Anatomy and Physiology* provides a "joy of understanding" of some truly fascinating biological systems and can be appreciated by all individuals with an interest in living things.

Sensory Systems: Anatomy and Physiology not only presents facts regarding the anatomy and function of sensory systems but it also provides interpretation and synthesis of our present understanding of the organization and function of these complex systems. The book covers areas that have not been extensively represented in other books such as the function of nonclassical sensory pathways and the input to the emotional brain. The book discusses *parallel processing* (processing of the same information in different populations of nerve cells) and *stream segregation* (processing of different kinds of information in different populations of nerve cells).

Scientists try to relate neural activity in the central nervous system (CNS) to sensory stimulation in order to understand how sensory stimuli reach our consciousness and how they become interpreted. This problem has been approached in two different ways: the first approach attempts to understand how individual neurons work and the second approach attempts to understand how different parts of the CNS are connected into systems and how such systems process information (systems approach). A *modern* analogy to these two approaches would be similar to studying how transistors work and how computers work. This book concerns a systems approach to understand how sensory systems work.

Sensory systems are also important for our general well-being. Sensory systems control basic bodily functions such as what we eat and how much. Whether we are going to be overweight or anorexic depends to some extent on our sensory systems. Under normal circumstances and during diseases, sensory systems are important for our mood. The connections to the limbic system are important in that respect. Sensory systems also provide input to other regions of the brain such as the motor systems, the cerebellum, and to the core of the brainstem (reticular formation) that control the degree of wakefulness.

This book covers the anatomy and function of the five senses: hearing, vision, somesthesia, taste, and olfaction. Pain is also included as a part of somesthesia. The book emphasizes the similarities between the function of receptors and between the ways that these different senses process very different physical stimuli. Specifically, the book discusses the information processing that occurs in the sensory organs and the nervous system.

Proprioception is not included in this book because activation of proprioception does not reach consciousness. The feeling of mental fatigue or the sensations of hunger and thirst are not included either because the feeling of fatigue, hunger, and thirst are caused by some internal processes and although they reach consciousness they are very different from the sensory systems that communicate information from the environment. Why include pain in this book? Although pain is not always caused by external events, the close association with the somatosensory system and the fact that it often communicates information from the environment to the conscious mind justifies inclusion.

Sensory Systems: Anatomy and Physiology emphasizes the similarities between the different sensory systems and their function. Therefore the different components of sensory systems are first discussed together in order to emphasize the similarities and differences rather than the more conventional way of treating the different sensory systems separately.

Why do we want to know about sensory systems and why write a book devoted to them? First of all, sensory systems are some of the most intriguing systems of our body. Exploring sensory systems is now more fascinating than ever before because technological advances have provided excellent tools for studying the function of sense organs and in particular to study the function of the nervous system.

Sensory Systems: Anatomy and Physiology is directed to physiologists who study sensory systems, to health professionals who are involved in diagnosis of disorders of sensory systems, and to any interested person who wants a broad understanding of how sensory systems function.

Chapter 1 provides a brief introduction to psychophysics. Chapter 2 is devoted to the anatomy and general function of sense organs. Chapter 3 discusses the anatomy and physiology of the ascending and descending sensory pathways. The anatomy and physiology of classical and nonclassical sensory systems are described and parallel processing and stream segregation important for processing sensory information are discussed. The different connections from sensory systems to limbic structures and other nonsensory parts of the CNS and their functional importance are also discussed in this chapter. Neural plasticity and its importance for development of sensory systems are discussed extensively in Chapter 3. How external and internal events can cause changes in the function of the nervous system by "rewiring" parts of the brain is described. Such changes can compensate for losses caused by injuries or diseases such as stroke. Plastic changes can also cause symptoms and signs of diseases that can manifest by chronic pain, hyperactivity, hypersensitivity, distension of sensory input, and emotional reactions to stimuli that normally do not elicit such reactions. Thus, there is both "good" and "bad" neural plasticity. Chapters 4 through 7 provide detailed descriptions of the anatomy and physiology of each of the five senses beginning with somesthesia (Chapter 4), followed by hearing (Chapter 5), vision (Chapter 6), and the chemical senses, taste and olfaction (Chapter 7).

ACKNOWLEDGMENTS

I have had help from many people in writing this book, especially Dr. George Gerken for his valuable comments on early versions of the manuscripts for many of the chapters of this book. Dr. Steve Lomber's suggestions about the chapter on vision were most valuable. Jan Nordmark commented on certain parts of the manuscripts. I also would like to thank many of my students at the University of Texas at Dallas for their valuable comments, and a special thanks to Pritesh Pandya for his comments on earlier versions of all the chapters. Many of my students at the School of Human Development, University of Texas at Dallas have provided valuable feedback and comments on earlier versions of the manuscript, as well as Dr. Karen Pawlowski. Phillip Gilley helped with the graphs and Karen Schweitzer typed the many revisions of the manuscripts.

I also want to thank Hilary Rowe and Cindy Minor at Academic Press in San Diego for their excellent work on the book, as well as Paul Gottehrer, Project Manager, for his dedicated work and professionalism. His copyeditor, Sarah Nicely Fortener meticulously copyedited this specialized book without changing its style or meaning. Debby Bicher worked on the artwork and supervised the redrawing of many of the illustrations. It has been a real pleasure to work with these professionals.

Without the support from the School of Human Development at the University of Texas at Dallas I would not have been able to write this book.

Last but not least I want to thank my wife, Margareta B. Møller, M.D., D. Med. Sci. not only for her patience with my occupation on this book and her encouragement during my writing of this book, but also for her comments and suggestions regarding earlier versions of the manuscripts.

PREFACE

The purpose of *Sensory Systems: Anatomy and Physiology* is to provide a comprehensive understanding of the anatomy and function of sensory systems. This book provides a systems approach to sensory systems and covers aspects of sensory systems not commonly found in textbooks such as the anatomy and function of nonclassical (nonspecific) sensory systems, parallel processing, stream segregation, and neural plasticity. The role of sensory input to nonsensory parts of the brain such as the limbic system (the emotional brain) and the physiology of various forms of pain are topics discussed extensively.

The book is written for all students of life sciences, for scientists who want a broad and comprehensive coverage of sensory systems, and for healthcare professionals dependent on sensory systems in one way or another, such as in restoring function after diseases that have impaired normal function of one or more of our senses. The book is based on a course I teach in the School of Human Development at the University of Texas at Dallas.

Sensory Systems: Anatomy and Physiology is suitable for anyone who wants to learn about the function of biological systems. I hope this book will encourage students to choose biology in one form or another for their career, be it clinical medicine, biomedical research, or other forms of life sciences.

I have enjoyed writing this book very much and hope the reader will have an equal enjoyment in acquiring insight to truly fascinating biological systems.

Aage R. Møller
Dallas, Texas
May, 2002

CHAPTER 1

Basic Psychophysics

ABBREVIATIONS

CFF: Critical fusion frequency
CNS: Central nervous system
dB: Decibel, one tenth of a logarithmic unit
HL: Hearing level. (Sound level above the normal threshold of hearing)
Hz: Hertz, (frequency in cycles per second)
MAF: Minimal audible field
MAP: Minimal audible pressure
msec: Millisecond
nm: Nanometer, 10^{-6} millimeter
SL: Sensation level. (Sound levels an individual person's threshold)
TTS: Temporary threshold shift
μm: Micrometers, (10^{-6} meter or 1/1000 millimeter)
μS: Microsecond

ABSTRACT

1. Threshold of detection and discrimination of specific features of sensory stimuli are assessed through psychophysics studies.
2. Detection of stimuli is dependent on:
   a. The intensity of the stimulus that reaches the receptor
   b. The sensitivity of the receptors for the stimulus in question
   c. The duration of the stimulation (temporal integration)
   d. Number of receptors that are stimulated (spatial integration)
   d. Background stimulation (masking)
   e. Prior stimulation (adaptation and fatigue)
   f. Attention
3. Discrimination of stimuli depends on temporal and spatial resolution of the sense.
4. Qualities of the stimuli that can be discriminated include:
   a. Intensity of stimuli
   b. Temporal properties
   c. Spatial properties
   d. Small difference in qualities (difference limen) such as intensity, visual contrast, frequency of sounds, concentration of odors and taste

I. INTRODUCTION

Psychophysics is a branch of psychology that deals with the relationship between a physical stimulus and the resulting sensation. Detecting the presence of a stimulus was probably the primary advantage of the evolution of such senses as vision and hearing when vertebrate species began to adapt to terrestrial life. Detecting odors was also important for many species. Early in the development of species, discrimination between different kinds of stimuli was of less importance. Much later in the evolution of terrestrial vertebrates it was still important to be able to detect the faintest sound or the weakest light, but the ability to discriminate between different stimuli became increasingly important as vertebrate species developed sensory systems adapted to these needs.

In this chapter, we will discuss such basic properties as the threshold of detection and the perception of strength and their relation to the physical properties of stimulation. The purpose is to provide the reader with a general overview of basic psychophysics in order to support the perspective on the main theme of this book, namely that of the anatomy and physiology of sensory systems. For more details on psychophysics, the readers are referred to standard texts on the subject.

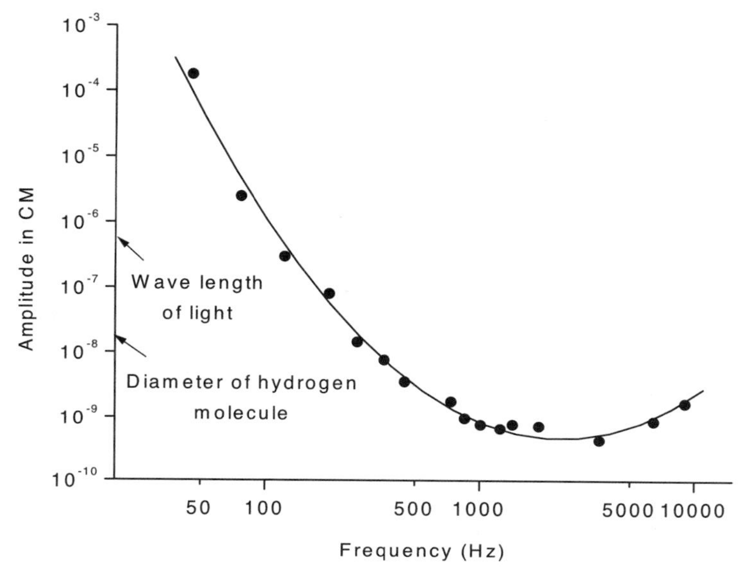

FIGURE 1.1 Threshold of hearing in a human subject. Circles indicate the amplitude of the displacement of the tympanic membrane at threshold (in cm);[42] the solid line is a curve fit to the data points.

II. THRESHOLD OF DETECTION

In psychophysics the threshold of detection of a physical stimulus refers to the minimum amount of the stimulus that is needed for the sensory system to elicit a behavioral response. The threshold varies between sensory systems, between animal species, and as a function of the properties of the stimulus.

The sensitivity of sensory systems is enormous and it surpasses most of our technical systems. Eyes of insects such as night moths, which are active when light is very weak, can detect one or two light quanta, and even our own eyes are very sensitive and can detect approximately 15 light quanta.[*,11] The threshold of hearing is no less impressive than that of vision. In the most sensitive frequency range the amplitude of vibration of the eardrum at the threshold of hearing is less than 10^{-9} cm (1/100 of one millimicron or *nanometer* [nm] or 10^5 *micrometers* [μm]), corresponding to less than 1% of the diameter of a hydrogen molecule (Fig. 1.1).[42]

*A light quantum is the smallest amount of light that can be produced. It refers to the quantum theory of light production, which assumes that light is both a particle and a wave.

Under the best circumstances, the sensitivity of the auditory system in humans is near its theoretical limits set by Brownian motion of the cochlear fluid. The sensitivity of the human ear is similar to that of other animal species, although the cat has an approximately 10-dB-lower threshold than humans between 1 kHz and 7 kHz and 20 dB or more between 10 and 20 kHz. Some animals can hear in a wider range of frequencies than humans.[15] For example the cat and the common laboratory rat hear up to 50 kHz, and some species of the flying bat, whales, and dolphins hear up to or above 100 kHz.

The sensitivity of the vibration of the skin is much less impressive, but nonetheless the stimulus required to elicit a conscious response is small. The greatest sensitivity for sinusoidal vibration applied to the palm is approximately 0.2 μm of skin displacement in humans. It is slightly higher for the monkey when the threshold is defined as a 50% correct response rate.[20] The sense of the skin is highly complex, involving several stimulus modalities, such as touch, pain, and thermal stimuli. The threshold of sensitivity in each of these modalities varies according to region of the body and species.

The sensitivity of the olfactory system also varies widely among species of mammals.[30] The sensitivity of the olfactory system in some mammals (for instance, dogs) is very high and only a few molecules of a substance are required for detection. The nose of humans is less sensitive, but the sensitivity of the human sense of smell varies within wide ranges for different odors. It also varies widely among different individuals and decreases with age.[16]

A. Factors That Affect Detection of Faint Stimuli

The sensitivity to a stimulus depends on many factors besides the type of the physical stimulus, such as the frequency of sounds or vibration, the wavelength of light, or the kind of taste and odor. Prior stimulation can decrease the sensitivity of sensory systems because of *adaptation* or *fatigue*. Simultaneous presentation of another stimulus can also increase the threshold to stimulation known as *masking*. Additionally, the threshold of detection decreases as the duration of the stimulus is increased up to a certain duration. This is known as *temporal integration*. Stimulation of a large number of receptors may decrease the threshold because of *spatial integration*. Finally, detection of faint stimuli as determined experimentally depends on the *criteria for detection* and how and where the stimulus is measured.

1. Type of Stimulation

The frequency range of hearing in different species of terrestrial vertebrates differs. Humans can hear in the frequency range from approximately

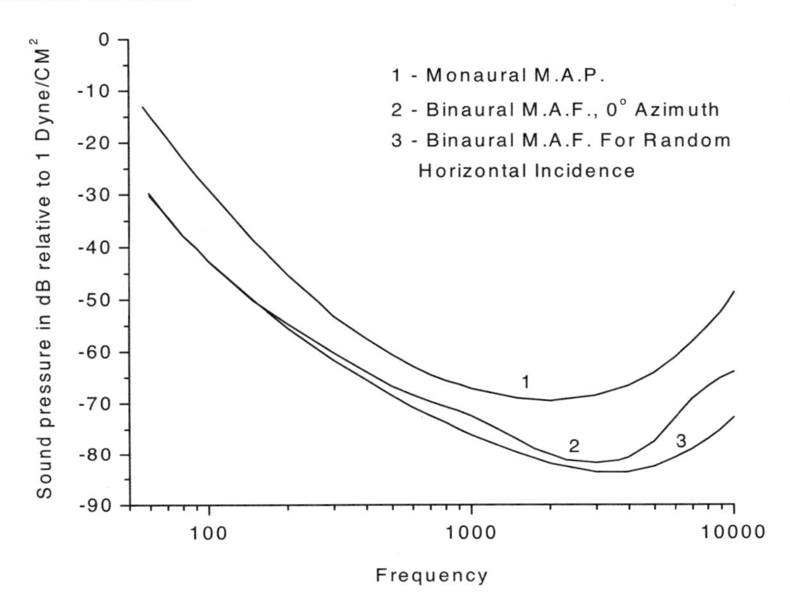

FIGURE 1.2 Threshold curves of hearing in three different situations: (1) monaural minimum audible pressure (MAP) determined using a calibrated earphone; (2) binaural minimum audible field (MAF) determined using a loudspeaker as sound source placed in front of the observer; and (3) similar as (2) but with the sound coming from a random azimuth.[32]

20 to 20,000 Hz (Fig. 1.1). The upper limit decreases with age. Some animals have hearing at higher frequencies than do humans. For example, the cat hears up to approximately 50 kHz, approximately the same as the rat. The flying bat, dolphins, and whales can hear sounds above 100 kHz.

The threshold of sensory systems depends on the type of stimulation, and the sensitivity of the ear is not uniform over the audible frequency range. The highest sensitivity of the human ear is in the frequency range from 500 to 6000 Hz (Fig. 1.1). The absolute value of the threshold depends on how the sounds are presented: monaural (sound to one ear only), binaural (sound to both ears at the same time) by earphones or in a free field at 0 degrees azimuth (angle in the horizontal plane) or at random incidence (Fig. 1.2).[32]

When measured in the ear canal, near the tympanic membrane, or at the entrance of the ear canal, hearing thresholds are usually referred to as the *minimal audible pressure* (MAP). If the threshold is referred to the sound pressure at the place where the person is located (without the person being present), yet other threshold values will be obtained. When the sound is measured in the place where the test subject is to be placed during the test, the threshold is known as the *minimal audible field* (MAF). Typically, the threshold for a MAP response is lower than that for a MAF response for sounds coming from a direction in front of the person being tested. The threshold for sounds that reach both ears is approximately 3 dB lower than when sounds reach only one ear when referred to the sound pressure at the entrance of the ear canal. However, when placed in a *free sound field*, the effect of the head on the sound pressure at the entrance of the ear canal is in most situations greater than that. The reason is the acoustic effects of the head on the sound pressure that reaches the entrance of the ear canal. Depending on the direction to the sound source, the head functions either as a baffle (that increases the sound, mostly for high frequencies) or as a shadow to the ear on the opposite side of where the sound source is located, which generally makes the sound at the ear canal lower than that measured in the place of the person in question.

Like the ear, the sensitivity of the eye depends on the stimulus, and the sensitivity of the eye is not uniform over the range of wavelengths to which it responds (Fig. 1.3). The range of vision is similar for most mammals but, for instance, insects can see ultraviolet light, which is outside the visible range for mammals. Rods have their best sensitivity at wavelengths of approximately 500 nm, corresponding to green, bluish light (Fig. 1.4). The color or wavelength of light to which the human eye is most sensitive therefore depends on its state of *adaptation* (see below). Daylight in the middle of the day has a broad spectrum with considerable energy emitted in the range of wavelengths from 420 to 700 nm, (i.e., from violet to deep red). The *light-adapted* eye using cones (*photopic vision*) is most sensitive to light of a wavelength of approximately 555 nm, corresponding to green light. The sensitivity of the eye also depends on

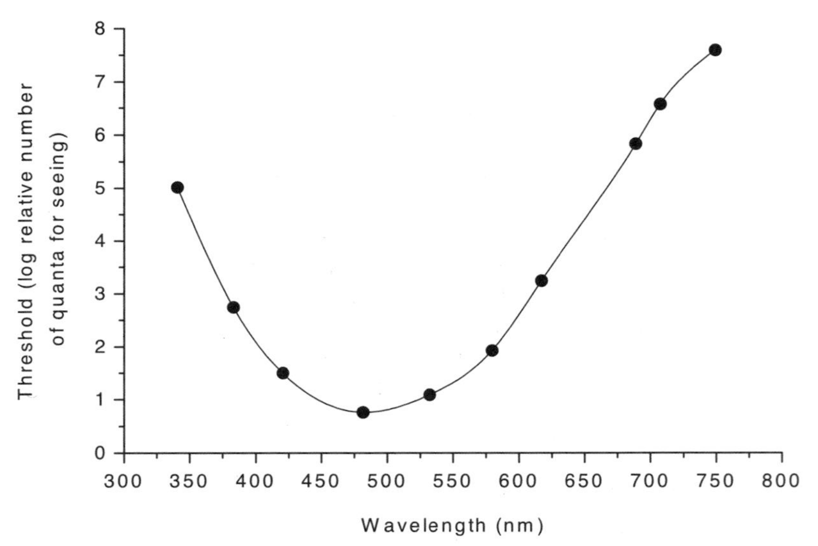

FIGURE 1.3 The sensitivity of the dark-adapted eye, peripheral to the fovea, expressed in the number of light quanta (left scale) and in log units (right scale) required to be detected as a function of the wavelength (in nm) of the test light.[5]

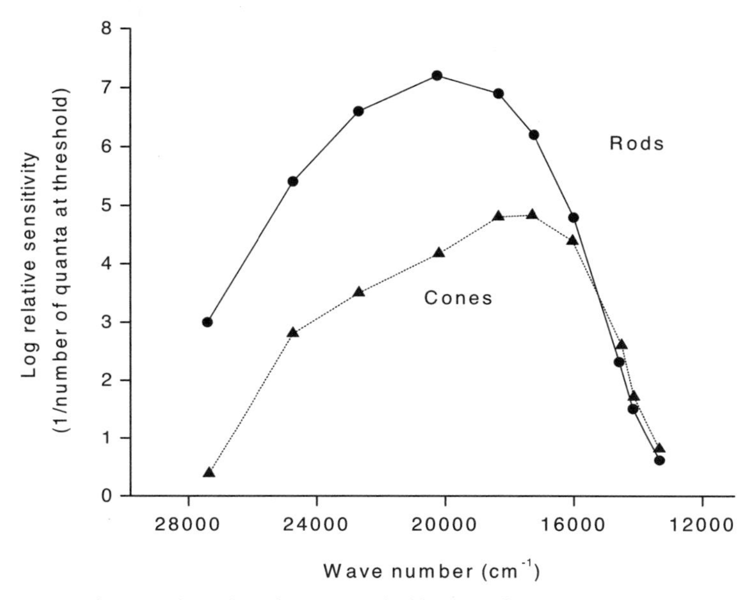

FIGURE 1.4 Photopic (cones) and scotopic (rods) spectral sensitivity, given in logarithmic measures, and shown as a function of the wave number (1/wavelength) (horizontal scale).[5] (Adapted from Wald.[39])

the location on the retina where the light is projected because the density of photoreceptors on the retina is not uniform (see Chapters 6, Fig. 6.1).

Mechanoreceptors in the skin are sensitive to deformation of the skin, and different types of mechanoreceptors have different sensitivity to such stimuli. The sensitivity to sinusoidal vibration is greatest around 200 Hz for some receptors known as *Pacinian corpuscles* (Fig. 1.5). Some receptors are most sensitive to rapid change in deformation of the skin compared with steady deformation of the skin, while in others the opposite is true.

While the olfactory receptors respond to a wide range of different odors, taste is limited to four categories, namely sweet, sour, salty, and bitter (and possibly a fifth, monosodium glutamate). The olfactory system has different sensitivity to different odors, and likewise the taste sense has different sensitivity to the four or five different substances to which taste receptors are sensitive.

2. Adaptation

The dependence of the threshold of sensory systems on prior stimulation is known as *adaptation*. Prior exposure decreases sensitivity of sensory receptors. For vision, the highest sensitivity is achieved when no light has

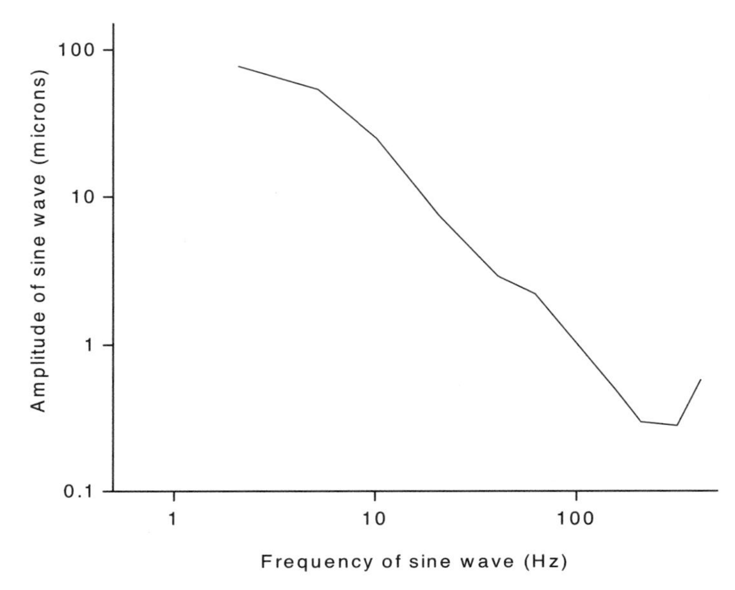

FIGURE 1.5 Threshold to mechanical vibration delivered to the skin of the hand for an experienced human observer. The amplitude that gave a response in 50% of the trials is shown as a function of the frequency of sinusoidal vibration.[21]

reached the eye for some time (the *dark adapted* state of the eye) (Fig. 1.6). Exposure to light reduces the sensitivity of the eye to an extent and for a period of time that depends on the exposure. In that state the sensitivity of the eye is determined by the sensitivity of the rods (*scotopic vision*).

Prior stimulation also affects the sensitivity of the ear, and exposure to loud sounds causes *temporary threshold shift* (TTS); exposure to even louder sounds causes permanent damage to the ear known as *permanent threshold shift* (PTS). The decrease in sensitivity that is caused by overexposure is also sometimes referred to as (auditory) *fatigue*. The reduction in the sensitivity of the ear depends on the intensity of the sound and its duration and frequency (spectrum). The TTS is largest approximately ½ octave (1400 Hz) above the frequency of the tone that caused the fatigue (1000 Hz) (Fig. 1.7); it accelerates as the intensity of the fatiguing sound is increased. In a similar way, prior stimulation of the chemical senses, olfaction and taste, affects the sensitivity of these senses, and prior stimulation of the skin causes a reduction in sensitivity.

3. Masking

The threshold of a test stimulus can be elevated by the concomitant presence of another stimulus of the same type, thereby *masking* the perception of the test

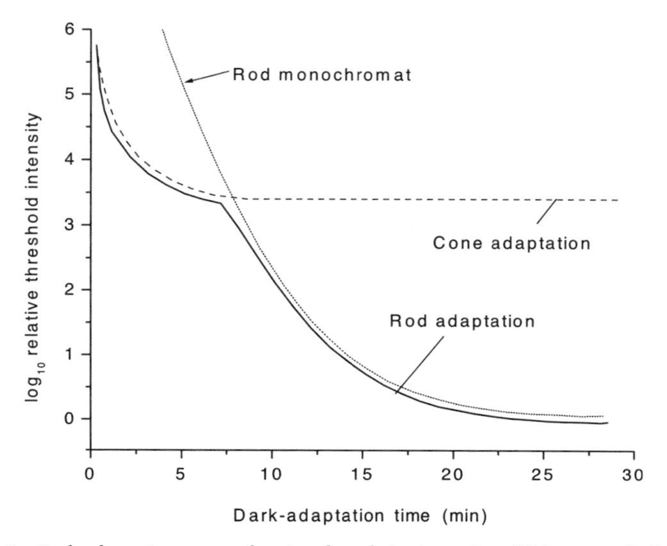

FIGURE 1.6 Dark-adaptation curves showing the relative intensity of light to reach threshold as a function of time after exposure to bright light. The dotted curve was obtained in a totally color-blind person assumed to have only rods. The dashed curve was obtained in response to red light illuminating the fovea, and the continuous curve was obtained using white light illuminating the extra foveal regions of the retina in a person with normal vision.[28]

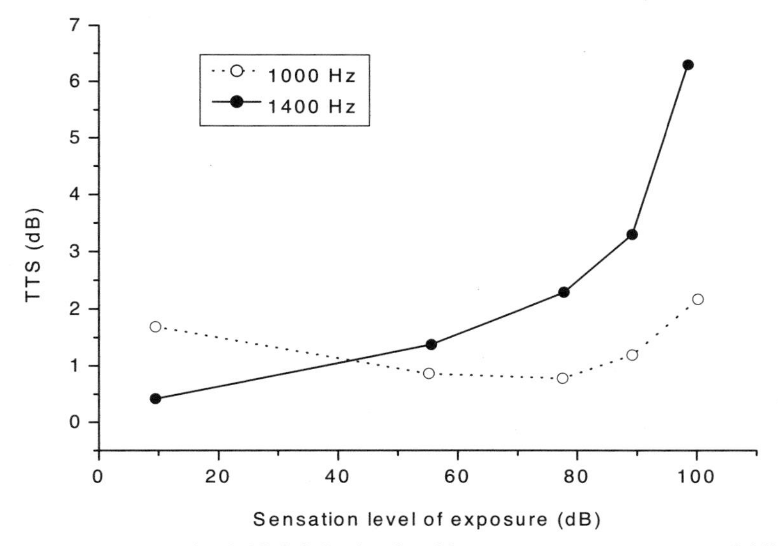

FIGURE 1.7 Temporary threshold shift (TTS) induced by exposure to 1000-Hz tones of different intensity given in sensation level (SL) (i.e., the level in decibels above the test subject's hearing threshold for two tones of different frequencies).[12,18]

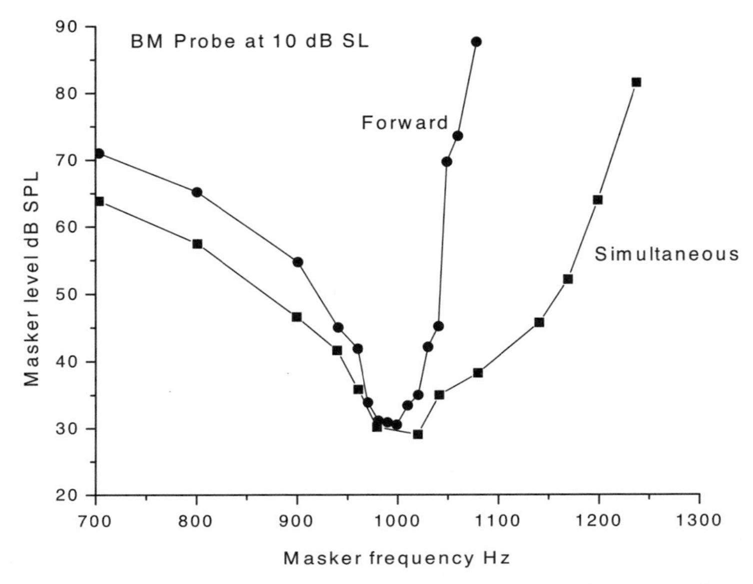

FIGURE 1.8 Comparison between simultaneous and forward masking of a weak (10-dB sensation level [SL]) 1000-Hz tone as a function of the frequency of the masker.[17]

stimulus. Masking is pronounced in hearing, where it has been studied extensively. It is well-known how stimulus parameters affect a sound's ability to mask another sound.

The efficiency of a sound in masking another sound depends on its intensity and its frequency relative to the test sound. In general, low-frequency sounds are more effective in masking high-frequency sounds rather than vice versa. Different sounds mask each other according to the width of the *critical band*.* Two sounds that are separated in frequency by less than one critical band are most efficient in masking each other, while sounds that are separated in frequency by more than one critical band are less effective in masking each other.

Stimuli do not have to occur simultaneously to mask each other. A stimulus that occurs just before another (second) stimulus, but does not overlap it in time, may change the threshold of the second stimulus. This temporal effect in masking is referred to as *forward* masking. Forward masking is less efficient than *simultaneous* masking (Fig. 1.8). A masker can also affect the threshold of

*Critical band is the band of frequencies over which the auditory system integrates the energy of a broadband sound. It is also a measure of the ability of one sound to mask another sound.

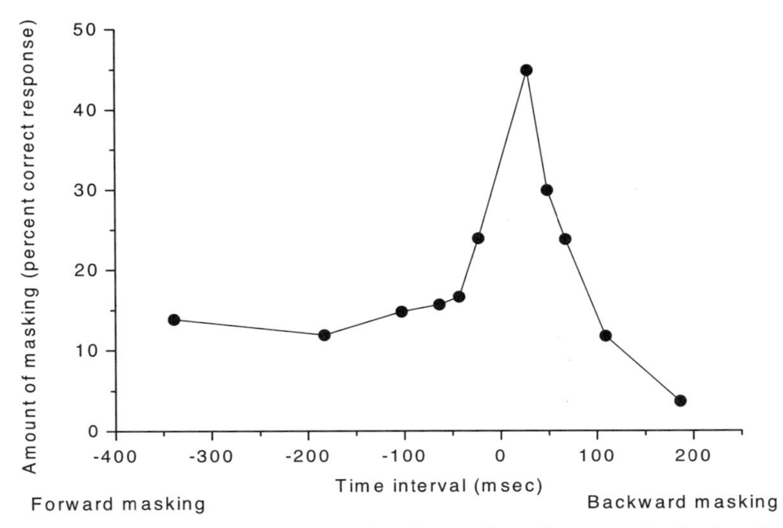

FIGURE 1.9 Stimulation of two points on the skin; masking of one stimulus by the other as a function of the temporal interval between the two stimuli.[6]

sensory systems to a stimulus that is presented after the masker, and this is known as *backward masking*. Backward masking is usually less efficient than forward masking, and its effect occurs during a shorter interval of time relative to the test stimulus.

For stimulation of the skin (touch), stimuli are most effective in masking each other if they occur within 100 msec relative to each other[6] (Fig. 1.9), but a masker can affect the threshold of a test sound that is presented as much as 1.2 sec prior to the test stimulus.

4. Temporal Integration

Temporal integration implies that energy is summed over a certain period of time. The result is that the threshold is lower for stimuli that last longer up to a duration, known as the *integration time*. For instance, the threshold of hearing decreases as the duration of the stimulus is increased up to approximately 300 msec[41,43] (Fig. 1.10), but the relationship between duration of sounds and the threshold is not an exact power function. Also, the temporal integration depends on the frequency of the sounds and it is different above threshold, where it is generally shorter. Temporal integration of mechanical stimulation of the skin is similar to that of sound in hearing, and it is different for different types of skin receptors.

Olfactory sensory organs integrate stimulus input over time; consequently, the sensitivity to short puffs of odors is less than it is to longer exposures when

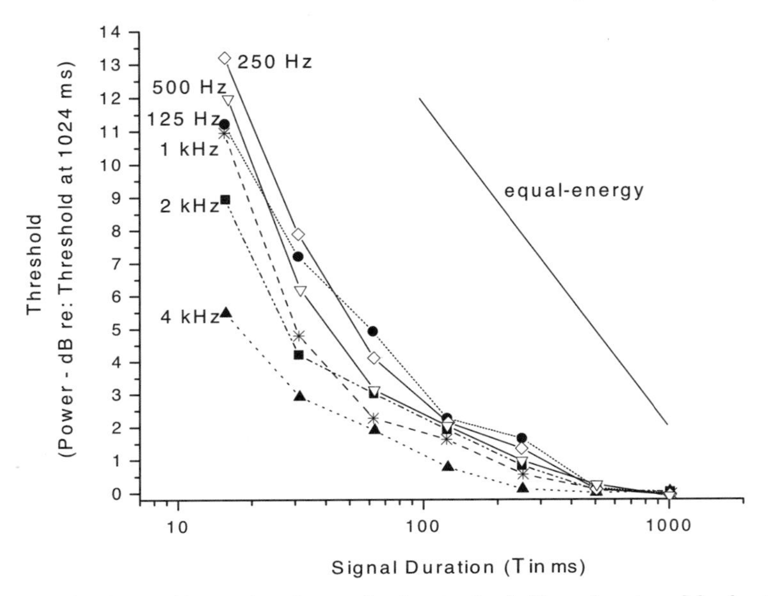

FIGURE 1.10 Temporal integration of tones showing the threshold as a function of the duration of tones of different frequencies.[43] The straight line shows the hypothetical threshold as a function of the duration for an ideal integrator of power.

the rate of delivery of the odor is constant. If the same amount of an odor chemical is delivered, the threshold will be the same independent of the time during which it is delivered, provided that the time is less than the integration time of the receptors.

The temporal integration in vision is often expressed in a slightly different way. The threshold of vision is related to the number of light quanta that the eye receives within approximately 100 msec. Even if the same number of light quanta is delivered during a very short time (for instance, 1 μS), the threshold measured in the number of light quanta is the same as if the light was delivered during a longer time up to 100 msec. This is a result of temporal integration, which means that the eye can add the light that is delivered during a time of approximately 100 msec. If the duration of light with a certain number of quanta is extended and the same number of light quanta is spread out over a time that is longer than the integration time, more quanta are required to reach threshold.

5. Spatial Integration

Stimulation of a small area of the skin has a higher threshold than stimulation of a larger area and is perhaps the most direct demonstration of spatial integration in a sensory system. The eye integrates light over a certain limited area of the

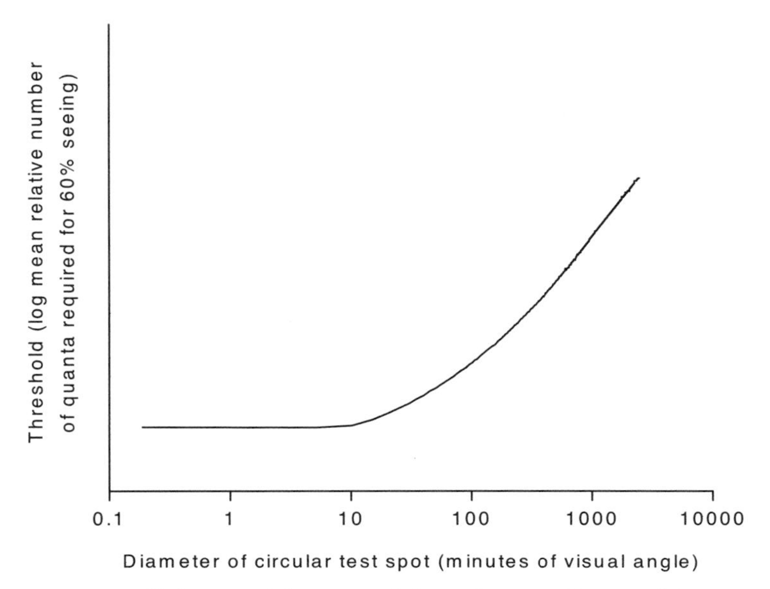

FIGURE 1.11 Total light required for seeing a flash as a function of the size of the test spot.[1,5]

retina, which means that the sensitivity of the eye depends on the size of the test spot of light that is shined on the retina. If the density of light of the spot is held constant then the threshold decreases when the size of the spot is increased, up to a certain size. The reason for that is that the light has been "diluted" by spreading the same light over a larger region of the retina and the energy per surface area decreases when the size of the light spot is increased. Consequently, the strength of the light that reaches an individual photoreceptor decreases when the illuminated areas increases. If, on the other hand, one would shine a spot of light with constant energy on the retina, the threshold will increase (more energy needed) when its size is increased beyond a certain size (Fig. 1.11).

The ear integrates sound over frequencies, and the threshold decreases when the spectrum of a test sound is increased up to a certain value, known as the *critical band*. This is a form of spatial integration. The width of the critical band depends on the (center) frequency of the band of sound (Fig. 1.12).

6. Criteria for Detection

Determination of the threshold of a sensory system is covered by *signal detection theory*, a discipline developed in engineering[29,37] that concerns the detection of signals and how the detection is affected by different circumstances. Detection

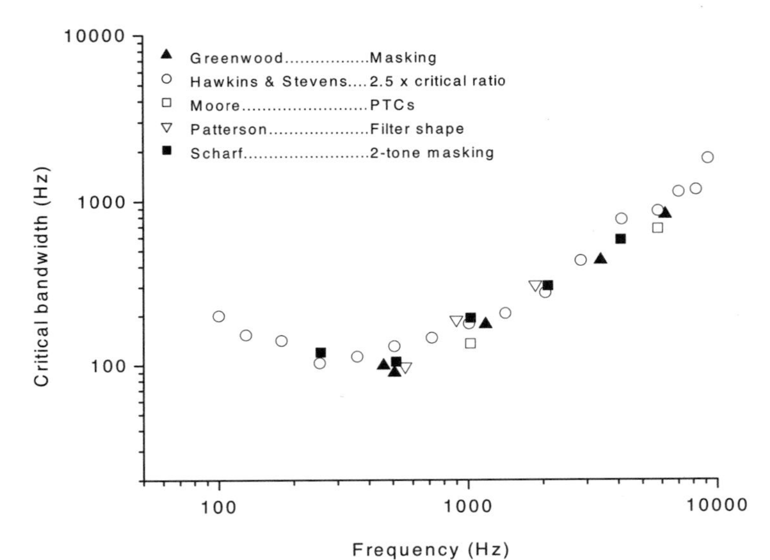

FIGURE 1.12 The width of the critical band as a function of the frequency. Five different sets of measurements are shown.[27]

of signals typically is characterized by a forced choice (i.e., the observer must make a decision, either yes or no). It can also be characterized by the observer making a yes, no, or do not know decision. This method results in a less sensitive threshold because it adds a conscious decision to the subconscious detection of stimulus that occurs in the former paradigm. The outcome of experiments to determine the sensitivity of a sensory system also depends on the definition of detection. In vision it is common to define the threshold as the intensity of the test light that results in correct detection of 60% of the presentations. If a higher percentage number is chosen, the sensitivity arrived at will be lower. Similar aspects can be applied to detection of stimuli by other sensory systems. In hearing, usually this criterion for threshold is defined as correct responses in 50% of the stimulus presentations.

Detection of the presence of a weak stimulus is associated with uncertainty, and this may be regarded as noise that is superimposed on the stimulus. This variability is an unavoidable property of the human observer that has been studied extensively. Repeating the threshold determination to many identical stimuli and then averaging the results reduces the variability. (In addition to this variability, there is an individual variability, which cannot be reduced by increasing the number of tests done on the same individual, and it has the character of an error or a deviation from the average threshold). When the stimulus intensity is varied in small steps near threshold, the correct responses

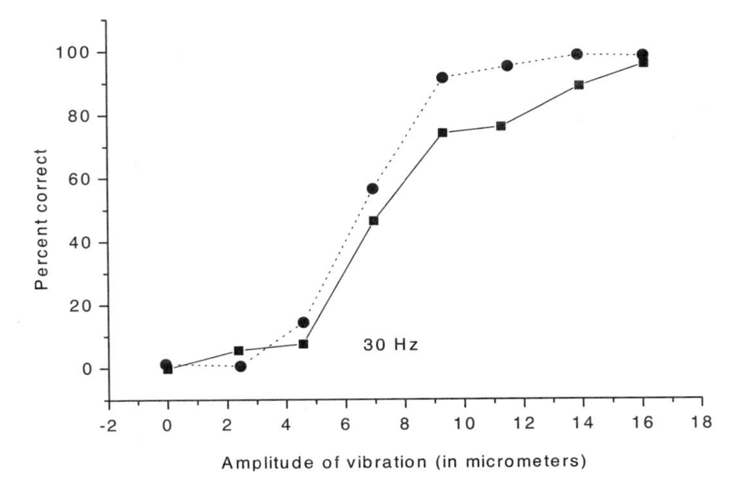

FIGURE 1.13 Percent correct response from an experienced human observer to vibration applied to the skin of the hand. The frequency of the vibration was 30 Hz. Each curve is the average of 6 separate runs, each of 256 trials distributed over 8 amplitude classes and delivered in random order.[21]

given by an individual shown as a function of the stimulus intensity follows a sigmoidally shaped curve (Fig. 1.13).[21]

The nature of the stimulus may contribute to the variability in the detection of a stimulus. Stimuli such as sounds applied to the ear or vibration applied to the skin can be presented at the same intensity repeatedly but this is not the case for light because light is a random process with an inherent variability because it is a series of quanta that are emitted randomly. This variability adds to the variability that is a property of the observer (Fig 1.14). The variability from the quantum nature of light is more pronounced if a weak stimuli can elicit a response (low threshold) of the eye because fewer quanta are required to reach the threshold, compared with a higher threshold of the eye. The stimulus response curves are therefore steeper when the threshold is high because of the lower variability of light that contains many quanta, compared with stimulus response curves of eyes with high sensitivity because fewer quanta are needed to reach the threshold (Fig. 1.14). That means that the steepness of the stimulus response curves can be used to determine how many quanta of light are needed to reach the threshold (Fig. 1.15).

The intensity of light is often given by the average number of quanta needed to elicit a response in a certain percentage of trials (Fig. 1.16). However, this is not a practical way of expressing the light intensity when the response to stimuli well above the threshold is concerned.

7. Individual Variation

The sensory threshold varies even between individuals who are regarded as having normal sensory functions. An example of that is the threshold of

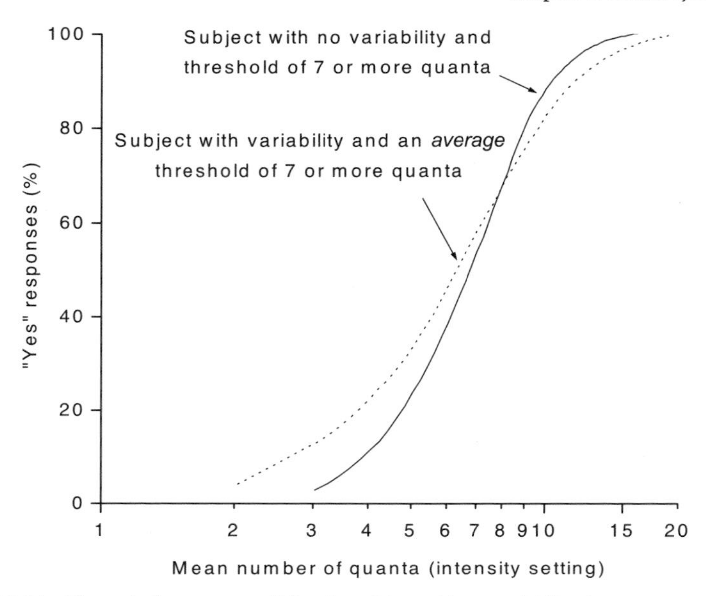

FIGURE 1.14 Theoretical response to light of a subject with no variability in response to seven or more quanta (solid lines). Dashed line shows average threshold of a subject with normal variability in response to seven or more quanta of light.[5]

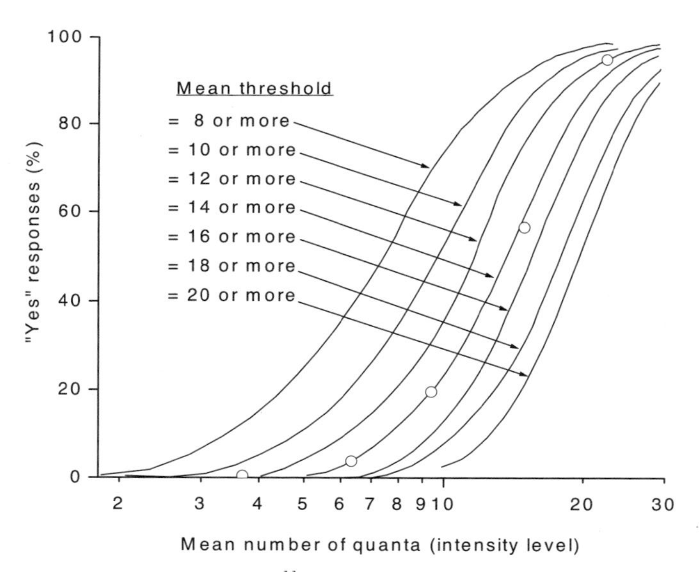

FIGURE 1.15 The data from Hecht *et al.*[11] fitted to a family of curves similar to the dotted line in Figure 1.14, assuming different thresholds from 8 or more quanta to 20 or more quanta.[5]

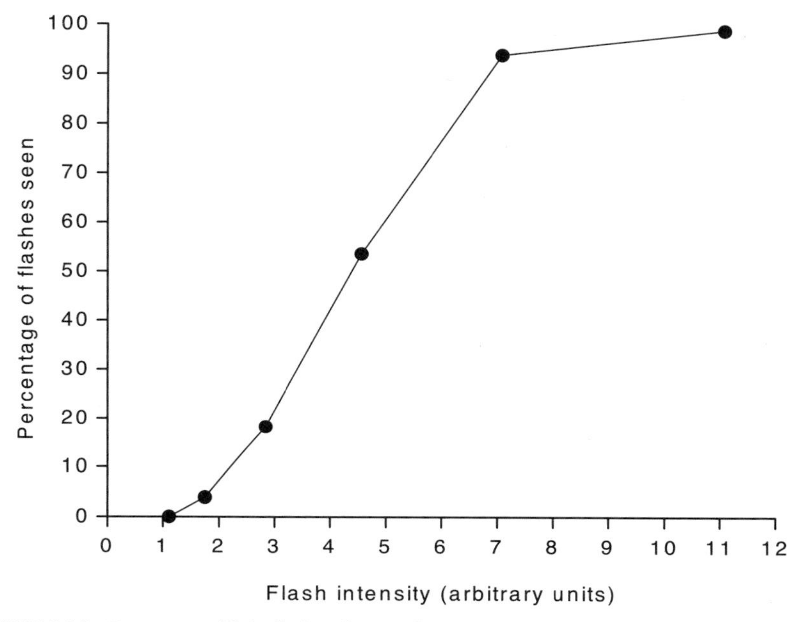

FIGURE 1.16 Percentage of light flashes that an observer sees as a function of the intensity of the flashes.[5,11]

hearing, which can be measured with great accuracy (Fig. 1.17). The thresholds for light, mechanical stimulation of the skin and the sensitivity to odors also vary widely from individual to individual.

The thresholds of sensory systems such as hearing and vision are usually measured in young individuals who have no known sensory defects of any kind. The sensitivity of most of the sensory systems decreases with age and certain diseases. Visual and hearing thresholds are typically higher in older individuals than in young individuals without known sensory disorders. It must therefore be remembered that the "normal" sensitivity that appears in the description of sensory systems applies to young healthy individuals, and thresholds in older individuals are likely to be higher.

8. Other Factors That Affect the Measured Threshold of Sensory Systems

Transmission of the stimulus to the receptors affects the intensity of the stimulus that reaches the receptors, and the measured threshold will depend on where the stimulus is measured. For hearing, the ear canal and the head transfer sound to the ear in a way that depends on its frequency (Fig. 1.2). The threshold of hearing

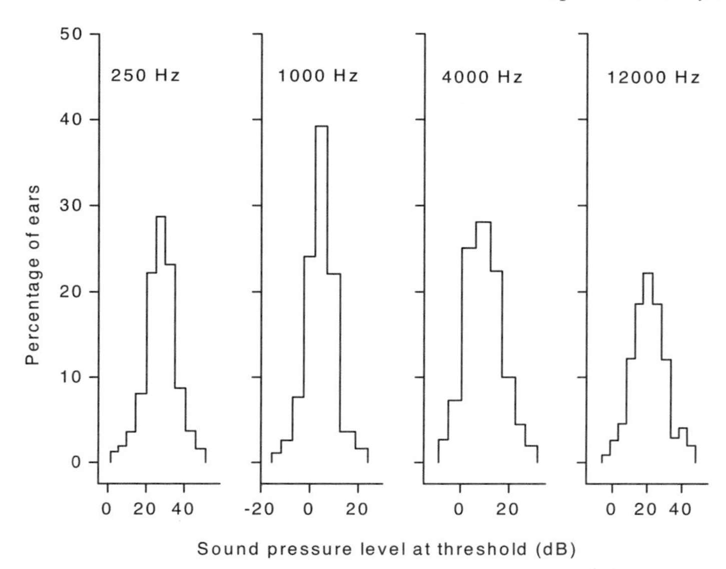

FIGURE 1.17 Individual variation in the threshold of hearing of 198 ears of individuals judged to have normal hearing. (Adapted from Dadson and King.[7])

therefore depends on where the sound is measured. The threshold values that refer to the sound at the entrance of the ear canal will be higher than the threshold that refers to the sound intensity at the tympanic membrane because of the resonance in the ear canal that increases the sound pressure. That increase is highest for sounds in the frequency range between 3 and 5 kHz, depending on the length of the ear canal.

Loss of light energy in the transmission from the environment to the receptors in the retina will result in a higher threshold than what would have been obtained if the light that reached the receptors were used as a measure of the threshold.

III. DISCRIMINATION BETWEEN DIFFERENT STIMULI

The ability to discriminate a large number of different stimuli evolved relatively late. For each sense a few basic features make it possible to discriminate among a large number of stimuli. For example, discrimination of the frequency (or spectrum) of sounds plays an important role in discrimination of stimuli such as speech sounds. While the time pattern of sounds is very important for discrimination of natural sounds, it plays a much less important role for visual stimuli. The important properties for discriminating among visual patterns are contrast and changes in contrast. Discrimination of colors naturally also plays a role in human vision.

Our senses can discriminate very small physical differences in sensory stimuli. For example we can discriminate small differences in visual images such as the differences in familiar faces. Likewise we can discriminate a large number of speech sounds and distinguish among many different speakers. Discrimination among a relatively few taste qualities and many different odors makes it possible to discriminate among a very large number of "tastes" of food.

The sensory organs of hearing and vision are paired organs, thereby making it possible to determine the direction of a sound source or a light source (visual object). These discriminatory abilities in turn contribute to the creation of the cognitive entities of auditory and visual space.

A. Perception of Stimulus Strength

The perception of the strength of a stimulus does not increase linearly with the physical intensity of the stimulus; rather, the apparent strength is more closely related to a power function of the intensity.[35] Estimates of strength of perception is done by having individuals indicate how much stronger a certain stimulus is compared with another stimulus. The results of such determinations are used to establish a *ratio scale*. Other methods make use of comparison of stimuli of different modalities. For example, the intensity of light may be equated to the strength of pushing a handle of a force meter (a hand dynamometer). Although seemingly subjective, these measures are remarkably reproducible in the same individual but vary among different individuals. The perceptions of the strength of sensations determined using such methods are *power functions*;[37] the perceived strength appears as straight lines when plotted on log–log scales as a function of the stimulus intensity. The exponent of the power functions is different for the different sensory modalities (Fig. 1.18).[28,36]

B. Discrimination of Small Differences Between Stimuli

Discrimination of small differences in stimuli is important for discricmination of natural stimuli such as speech sounds. Small differences in the contrast of visual images make it possible to discriminate among different patterns such as those of faces. These matters have been studied in laboratories using simple stimuli such as pure tones where the intensity or the frequency is varied. The smallest difference between such properties of sensory stimuli that an observer

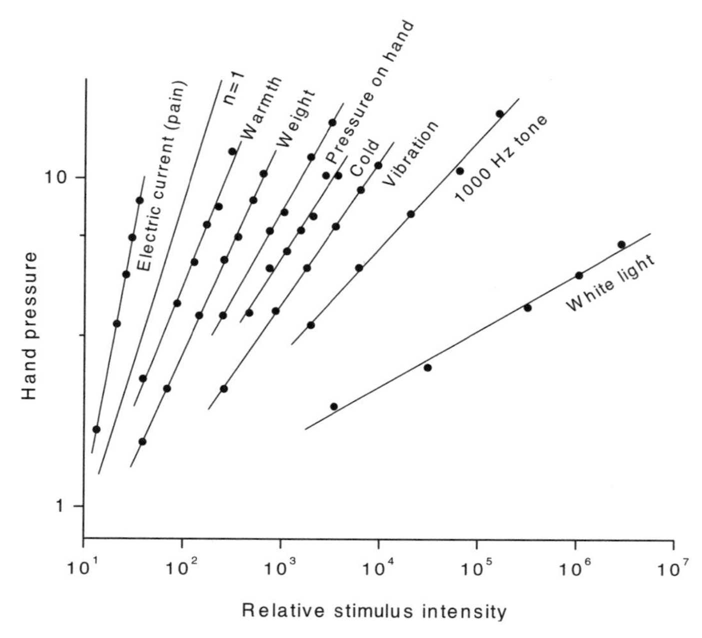

FIGURE 1.18 Perceived intensity of sensation determined by cross modality intensity comparison using the force exerted upon a hand dynamometer as ordinate. Both scales are logarithmic.[28,36]

can discriminate is known as the *difference limen* which describes the ability to discriminate between two stimuli that differ slightly.

The difference limen for intensity of sounds depends on the intensity of the sound (and its frequency) (Fig. 1.19) and the difference limen for the frequency of a tone depends on the frequency of the tone (and its intensity) (Fig. 1.20). The difference limen for intensity at moderate sound intensities is approximately 1 dB, and the difference limen for frequency of pure tones is approximately 1 Hz for frequencies below 1000 Hz and approximately 0.1% of a given frequency for frequencies above 1000 Hz.

C. Temporal Resolution

Temporal resolution of sensory systems can be studied in several ways, and the results often differ depending on how it is studied. Discrimination of the temporal order of two different sounds of the same duration (for example 500 msec) is one such method, and the smallest time interval where it is possible to determine the order of such two sounds presented after each other is 20 to 30 msec.[13] A similar experiment using two different light flashes

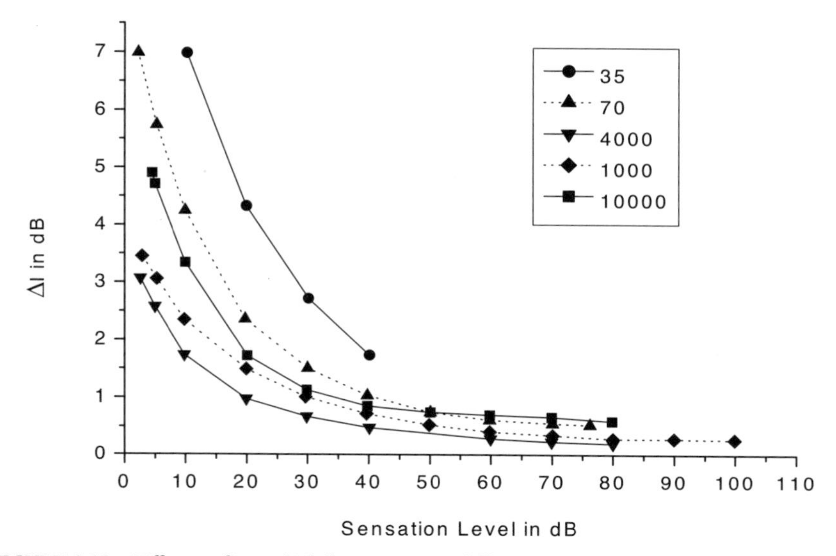

FIGURE 1.19 Difference limen (ΔI) for intensity at different sensation levels (in dB) for tones of different frequencies (Hz).[24,34]

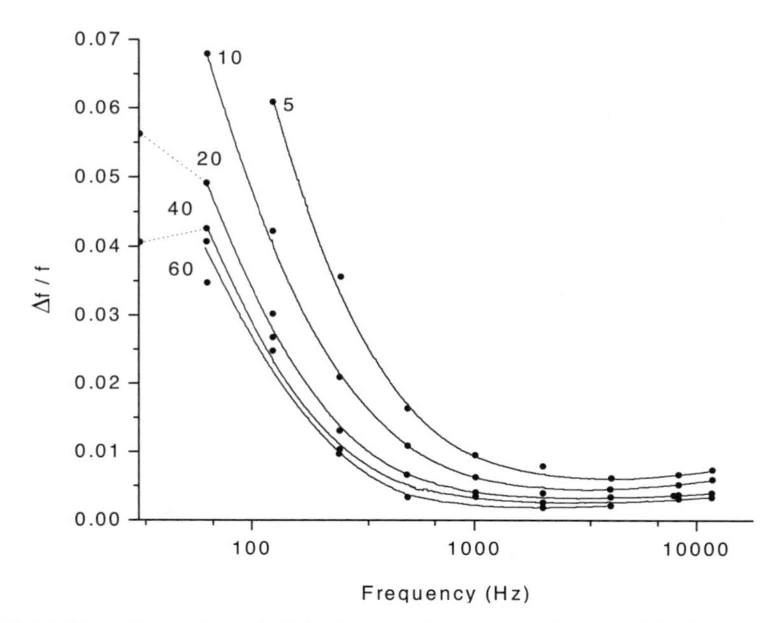

FIGURE 1.20 Difference limen (Δf) for frequency for tones, as a function of the frequency of the tones, for different sensation levels (dB).[31,34]

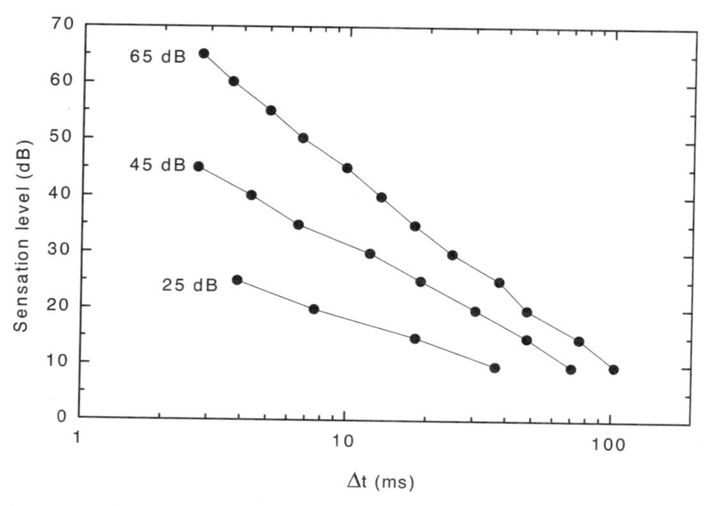

FIGURE 1.21 Gap detection. The minimum detectable silent interval between two sounds as a function of the sensation level of the second sound for three different sensation levels of the first sound.[19,22]

showed that the threshold for determining which of the light flashes was presented first is also 20 to 30 msec.[14] The discrimination of temporal order increases with training. The ability to discriminate the order of two sounds also increases when the duration of the sounds is decreased.

Temporal acuity has been studied extensively using sounds with a gap between them (Fig. 1.21).[22] Gap detection is little affected by many stimulus parameters, but when intensity of the sound that follows the gap is different from that preceding the gap a distinct dependence on the sound intensity becomes apparent (Fig. 1.21). The threshold for detecting the difference between an amplitude-modulated sound and an unmodulated sound is yet another measure of temporal acuity (Fig. 1.22). The smallest modulation that can be detected is approximately 3% at a modulation frequency of 50 Hz. The threshold increases for frequencies above 50 Hz and reaches a constant level of approximately 40% above 1000 Hz (Fig. 1.22). The cut-off frequency for detection of modulation at 50 Hz can be translated into a time constant of 2 to 3 msec if detection of temporal patterns is assumed to follow the rules of a simple first-order integrator.[8]

In the visual system, temporal resolution can be measured by displaying a pattern of stripes moved at different rates until the pattern fuses (Fig. 1.23).[5] Temporal resolution can also be determined by measuring the response to light the intensity of which is modulated (Fig. 1.24). The threshold for detection of the modulation of sinusoidally modulated light is 5% modulation and is known as

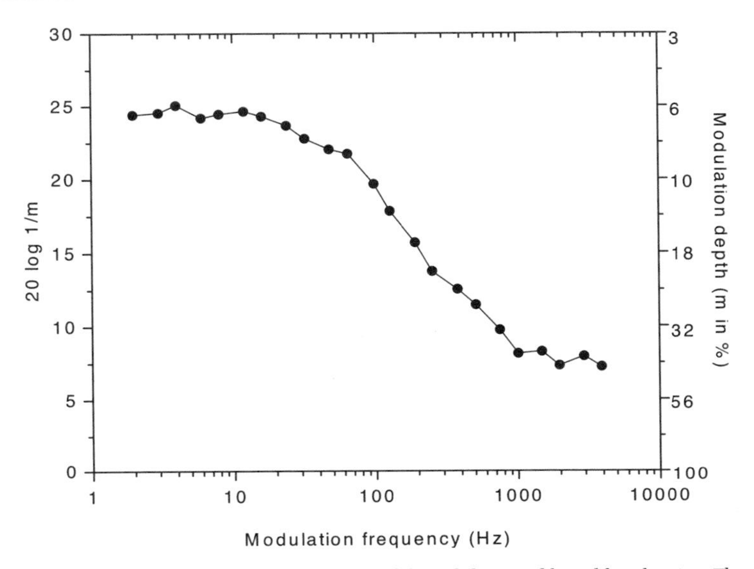

FIGURE 1.22 Threshold for detection of sinusoidal modulation of broad-band noise. The vertical scale is 20 log $1/m$, where m is the modulation degree.[38]

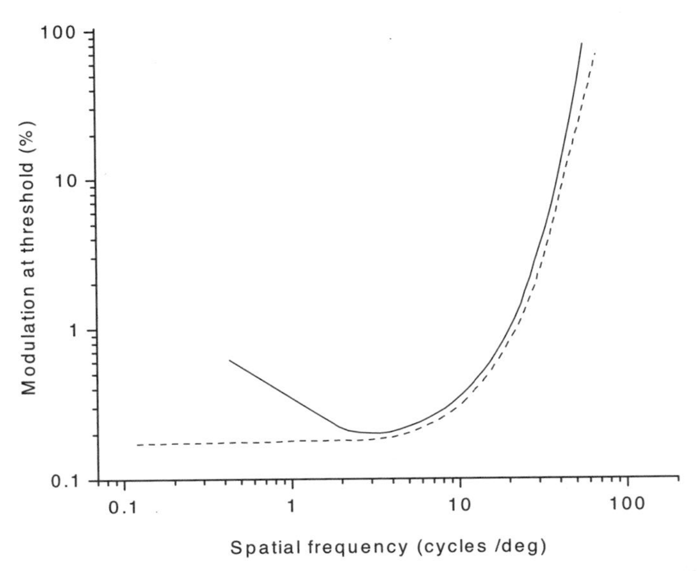

FIGURE 1.23 *Solid line*: Threshold for spatial modulation as a function of the spatial modulation in cycles/degree obtained in the human visual system. *Dashed line*: Expected modulation transfer function of a lens under the same conditions.[5]

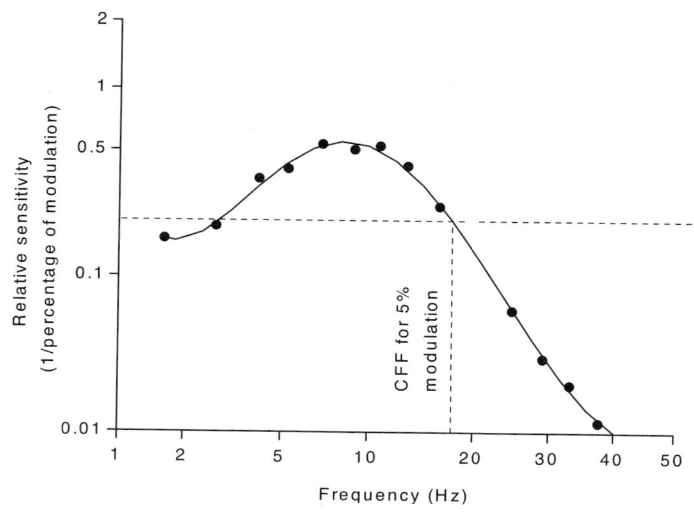

FIGURE 1.24 Critical fusion frequency (CFF) determined using sinusoidally intensity modulated light.[5]

the *critical fusion frequency* (CFF). The CFF is approximately 20 Hz, and the threshold to modulation above that modulation frequency increases at a steep rate. This means that the eye does not follow intensity variations above 25 Hz, well below the upper rate of modulation frequencies where the ear can detect amplitude modulation (approximately 500 Hz).

D. Spatial Resolution

Spatial resolution is the ability to discriminate between stimulations of two different points, such as two points on the skin or different locations on the retinal such as can be done by projecting vertical or horizontal stripes onto the optical field. The spatial resolution of the skin, known as the *two-point discrimination*, is defined as the minimum distance between two stimuli that can be perceived as separate stimuli. The test is typically carried out by placing a compass with sharp tips on the skin. The distance between the two points is varied until the observer perceives the stimulation as two distinct stimulations. The two-point discrimination determines the size of objects that can be discriminated by the somatosensory system. It is a function of the density of the innervation of the skin and is therefore different for different locations on the skin (Fig. 1.25). On palm skin, the distance between two points pressed on the skin that individual observers judged as being two separate stimuli was approximately 7 mm for 50% correct response (Fig. 1.26).[4]

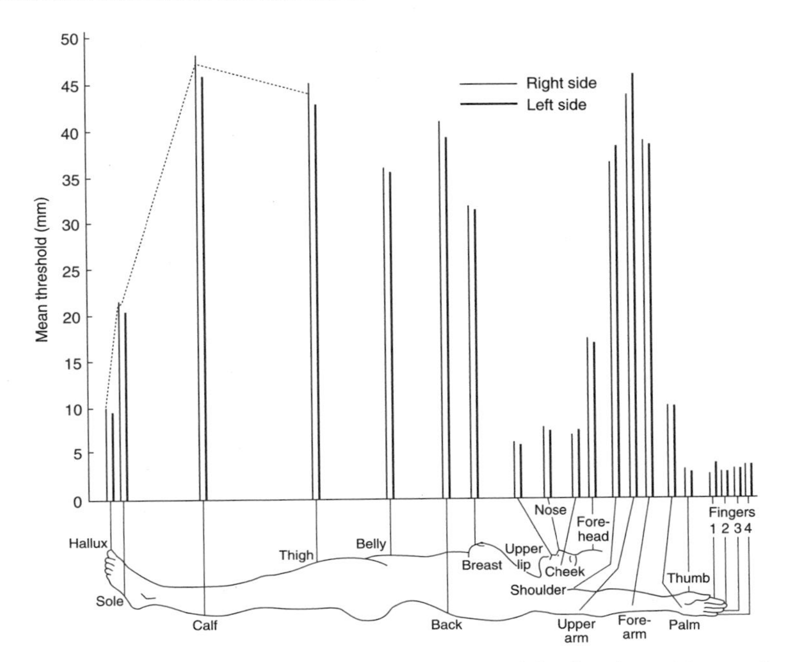

FIGURE 1.25 Two-point discrimination on different points of the skin. (From Mountcastle, V.B., ed., *Medical Physiology*, St. Louis: C.V. Mosby, 1974.)

The spatial resolution in vision is measured by projecting stripes on a screen. The distance between the stripes is decreased until individual stripes can no longer be discerned. The spatial resolution is expressed as the spatial frequency, which is the number of repetitions of stripes per degree of visual angle.

These measures of spatial resolution may be comparable to the frequency discrimination of hearing, which is a measure of how much the frequency of two tones must differ in order for them to be perceived as being different. Frequency discrimination in hearing is more complex because it depends not only on the (spatial) representation of frequency along the basilar membrane but also on the time pattern of the sound (see Chapter 5).

1. Lateral Inhibition (Contrast Enhancement)

Spatial resolution, such as two-point discrimination in somesthesia, depends on the size of the *receptive field** of the sensory receptors in the skin.

*The receptive field of a receptor such as a mechanoreceptor in the skin is the region of the skin where mechanical stimulation exceeds the threshold of the receptor. Receptive fields of other receptors such as photoreceptors are defined in the same way. The receptive field of auditory receptors would be the range of frequencies of a pure tone that elicit a response.

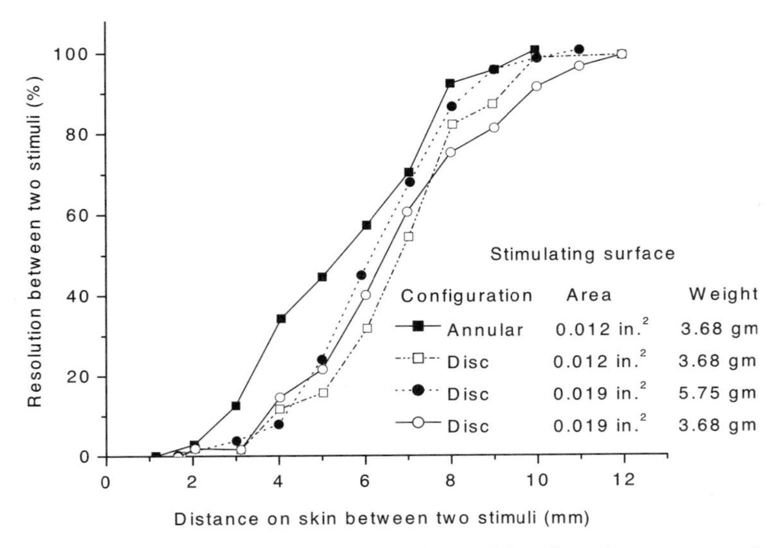

FIGURE 1.26 Correct responses to two-point stimulation of the palmar skin as two sensations as a function of the separation of the stimulations.[4]

Stimulation within a receptive field increases the discharge rate in the nerve fiber that innervates the receptor in question. Stimulation of the skin at a single point elicits response from a population of neurons, the size of which depends on the size of the receptive fields of the receptors in question and of the intensity of the stimulation. Stimulation of two points that are close together elicits responses from two populations of sensory nerve fibers that overlap. The ability to distinguish two locations of stimulation depends on the degree of separation of these two populations of neurons which depends on the size (width) of each of these populations. In vision, the separation between such two populations of neurons determines the *contrast* between a light and a darker area.

The contrast between lighter and darker parts of a visual image or between stimulation of two points on the skin is enhanced by *lateral inhibition*.[23] The physiologic basis for lateral inhibition is an arrangement of inhibitory receptive fields that are located on each side of (excitatory) receptive fields. Stimulation within these inhibitory fields decreases the excitation caused by stimulation within the receptive field (the excitatory area). Lateral inhibition was first described in the eye of the horseshoe crab *limulus*.[10] Lateral inhibition increases contrast by making the population of neurons that is activated narrower, and it can be demonstrated by observing dark stripes adjacent to less dark stripes (Fig. 1.27).[23] The stripes in Fig. 1.27 have the same density but the part of a lighter stripe that is adjacent to a darker stripe appears lighter than the parts of the stripe that are farther away from the transition. This demonstrates the enhancement of contrast

FIGURE 1.27 Homogeneous black bands with different density placed together to illustrate the enhancement of contrast.[23]

that is a result of lateral inhibition. Similar enhancement of contrast can be demonstrated in somesthesia where lateral inhibition enhances two-point discrimination. The response areas of cochlear sensory cells (and auditory nerve fibers) also have inhibitory areas for frequencies below and above that to which they respond best, but the effect on (frequency) contrast is more complex than that of vision and somesthesia (see Chapters 3 and 5).

2. Discrimination of Direction to the Source of a Stimulus

Paired sensory organs such as hearing and vision have the ability to discriminate the direction of a stimulus to the stimulus source on the basis of the difference in the stimuli arriving at the two sensory organs. These two sensory systems also have this ability independent of the other system. Stereoscopic vision is the basis for perception of depths and for creation of the sensation of space (three-dimensional images). Binaural hearing is the basis for the sensation of (auditory) space (stereophonic sound).[3]

Binaural hearing provides information about the direction to a sound source which has been important for many species during evolution for locating prey and for avoiding becoming prey. The use of binaural hearing to create a three-dimensional image of the environment is perhaps most developed in certain animals that are active in low-light environments such as the flying bat, which uses echolocation for navigation and hunting for food, or the owl, which uses binaural hearing to locate sound-emitting prey (mice).

The physical basis for directional hearing is the difference in arrival time of sounds at the two ears and the difference in the sound intensity. The difference in time is created by the difference in traveling time from a sound source to the two ears and is a direct function of the azimuth to the sound source. The difference in the sound intensity at the two ears is a result of the acoustic properties of the head (shadow and baffle effects; see Chapter 5).

These considerations refer to a person who is located in a free sound field, which means that sounds travel in a space where there are no obstacles. The reflections of sounds from the walls of rooms make the sound field in common rooms much more complex and sounds that arrive at the ears of an observer in such a situation consist of the direct wave; later, the reflected waves arrive. It was therefore a surprise to find that studies of sound localization in reverberant rooms could be done accurately.[40] The reason that we are able to determine the direction of a sound source under such circumstances is assumed to be the *precedence effect*,[9,44] which means that it is the sound that arrives first that is used by the central auditory nervous system to determine the direction to a sound source (see page 343).

The precedence effect can be demonstrated by having an observer sitting in an anechoic chamber with two loudspeakers placed at a certain distance in front of the observer with an angle between the directions to the loudspeakers of, for instance, 80 degrees. Simultaneous presentation of sounds from the two loudspeakers gives the impression of a (phantom) sound source that is located in between the two loudspeakers. Tone bursts presented to the loudspeakers with a small delay (less than 1 msec) give the impression of a sound source located somewhere in between the two loudspeakers, closer to the loudspeaker with the leading sound.[3] The direction depends on the delay between the sounds. When the delay is increased to between 1 and 30 msec, the observer gets the impression that the sound source is located at the lead loudspeaker independent of the sound that is generated by the second loudspeaker. This is assumed to be the phenomenon that suppresses echoes from the walls and ceiling of reverberant rooms and makes it possible to accurately determine the direction to a sound source independent of the echoes. This determination seems to be a result of the sound that first arrives at the observer's ears.

Echolocation using sound is not restricted to bats. Some sea animals use sound for navigation. The best known example is the dolphin, which emits sounds of various kinds and is known to use sounds for communication and navigation (echolocation). The bottlenose dolphin has been extensively studied in this respect.[1]

Determining the direction to a visual object depends on the location on the retina where the image is projected. Stereoscopic vision is based on the angle between the optical axis of the two eyes when the image is fused. Stereoscopic vision is useful for determining the distance to an object only for short distances. The distance to objects that are far away is based on knowledge about the size of the object.

Determining the direction to the source of odors is mainly a result of change in the concentration when the observer moves. Theoretically, it might be possible to determine the direction based on the different concentrations that arrive at the two nostrils, but the practical importance of that is doubtful.

3. Determination of Movement

Presenting stimuli to each of paired sense organs with a small time difference creates a sense of motion. Two brief flashes of light presented with a small interval are perceived as a single spot of light that moves. In the same way, presenting sounds to the two ears in such a way that the sounds reach the two

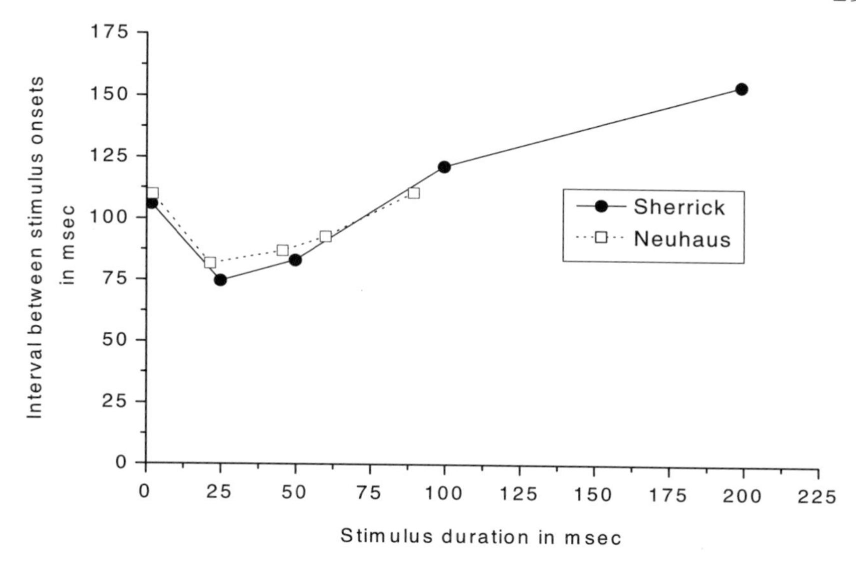

FIGURE 1.28 The threshold of feeling an apparent motion from presentation of two stimuli to the skin shown as the interval between the stimuli as a function of their duration.[33]

ears one after the other with a short interval can give the perception of a moving sound source. It is also possible to create a sensation of movement by stimulating the skin at two locations within a small time interval. Two vibratory stimuli presented to the skin at two different locations of the skin with a certain small time interval between them can create a sensation of motion (*apparent motion*) from the first stimulated location to the later stimulated location (Fig. 1.28).[33]

4. Specialized Systems for Directional Hearing in Animals

Perhaps the simplest example of binaural hearing systems in animals is that of the night moth. The night moth has four cells (two sets of two, with each set being located on each side of its belly)[2,25,26] that enable it to determine the direction of sound emitted by the echolocating bat that has the moth as it prey. The two pairs of cells have different sensitivities, making it possible for the night moth to determine the best escape route. The spectral sensitivity of the cells matches the sound emitted by bats. If the most sensitive set of receptors is activated, such as will occur when a bat approaches the moth at a distance of 30 to 40 meters, the moth will turn away from the bat to escape. If the moth is surprised by a bat at close range (2 to 3 m), then the two less-sensitive cells are activated, which causes the moth to use all means available to escape, such as power diving, flying in loops, or folding its wings together to fall toward the ground.[25] It is interesting that the moth has developed such a sophisticated escape tool in view of the fact that the moth is phylogentically much older than the bat.

REFERENCES

1. Barlow, H.B., Temporal and spatial summation in human vision at different background intensities. *J. Physiol. (London)*, 141: 337–350, 1958.
2. Bennett, T.L., *The Sensory World*. Monterey, CA: Cole Publishing Company, 1978.
3. Blauert, J., *Spatial Hearing: The Psychophysics of Human Sound Localization*. Cambridge, MA: MIT Press, 1983.
4. Carmon, A., Stimulus contrast in tactile resolution. *Percept. Psychophysiol.*, 3: 241, 1968.
5. Cornsweet, T.N., *Visual Perception*. New York: Academic Press, 1970.
6. Craig, J.C., Capacities and limitations of tactile processing. In: *Information Processing in the Somatosensory System* edited by O. Franzén and J. Westman, London: MacMillan, 1991.
7. Dadson, R.S. and King, J.H., A determination of the normal threshold of hearing and its relation to the standardization of audimeters. *J. Laryngol. Otol.*, 66: 366–378, 1952.
8. Eddins, D.A. and Green, D.M., Temporal integration and temporal resolution. In: *Hearing*, edited by B.C.J. Moore, San Diego: Academic Press, 1995, pp. 207–242.
9. Gardner, M.B., Historical background of the Haas or precedence effect. *J. Acoust. Soc. Am.*, 43: 1243–1248, 1968.
10. Hartline, H.K., Response of single optic nerve fibers of the vertebrate eye to illumination of the retina. *Am. J. Physiol.*, 121: 400–415, 1938.
11. Hecht, S., Schlaer, S. and Pirenne, M.H., Energy, quanta, and vision. *J. Gen. Physiol.*, 25: 819–840, 1942.
12. Hirsh, I.J. and Bilger, R.C., Auditory threshold recovery after exposures to pure tones. *J. Acoust. Soc. Am.*, 27: 1186–1194, 1955.
13. Hirsh, I.J., Auditory Perception of temporal order. *J. Acoust. Soc. Am.*, 31: 759–767, 1959.
14. Hirsh, I.J. and Sherrick, C.E., Percieved order in different sense modalities. *J. Exp. Psychol.*, 62: 423–432, 1961.
15. Lewis, D.B. and Grower, D.M., *Biology of Communications*. New York: Wiley, 1980.
16. Meisami, E., Mikhail, L., Baim, D. and Bhatnagar, K.P., Human olfactory bulb: aging of glomeruli and mitral cells and a search for the accessory olfactory bulb. *Ann. N.Y. Acad. Sci.*, 855: 708–715, 1998.
17. Moore, B.C.J., Psychophysical tuning curves measured in simultaneous and forward masking. *J. Acoust. Soc. Am.*, 63: 524–532, 1978.
18. Moore, B.C.J., *An Introduction to the Psychology of Hearing*. London: Academic Press, 1989.
19. Moore, B.C.J., *Hearing*. San Diego: Academic Press, 1995.
20. Mountcastle, V.B., LaMotte, R.H. and Carli, G., Detection thresholds for vibratory stimuli in humans and monkeys: comparison with threshold events in mechanoreceptive first order afferent nerve fibers innervating monkey hands. *J. Neurophysiol.*, 35: 122, 1972.
21. Mountcastle, V.B., ed., Neural mechanisms in somesthesia. In: *Medical Physiology*, St. Louis: C.V. Mosby, 1974.
22. Plomp, R., Rate of Decay of Auditory Sensation. *J. Acoust. Soc. Am.*, 36: 277–282, 1964.
23. Ratliff, F., *Mach Bands. Quantitative Studies on Neural Networks in the Retina*. San Francisco: Holden-Day, 1965.
24. Riesz, R.R., Differential intensity of the ear for pure tones. *Phys. Rev.*, 31: 288–306, 1928.

25. Roeder, K.D. and Treat, A.E., Detection and evasion of bats by moths. *Am. Sci.*, 49: 135–148, 1961.
26. Roeder, K.D., Acoustic alerting mechanisms in insects. *Ann. N.Y. Acad. Sci.*, 188: 63–79, 1971.
27. Scharf, B., *Critical Bands*. New York: Academic Press, 1970.
28. Schmidt, R.F., *Fundamentals of Sensory Physiology*. New York: Springer-Verlag, 1981.
29. Shannon, C.E., *The Mathematical Theory of Communication*. Urbana: University of Illinois Press, 1949.
30. Shepherd, G.M., *Neurobiology*. New York: Oxford University Press, 1994.
31. Shower, E.G. and Biddulph, R., Differential Pitch Sensitivity of the Ear. *J. Acoust. Soc. Am.*, 3: 275–287, 1931.
32. Sivian, L.J. and White, S.D., On minimum audible sound fields. *J. Acoust. Soc. Am.*, 4: 288–321, 1933.
33. Srinivasan, M.A. and LaMotte, R.H., *Encoding of Shape in the Responses to Cutaneous Mechanoreceptors*. New York: Macmillan, 1991.
34. Stevens, S.S. and Davis, H., *Hearing: Its Psychology and Physiology*. New York: Wiley, 1938.
35. Stevens, S.S., On the psychophysical law. *Psychol. Rev.*, 64: 153–181, 1957.
36. Stevens, S.S., The Psychophysics of Sensory Function. *Am. Sci.*, 48: 226–253, 1960.
37. Swets, J.A., *Signal Detection and Recognition by Human Observers*. New York: Wiley, 1964.
38. Viemeister, N.F., Temporal modulation transfer function based upon modulation thresholds. *J. Acoust. Soc. Am.*, 66: 1364–1380, 1979.
39. Wald, G., Human vision and the spectrum. *Science*, 101: 653–658, 1945.
40. Wallach H., Newman, E.B. and Rosenzweig, M.R., The precedence effect in sound localization. *Am. J. Psychol.*, 62: 315–336, 1949.
41. Watson, C.S. and Gengel, R.W., Signal duration and signal frequency in relation to auditory sensitivity. *J. Acoust. Am.*, 46: 989–997, 1969.
42. Wilska, A., Eine Methode Zur Bestimmung Der Horschwellenamplituden Des Trommelfells Bei Verschiedenen Frequenzen. *Skand. Arch. F. Physiol.*, 72: 161–165, 1935.
43. Yost, W. and Nielsen, D.W., *Fundamentals of Hearing*. New York: Holt, Rinehart & Winston, 1985.
44. Zurek, P.M., The precedence effect. In: *Directional Hearing*, edited by W. Yost and G. Gourevitch. New York: Springer-Verlag, 1987.

CHAPTER 2

Anatomy and Physiology of Sensory Organs

ABBREVIATIONS

- **CN** : Cranial nerve
- **CNS** : Central nervous system
- **dB** : Decibels
- **EPSP** : Excitatory postsynaptic potential
- **FTC** : Frequency threshold curves of frequency tuning curves
- **HL** : Hearing level (sound level in decibels above normal threshold)
- **IHC** : Inner hair cells
- **IPSP** : Inhibitory postsynaptic potential
- **OHC** : Outer hair cells
- **RA** : Rapid adapting (mechanoreceptors)
- **SA** : Slow adapting (mechanoreceptors)
- **μm** : Micrometer (1/1000 of one millimeter)

ABSTRACT

1. Sensory organs contain sensory receptors and structures that conduct the physical stimulus to the receptors and which ultimately provide conscious awareness of physical stimuli.
2. The ear consists of the outer ear, the middle ear, and the cochlea (part of the inner ear). The cochlea contains the auditory sensory cells, namely the inner and outer hair cells.
3. The eye consists of the pupil, the lens, and the retina. The retina contains the photoreceptors (rods and cones) and a neural network that processes visual information and conducts it to the optic nerve.
4. The skin has mechanoreceptors and thermoreceptors that respond to innocuous mechanical stimulation and to temperatures varying within a narrow range from the body temperature (somesthesia). Nociceptors in the skin, joints, muscles, and viscera respond to noxious stimuli such as certain chemicals, tissue injury, and heat or cold.
5. The chemical senses, smell and taste, respectively, have receptors in the olfactory epithelia in the nose and in the tongue, soft palate, and glottis.
6. Sensory receptors convert a physical stimulus into a code of nerve impulses in the fibers of the afferent nerve that innervates the sensory cells.
7. All receptors are specific in regard to the modality with which they respond (except polymodal pain receptors).
8. Photoreceptors (cones) respond to light of different wavelengths, and olfactory and taste receptors respond selectively to different chemicals (odors and tastes).
9. Sensory receptor cells are of two principally different types:
   a. Type I: A specialized membrane on the distal portion of the afferent nerve fiber that projects to the central nervous system.
   b. Type II: A specialized receptor cell that connects to one or more afferent nerve fibers by synapse-like structures.
10. Sensory transduction occurs when the specialized region of the cell membrane is activated by the physical (including chemical) stimulus to which it is sensitive and causes a change in the membrane conductance and a subsequent change in the membrane potential (generation of a receptor potential).
11. In receptors where the specialized membrane is located on the axon of the afferent nerve fiber (Type I), the receptor potential is conducted electrotonically to the site of impulse generation (the first node of Ranvier).

12. In receptors where the afferent axon makes synaptic contact with the receptor cell (Type II), the receptor potential is conducted electrotonically to the synapse.
    a. In hair cells of the cochlea and taste receptor cells, impulse generation occurs in the axon that connects to the synapse.
    b. The synapses of photoreceptors and olfactory receptors connect to other cells in which the first impulse generation occurs.
13. Receptors have been characterized by their receptive field, which describes the area of a receptor surface where stimulation can cause the afferent fiber of the receptor cell to respond.
14. Adaptation, temporal and spatial integration, and compression of amplitude are common properties of all sensory organs.
    a. Adaptation attenuates steady-state and slowly varying stimuli and enhances rapid changes in stimulus intensity.
    b. Temporal integration decreases the threshold for repeated stimuli and stimuli of long duration.
    c. Spatial integration decreases the threshold of stimuli that activate a large group of receptors.
15. The medium that conducts the stimulus to the receptors modifies the stimulus in different ways by providing adaptation (skin), frequency selectivity (cochlea), and amplitude compression (ear and eye).

I. INTRODUCTION

The main function of sensory receptors is to transform physical stimuli into trains of nerve impulses in the afferent nerve fibers that innervate the receptors. Receptors only respond to physical stimuli of certain types. The properties of the receptors and the media that transmit the physical stimuli to the receptors determine the range of stimuli that are coded in the discharge pattern of the afferent nerve fibers. The frequency and intensity range of hearing, the wavelength range of visible light, and the range of vibrations that can be sensed by skin receptors, along with the range of chemicals that can be sensed by the chemical senses, depend on the properties of the receptors and the media that transmit the physical stimuli to the receptors.

Sensory organs that respond to innocuous stimulation perform two important kinds of tasks for the organism: (1) detect a physical stimulus that reaches one of its sensory organs and (2) communicate that information to the sensory nervous system where extraction of useful information occurs. Sensory receptors that respond to noxious stimulation warn and protect the animal from trauma and other dangers to their existence. Early in the evolution of sensory systems, the ability to detect weak stimuli was probably the most important task of sensory systems, and most sensory organs can respond to very weak physical stimuli. Later, interpretation of the messages that physical stimuli carried became important for avoiding dangers and for detecting prey, leading to the development of the sensory nervous systems. Some animals developed specialized sensory organs and neural processing schemes for specific purposes.

The sensitivity of sensory systems depends on the sensitivity of sensory cells and the efficiency with which the physical stimulus is conducted to the sensory cells. Many sensory cells such as photoreceptors and mechanoreceptors in the ear (*hair cells*) have a very high sensitivity that is accomplished by a very efficient transduction process. The efficiency of the structure that conducts the stimulus to the receptors can also improve the sensitivity of sense organs. The middle ear and the cochlea improve transmission of a physical stimulus to the receptors many-folds, and in some animals the backside of the eye reflects light so that light passes twice through the receptors, which increases the sensitivity of the eye in such animals.

Interpretation of sensory input depends on the processing of information that occurs in the different sense organs and in the sensory nervous system. The medium that conducts the stimulus to the mechanoreceptors in the skin can suppress slow changes in deformation of the skin (adaptation) before such stimuli reach the receptors. The processing that occurs before sound reaches the mechanoreceptors in the ear is more extensive and complex than what occurs in other sensory systems. The cochlea, which conducts sounds to the receptors (hair cells), possesses specific selectivity with regard to the spectrum of sounds, and the basilar membrane of the cochlea performs a form of spectral analysis by separating the audible spectrum into narrow bands before the sound reaches the receptors. The cochlea is an extreme example of a sensory organ in which selectivity of the medium that conducts the stimulus to the receptors provides extensive signal analysis of sensory stimuli before the sensory cells are activated.

Some sensory cells have properties that can separate physical stimuli according to their properties. Some photoreceptors (*cones*) respond specifically with regard to the wavelength of light, they separate light according to its wavelength serving as the basis for color vision. Olfactory and taste receptors are likewise selective with regard to the chemical substances to which they respond, and that selectivity is important for discriminating different tastes and odors.

It is interesting to note that the spectral selectivity in the ear is a result of the properties of the medium conducting the stimulus to the receptors, while in the eye a similar spectral selectivity (for color vision) is a result of the properties of the receptor cells. The cochlear hair cells have no specific selectivity in sensing the vibration of the basilar membrane. Likewise, the mechanoreceptors of the skin have little or no specificity regarding the stimuli but the medium that conducts the stimulus to the receptors provides adaptation, which suppresses stimuli that changes slowly.

The most extensive signal processing of sensory information occurs in the Central Nervous System (CNS). Interpretation of processed neural activity that is evoked by sensory stimulation is a result of the extensive signal processing that occurs in the CNS and will be discussed in the following chapters.

This chapter will discuss the similarities and the differences between various sensory organs and sensory cells and will provide an overview of the anatomy and function of sensory organs and the commonalties regarding the signal processing that occurs in sensory organs for the different senses.

This chapter first provides a brief description of the anatomy of the five sense organs and their receptors, which is followed by discussions of the basic physiology of the sensory organs and the mechanisms of sensory transduction. The detailed anatomy and physiology of the different sensory systems will be covered in subsequent chapters (4 through 7), each of which is devoted to an individual sense.

II. ANATOMY OF SENSORY ORGANS

The sensory receptors of the five primary senses are located in the special sensory organs of hearing, vision, olfaction, and taste and in the somatosensory receptors that are distributed in the skin over the entire body. A physical stimulus that reaches a sense organ is conducted to the receptors by a medium that transforms the stimulus in various ways. The eye also contains a neural network that processes visual information before it enters the CNS through the optic nerve. The sense organs communicate sensory information to the CNS through sensory nerves, several of which are cranial nerves or parts of cranial nerves (olfaction, CN I; vision, CN II; taste, CN VII, IX, and X; hearing, CN VIII). Skin receptors are innervated by spinal nerves (dorsal roots) and the sensory portion of the trigeminal nerve (CN V) for the face. Skin receptors include receptors that respond to noxious stimuli (nociceptors) and are also found in viscera.

A. The Ear

The ear consists of three parts, the *outer ear*, the *middle ear*, and the *cochlea* (Fig 2.1). The acoustic effect of the head and the ear canal affects the sound that reaches the tympanic membrane in a way that depends on the direction of a sound source. The cochlea, in which the sensory cells for hearing are located, is fluid filled. Sound is converted to vibrations of the cochlear fluid by the middle ear.

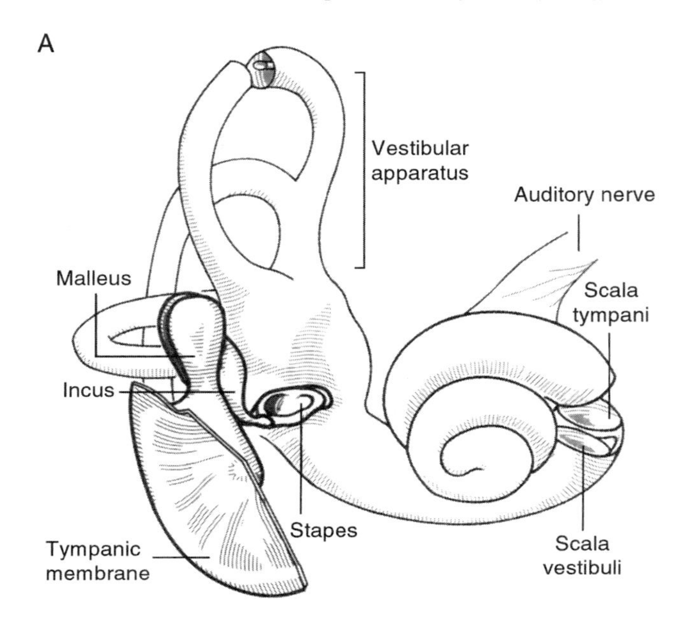

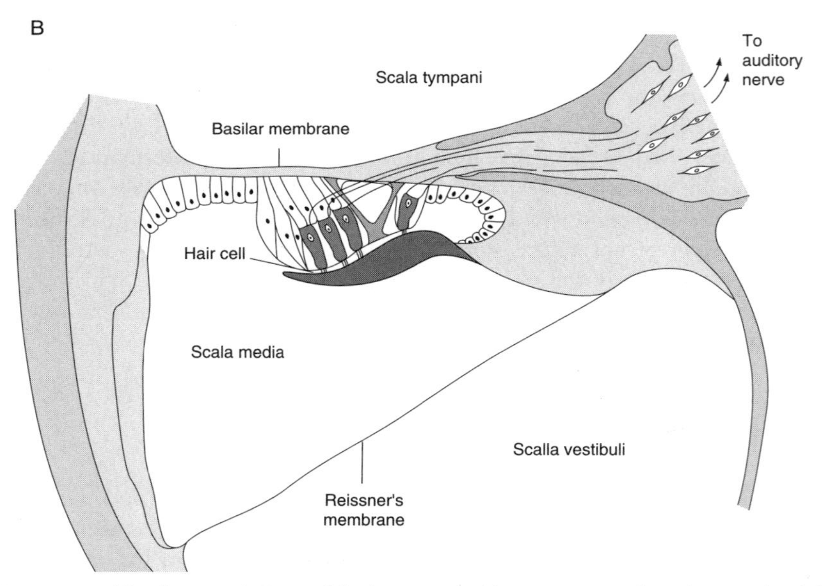

FIGURE 2.1 (A) Schematic drawing of the human ear. (B) Cross-section through one side of the second turn of a guinea pig cochlea, as indicated in (A). (Adapted from Møller, A.R., *Ambio*, 4: 6–13, 1975.)

1. The Middle Ear

The middle ear consists of the *tympanic membrane*, which converts sound from the air into vibrations of the three small bones (ossicles: *malleus, incus*, and *stapes*). The outer ear consists of the pinna and the ear canal, which modify the spectrum of the sound that is conducted to the cochlea in a way that depends on the direction of the sound source (see Chapter 5 and Møller[23]).

The *manubrium of the malleus* is imbedded in the tympanic membrane (Fig. 2.1A). The malleus is firmly connected to the incus by a joint that does not allow movement. The incus connects to the stapes with a joint that allows motion in a direction that is perpendicular to the normal direction of the motion induced by sound. The *footplate* of the stapes rests in the *oval window*, which is one of the two openings in the cochlear capsule. The other opening, the *round window* is covered by a membrane. The cochlear capsule is rigid but the flexible membrane of the round window allows the cochlear fluid to move in response to the in- and outward motion of the stapes footplate. When the stapes footplate moves inward, the round window moves outward. The middle ear acts as an *impedance transformer* that improves the transmission of sound to the cochlear fluid. This is mainly accomplished due to the difference in the surface areas of the tympanic membrane and the footplate of the stapes. The transformer action of the middle ear improves sound transmission from air to vibrations of the fluid in the cochlea by slightly less than 30 dB.[32] However, the real gain, compared with a situation where sound reaches both the oval and the round windows of the cochlea, is much larger because it is the difference in the sound reaching the two windows that sets the cochlear fluid in motion and acts as the efficient acoustic stimulation of the ear. Sound that reaches both windows of the cochlea with the same intensity and in the same phase will not cause any motion of the cochlear fluid.[23] The increased sound transmission to the oval window that is normally accomplished through the transformer action of the middle ear provides a large difference between the force that acts on the two windows of the cochlea, making sound transmission to the cochlear fluid more efficient.

Two small muscles are located in the middle ear. One, the *tensor tympani muscle*, is attached to the manubrium of the malleus and pulls the tympanic membrane inward when it contracts. The other muscle, the *stapedius muscle* (the smallest striate muscle in the body), is attached to the stapes and pulls the stapes in a direction that is perpendicular to its normal piston-like movement. The tensor tympani muscle is innervated by the trigeminal nerve (CN V) and the stapedius muscle is innervated by the facial nerve (CN VII). In humans, the stapedius muscle contracts as an *acoustic reflex* in response to strong sounds (above approximately 85 dB hearing level [HL]). In many animals both muscles contract as an acoustic reflex. Contraction of either one of these two muscles decreases the sound transmission to the cochlea, but the stapedius muscle is the most effective one in reducing the sound transmission to the cochlea (for more details, see Møller[23]).

2. The Cochlea

The snail-shaped cochlea has 2¼ turns in humans (Fig. 2.1A). The length of the cochlea is approximately 3.5 cm when unrolled. The outer ear and the middle ear are usually regarded as the sound-conducting parts of the ear but the cochlea also participates in conducting the physical stimulus to the receptor cells and it may therefore be regarded as a part of the sound-conducting apparatus.

The fluid-filled space of the cochlea is separated longitudinally to form three separate compartments known as *scalae*: *scala tympani* and *scala vestibuli* and, in the middle, *scala media* (Fig. 2.1B). The scala media is separated from the two other scalae by *Reissner's membrane* and the *basilar membrane*. The fluid in the scala media, *endolymph*, has a different ionic composition than that of the two other scalae, which contain *perilymph*. Endolymph has an ionic composition similar to that of intracellular fluid (rich in potassium and poor in sodium), and the perilymph has a composition similar to extracellular fluid (rich in sodium and poor in potassium).

The receptor cells, known as hair cells, are located in the *organ of Corti*, which is located along the *basilar membrane* and divides the cochlea longitudinally (Fig. 2.1B). The two types of hair cells are *inner hair cells* and *outer hair cells*. The hair cells are located along the basilar membrane in a row of inner hair cells and three to five rows of outer hair cells (Fig. 2.1B).

The auditory nerve, which innervates the hair cells, is a part of CN VIII. It passes through the *internal auditory meatus* into the skull cavity, together with the vestibular portion of the CN VIII and the facial nerve, CN VII (and *nervus intermedius*).

B. The Eye

The eye globe is formed by the *sclera* outside and by the *choroids* inside (Fig. 2.2). In between these two layers (tunics) is a *vascular layer* that is pigmented in humans (Fig. 2.2). In the front, the sclera continues in the *cornea* and the choroids become the *ciliary body* and the *iris*—so named because of its color. The *pupil*, an opening in the iris located in front of the eye, can regulate the light that reaches the *retina*. The retina, which contains the photoreceptors and a neural network that processes the output of the photoreceptors, is located at the inside of the back wall of the eye.

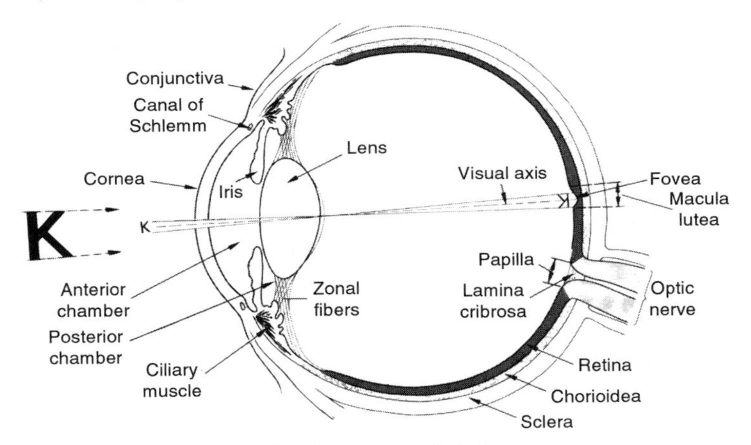

FIGURE 2.2 Cross-section of the human eye.

In the human eye, the wall behind the photoreceptors is covered with a dark (pigmented) layer. In many animals, the wall behind the photoreceptors reflects light, which causes light to travel through the layer of photoreceptors twice. Such a reflecting layer is present in the eyes of mammals that are mainly nocturnal, such as the cat and the cow, causing the eyes to glow in the dark when hit by light. The fact that light passes twice through the photoreceptors increases the sensitivity of the eye, but the slight difference between the direct and the reflected image can blur the image and reduce the spatial resolution of such eyes. The pigmented layer behind the photoreceptors in humans and other primates prevents light from being reflected after it has passed the photoreceptor and prevents the photoreceptors from being activated twice.

The eye globe is filled with the *vitrous body*, a semigelatinous transparent substance. A *lens* in front of the eye projects an (inverted) image on the retina. Small muscles (the *ciliary muscles*) can change the focal distance of the lens to focus images on the retina. When these muscles contract, they decrease the tension on the lens capsule from the zonal fibers, and the curvature of the lens increases, which decreases the distance to objects that are focused on the retina, thus the occurrence of near accommodation.*

*The focal distance (distance beyond a lens where incoming parallel light rays are focused to a single point) is a measure of the refractive power of a lens and is expressed in diopters (one divided by the focal length). Most of the refraction of the eye is a result of the interface between air and the cornea, and the lens contributes only a small portion of the refractive power. The optical system must be refocused to view objects at a short distance after viewing objects at a long (infinite) distance. In cameras, such refocusing is normally done by shifting the position of the lens relative to the film, but in the mammalian eye it is done by changing (increasing) the refractive power of the lens by making the lens more spherical. This is called *near accommodation*.

1. Retina

The sensory cells (*photoreceptors*) are located in the retina, which covers the inside back wall of the eye. The human eye has two kinds of photoreceptors: *rods* and *cones*. Rods are more light sensitive than cones, therefore, rods are used in detection of dim light. Cones provide color discrimination and fine resolution of vision. The neural network that is located in front of the sensory cells in the retina processes visual information before it reaches the optical nerve. Because the neural network is located in front of the photoreceptors, light must pass through the neural network in order to reach the photoreceptors, and it must even pass through the cell bodies of the photoreceptors to reach the light-sensitive part of the photoreceptors. That means that these structures must be transparent.

It has recently been shown that other light-sensitive cells, besides rods and cones, are found in the retina. These other light sensors in the retina are involved in resetting the body's clock,[30] which is important for maintaining correct circadian rhythm. This means that individuals whose normal retinal cones and rods are destroyed and are therefore blind can still maintain their circadian rhythm. These photosensitive cells, which were identified by Berson et al.,[3] use *melanopsin*[28] as the light-sensitive substance.

C. Skin

The sensory receptors of the skin are distributed over the entire body surface. There is some difference between the receptors located in *hairy skin* compared with those located in the *glabrous* (nonhairy) skin (Fig. 2.3).[14,24] Some receptors are located near the surface of the skin while others are located deeper in the skin which causes mechanical stimuli of the skin to be transformed differently before activating the receptors. The sensory receptors in the skin of the body are innervated by spinal nerves that travel in dorsal roots, and the receptors of the face and mouth are innervated by the sensory portion of the trigeminal nerve (CN V).

D. The Chemical Senses

The taste receptors are located in *papillae* on the surface of the tongue, the soft palate, and the glottis. The papillae vary in different parts of the tongue (Fig. 2.4). The olfactory epithelium, which contains the olfactory receptors (Fig. 2.5A), covers the surface of a complex system of *turbinates* located deep in the nose (Fig. 2.5B). Air is brought in contact with the olfactory receptors through turbulence over the turbinates (inferior, middle, and superior).

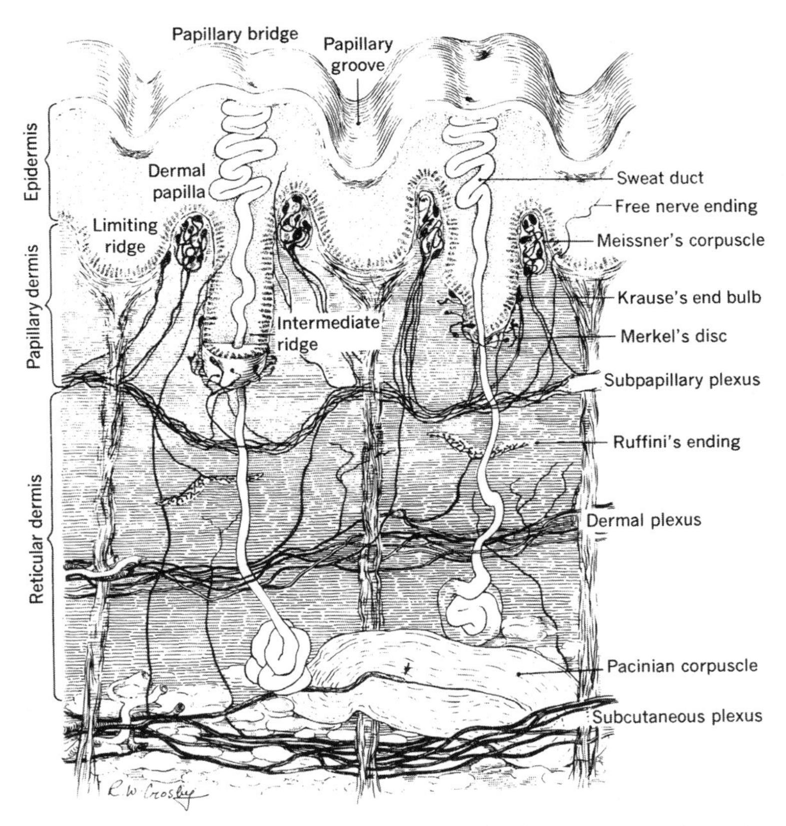

FIGURE 2.3 Section of glabrous (hairless) skin, showing the different receptors that are located in the three layers: epidermis, papillary dermis, and reticular dermis. (Adapted from Mountcastle, V.B., *Medical Physiology*, St. Louis: C.V. Mosby, 1974.)

III. SENSORY RECEPTORS

Sensory receptors transduce physical stimuli into a neural code that is conveyed to the CNS by the afferent nerve fibers that innervate the receptors. Sensory receptors consist of structures (specialized membranes) that are sensitive to specific physical stimuli, such as mechanical deformation, light, temperature, and specific chemicals. All receptors of innocuous stimulation are specific with regard to the modality of physical stimuli to which they respond. Some somatosensory receptors that respond to noxious stimuli respond to more than one modality of stimulation (*polymodal receptors*). These receptors belong to the nonclassical somatosensory pathways (see page 205).

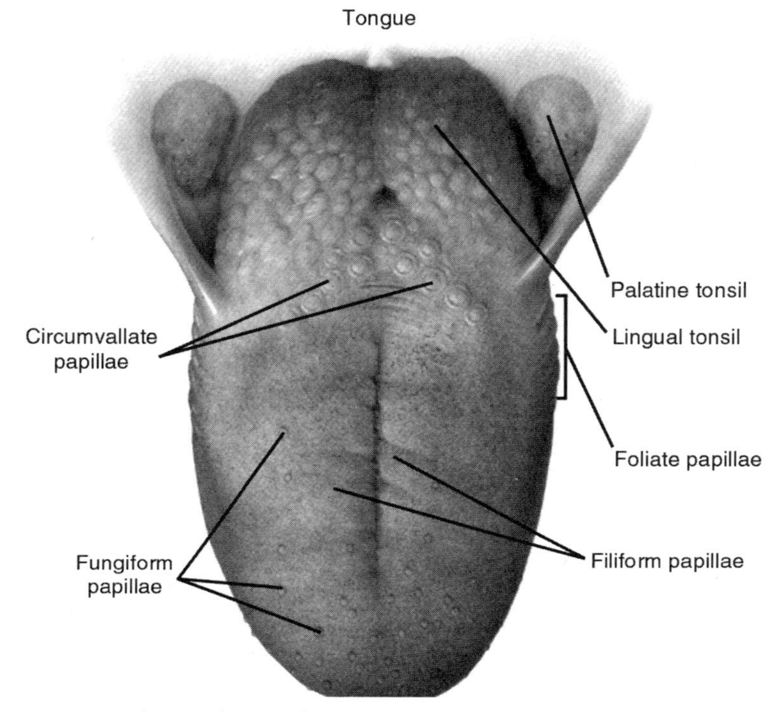

FIGURE 2.4 Distribution of taste cells (gustatory papillae) in the tongue and their innervation. (Adapted from Smith and Margolskee.[38])

Receptors of hearing and touch are mechanoreceptors, while chemical receptors are found in the sense organs of taste and smell. Light receptors (photoreceptors) are found in the eye. Thermoreceptors are found in the skin. Receptor cells that mediate the sensation of pain respond to chemicals, heat, and cold. Similar receptors are involved in the sensation of itch.

The anatomy and general function of sensory receptors are discussed below, and some important similarities are emphasized. The sensory receptors of the different senses will be discussed in more detail in the following chapters that deal with the individual senses (Chapters 4 to 7).

A. Anatomy of Sensory Receptors

Receptors that respond to the different kinds of stimuli have many morphologic and functional similarities, although the receptor cells at first glance may appear to have great differences morphologically (Fig. 2.6). The specialized region of

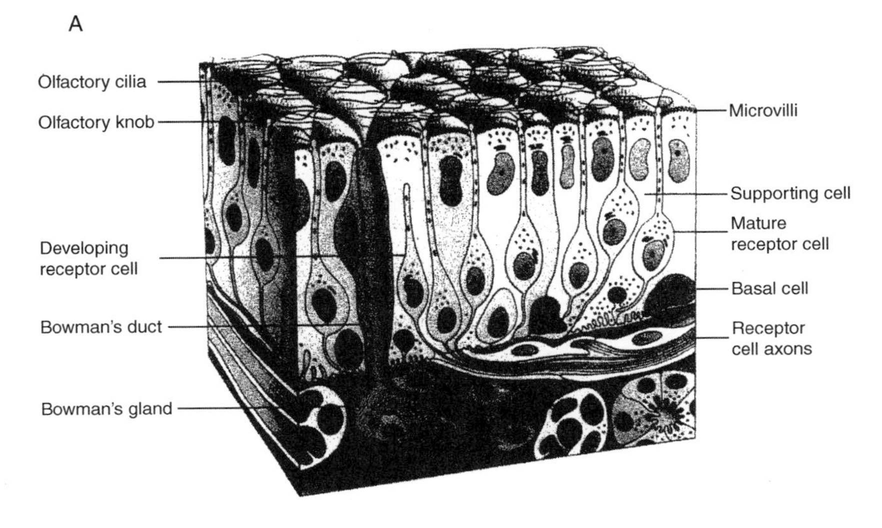

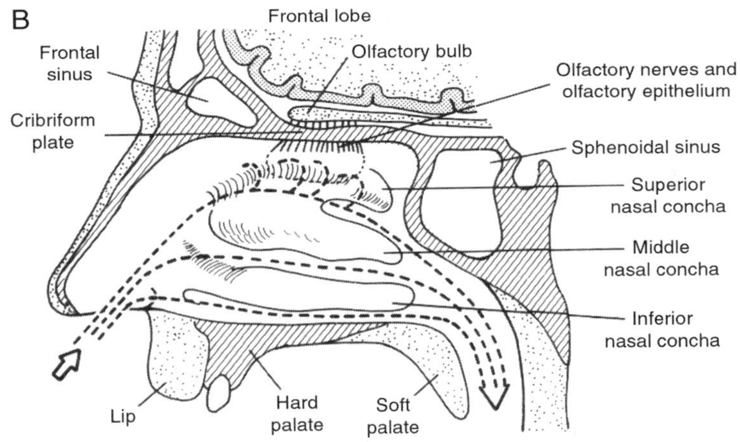

FIGURE 2.5 (A) Olfactory sensory epithelium. (Adapted from Warwick, R. and Williams, P.L., *Gray's Anatomy*, Philadelphia: W.B. Saunders, 1973.) (B) Human nose showing the location of the olfactory organ. Dashed lines indicate airflow. (Adapted from Shepherd, G.M., *Neurobiology*, New York: Oxford University Press, 1994.)

a membrane where the transduction of the physical stimulus occurs can be located on the distal portion of an axon that continues as the afferent nerve fiber that leads the information to the CNS. We will call such receptors *Type I* receptors. Type I receptors include mechanoreceptors, warm and cool receptors in the skin, and various kinds of nociceptors. Other receptors consist of specific cells to which the afferent nerve fiber connects via a synapse or synapse-like

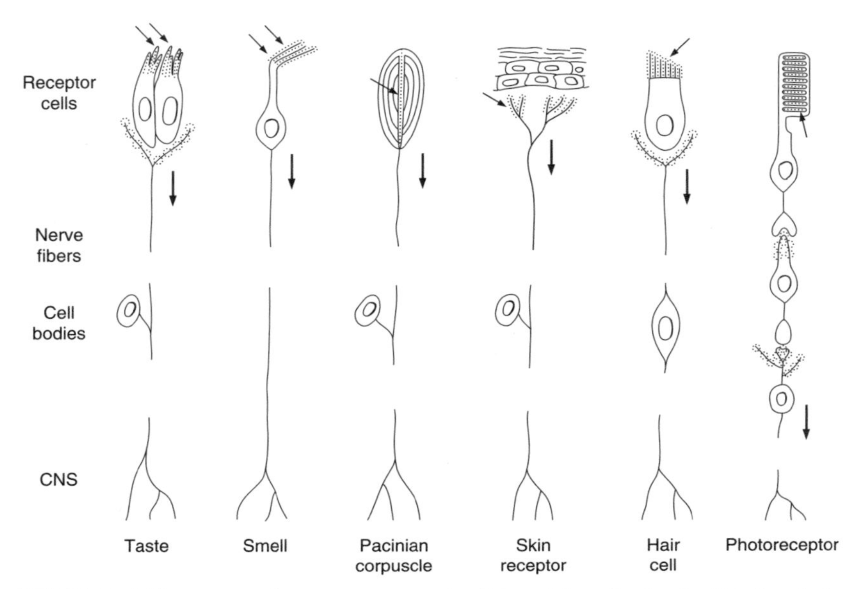

FIGURE 2.6 Different types of sensory receptor cells. (Adapted from Shepherd, G.M., *Neurobiology*, New York: Oxford University Press, 1994.)

structure. We will call such receptors *Type II* receptors. Type II receptors are found in the ear and the eye, and they serve as receptors in the chemical senses (olfaction*) and taste. The specialized membrane that is sensitive to physical stimuli is located in cilia (hair cells and photoreceptors) and villi (taste and olfactory).

1. Type I Receptors

The sensitive area of several kinds of mechanoreceptors found in the skin is the bare part of an axon that is surrounded by different kinds of tissue, which give the receptors their individual morphologic shapes. The properties of the tissue in which the sensitive region is imbedded modify the mechanical stimulus before it reaches the receptors (see pages 192 and 222).

The mechanoreceptors in the skin can be divided into five different types according to their morphology:[14] *Meissner's corpuscles* and *Merkel's disks*, which are located superficially in the skin, and *Pacinian corpuscles*, *Ruffini endings*, and *hair follicle receptors*, which are located deep in the skin. The receptors in the

*Olfactory cells may be regarded as Type II cells (as the distal axon is regarded as a part of the cell) or, if their axons are regarded to be the afferent fibers of the receptor cells, these cells may be regarded to be bipolar cells similar to the mechanoreceptors of the skin (Type I receptor cells) (see Chapter 7).

glabrous (hairless) skin are different from those found in hairy skin. Three different types of receptors are located in the glabrous skin: Meissner's corpuscles, Merkel's disk (Ruffini endings), and Pacinian corpuscles. Hairy skin contains Pacinian corpuscles, hair follicle receptors, tactile disks (*pincus domes*) and Ruffini endings (see Chapter 4 for details). The free nerve endings in the Pacinian corpuscles are surrounded by onion-shaped structures of alternating concentric layers of cellular membranes and fluid-filled spaces.

In addition to these mechanoreceptors, thermal receptors and nociceptors are also found in the skin. There are four kinds of thermal receptors: warmth, cool, heat, and cold.[13,17] The cold and heat receptors are regarded to be nociceptors because they mediate the sensation of pain. Little is known about the morphology of thermal receptors, but it has been assumed that they are bare axons with a specialized membrane for sensing temperature.

Other nociceptors are mainly free nerve endings that are located in different layers of the skin mucosa, tendons, joints, and viscera. Similar receptors located in the skin or mucosa mediate the sensation of itch. Some of these receptors are polymodal nociceptors that respond to several modalities of noxious stimuli, and they are mostly bare axons (see Chapter 4). Some nociceptors mediate the sensation known as itch, and some nociceptors are sensitive to *capsaicin*, the ingredient in red pepper.

2. Type II Receptors

Type II receptors are more complex than Type I receptors because transmission of sensory excitation involves a synapse and it is possible to modulate the sensitivity by efferent (descending) neural activity. Some of the receptors in this group are mechanoreceptors, such as those in the ear; others are sensitive to specific chemical substances such as the receptors for olfaction and taste. The photoreceptors also belong to this group of receptors.

The specialized region of such Type II receptor cells may be *cilia* (as in olfactory receptors), *stereocilia* (as in the hair cells of the cochlea), or *microvilli* (as in taste receptors). The outer segments of photoreceptors are specialized to be sensitive to light by the *photopigment* that is located in modified cilia.

a. Cochlear Receptors

The receptors (hair cells) in the cochlea are mechanoreceptors that sense the vibration of the basilar membrane. These cells are similar to the receptor cells in the *vestibular apparatus*. The hair cells in the vestibular apparatus have both *kinocilia* and *stereocilia*, but hair cells are stereocilia only (Fig. 2.7). The hair cells in the cochlea lose their kinocilia as they mature. A basal body can be identified in the place of the kinocilium in the cochlear hair cells.

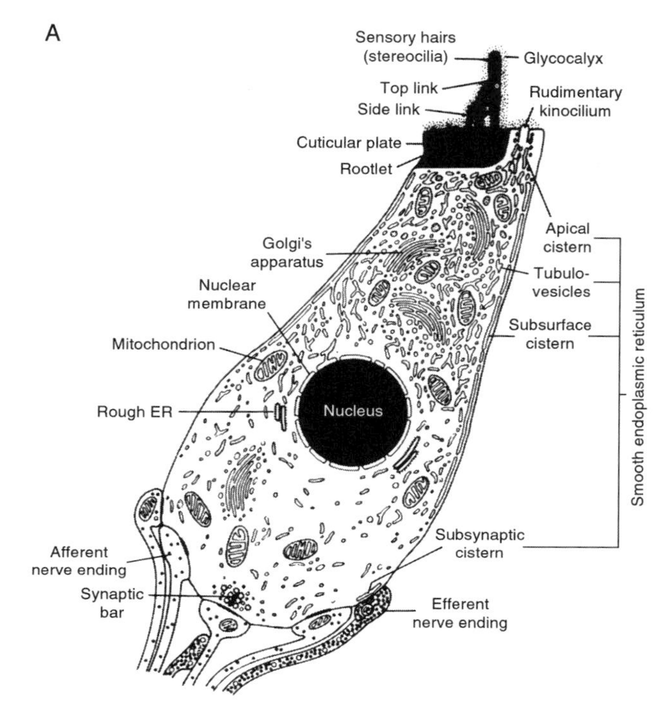

FIGURE 2.7 (A) Schematic drawing of an inner hair cell and (B) schematic drawing of the cross-section of an outer hair cell. (Adapted from Lim, D.J., *Am. J. Otolaryngol.* 7: 73–99, 1986.)

The morphology of inner hair cells and outer hair cells is slightly different. The inner hair cells are flask shaped (Fig. 2.7A), whereas the outer hair cells are cylindrical in shape (Fig. 2.7B). In both kinds of hair cells, the stereocilia emerges from a *cuticular plate*.

The stereocilia of the outer hair cells are arranged in W-shaped formations, whereas the cilia of the inner hair cells assume the shape of a flattened U. Inner hair cells (Fig. 2.7A) have approximately 60 stereocilia. Outer hair cells (Fig. 2.7B) have 50 to 150 stereocilia, the length of which are less than 2 μm (μm = micrometer, 1/1000 of one millimeter) in the basal end of the cochlea and approximately 8 μm in the apical end of the cochlea. (Compare that to the diameter of a human hair, which is approximately 100 μm, a red blood cell is approximately 7 μm in diameter.) The tallest tips of the stereocilia of the outer hair cells are imbedded in the overlying membrane (the *tectorial membrane*) (Fig 2.1B). The tips of the stereocilia of the inner hair cells are not embedded in the tectorial membrane.

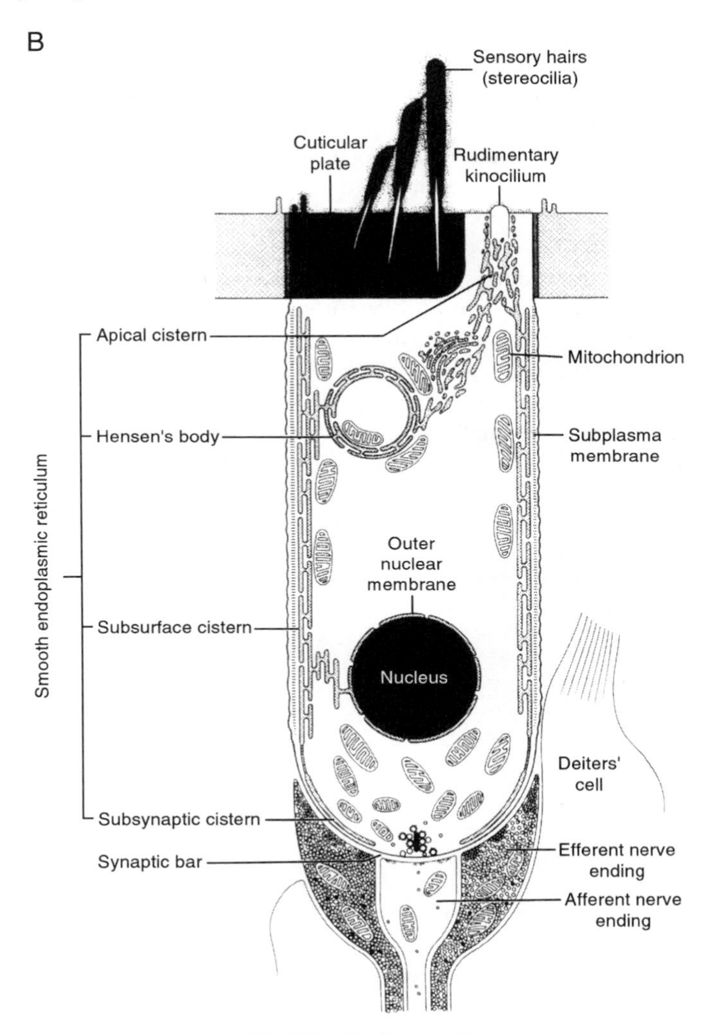

FIGURE 2.7 *(continued)*

Similarities in the anatomy of sensory receptors do not necessarily mean that there are similarities in their function. For instance, the two types of hair cells in the cochlea that are morphologically similar have completely different functions in that only the inner hair cells act as sensory cells that convert the motion of the basilar membrane into a neural signal. The outer hair cells are motile, participate in the motion of the basilar membrane, and act as "motors" that amplify the motion of the basilar membrane[4] (see page 301).

b. Photoreceptors

Two types of photoreceptors are found in the retina of vertebrates: cones and rods (Fig. 2.8). The outer segment of both types of receptors is the region that is sensitive to light. These cilia consist of stacks of disc membranes formed by folded plasma membranes. The photosensitive substance in the rods, which are the most sensitive photoreceptors, is *rhodopsin*. The eyes of mammals, which can discriminate color, have three different types of cones, each of which contains one of three different but related molecules of photopigment. The different pigments absorb light of different wavelengths, which makes the cells sensitive to different parts of the light spectrum—the basis for color discrimination. The inner segment of the photoreceptors contains the nucleus and the energy-producing apparatus of the cells (mitochondria). The photoreceptors end in a small stalk; the synaptic terminal that makes contact with cells in the neural network is located in the retina in front of the photoreceptors (Fig. 2.8). (See chapters 6 for details).

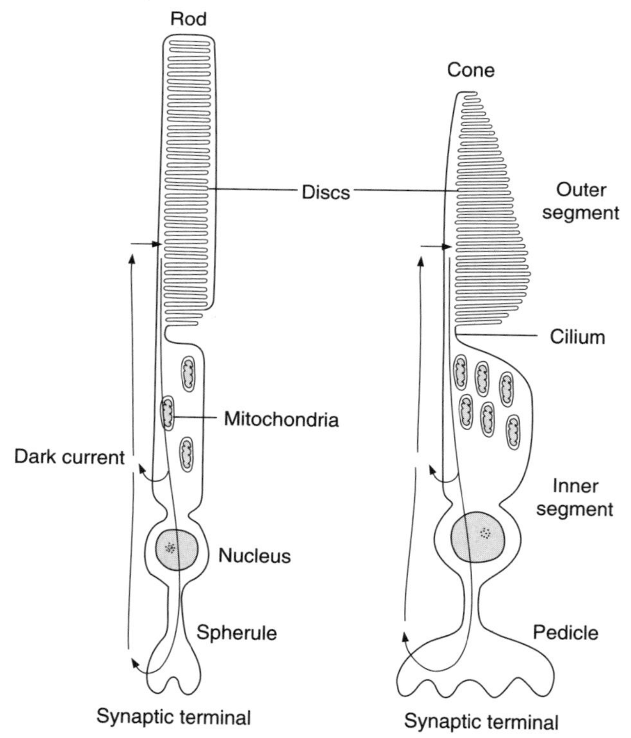

FIGURE 2.8 Schematic drawings of photoreceptors. (Adapted from Shepherd.[37])

c. *Taste Receptors*

Taste cells are transformed epithelium cells where the sensory transduction takes place in the *microvilli* that are parts of the taste cells. Various taste cells being sensitive to different chemicals is the basis for discrimination of taste. Taste cells are clustered in *taste buds*, which are located in papillae[26] (Fig. 2.9A) in the epithelium of the tongue. In humans, three different kinds of papillae are found in different regions of the tongue (Fig. 2.9B). Each taste bud contains 50 to 150 receptor cells (Fig. 2.9A). Saliva comes in contact with the microvilli on the taste cells through the *taste pore* (see Chapter 7). Taste cells undergo constant renewal; therefore, at any one time, cells can be found at different stages of their life cycle.

d. *Olfactory Receptors*

Olfactory sensory cells may be regarded as bipolar cells with dendrites that extend toward the surface of the olfactory epithelium and unmyelinated axons that extend superior towards the olfactory bulb. Cilia are located on a small knob at the end of these long, thin dendrites. The knobs have a diameter of 0.1 to 0.2 μm and the cilia may be up to 200 μm long. These slender dendrites extend to the surface of the epithelium and send out tufts of fine processes known as olfactory hairs (Fig. 2.10).[6] The axons collect in bundles and pass through the holes in the cribiform plate of the ethmoid bone (Fig. 2.10). The cilia that are located on top of these receptor cells are sensitive to different odors and thus provide the basis for discrimination of odors. The olfactory epithelium is remarkable because it undergoes a constant renewal where new dendrites and new knobs with cilia constantly develop and older ones degenerate. The axons of the receptor cells in the olfactory epithelium form the *olfactory nerve* (CN I) and make contact with nerve cells in a neural network in the olfactory bulb that is similar to that of the retina.[36]

B. Innervation of Sensory Cells

The afferent nerve fibers of Type I receptors are the continuation of the axons that contain specialized regions that serve as the receptors. The innervation of Type II receptors is more complex because the afferent nerve fibers connect to the receptor cells via a synapse-like structure. This makes it possible for more than one afferent fiber to innervate a receptor cell, and it also allows efferent innervation to modulate the afferent flow of information and to affect the function of the receptor cell.

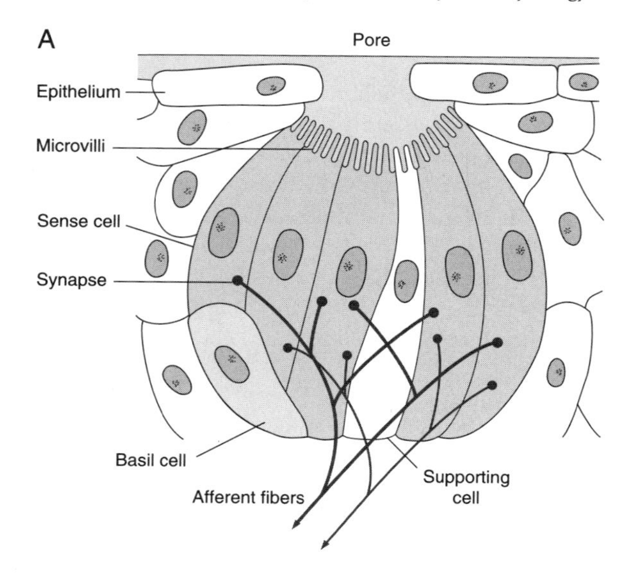

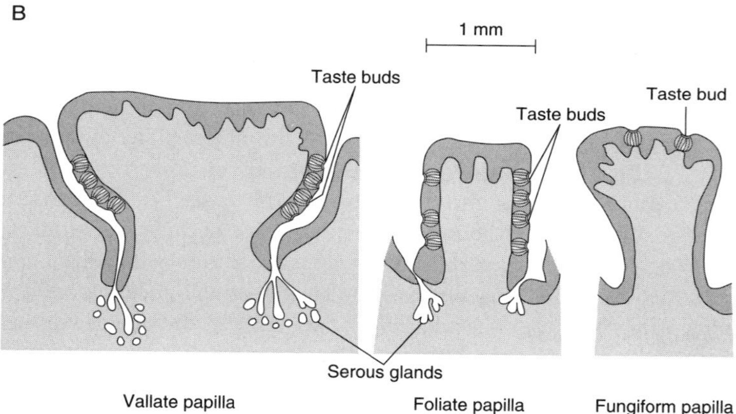

FIGURE 2.9 (A) Structure and innervation of a taste bud and (B) location of taste buds in three types of gustatory papillae. (Adapted from Schmidt, R.F., *Fundamentals of Sensory Physiology*, New York: Springer-Verlag, 1981.)

1. Type I Receptors

Mechanoreceptors that mediate touch, vibration, and deformation of the skin are innervated by myelinated nerve fibers of the Aβ type with diameters of 6 to 12 μm and conduction velocities between 35 and 75 m/sec (Table I). Warmth and cool receptors are regarded receptors for innocuous stimuli and are innervated by Aδ-type nerve fibers with a diameter of 1 to 5 μm and a

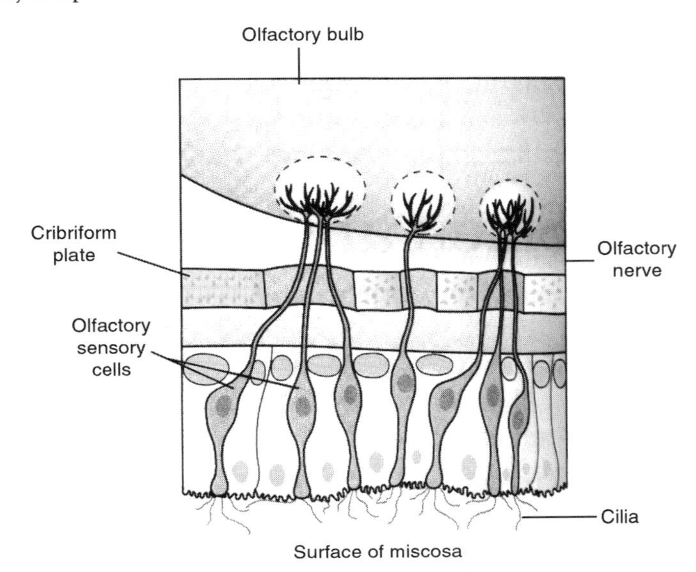

FIGURE 2.10 Schematic drawing of olfactory sensory cells. (From Buck, L.B., *Principles of Neural Science*, 4th ed., edited by E.R. Kandel *et al.* New York: McGraw-Hill, 2000, pp. 625–647. With permission of the McGraw-Hill Companies.)

conduction velocity of 5 to 30 m/sec. Some nociceptors that respond to stimuli that elicit pain sensation, including heat and cold receptors, are also innervated by Aδ-type fibers while some polymodal nociceptors are innervated by unmyelinated C-fibers with a diameter of 0.2 to 1.5 μm and a conduction velocity of 0.5 to 2 m/sec. The area of the specialized axon membrane that acts as the sensor in these receptors differs, which makes them sensitive to different modalities of stimulation (chemical, heat, etc.). Some pain receptors are bare axons.

The nerves that innervate mechanoreceptors, thermoreceptors, and nociceptors in the skin, joints, tendons, and muscles form dorsal roots of the spinal cord. In the face, innervation of the same kinds of receptors is mainly by parts of

TABLE I Fiber Diameters and Conduction Velocity of Nerve Fibers

Name	Diameter (μm)	Conduction velocity m/sec
Aα	13–20	80–120
Aβ	6–12	35–75
Aδ	1–5	5–30
C (unmyelinated)	0.2–1.5	0.5–2

the trigeminal nerve (CN V), but CN IX and CN X also contain nerve fibers that innervate skin receptors in the face and receptors in the mouth and the pharynx.

> Polymodal nociceptors that are innervated by Aδ and C fibers respond to heat and cold and high-intensity mechanical stimulation. The polymodal nociceptors that are innervated by C-fibers also respond to capsaicin, which causes a burning sensation. Nagy and Rang[27] identified two types of dorsal root ganglion cells that responded to heat. One has a high threshold, is not sensitive to capsaicin, and is innervated by large-diameter ganglion cells. The other has a lower heat threshold, is sensitive to capsaicin, and is associated with small-diameter ganglion cells. Itch is likewise mediated by C-fibers that innervate bare axons. These receptors may be slightly different from those mediating pain.

2. Type II Receptors

Innervation of Type II receptors, such as those in the ear and the eye and those of chemical senses, is more complex than that of Type I receptors, where the receptor is just a continuation of the axon of the afferent nerve fiber. In Type II receptor cells, one or more afferent fibers connect to the body of the cell by synaptic contacts, and efferent fibers connect to the cell body or to the afferent nerve fibers by synaptic contacts.

The nerve fibers that carry information from receptors of hearing, vision, smell and taste are all cranial nerves or parts of cranial nerves. Olfactory information is carried in cranial nerve I, vision in cranial nerve II, and taste in cranial nerve VII, IX, and X. Auditory information (together with information from the balance organ) is carried in cranial nerve VIII.

a. The Ear

The outer and inner hair cells (Fig. 2.7) have different functions and differ in their innervation. The inner hair cells mediate information about the vibration of the basilar membrane to the CNS, whereas the outer hair cells are mechanically active and participate in the motion of the basilar membrane. The inner hair cells connect to the CNS by multiple afferent nerve fibers. Both inner and outer hair cells are innervated by efferent nerve fibers that originate in the CNS.

Each inner hair cell is innervated by many afferent auditory nerve fibers (Fig. 2.7A), known as Type I auditory nerve fibers. The first part of the afferent fibers is unmyelinated (and may be regarded as a dendrite). Efferent nerve fibers terminate on these unmyelinated fibers and also to a lesser extent on the body of the inner hair cells.[39]

The outer hair cells are also innervated by afferent fibers of the auditory nerve (Type II fibers), but each auditory nerve fiber innervates many outer hair cells. The efferent innervation of the outer hair cells, which is abundant, mainly terminates on the cell bodies (Fig 2.7B).[39] The outer hair cells of the cochlea function as mechanically active elements[4] that are integral parts of the basilar membrane, and they can therefore affect the vibration of the basilar membrane (see Chapter 5). The efferent nerve fibers that terminate on these hair cells can therefore control the mechanical properties of the outer hair cells from the CNS and thereby the mechanical property of the medium that transmits the physical stimulus to the inner hair cells.

b. The Eye

Photoreceptors are not innervated directly by fibers of the optic nerve but rather make synaptic connection with cells of the neural network that is located in the eye. *Bipolar cells* that make synaptic contact with the photoreceptor cells are intermediate cells between photoreceptors and the *ganglion cells* that give rise to the fibers of the optic nerve. Other cells in that network, *horizontal cells* and *amacrine cells*, provide connections between photoreceptors cells and between bipolar cells (see Chapter 6). The first impulse generation occurs in the ganglion cells. The retina performs complex processing of the output of the photoreceptors, and the axons of the ganglion cells of the retina form the *optic nerve* (CN II). This means that a part of the visual nervous system is located in the eye. The optic nerve emerges from the retina and passes through the *lamina cribrosa* of the *sclera* before it enters the skull cavity through the *orbital plate* of the frontal bone, together with blood vessels and some of the *extraocular muscles*. The neural network of the retina is regarded as a part of the visual nervous system and will be discussed below (see page 377).

c. Gustation and Olfaction

Nerve fibers that innervate taste receptors connect to the receptor cells via synapse-like endings like other Type II receptors. Taste receptors of the frontal part of the tongue are innervated by the lingual nerve, which becomes the chorda tympani, which is a part of CN VII, and those of the posterior part are innervated by fibers of the glossopharyngeal nerve (CN IX). Taste receptors in the pharynx and on the glottis are innervated by CN IX and the vagus nerve (CN X).

The innervation of the sensory cells of olfactory is similar to that in the retina. Olfactory receptors connect to nerve cells in a neural network in the *olfactory bulb*. The axons of the bipolar receptor cells are unmyelinated fibers that form the olfactory nerve (CN I). The fibers collect into approximately 20 bundles that pass through the openings of the cribiform plate of the ethmoid bone (Fig. 2.10). These fibers make synaptic connection with *mitral cells* in the *glomeruli* of the *olfactory bulb* (see Chapter 7). This means that the olfactory nerve (CN I) actually consists of axons that are part of the olfactory receptor cells. The network of nerve cells to which the olfactory nerve fibers connect is similar to that of the retina (see Chapter 3).[36]

IV. CONDUCTION OF THE PHYSICAL STIMULUS TO THE RECEPTORS

In the early evolution of sensory systems, sensitivity was of primary importance because detection of a stimulus was the most important factor for survival and for detecting prey. This means that optimization of energy transfer to the receptor cells was the most important task of sensory systems. Later in evolution, discrimination of stimuli became important, and this task became served primarily by neural processing in the ascending sensory pathways. The structures that conduct the physical stimulus to the receptors also developed to serve that purpose in an optimal way. The structures that conduct the physical stimuli to the receptors can improve the transmission of the physical stimulus to the receptors and thereby increase the sensitivity of a sense.

The structures that conduct the stimulus to the receptors modify the stimulus before it reaches the sensory receptors and play a role in the information processing of sensory stimuli. The analysis that occurs in the sense organs aid the subsequent analysis that occurs in the CNS.

A. Improvement of Transmission to the Receptors

The greatest improvement of transmission of a physical stimulus to the receptors occurs in the ear. The middle ear improves the transmission of sound to the fluid-filled cochlea by acting as an impedance transformer that increases the efficiency of the transfer of sound from air to the fluid of the cochlea by approximately 30 dB compared with applying the sound directly to one of the two windows of the cochlea.[32] The impedance-matching action of the middle ear causes the force on the oval window to be much greater than that of the round window, and this ensures a large difference between the forces that act upon the two windows of the cochlea. If sound is allowed to reach both windows of the cochlea in an identical way there would be no motion of the fluid in the cochlea because it is the difference between the forces at the two windows that causes the cochlear fluid to move. The gain in sound transfer to the cochlear fluid by the action of the middle ear is therefore much larger than the 30-dB increase in sound transmission to the cochlear fluid that results from the improved impedance matching.

Because the properties of air and (cochlear) fluid differ greatly, most of the energy is reflected at the interface between air and fluid, and only 0.1% is transmitted to the fluid as vibration. This is caused by the mismatch of the impedance of air and that of the cochlear fluid. The middle ear acts as an impedance transformer that reduces the reflection and improves the transmission of sound into vibration of the cochlear fluid (see Chapter 5).

The vibration of the cochlear fluid is transferred to the basilar membrane, which activates both inner and outer hair cells. As was mentioned above, the outer hair cells do not participate in sensory transduction; rather, the deflection of the cilia of the outer hair cells, which occurs as a result of the motion of the basilar membrane, causes the outer hair cells to contract or elongate.[5] This mechanical motion of the outer hair cells is transferred to the basilar membrane and amplifies the motion of the basilar membrane and compensates for the frictional losses that otherwise would have decreased the motion of the basilar membrane. The action of the outer hair cells increases the sensitivity of the ear by approximately 50 dB. The active process of the outer hair cells also makes the basilar membrane more frequency selective, which is important for detection of weak sounds in a noise. This "motor" action of the outer hair cells is most effective at low sound intensities, and the amplitude dependence provides amplitude compression (see page 304). The light-reflecting surface behind the photoreceptors in some animals causes light to pass twice through the photoreceptors, which increases the probability that light activates a photoreceptor, thereby increasing the sensitivity of the eye.

A peculiar pattern of submicroscopic protuberances has been found on the surface of the corneas of some species of night moths[1] which provides antireflection properties, which in turn increase transmission of light to the photoreceptors. The light reflection of a plain surface of the corneas of facet eyes is 4%. This pattern of small cone-shaped protuberances reduces light reflection from the surface of the facet eyes of these insects[1,2] and enhances transmission of light to the receptors by approximately 4%. It is interesting to speculate if such a small gain has been sufficient for the development of such an elaborate structure on the surface of the cornea of certain insects. The antireflection action of these protuberances also reduces the risk of these nocturnal insects being discovered by predators when hiding in the day in bright sunlight, a factor that may have been a more important promoter of the development of these structures.

B. Modification of the Physical Stimulus

Transfer of the physical stimulus to the receptors involves various degrees of modification of the stimulus before it reaches the receptors. The adaptation and frequency selectivity that are properties of the medium that conducts the stimulus to mechanoreceptors in the skin and ear are examples of modifications of physical stimuli that occur before the stimuli reach the receptors.

1. Adaptation

Adaptation implies that the response to steady stimulation decreases gradually after the onset of a steady stimulation, which results in attenuation of slow components of a stimulus. Adaptation may be regarded as a spectral filter that attenuates low frequencies and thereby emphasizes fast components, which often contain more important information than slow components. The attenuation of slow stimuli also acts to increase the range of stimulus intensities where the receptors can operate efficiently.

The effect of adaptation can be demonstrated by observing the discharge in the nerve fiber that innervates the receptor in response to deformation of the skin that rises rapidly and then stays constant. The discharge rate of afferent nerve fibers in response to such stimuli is an initial increase in the discharge rate and then a decrease to asymptotically reach a constant value that is maintained during the steady phase of the stimulation (Fig. 2.11). The decrease in discharge rate is approximately an exponential function.

Receptors with different degrees of adaptation such as those found in the skin communicate different components of complex skin deflections to the CNS, and mechanoreceptors have been classified according to their adaptation (Fig. 2.11). *Rapid adapting* (RA) mechanoreceptors of the skin respond best to rapid changes in deformation of the skin, while *slow adapting* (SA) receptors respond to slow deformations as well as fast stimuli.[15] Thus, SA receptors respond to displacement of the skin, while RA receptors mainly respond to the *rate* of displacement of the skin (*velocity*). Some receptors such as the Pacinian corpuscles have very rapid adaptation and mainly respond to the *acceleration* of skin displacement.

While the majority of adaptations of mechanoreceptors are caused by the properties of the media that conduct the stimuli to the receptors, the receptors also possess an adaptation that is a part of the transduction process (page 68). The responses of mechanoreceptors to deformation of the skin, such as those illustrated in Fig. 2.11, thus also include the effect of adaptation in the receptor cells in addition to the effect of the media conducting the stimuli to the receptors.

2. Frequency Selectivity

The media that conduct the stimulus to the receptors may provide *frequency selectivity*. For example, Pacinian corpuscles respond best to sinusoidal vibration within a narrow frequency range, and their response to vibrations of higher and lower frequencies are attenuated. Pacinian corpuscles may be regarded as being "tuned" to a certain frequency of vibration (Fig. 2.12). This tuning is a result of the properties of the onion-shaped structures that surround the bare axons.

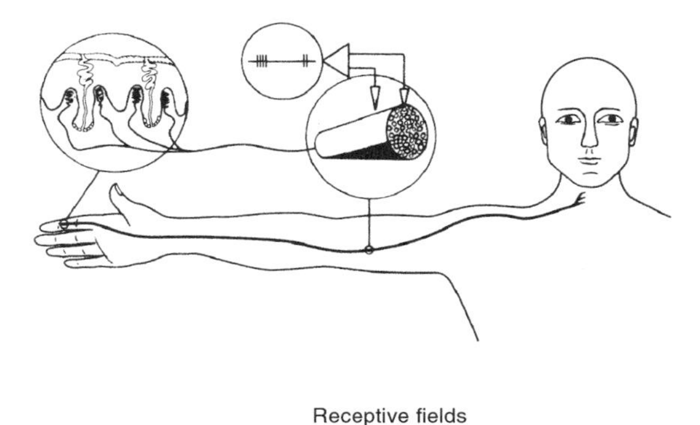

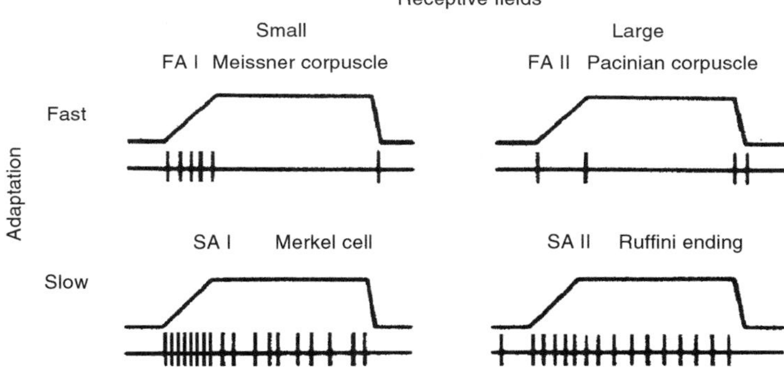

FIGURE 2.11 Schematic illustration of adaptation in different types of mechanoreceptors. The illustrations are based on recordings from single nerve fibers in a peripheral nerve on the arm of a human subject during stimulation of mechanoreceptors in the finger. (Adapted from Vallbo, A.B., and Johansson, R.S., *Hum. Neurobiol.*, 3, 3–14, 1984 by Springer-Verlag.)

More extensive transformation of the physical stimulus occurs in the ear where the medium (the cochlea) that transfers the stimulus to the receptors separates sounds according to their frequency (or spectrum) before being converted into a neural code. The transfer of the vibration of the cochlear fluid to the deflection of the cilia of the hair cells involves the amplitude of motion of a small segment of the basilar membrane being not only a function of the intensity of the sound that reaches the ear but also a direct function of the frequency of the sound. In fact, each segment of the basilar membrane can be regarded as being tuned to a certain frequency of sounds (Fig. 2.13) (see Chapter 5). This frequency selectivity is a result of the interaction between the basilar membrane and the surrounding fluid, and the properties of the outer hair cells.

Hair cells are located along the basilar membrane, and, because different segments of the basilar membrane are tuned to different frequencies, different

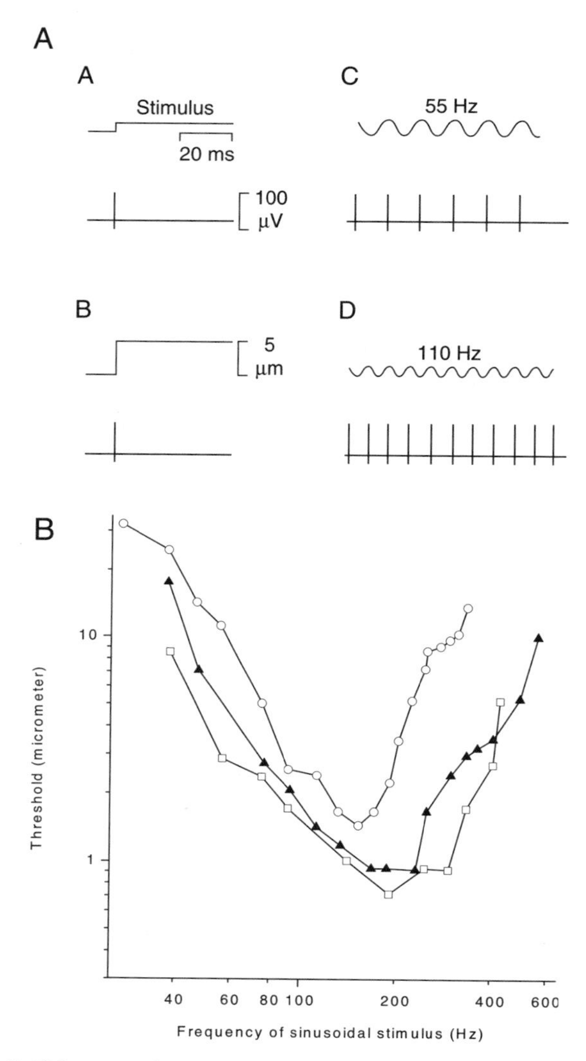

FIGURE 2.12 (A, B) Response of a Pacinian corpuscle to step stimulus of different intensities. (C, D) Sinusoidal vibration of 55 and 110 Hz. (E) Threshold of three different Pacinian corpuscles in response to sinusoidal stimulation. (C, D, and E adapted from Schmidt, R.F., *Fundamentals of Sensory Physiology*, New York: Springer-Verlag, 1981.)

populations of hair cells will be excited according to the frequency of a sound. Individual receptor cells are therefore activated by a narrow band of sound frequencies. The hair cells themselves do not possess any frequency selectivity within the audible frequency range. The frequency tuning of the basilar

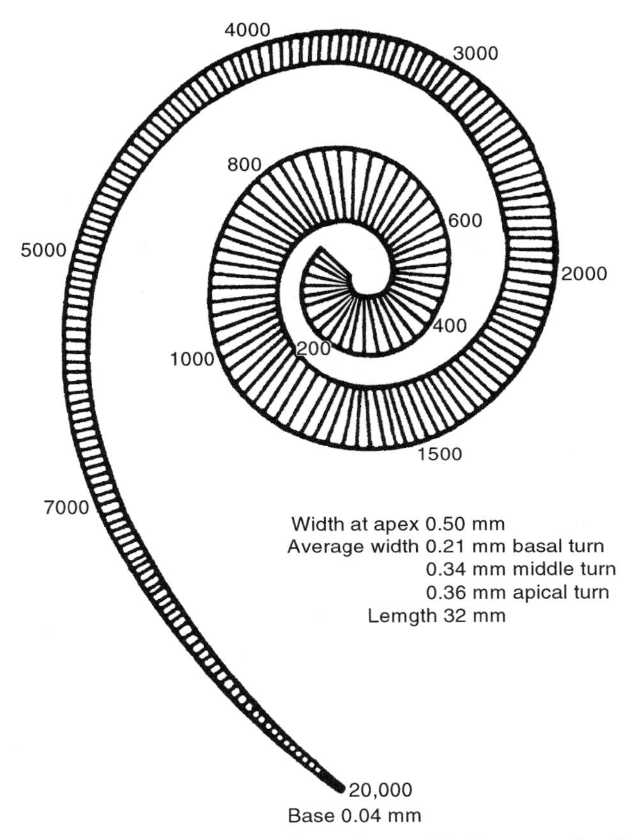

FIGURE 2.13 Schematic drawing of the basilar membrane of the human cochlea showing that the width of the basilar membrane increases from the base to its apex and that high frequencies are represented at the base of the cochlea and low frequencies at the apex. (Adapted from Stuhlman.[40])

membrane is reflected in most nerve cells throughout the auditory nervous system (see Chapters 3 and 5).

Inner hair cells do not directly respond to the motion of the basilar membrane rather, the excitation of inner hair cells from the motion of the basilar membrane is affected by the coupling between the *tectorial* membrane and the cilia of the inner hair cells.[10] Thus, the transfer of vibrations of the basilar membrane to the deflection of the hair cells is complex; some hair cells respond to the velocity of the basilar membrane motion, other hair cells respond to the acceleration of the basilar membrane, and yet other hair cells respond to the displacement of the basilar membrane.[18,43] This specificity of the response of inner hair cells to the mode of vibration of the basilar membrane is an additional property to their frequency selectivity.

3. Amplitude Compression

Amplitude compression is important in sensory systems because of the relatively narrow dynamic range of neural coding. The high dynamic range of sensory systems is achieved by various kinds of amplitude compression, some of which is accomplished by the media that conduct the stimuli to the receptors and some of which is a result of the function of receptor cells. In the ear, the acoustic middle ear reflex attenuates sounds through the middle ear above the approximately 85-dB hearing level (HL),[20,23] and the amplification in the cochlea decreases with increasing sound intensity for low-intensity sounds, acting as an automatic gain control. This is mostly a result of the active role of outer hair cells[4] (see page 304) which also makes the frequency tuning of the basilar membrane become intensity dependent.[16,22,29] The light reflex has a similar action for vision as the acoustic middle ear reflex. Because the acoustic middle ear reflex is relatively slow, the resulting amplitude compression affects steady and slowly varying sounds more than transient sounds. The amplitude compression that occurs in the cochlea is nearly instantaneous and covers a much larger range of sound intensities than the acoustic middle ear reflex.

In the eye, the pupil constricts in response to light that reaches the retina. That light reflex reduces the amount of light that reaches the retina and acts as amplitude compression. The degree of that compression is relatively small in humans and much larger in animals such as the cat, which has pupils in the shape of a slit that can vary the light that passes through them to a greater extent than the circular pupils in humans.

4. Temporal Integration

The media that conduct the stimulus to the receptors can cause *temporal integration*.* Temporal integration of physical stimuli by the media that conduct the stimuli to the receptors is greatest in thermal senses where the thermal inertia and storage capacity make the activation of thermal receptors dependent upon the duration of the exposure to change in the temperature of the skin surface. Temporal integration in the media that conduct light or mechanical stimuli, including sound, to their respective receptors is negligible. The chemical senses have some storage capacity for odors and tastes which provides temporal integration. The temporal integration that occurs in receptors and in particular in the CNS is more important than that which occurs in the media that conduct the stimuli to the receptors.

*Temporal integration involves a summation of the response over time in such a way that a larger response is obtained to a stimulus of long duration compared with a short duration stimulus.

5. Delay of Response

Transmission of chemical substances to taste receptors depends on how soluble the substances are. Different water-soluble substances that reach the tongue diffuse through the pores of the taste buds to reach the sensory cells in similar ways, but solid substances may activate taste receptors in different ways depending on their solubility. This activation takes some time so the sensation of such substances is delayed. Enzymes in the saliva can break down substances and change how they stimulate taste receptors. For example, starch has no taste but enzymes in the saliva break down starch into sugars, giving a taste of sweetness. That sensation builds up slowly because the conversion of starch takes time. The fact that sensory cells that are sensitive to the different taste qualities are located at different regions of the tongue may affect the time course of activation of taste receptors.

Odorants must interact with the mucus on the ciliated surface to activate the receptor cells in a yet-unknown way. It is, however, known that thickening of this mucus layer, such as may occur during upper airway infections by a cold or as a result of allergies, reduces the ability to detect odors.

6. Change in Receptive Fields

The medium that conducts the physical stimulus to the receptors can affect the *receptive field** (see page 129) of receptors because of spread of the stimulus to adjacent receptors. For example, only a few mechanoreceptors in the skin will respond to a pin prick of low intensity. When the strength of the stimulation is increased, the mechanical deformation of a larger region of the skin will exceed the threshold of an increasing number of receptors. This means that the receptive field of receptors in the skin will become broader when the stimulus intensity is increased.

V. PHYSIOLOGY OF SENSORY RECEPTORS

All sensory receptors convert physical stimuli into the discharge pattern of afferent nerve fibers that innervate the receptors. The properties of the receptor cells determine the sensitivity and the sensory modality to which the afferent fibers respond. Some sensory receptors (photoreceptors and olfactory and taste receptors) have specific selectivity regarding stimuli within their modality.

*The receptive field of a receptor is the area of a receptor surface to which stimulation gives a response in the afferent nerve fiber of a receptor (see also page 129).

A. General Principles for Sensory Transduction

The principles of sensory transduction in the different senses have many similarities despite the fact that the physical stimuli that activate receptor cells and the morphology of the receptor cells differ among the senses. In all sensory cells, transduction of a physical stimulus involves a change in the electrical conduction of a specific region of the membrane of the receptor. The intracellular potential changes when that specialized part of a membrane is exposed to a stimulus of the modality of the particular receptor. This change in the membrane potential is known as the *receptor potential* (or *generator potential*) (Fig. 2.14). The receptor potential is a graded potential, the size of

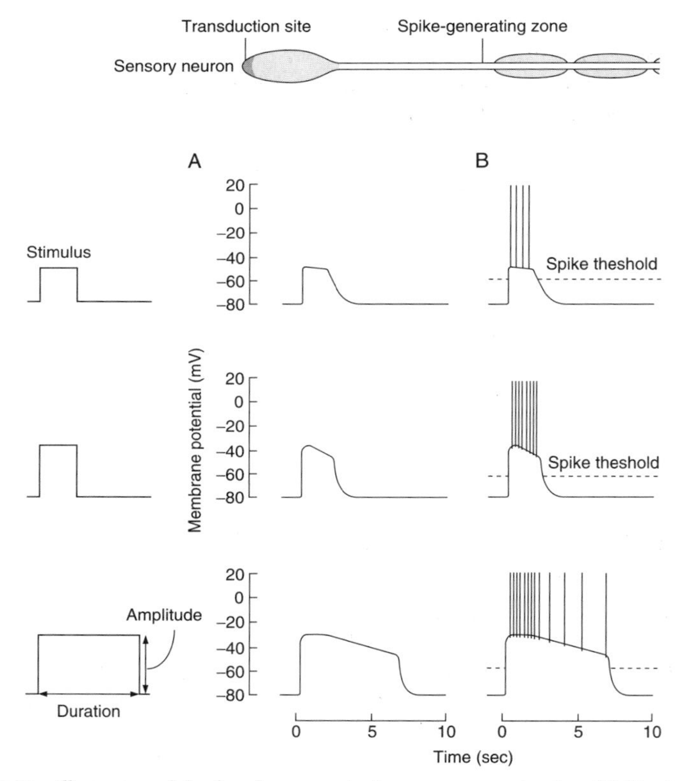

FIGURE 2.14 Illustration of the first four steps in the sensory transduction. (A) Physical stimulus causes the membrane potential to change (generate a receptor potential). (B) The receptor potential, transferred to the first node of Ranvier of the nerve fiber that innervates the receptors, causes the generation of action potentials (spikes or discharges).

which is related to the strength of the stimulation. In all receptors (except photoreceptors) a stimulus depolarizes the receptor cell corresponding to a positive receptor potential. In photoreceptors, light instead hyperpolarizes the cell membrane (this will be discussed in more detail in Chapter 6).

In Type I receptors (such as touch receptors in the skin), the specialized part of the cell membrane where the stimulus acts is a part of an axon. In Type II receptors, the active membrane is located on specific structures of the receptor cell such as microvilli (taste), cilia (smell and vision), or stereocilia (hearing).

Receptor potentials are generated by different means, depending on receptor type. The transduction of physical stimuli into receptor potentials in chemoreceptors involves secondary messengers that are activated when a chemical stimulus binds to a specific cell membrane receptor that is coupled to G-proteins. Transduction in mechanoreceptors, such as the hair cells of the inner ear, occurs when deformation of the receptor (i.e., stereocilia) causes membrane "gates" to open and allow current from a source in the cochlea to flow through the open gate into the cell, generating the receptor potential.

While mechanoreceptors of the skin respond to mechanical deformation in all directions, the mechanoreceptors in the ear (hair cells) only respond to mechanical deflection of the hairs in one direction. Deflection of the stereocilia of inner ear hair cells generates a receptor potential, the polarity of which is positive (depolarizing the cell) when the cilia are deflected in the direction of the place where the kinocilium normally is located (toward the basal body), as shown in studies of the lateral line organ in fish (Fig. 2.15) and later confirmed in studies of cochlear hair cells in the guinea pig.[33] Deflection in the opposite direction generates a negative receptor potential that hyperpolarizes the cell.

The receptor potential controls the discharges in the afferent nerve fiber of the receptor in different ways in the Type I and Type II receptors. In Type I receptors, where the specialized membrane is a part of the axon of the afferent fiber of the receptor, the receptor potential is conducted *electrotonically** in the unmyelinated portion of the axon to the site of impulse generation, which occurs at the first *node of Ranvier* of the afferent fiber (Fig. 2.14). (While the receptor potential is a graded potential, the nerve impulses in axons [action potentials or neural discharges] are *all-or-none* potentials.) This is the final and only step in sensory transduction in Type I receptors, and the receptor potential controls the neural discharge in the afferent nerve fiber directly.

In Type II receptor cells, the sensory transduction occurs in a cell that is separate from the afferent nerve fiber but is connected to the afferent nerve fiber

*Electrotonic transmission means conduction of electrical current in the same way as in a wire, for instance. This is in contrast to *propagated* neural activity such as occurs in a nerve fiber. While electrotonic conduction occurs instantaneously, neural activity in a nerve fiber propagates slowly.

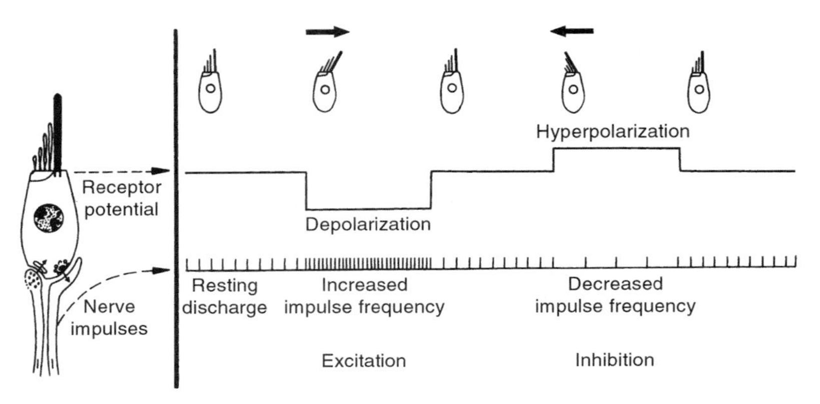

FIGURE 2.15 Illustration of the directional sensitivity of cochlear hair cells. (Adapted from Flock, A., *Cold Spring Harbor Symp. Quart. Biol.*, 30: 133–146, 1965.)

by a synapse-like structure where it elicits packages of a specific *transmitter substance* that are conducted over the *synaptic cleft* to the opposite side of the synapse, where a *postsynaptic potential* is generated (*excitatory postsynaptic potential,* EPSP).

In Type II receptors, the receptor potential is conducted to the afferent fiber (or fibers) by synaptic transmission. In cochlear hair cells and in taste receptors, the EPSP is conducted electrotonically to the first node of Ranvier of the afferent nerve fiber that connects to that synapse. The impulse generation occurs in the first node of Ranvier of that nerve fiber.

Impulse generation in Type II receptor cells of hearing and taste is affected by *inhibitory* input from the CNS that can act on the postsynaptic fiber or by synaptic contact with the cell body of the receptor. The inhibitory input may affect the membrane potential of the receptor cell by generating an *inhibitory postsynaptic potential* (IPSP), or by modulating the membrane potential in the postsynaptic fiber of the receptor cell.

In the eye and in olfaction, the impulse generation occurs after several steps of synaptic transmission that do not involve impulse generation (see page 84). The postsynaptic potential is conducted electrotonically as a graded potential to another cell. Those cells communicate by graded potentials rather than generating nerve impulses. In the retina, both the first and the second cell in the network of nerve cells conduct graded potentials, and the earliest spike-generating cells are the ganglion cells, the axons of which leave the eye and form the optic nerve (CN II). Extensive signal processing takes place in the network through complex synaptic connections. The function of this network will be discussed in Chapters 3 and 6 because it is regarded as a part of the nervous system.

The function of the neural network in the olfactory bulb is in many ways similar to that of the retina; for example, nerve cells of a neural network communicate with graded potentials. But, it is the axons of the (bipolar) receptor cells that form the olfactory nerve (CN I), and the neural network is located in the olfactory bulb.

1. Sensitivity of Receptors

The generation of the receptor potential, its (electrotonic) transmission to the impulse generation site, and the threshold of the subsequent impulse generation determine the threshold of Type I receptors. In Type II receptors, generation of the receptor potential and its conduction to the synaptic site, the efficacy of the synapse, and the threshold of impulse generation set the threshold. The size of the receptor potential that a given stimulus elicits depends on the sensory transduction properties of the receptor cell. The properties of the receptor cells and the media that conduct the stimuli to the receptor cells also contribute in different ways to setting the sensitivity of sensory systems. Sensory transduction in some receptor cells implies amplification, and conduction of the stimulus to the sensory cells also implies amplification in some sensory systems.

The ear is an example of a sensory organ with a very high sensitivity, as are the eyes of some insects, which can detect one or two quanta of light. The human eye is slightly less sensitive, responding to approximately 15 quanta of light.[12] The high sensitivity is mainly a result of amplification that occurs in the photoreceptors.

The sensitivity of many sense organs is enhanced by amplification in the transduction process that is caused by positive feedback mechanisms. These mechanisms amplify the weak signals that are elicited by the physical stimulus acting on the specialized region of the receptor membrane. The receptors in the ear (hair cells) do not convert acoustic energy into neural energy, but the deflection of the cilia gates (modulates) an electrical current from a source (battery) in the cochlea, which generates the receptor potential. This represents amplification, because the energy needed to operate this gate is much less than what would have been needed to generate a receptor potential of the same magnitude by converting acoustic energy into electrical energy.

Amplification may also occur in the synaptic transmission between a receptor cell and the afferent nerve fiber. Synaptic transmission may amplify sensory signals because a single synaptic vesicle may contain thousands of molecules of the transmitter substance and it may take only two molecules to open a postsynaptic ion channel.

The *spatial integration*, which is accomplished by adding the output of many receptors, also increases the sensitivity and the *signal-to-noise ratio*[7] (see Chapter 3), which often is the limiting factor in detecting a weak stimulus.

External factors can affect the sensitivity of sensory receptors. Although *adrenergic* innervation of sensory organs is often assumed to affect only blood vessels, evidence suggests that the sensitivity of pain receptors can change (increase) because of secretion of *norepinephrine* from adrenergic nerve fibers that are located near mechanoreceptors in the skin.[31] Also, in the cochlea adrenergic nerve endings are located close to the hair cells that may affect their sensitivity by secretion of norepinephrine.[8]

B. Information Processing at the Receptor Level

Information processing at the receptor level includes the effects of adaptation, temporal integration, and selectivity with regard to different properties of stimuli within the modality of the sense. The information processing at the receptor level is one of the reasons why the discharge pattern of afferent sensory nerve fibers is not an exact replica of the physical stimulus that reaches the receptor cell.

1. Adaptation

Adaptation is an almost universal feature of sensory transduction. As mentioned earlier, mechanoreceptors in the skin are often classified according to their degree of adaptation. That adaptation is a combination of the properties of the medium that conduct the sensory stimulus to the receptors and the properties of the receptors themselves.

Adaptation in receptors is caused by properties of the receptor membrane, and it manifests as a gradual decrease of the receptor potential after application of a constant stimulation. The decreased receptor potential results in a decrease in discharge rate of the nerve fiber that innervates the receptors. The rate of adaptation varies among receptors. Adaptation at the receptor level is most pronounced in thermoreceptors. When referring to the adaptation of the eye to light, it is usually the *dark* adaptation that is discussed. Dark adaptation is the recovery to normal sensitivity after light exposure.

The molecular mechanism of adaptation is not known, but one hypothesis links it to the effect of a final reservoir of some ions transported over a cell membrane causing cellular excitation. Other hypotheses support a mechanism of adaptation that is based on the internal concentration of free Ca^{++} ions, and yet another hypothesis concerns micromechanical properties of the cell membrane or adjacent structures.

2. Temporal Integration

Temporal integration is a common property of sensory cells and the receptor potential. Temporal integration in sensory systems implies that the sensitivity is greater to a stimulus of long duration compared with that of a short duration, or that the threshold of a stimulus that follows after another stimulus is lower than that of the first stimulus. Temporal summation of sensory stimuli is noticeable for all sensory modalities but most of that is a result of processing in the CNS (see Chapter 3).

Temporal integration implies a slow build-up of the output. Temporal integration is often described by a time constant, a description borrowed from engineering that assumes that the integration is similar to that of a simple integrator such as that which can be realized by an electrical circuit consisting of a single resistor and a single capacitor. Many receptors and synaptic transmissions work as "leaky integrators." An ideal integrator keeps adding the input to the previous input, but a leaky integrator loses its stored potential at a certain rate. When two impulses are applied within a short interval to a receptor that functions as a leaky integrator, the response to the second impulse will add to that of the first impulse, but the contribution from the first response will depend on how much the response to the first applied impulse has decayed when the second impulse is applied, which depends on the interval between the two impulses and the rate of the decay. Assuming that synaptic integration works as a simple leaky integrator, the rate with which the integrator loses its stored energy is determined by an exponential function which can be described by its time constant, defined as the time it takes for the output to decrease to $1/e$ of its value, where $e = 2.718$ is the basis for the natural logarithm ($1/e = 37\%$).

The temporal integration that occurs in the CNS is much greater than that occurring in the sensory organs. As an example, the temporal integration in the cochlea is on the order of a fraction of a millisecond, depending on the frequency of the stimulus, while the temporal integration that is associated with perception of sounds is approximately 200 msec at threshold and approximately 100 msec at physiologic sound levels (see Chapter 1).

C. Amplitude Compression

Compression of the range of the intensities of sensory stimuli that reach the sense organs is another important property of sensory systems. Such compression is essential for neural coding of stimuli that have a wide range of intensities. Sensory cells convey information about the properties of stimuli over a very large range of stimulus intensities, but nerve impulses in a single nerve fiber that innervates a receptor can only code the intensity of physical stimuli within a narrow range. Amplitude compression commonly occurs in the receptors as a part of the transduction process. In Type II receptors, amplitude compression may also occur in the synaptic transmission that occurs from the receptor cell to the axon or to the cells in the neural network in the retina and the olfactory bulb before impulse initiation occurs.

D. Selectivity of Receptors

Perhaps the most distinct processing that occurs at the receptor level is the selectivity of individual receptors. While all sensory receptors are selective with regard to the modality of stimulation, some receptors are selective regarding specific qualities within their modality. Some receptors such as photoreceptors (cones) and taste and olfactory receptors typically respond best to specific stimuli within the range to which the respective sensory systems respond. The three types of cones in the retina are red, green, and blue sensitive. Red cells will respond better to red light than to green or blue light, and so on (Fig. 2.16).[35] In a similar way the taste cells in the tongue and the olfactory cells in the nose will respond to several different taste substances and odors but to different degrees. Inner hair cells in the cochlea also exhibit selectivity, namely to the frequency of sounds. That selectivity, although very similar to that of photoreceptors, is not a result of selectivity of the inner hair cells but rather a result of the selectivity of the basilar membrane (see page 291), thus the medium that conducts the stimulus (sound) to the receptor. Recordings from inner hair cells or from their nerve fibers describe this selectivity to the frequency of sounds in a similar way as the curves in Fig 2.16 describe the selectivity of photoreceptors to light as a function of wavelength. Curves showing the frequency selectivity of auditory nerve fibers are known as *frequency threshold curves* (FTC) or *frequency tuning curves*.

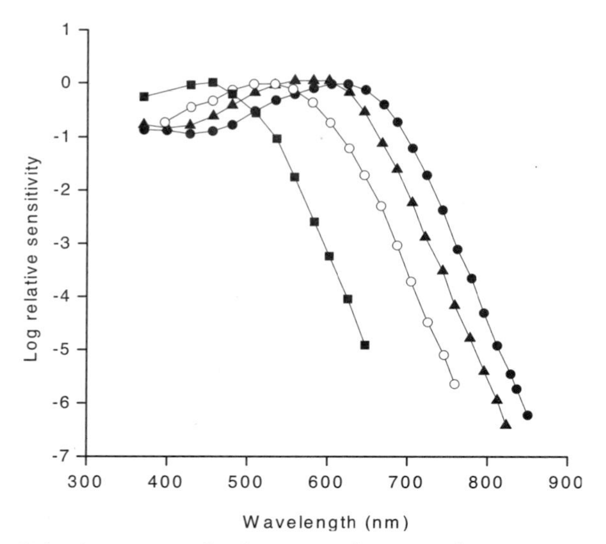

FIGURE 2.16 Light absorption in the three types of cones in the primate retina vs. light as a function of the wavelengths. (Data from Schnapf et al.[35])

Because our perceptions of the different qualities of sensory stimuli are distinct, there must be mechanisms in the CNS that can unravel the neural code from many different sensory cells and produce a "clean" interpretation of sensory stimuli.

The selectivity of cones in the retina with regard to specific colors (wavelengths) of light is similar to the frequency selectivity in the cochlea, but with the difference that in the eye the selectivity is caused by the properties of the sensory cells themselves, while the distinct frequency selectivity in the ear is a result of the mechanical properties of the medium that conducts the physical stimulus to the receptors (the basilar membrane of the cochlea).

All receptors have a threshold below which they do not respond. Sensory receptors generally code the intensity of the stimulation in the discharge rate of their afferent nerve fibers as a rate increase with increasing stimulus intensity. Some thermoreceptors, however, respond selectively to temperature, which means that above threshold the discharge rate increases with increasing temperature for warmth receptors. This occurs only up to a certain temperature above which the discharge rate again decreases. For cold receptors, the discharge rate increases with decreasing temperature, down to a certain temperature. This may be interpreted as a specificity regarding stimulus strength. Also some olfactory cells have a similar selectivity.

E. Receptive Fields

Sensory receptors in the skin, for example, respond only to stimulation of a limited area of the skin. This can be viewed as a certain receptor responding selectively to stimulation within a certain region of skin. This form of specificity is commonly referred to as the *receptive field** of the particular receptor. The concept of receptive field was introduced by Hartline[11,**] and is defined as the "region of the retina that must be illuminated in order to obtain a response from a given fiber." The "fiber" refers to an axon of a retinal neuron. This definition of receptive fields applies to any nerve fiber that innervates a sensory organ.

*The receptive field of nerve fiber that innervates a receptor cell is the area of a receptor surface such as the skin or the retina where stimulation gives rise to a response. The original use of the term was for describing the response areas of fibers of the optic nerve. Now it is also used to describe the area of response of nerve cells and fibers in the somatosensory and visual nervous systems.

**Hartline studied the eyes of the horseshoe crab (*Limulus*), which are compound eyes, each of which consists of about 300 *ommatidia* that are arranged in a hexagonal pattern. *Limulus* is a very old animal phylogentically. The eyes of insects, which date to approximately 250 million years ago, resemble those of the *Limulus*.

The area of skin from which a response (increase in discharge rate) can be elicited is known as the receptive field of that particular receptor. The size of the receptive field increases when the stimulus intensity is increased because of the spread of stimulation in the medium that conducts the stimulus to the receptor (the skin). The degree of the spread depends on the properties of the skin. The size of the receptive fields of skin receptors therefore depend upon the properties of the medium that conducts the stimulus to the receptors.

While the receptive field of a nerve fiber of the optic nerve is the visual field within which light causes the fiber to respond, the receptive field of auditory receptors (cochlear inner hair cells) is the range of frequencies to which an auditory nerve fiber responds. The receptive field of auditory nerve fibers can be determined by presenting tones of various frequencies and intensities to the ear of experimental animals while recording from single auditory nerve fibers. Such receptive fields are the areas under frequency threshold curves or frequency tuning curves.

In conclusion, the sensory organs of the five senses that provide conscious awareness of physical stimuli from the environment have many similarities. Studying their similarities as well as the ways in which they differ can aid in understanding the function of the sensory systems.

REFERENCES

1. Bernhard, C.G., Miller, W.H., and Møller, A.R., Function of the Corneal Nipples in the Compound Eyes of Insects. *Acta Physiol. Scand.*, 58: 381–382, 1963.
2. Bernhard, C.G., Gemne, G., and Møller, A.R., Modification of Specular Reflection and Light Transmission by Biological Surface Structures: To See, To Be Seen or Not To Be Seen. *Quart. Rev. Biophys.*, 1: 89–105, 1968.
3. Berson, D.M., Dunn, F.A., and Takao, M., Phototransduction by Retinal Ganglion Cells That Set the Circadian Clock. *Science*, 295: 1070–1073, 2002.
4. Brownell, W.E., Observation on the Motile Response in Isolated Hair Cells. In: *Mechanisms of Hearing*, edited by W.R. Webster and L.M. Aiken. Melbourne: Monash University Press, 1983, pp. 5–10.
5. Brownell, W.E., Bader, C.R., Bertrand, D., and de Ribaupierre, Y., Evoked Mechanical Responses of Isolated Cochlear Hair Cells. *Science*, 227: 194–196, 1985.
6. Buck, L.B., Smell and Taste: The Chemical Senses. In: *Principles of Neural Science*, 4th ed., edited by E.R. Kandel et al. New York: McGraw-Hill, 2000, pp. 625–647.
7. Büki, B., Avan, P., and Ribari, O., The Effect of Body Position on Transient Otoacoustic Emission. In: *Intracranial and Intralabyrinthine Fluids*, edited by A. Ernst, R. Marchbanks, and M. Samii. Berlin: Springer-Verlag, 1996, pp. 175–181.
8. Densert, O., Adrenergic Innervation in the Rabbit Cochlea. *Acta Otolaryngol. (Stockholm)* 78: 345–356, 1974.
9. Flock, A., Transducing Mechanisms in Lateral Line Canal Organ. *Cold Spring Harbor Symp. Quant. Biol.*, 30: 133–146, 1965.
10. Geisler, C.D., *From Sound to Synapse.* New York: Oxford Press, 1998.

11. Hartline, H.K., Response of Single Optic Nerve Fibers of the Vertebrate Eye to Illumination of the Retina. *Am. J. Physiol.*, 121: 400–415, 1938.
12. Hecht, S., Schlaer, S., and Pirenne, M.H., Energy, Quanta, and Vision. *J. Gen. Physiol.*, 25: 819–840, 1942.
13. Hensel, H., Correlations of Neural Activity and Thermal Sensation in Man. In: *Sensory Functions of the Skin in Primates*, edited by Y. Zotterman. Oxford: Pergamon Press, 1976, pp. 331–353.
14. Iggo, A. and Andres, K.H., Morphology of Cutaneous Receptors. *Annu. Rev. Neurosci.*, 5: 1–31, 1982.
15. Johansson, R.S., and Vallbo, A.B., Tactile Sensory Coding in the Glabrous Skin of the Human Hand. *Trends Neurosci.*, 27–32, 1983.
16. Johnstone,, B.M., Patuzzi, R., and Yates, G.K., Basilar Membrane Measurements and the Traveling Wave. *Hear. Res.*, 22: 147–153, 1986.
17. Kenshalo, D.R., Correlations of Temperature Sensitivity in Man and Monkey, a First Approximation. In: *Sensory Functions of the Skin in Primates*, edited by Y. Zotterman. Oxford: Pergamon Press, 1976, pp. 305–330.
18. Konishi, M. and Nielsen, D.W., The Temporal Relationship Between Motion of the Basilar Membrane and Initiation of Nerve Impulses in Auditory Nerve Fibers. *J. Acoust. Soc. Am.*, 53: 325, 1973.
19. Lim, D.J., Effects of Noise and Ototoxic Drugs at the Cellular Level in the Cochlea. *Am. J. Otolaryngol.*, 7: 73–99, 1986.
20. Møller, A.R., The Acoustic Reflex in Man. *J. Acoust. Soc. Am.*, 34: 1524–1534, 1962.
21. Møller, A.R., Noise as a Health Hazard. *Ambio*, 4: 6–13, 1975.
22. Møller, A.R., Frequency Selectivity of Single Auditory Nerve Fibers in Response to Broadband Noise Stimuli. *J. Acoust. Soc. Am.*, 62: 135–142, 1977.
23. Møller, A.R., *Hearing: Its Physiology and Pathophysiology*. San Diego: Academic Press, 2000.
24. Mountcastle, V.B., LaMotte, R.H., and Carli, G., Detection Thresholds for Vibratory Stimuli in Humans and Monkeys: Comparison with Threshold Events in Mechanoreceptive First Order Affernt Nerve Fibers Innervating Monkey Hands. *J. Neurophysiol.* 35: 122, 1972.
25. Mountcastle, V.B., *Medical Physiology*. St. Louis: C.V. Mosby, 1974.
26. Murray, R.G., The Ultrastructure of Taste Buds. In: *The Ultrastructure of Sensory Organs*, edited by I. Friedmann. New York: Elsevier, 1973, pp. 1–81.
27. Nagy, I. and Rang, H., Noxious Heat Activates All Capsaicin-Sensitive and Also a Subpopulation of Capsaicin-Insensitive Dorsal Root Ganglion Neurons. *Neuroscience*, 88: 995–999, 1999.
28. Provencio, I., Rodriguez, I.R., Jiang, G., Hayes, W.P., Moreira, E.F., and Rollag, M.D., A Novel Human Opsin in the Inner Retina. *J. Neurosci.*, 20: 600–605, 2000.
29. Rhode, W.S., Observations of the Vibration of the Basilar Membrane in Squirrel Monkeys Using the Mossbauer Technique. *J. Acoust. Soc. Am.*, 49: 1218–1231, 1971.
30. Rimmer, D.W., Boivin, D.B., Shanahan, T.L., Kronauer, R.E., Duffy, J.F., and Czeisler, C.A., Dynamic Resetting of the Human Circadian Pacemaker by Intermittent Bright Light. *Am. J. Physiol. Regul. Integr. Comp. Physiol.*, 279: R1574–1579, 2000.
31. Roberts, W., A Hypothesis on the Physiological Basis for Causalgia and Related Pains. *Pain*, 24: 297–311, 1986.
32. Rosowski, J.J., The Effects of External- and Middle-Ear Filtering on Auditory Threshold and Noise-Induced Hearing Loss. *J. Acoust. Soc. Am.*, 90: 124–135, 1991.
33. Russell, I.J., and Sellick, P.M., Low Frequency Characteristic of Intracellularly Recorded Receptor Potentials in Guinea-Pig Cochlear Hair Cells. *J. Physiol.* 338: 179–206, 1983.
34. Schmidt, R.F., *Fundamentals of Sensory Physiology*. New York: Springer-Verlag, 1981.
35. Schnapf, J.L., Kraft, .T.W., Nunn, B.J., and Baylor, D.A., Spectral Sensitivity of Primate Photoreceptors. *Vis. Neurosci.*, 1: 255–261, 1988.
36. Shepherd, G.M., Microcircuits in the Nervous System. *Sci. Am.*,. 238: 92–103, 1978.

37. Shepherd, G.M., *Neurobiology*. New York: Oxford University Press, 1994.
38. Smith, D.V., and Margolskee, R.F., Making Sense of Taste. *Sci. Am.* 284: 32–39, 2001.
39. Spoendlin, H., Structural Basis of Peripheral Frequency Analysis. In: *Frequency Analysis and Periodicity Detection in Hearing*, edited by R. Plomp and G.F. Smoorenburg. Leiden: A.W. Sijthoff, 1970, pp. 2–36.
40. Stuhlman, O., *An Introduction to Biophysics*. New York: Wiley, 1943.
41. Vallbo, A.B. and Johansson, R.S., Properties of Cutaneous Mechanoreceptors in the Human Hand Related to Touch Sensation. *Hum. Neurobiol.*, 3: 3–14, 1984.
42. Warwick, R. and Williams, P.L., In: *Gray's Anatomy*. Philadelphia: W.B. Saunders, 1973.
43. Zwislocki, J.J. and Sokolich, W.G., Velocity and Displacement Responses in Auditory Nerve Fibers. *Science* 182: 64–66, 1973.

CHAPTER 3

Sensory Nervous Systems

ABBREVIATIONS

AAF : Anterior auditory fields
ACTH : Adrenocorticotropic hormone
AI : Primary auditory cortex
AII : Secondary auditory cortex
BIC : Brachium of the inferior colliculus
CN : Cranial nerve
CNS : Central nervous system
DCN : Dorsal column nuclei
dMGB : Dorsal division of the medial geniculate body
EPSP : Excitatory postsynaptic potentials
fMRI : Functional magnetic resonance imaging
IC : Inferior colliculus
ICC : Central nucleus of the inferior colliculus
ICX : External nucleus of the inferior colliculus
IPSP : Inhibitory postsynaptic potential
LGN : Lateral geniculate nucleus
LL : Lateral lemniscus
MD : Mediodorsal (nucleus of the thalamus)
MEP : Magnetic evoked potentials
MGB : Medial geniculate body
dMGB : Dorsal division of the medial geniculate body
mMGB : Medial division of the medial geniculate body

ML : Medial lemniscus
MN : Mammillary nucleus
NST : Nucleus of the solitary tract
OCB : Olivocochlear bundle
PAF : Posterior auditory field
PAG : Periaqueductal gray
PET : Positron emission tomography
PIN : Posterior intralaminar nucleus
PO : Posterior complex (of the thalamus)
PSP : Postsynaptic potentials
RA : Rapid adapting
RE : Reticular nucleus
SA : Slow adapting
SC : Superior colliculus
SG : Suprageniculate nucleus
SII : Secondary somatosensory field
SNR : Signal-to-noise ratio
SOC : Superior olivary complex
SPECT : Single photon emission computed tomography
SPL : Sound pressure level
TENS : Transdermal electrical nerve stimulation
TN : Trigeminal nucleus
VB : Ventrobasal (nucleus of the thalamus)
VPL : Ventral posterior lateral (nucleus of the thalamus)
VPM : Ventral posterior medial (nucleus of the thalamus)

ABSTRACT

Anatomy

1. Two parallel pathways known as the classical and nonclassical pathways convey information from somatosensory, auditory, visual, and taste receptors to higher central nervous system centers. The classical ascending pathways of these four senses project to the primary sensory cortices.
2. The pathways for olfaction are different from those of the other sensory systems, and they do not have distinct classical and nonclassical parts.

Classical Ascending Sensory Pathways

3. The fibers of the ascending somatosensory, auditory, and taste pathways all make synaptic connections with the cells of a nucleus before crossing

the midline (dorsal column nuclei, cochlear nuclei, and nucleus of the solitary tract).

4. The retina contains a neural network with two synapses before it enters the optic nerve.
5. All fibers of the somatosensory, auditory, visual, and taste fibers are interrupted by synaptic transmission in the ventral part of the thalamus, which may be regarded as the gateway to the primary sensory cerebral neocortex where processing of information occurs. The thalamus is the common pathway for sensory information.
6. Olfactory pathways project to allocortex and structures of the limbic system such as the medial and the central nucleus of the amygdala.
7. The first nuclei of the somatosensory pathways (dorsal column nuclei) and the taste pathways (nucleus of the solitary tract) project directly to the thalamic nucleus (ventral posterior lateral nucleus), while all fibers from the first auditory relay nucleus (cochlear nucleus) make synaptic connections with cells in a midbrain nucleus before reaching the thalamic relay nucleus (medial geniculate body).
8. The optic nerve and the optic tract proceed uninterrupted to the thalamic nucleus (lateral geniculate nucleus), which projects to the primary visual cortices.
9. The primary cerebral sensory cortices relay information to other (secondary and association) cortices and to subcortical structures (descending system). The primary sensory cortices may be regarded as the second common pathways for sensory information.
10. The cerebral sensory cortex has six layers and is organized in columns where the neurons of a particular column represent the same sensory input.
11. Each layer of the cerebral cortex is specific with regard to receiving and sending information in similar ways for the different senses. Interconnections between neurons in different layers of the cerebral cortices are the anatomical basis for information processing in the sensory cerebral cortices.
12. The corpus callosum connects corresponding cortical areas on both sides with each other.

Descending Pathways

13. The four sensory systems (auditory, somatosensory, visual, and taste) have descending pathways, the extent of which differs among the different senses.

14. The descending pathways have been described as separate systems by some authors while other authors prefer to regard the descending systems as reciprocal connections along the ascending pathways.

Nonclassical Pathways

15. The nonclassical pathways of the somatosensory, auditory, and visual fibers are interrupted by synaptic transmission in the dorsal and medial parts of the thalamus which project to secondary and association cortices and several other regions of the brain such as motor regions and structures of the limbic system.
16. The nonclassical pathway of the somatosensory system is the anterior-lateral system consisting of the spinothalamic, spinomesencephalic, and spinoreticular systems that receive their input from specific receptors (such as nociceptors) as well as from the same receptors that provide input to the classical pathways
17. The nonclassical pathways of the visual system are the superior colliculus pathways and one that uses the pretectal and pulvinar of the thalamus.
18. Many neurons of the nonclassical auditory pathways receive input from the somatosensory and visual system in addition to auditory input.

Information Processing in Ascending Sensory Pathways

Classical Pathways

19. Sensory information that is coded in the afferent fibers of receptors is transformed, analyzed, rearranged, and restructured in the nuclei of the ascending sensory pathways.
20. Neurons in the classical sensory systems respond distinctly to different qualities of sensory stimuli, and the different qualities of sensory stimuli are coded in the discharge pattern of the neurons throughout the ascending sensory pathways, including the cerebral cortices.
21. Generally, small changes in sensory stimuli are enhanced as the information ascends in the ascending sensory pathways toward the primary sensory cerebral cortices.
22. Adaptation contributes to enhancement of the responses to rapid changes in the stimulus. Adaptation occurs at all levels of sensory systems including the receptors and the structures that conduct the stimulus to the receptors.

23. Synaptic interaction between inhibition and excitation increases contrast of visual images, of stimulation at different points of the skin and between different frequencies of sounds in hearing. Interaction between inhibition and excitation also enhances the neural representation of rapid changes in sensory stimuli.
24. Sensory stimulation creates two-dimensional neural maps that are projections of (physical) receptor surfaces such as the skin and the retina. The basilar membrane in the cochlea can be regarded as a one-dimensional receptor surface regarding frequency of sounds. The frequency of sounds is mapped throughout the ascending auditory pathways, including the auditory cortices.
25. Spatial sensory information from paired (bilateral) sensory organs creates three-dimensional maps in the central nervous system, which is the basis for directional hearing and stereoscopic vision at short distance.
26. The physical basis for directional hearing (creation of auditory space) is the difference between the sounds that reach the two ears. The resulting neural maps in the auditory nervous system are computational maps rather than projections of receptor surfaces.
27. Spatial discrimination is enhanced by interaction between inhibitory and excitatory receptive fields (such as lateral inhibition).
28. Flow and processing of information in the sensory systems are controlled by:
    a. Descending systems (reciprocal innervation)
    b. Arousal that exerts a general control of excitability of cerebral cortices
    c. Selective attention, which is controlled from high central nervous system centers and which selects and suppress sensory specific messages
29. Parallel processing is prominent in sensory systems where the same information is processed in different ways in anatomically different parts of the central nervous system.
30. Stream segregation, where different kinds of information such as object ("what") and spatial ("where") information is processed in anatomically different parts of the central nervous system, has been demonstrated in the visual and auditory systems and probably also exists in the somatosensory system.
31. Neural plasticity can change the way sensory stimuli are processed for shorter or longer times by several means:
    a. Unmasking dormant synapses to open normally inactive connections
    b. Sprouting of axons to establish new connections
    c. Masking of normally functioning synapses that close connections that are normally open

d. Anatomical elimination of synapses, axons, dendrites, and cell bodies

Nonclassical Pathways

32. Neurons in the nonclassical pathways respond in a less distinct way to sensory stimuli than neurons in the classical pathways.
33. Neurons in the nonclassical pathways often respond to broad ranges of stimuli, including stimuli of other modalities.

Connections from Sensory Systems to Nonsensory Regions of the Central Nervous System

34. Sensory information can reach motor systems and elicit contractions of muscles.
35. Sensory information reaches the reticular formation where it controls arousal. Nonclassical pathways, in particular, supply arousal by activating the brainstem reticular formation and provide subcortical input to structures of the limbic system.
36. Sensory information can reach structures of the limbic system (such as the amygdala) through a cortical and a subcortical route:
    a. Primary sensory cortices, secondary cortices and association cortices (the "high route")
    b. Dorsal and medial thalamic nuclei directly to the amygdala (the "low route").

I. INTRODUCTION

The complexity of the *central nervous system* (CNS) is overwhelming, compared to any manmade structure, with regard to the number of elements (neurons) and the number of connections between neurons. The exact number of neurons in the human brain is unknown, but it has been estimated to be on the order of 100 billion neurons. Many individual neurons receive 1000 or more inputs. The total number of connections has been estimated to be on the order of 100 trillion.[1]

The brain consists of many seemingly independent parts that are assigned different functions; however, there are ample interconnections among different parts of the brain that make possible many forms of interactions among the various parts of the brain. As our understanding of the function of the CNS progresses, the distinction among the functions of different parts of the brain becomes more and more diffuse. For example, the cerebellum was earlier regarded mainly as the center of motor control, but more recently it has been shown that the cerebellum has many other functions — including memory.

Despite the overwhelming complexity of the brain, studies of isolated, well-defined parts of the CNS have provided important insights with regard to how individual neurons function and how certain well-defined functions are performed by various systems of the CNS. Studies of the CNS have been mainly of two types, namely studies of how individual elements (neurons) function and studies of how systems function (*systems neurophysiology*).

This chapter concerns systems-level neurophysiology. Those readers who are interested in more details regarding the function of individual neurons and specifically the biology of the synapse are referred to books such as those by Shepherd.[195,196]

In this chapter, we will discuss the anatomy and basic physiology of the neural pathways between sensory organs and the cerebral cortices, emphasizing the similarities and the differences of the anatomy and the physiology of the different sensory systems. We will also discuss how sensory information is processed in the nervous system in general, again emphasizing similarities and differences between the five senses. More specific aspects on the five sensory systems will be discussed in detail in this chapter and in the chapters on individual sensory systems (Chapters 4 to 7).

II. ANATOMY OF SENSORY NERVOUS SYSTEMS

The first part of this chapter discusses the organization and connections of the ascending sensory pathways. The main anatomical organization of the cerebral cortices and their connections with other parts of the brain are also discussed.

It must be emphasized that the anatomical connections that are described in this chapter are those that have been verified using anatomical methods. Anatomically verified connections may not all be functional because the synapses with which they are connected may not be open (may become *dormant*[224], see page 235). Furthermore, connections that are normally not functional because of dormant synapses may become functional under certain circumstances because of *unmasking*[224] of dormant synapses.

A. Ascending Sensory Pathways

For our discussion of the nervous system of the senses, we may assume that sensory information ascends in mainly two parallel pathways through a series of nuclei where the fiber tracts are interrupted and information is processed by synaptic transmission before it proceeds to the following nuclei and finally reaches the sensory cerebral cortices, from which the information can reach many other regions of the CNS. This organization of complex networks of neurons in the nuclei of the ascending pathways and in the cerebral cortices is the anatomical basis for the *hierarchical* and *parallel processing* of sensory information.

Two main separate ascending neural pathways known as the *classical ascending sensory pathways* and the *nonclassical ascending pathways** have been identified in the senses of hearing, vision, and somesthesia. The classical ascending pathways project to the primary sensory cortices, which in turn connect to other parts of the cerebral cortex. The classical ascending pathways are phylogenetically younger than the nonclassical pathways that ascend parallel to the classical pathways. There are also indications of two separate ascending pathways for taste, but the olfactory nervous system has no distinct divisions to separate into classical and nonclassical pathways, and its organization is in many ways different from that of the other four senses.

In addition to the classical and the nonclassical ascending pathways there are extensive *descending pathways*, which are mainly *reciprocal connections* to the ascending pathways. There are also subcortical connections from sensory pathways to parts of the CNS other than those that are normally regarded as belonging to sensory systems.

Most studies have centered on the classical ascending pathways, which convey information to the primary cerebral cortex from the sensory organs. The neurons in these pathways perform extensive processing of sensory information, which will be discussed later in this chapter. The analysis that occurs in the classical ascending pathway is to a great extent responsible for our ability to discriminate sounds, visual images, tactile stimulation, odors, and tastes. The analysis that occurs in the sense organs (discussed in the previous chapter) can be regarded as a pre-analysis that prepares the sensory information for the further analysis that occurs in the nervous system.

Less is known about the morphology and function of the nonclassical pathways than the classical pathways, and the role of nonclassical pathways have received little attention in the past. We will show in this chapter (and in the following chapters concerning individual senses) that the nonclassical pathway may indeed play important roles in the function of the normal sensory systems, and the nonclassical pathways may play a role in generating symptoms and signs of some disorders.

*Various authors have used different names for these two types of pathways. Some investigators refer to these pathways as the "lemniscal" and the "extralemniscal;" other commonly used names are "specific" and "nonspecific". Some investigators divide the nonclassical pathways into "diffuse" and "polysensory." pathways. In many textbooks, only the classical pathways are considered, or the nonclassical pathways are discussed only briefly.

The anatomy of descending systems are relatively well known but little is known about their function in general, and only a few studies have addressed their role in sensory processing.

1. Sensory Systems Project to the Conscious and the Unconscious Brain

Our five senses of hearing, vision, touch, smell, and taste can discriminate a very large number of different stimuli. Moreover, minute differences in the stimuli can evoke distinct conscious responses. It is therefore natural that researchers have focused on the conscious brain when studying these five senses. However, there is a considerable amount of connections from the sensory pathways to nonconscious parts of the CNS, and these connections may have importance that has not been fully recognized because they are not associated with conscious awareness. *Proprioception** is sometimes included in discussion of sensory systems, but it is outside the scope of this book (for details, see, for example, Zigmond *et al.*[247]). Proprioception is not usually associated with awareness.

The nonclassical pathways do not project to primary cerebral cortical areas as the classical pathways do. Instead, they project to secondary and association sensory cortices and make subcortical connections to other parts of the brain that are regarded as belonging to the old brain, such as structures of the *limbic system.*** This means that many different parts of the brain may receive sensory input from the same sensory event. Because the nonclassical pathways provide little input to primary sensory cortices and project to many nonsensory parts of the CNS, the activity in nonclassical pathways may not elicit immediate awareness but may affect vital functions such as memory, emotions, and endocrine and other autonomous functions.

The senses of taste and olfaction have the strongest indications of providing input to the unconscious and autonomic brain. Taste can result in vomiting and can cause either a strong feeding desire or the opposite. The same is the case for olfaction, where specific odors may affect behavior and bodily functions without causing awareness of such.

*Proprioception monitors muscle contractions and movements and the position of the body. Proprioception uses sensory organs that are similar to those of conscious senses. Proprioception gets its main input from muscles and tendons and joints. This information is integrated with input from the vestibular apparatus.

**The limbic system is a collective name used for a group of structures that are located deep in the brain near the edge of the medial wall of the cerebral hemisphere. Usually the *hippocampus*, *amygdala*, and *fornicate gyrus* are included in this term, but often structures to which they interconnect, such as the *septal area,* the *hypothalamus*, and a medial part of the *mesencephalic tegmentum*, are regarded as belonging to limbic structures. Limbic structures are sometimes referred to as the *visceral brain*.

Perhaps a clearer example of sensory information that does not reach awareness is mediation by the *vomeronasal system*, which does not project to cerebral cortices but instead projects exclusively to structures of the limbic system. The vomeronasal system that responds to *pheromones** has many similarities with a nonclassical sensory systems and may be regarded as the nonclassical pathway of olfaction.

2. Innervation of Sensory Receptors

Afferent nerve fibers communicate sensory input from the receptors to the CNS. The nerve fibers that carry information from receptors of hearing, vision, smell, and taste are all *cranial nerves* or parts of cranial nerves. Olfactory information is carried in cranial nerve(CN) I, vision in CN II, and taste fibers in CN VII, IX, and X. Auditory information is carried in the cochlear part of CN VIII. The other part, the vestibular nerve, carries information from the balance organs in the inner ear. The nerve fibers that innervate somatosensory receptors of the skin in the face and in the mouth travel in CN V. The fibers that innervate skin receptors in the rest of the body travel in peripheral nerves and enter the spinal cord through the *dorsal roots* of spinal nerves.

Innervation of Type I and Type II sensory organs (see Chapter 2) is different. The nerve fibers that innervate Type II receptors such as those in the cochlea, the eye, and the taste organ connect to receptors through synapse-like connections. Some of these Type II receptors also receive inhibitor input from efferent fibers originating in the CNS. These efferent fibers may connect to the cell bodies of the receptor cells or to the axons of the afferent fibers of the receptor cells. The receptor cells of olfaction may also be regarded as Type II receptors but are *bipolar cells* (see Chapter 2, page 46) where the dendrites contain the specialized membranes that are sensitive to odors. The axons of these bipolar cells form the olfactory nerve, the fibers of which make synaptic connections with cells in the olfactory bulb.

The photoreceptors of the eye do not connect directly to the optic nerve but visual information reaches the optic nerve through an elaborate neural network located in the retina. This network consists of two perpendicular paths: one consisting of the bipolar cells that lead from the photoreceptors toward the *ganglion cells,* the axons of which are fibers of the optic nerve, and the other consisting of the *horizontal cells* and *amacrine cells* that connect between the photoreceptors and the ganglion cells, respectively (Fig. 3.1).[194]

The neural connections of the olfactory sensory cells that are located in the olfactory bulb have similarities with those of the retina (Fig. 3.1).[194] As in

*Pheromones are chemicals that act as messengers between members of the same species. Pheromones are important for sexual behavior and are best known from studies of insects.

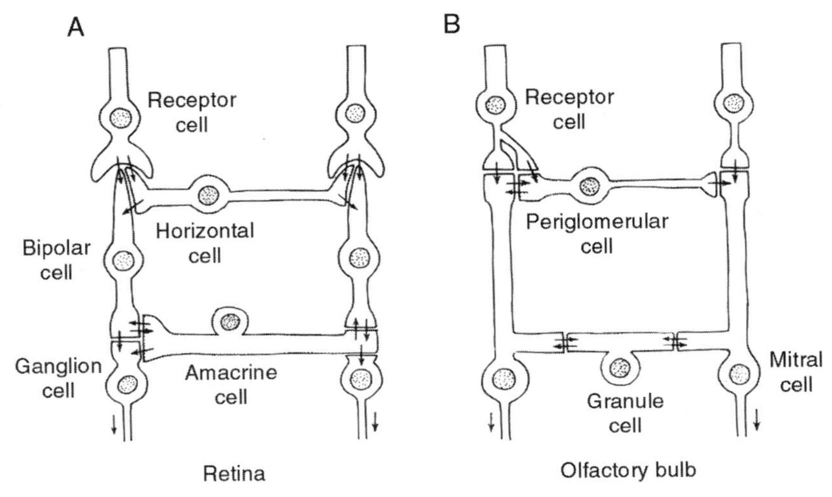

FIGURE 3.1 Illustration of the similarity between the neural network in the retina (A) and the network in the olfactory bulb (B). (Adapted from Shepherd, G.M., *Sci. Am.*, 238: 92–103, 1978.)

the retina, two different paths can be identified: a vertical path that leads from the olfactory receptors to the neurons (*mitral cells*) that are the output neurons of the olfactory bulb and which connect to the CNS and a horizontal path that consists of the *periglomerular cells* that connect between the receptor cells, and the *granular cells* that connect between the output cells (mitral cells).

The efferent innervation of Type II sensory cells makes it possible to modify the output of the receptors from the CNS. The efferent innervation can control the sensitivity of the receptors, and efferent activity mainly decreases the neural activity in the afferent fibers of the receptors. In the ear, the effect of efferent innervation is more complex than in other sensory organs because some of the receptors in the cochlea, the *outer hair cells*, function as mechanically active elements, while the *inner hair cells* are the sensory cells that communicate auditory information to afferent fibers of the auditory nerve (Chapter 2, pages 40 and 48). Efferent fibers that terminate on the afferent fibers of inner hair cells can therefore affect the flow of activity from the sensory cells to the afferent auditory nerve fibers. Efferent fibers terminate directly on the bodies of the outer hair cells, and efferent activity can therefore control the mechanical properties of the outer hair cells through commands from other structures of the CNS. Because outer hair cells are integral parts of the mechanical system of the cochlea (basilar membrane),[23] the mechanical properties of the media that transmit the physical stimuli to the receptors (the inner hair cells) are under the control of the CNS.[70,154]

3. Ascending Sensory Neural Pathways

Many similarities exist between the organization of the neural pathways of hearing, somesthesia, vision and, to some extent, taste. The primary sensory cerebral cortices that are the target of the ascending sensory pathways also have many similarities for these four senses. While there are similarities between the organization of the olfactory bulb and the retina as mentioned above, the pathways of olfaction are different from those of the other four senses. In the following section, we will discuss these similarities and the differences. A more detailed description of the sensory pathways of the different sensory systems is found in subsequent chapters that are devoted to specific senses (Chapters 4 to 7).

The information from the ear, the eye, and receptors in the skin (touch) ascends toward the cerebral cortex in the classical and nonclassical pathways (Fig. 3.2). In general, the classical ascending sensory systems perform a high degree of analysis and project to the *neocortex,* where specific features of stimuli activate specific populations of neurons. The classical pathway represents the phylogenetically newer part of the brain compared with the nonclassical sensory

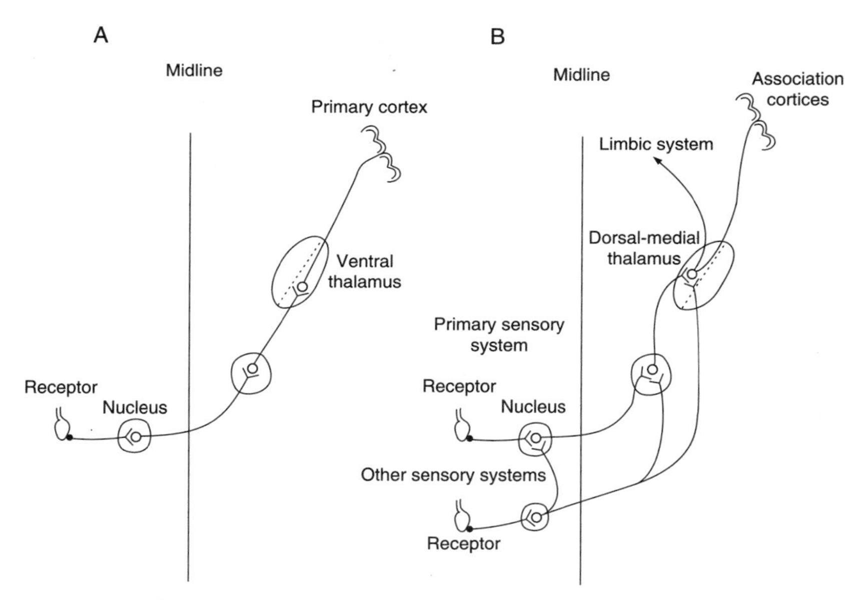

FIGURE 3.2 Schematic drawing of the general outline of the ascending pathways of a sensory system emphasizing the differences and similarities between the classical (A) and the nonclassical (B) pathways. Note that the two receptors in (B) are from two different sensory systems—for instance, auditory and somatosensory.

systems, and the classical pathways perform detailed processing of sensory information and provide the basis for fine discrimination of sounds, fine touch, high visual acuity, and perhaps discrimination of taste. The nonclassical pathways are parts of the old brain, where sensory systems were presumed to be involved in detection of faint stimuli with less ability to discriminate stimuli than within the classical sensory systems. It seems as though the classical sensory pathways have developed as species have become more sophisticated and more and more dependent on being able to interpret complex sensory stimuli for identification of prey, to distinguish between friend and foe, and, to a limited extent, for intraspecies communication. The connections from the nonclassical system to *allocortical** regions and structures of the limbic system fit the hypothesis that sensory input mainly served as a warning device at early stages of the phylogenetic development of terrestrial vertebrate species. The classical pathways that developed in terrestrial species are assumed to have developed as the demands on the use of sensory input shifted from detecting warning signals to identifying a friend, searching for food, or communicating between individual members of species, which required an increasing sophistication of analysis of sensory signals. The fine discrimination by olfaction that many animals (such as the dog) can perform is comparable to the discrimination that is performed by the classical sensory systems of hearing, vision, and somesthesia. This may seem to be a contradiction of the rules that phylogenetically old systems are inferior to newer systems regarding discrimination of fine details of sensory stimuli.

The emphasis on nonclassical pathways is assumed to have decreased concomitantly during phylogenetic development. Decrease of input could occur as a result of the normal *pruning* of anatomical connections and *apoptosis,*** which occur during ontogeny, but it could also be due to decreased input to these systems caused by decreased efficacy of the synapses that connect sensory information to the nonclassical systems, without a change in the anatomical organization of these systems. It is also possible that the nonclassical pathways during development may have provided a substrate for the development of the more advanced analysis capabilities in the classical sensory systems.

Most studies of the auditory system have focused on the classical pathway, but the nonclassical pathway was studied relatively early and it was often referred to as the extralemniscal pathway.[7,69,152,200] The classical ascending pathway of somesthesia is the dorsal column system and the nonclassical pathway is the *anteriorlateral system,* which is the pathway for pain and crude touch. It consists of at least four different and separate pathways.[19,156,158]

**Allocortical regions* is Vogt's term denoting several regions of the cerebral cortex, in particular the olfactory cortex and the hippocampus, characterized by fewer cell layers than the neocortex (isocortex).

***Apoptosis* is programmed cell death.

The classical visual system is the *retinogeniculocortical pathway,* and the nonclassical is the *superior colliculus* (SC), or *tectal pathway,* including the *pulvinar* of the *thalamus.*[19,34,163] Connections to the *suprachiasmatic nucleus* (SCN) and the hypothalamus may also be regarded as a nonclassical visual pathway. These connections are important for resetting the body clock that controls the circadian rhythm.[13,182]

The taste pathways have two parallel pathways,[19,196] one specific and one less specific, which may be regarded as the classical and nonclassical pathways, respectively. The vomeronasal pathway[71] of the olfactory system may be regarded as the nonclassical olfactory system.

Two important differences between the classical and the nonclassical pathways are (1) the classical pathways are interrupted by synaptic transmission in nuclei in the ventral thalamus, and these neurons project to primary cortices; and (2) the nonclassical pathways use the dorsal and medial thalamic nuclei, and the axons of these neurons project to many different parts of the CNS, most notably the *secondary cortices* and *association cortices*, the cells of which receive input from more than one sensory system.

4. Organization of the Classical Ascending Sensory Pathways

The classical ascending sensory pathways consist of a string of well-defined nuclei connected by fiber tracts (Fig. 3.3A–D). The last nucleus in this chain is located in the ventral thalamus, and fibers from these neurons connect to neurons in the primary sensory cerebral cortices. The classical ascending sensory pathways for hearing, vision, touch, and taste have many similarities in organization but differ with regard to the complexity of the pathways. All fibers of the classical ascending sensory pathways of these four senses are interrupted by synaptic connections in cells of the ventral portion of the thalamus. The auditory and the visual systems are the most complex, and those of somesthesia and taste are the least complex. The ascending auditory pathways have many connections between the two sides while the other pathways do not have such connections. In vision, the neural network in the retina carries out many of the functions that are carried out by the first several levels of the auditory and somesthesia pathways.

The basic organization of the classical ascending pathways for the four sensory systems, hearing, somesthesia, taste, and vision (Fig. 3.3A–D), can be summarized as:

1. The nerve fibers from the sensory organs terminate in relay neurons in nuclei that are located on the same side of the head or body (in vision, the ganglion cells in the retina may be regarded as the first relay neurons because they are the first cells where impulse activity occurs).

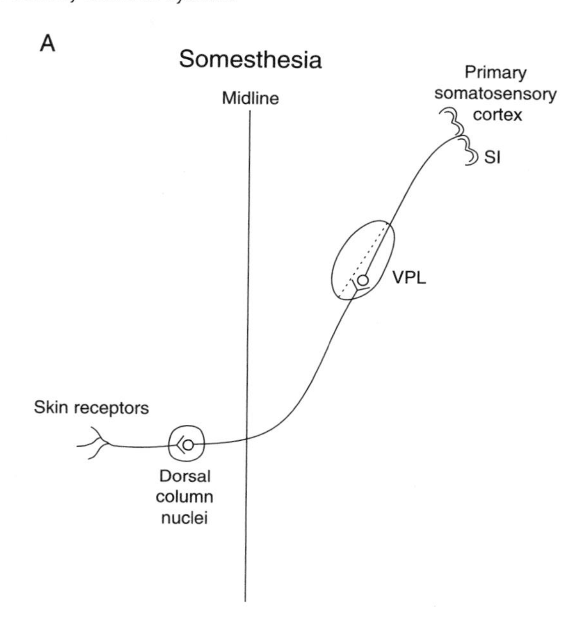

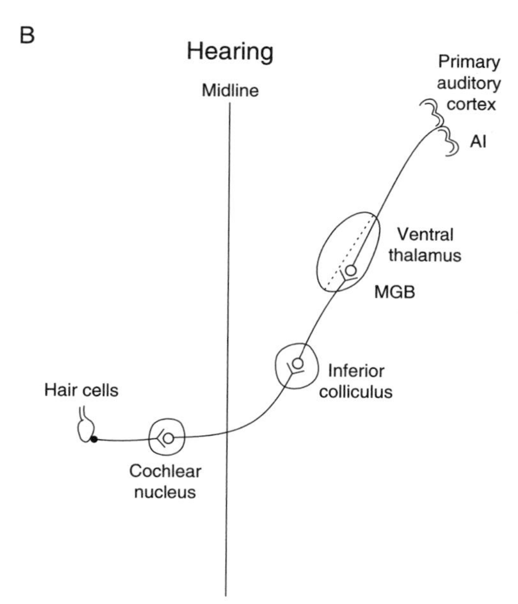

FIGURE 3.3 Schematic drawing showing the principles of the organization of the classical ascending pathways of the different senses. (A) Somesthesia, (B) hearing, (C) vision, (D) taste, and (E) olfaction.

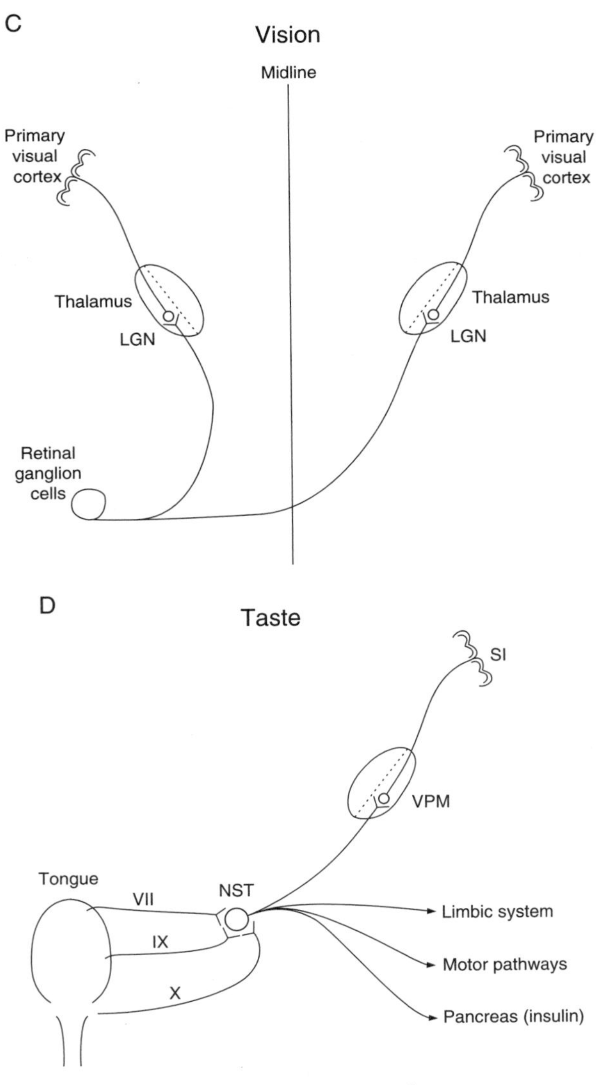

FIGURE 3.3 *(continued)*

2. From the first nucleus, fiber tracts cross the midline.
3. In somesthesia, all fibers cross and continue uninterrupted to the thalamus.
4. In the auditory pathways, most of the fibers that originate in the first relay nucleus cross the midline and are interrupted by synaptic

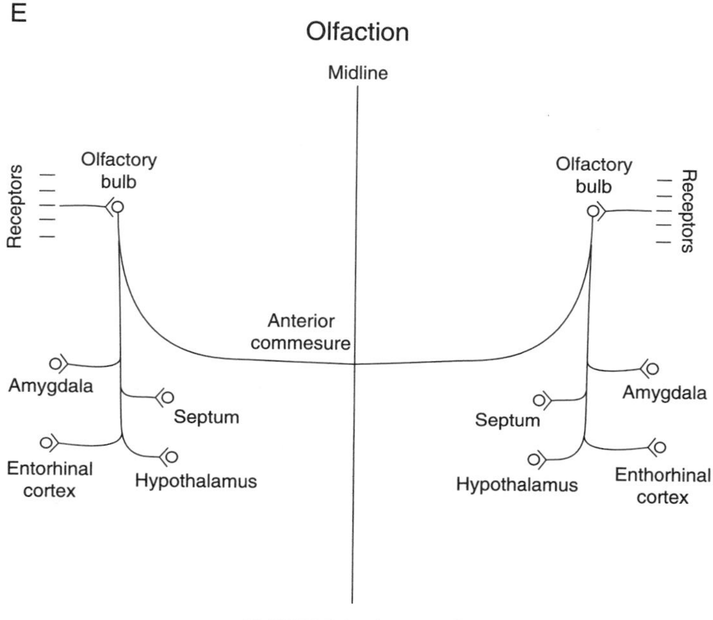

FIGURE 3.3 *(continued)*

transmission at least once (in the *inferior colliculus,* IC) before reaching the thalamus. Many fibers are also interrupted in other nuclei. Some fibers from the first nucleus ascend on the same side and there are ample connections between the pathways on the two sides.

5. Somesthesia and taste have no uncrossed connections.
6. Vision in humans and other animals with forward-pointing eyes has symmetrical crossed and uncrossed connections. The vision of animals with eyes pointing sideways has mainly crossed connections.
7. All fibers of the classical ascending pathways from the visual, auditory, and somatosensory nuclei are interrupted by synaptic transmission in the ventral thalamus, and the axons of these neurons terminate in specific primary sensory cortical regions from where connections are made to the other side (through the *corpus callosum*) and to secondary and association cortices. Most of these connections are reciprocal.
8. Connections from sensory cortices to secondary and association cortices also reach other structures of the CNS such as the limbic system, and this long pathway (the "high route"[104]) provides a cortical route to the amygdala nuclei. The connections between cortical areas and limbic structures are mostly reciprocal.

The first nuclei of the somatosensory system, where all fibers that carry information from the skin of the body are interrupted by synaptic transmission, is the *dorsal column nuclei* (DCN), consisting of the *gracilis* and *cuneatus nuclei* located in the *medulla oblongata*.[19,158] The *trigeminal* (sensory) *nucleus* (TN) interrupts the information from receptors in the skin of the face and inside the mouth.[87] The fibers from the neurons in the TN join those of the DCN to form the *medial lemniscus* (ML), which ascends to the somatosensory nuclei of the thalamus, namely the *ventrobasal* (VB) nuclei, the *ventral posterior medial* (VPM) nucleus for the face region, and the *ventral posterior lateral* (VPL) nucleus for the body senses.

Auditory information is first interrupted by synaptic transmission in the *cochlear nucleus* located in the *pons* (Fig. 3.3B). The cochlear nucleus has three divisions: the *anterior ventral*, *posterior ventral,* and *dorsal*. Auditory nerve fibers bifurcate twice, and one of the resulting three branches terminates in each of the three divisions of the cochlear nucleus. Each of the three divisions sends fibers that cross the midline and then form the *lateral lemniscus* (LL).[48,151,226,227] All information* in that pathway is interrupted by synaptic transmission in the *central nucleus of the inferior colliculus* (ICC) that is located in the *midbrain*. Another pathway from the cochlear nucleus ascends on the ipsilateral side, forming an uncrossed pathway to the ICC. From the ICC, a fiber tract, the *brachium of the inferior colliculus* (BIC) conducts the information to the thalamic auditory nucleus, the *medial geniculate body* (MGB). In addition, some of the fibers of the ascending auditory pathways are interrupted and make synaptic connections to cells in several other nuclei. Many fibers give off collaterals that make synaptic connections to cells in different nuclei. Numerous connections also exist between the two sides, most notably at the midbrain level, but there are also connections between the two sides at the pontine level. No connections between the MGBs of the two sides have been identified.

Information in the visual system is first interrupted by synaptic conduction in the nerve cells in the neural network that is located in the retina. The information that travels in the optic nerve and optic tract is interrupted by synaptic transmission in the thalamic nucleus (the *lateral geniculate nucleus*, LGN) before the information reaches the cells of the *primary* (striate) *visual cortex* (Fig. 3.3C).[19] In animals with forward-pointing eyes, the portion of the optic nerve fibers that innervates the medial (nasal) part of the retina (corresponding to the temporal field) crosses the midline and progresses toward the contralateral LGN at the *optic chiasm*, while the other portion (that innervates the lateral retina) continues on the same side to the ipsilateral geniculate nucleus.

* It has recently been shown that some fibers from the dorsal cochlear nucleus, in fact, bypass the ICC and project to neurons in the medial division of the MGB.[113] These fibers may be regarded as belonging to the nonclassical pathway (see pages 311, 345, and 347).

The classical sensory pathways for each of the visual, hearing, and somesthesia senses all project to regions of the cerebral cortex that are specific for each sense, and are known as the primary sensory cortices. From there, information progresses to secondary and association cortices. While neurons in the classical ascending pathways, including the primary sensory cortices, receive input from only one sensory modality, neurons in association cortices (and some in secondary cortices) respond to more than one sensory modality.

The fibers of taste pathways (Fig. 3.3D) are first interrupted and make synaptic connections with cells in the *nucleus of the solitary tract* (NST).[19,196] The main input to that nucleus comes from the *chorda tympani* (part of cranial nerve VII), the *glossopharyngeal nerve* (part of cranial nerve IX), and the *vagus nerve* (cranial nerve X). There are considerable species differences. In the rat, the NST connects to a pontine taste nucleus, which projects both to the VPM of the thalamus and structures of the limbic system, including the amygdala. In the monkey, neurons of the NST connect directly to the neurons of the VPM. The pontine taste nucleus in the monkey relays information only to the limbic structures.[162,196] It is assumed that the taste pathways in humans resemble those of the monkey. Cells in the VPM nucleus project to areas of the *insular cortex* and somatosensory areas. There are also connections from the NST to motor systems and parts of the digestive tract such as glands that secrete saliva and other glands that secrete digestive fluid. Connections to the pancreas make taste stimulation affect the excretion of insulin. In the rat, taste pathways also project directly to the nuclei of the amygdala.[162,196]

Information from the olfactory epithelium reaches the *olfactory bulb*, where the first synapse is located (Fig. 3.3E). Shepherd[194] has pointed out that the circuit diagram of the peripheral part of the olfactory bulb has similarities with that of the retina (Fig. 3.1). Aside from these similarities, the neural circuitry of the ascending olfactory pathways shares few similarities with the ascending pathways of any of the other four sensory systems.

The ascending olfactory pathway reaches parts of the CNS that are different from those of the four other senses, and the olfactory pathways mainly project to *allocortex*. The olfactory pathways have abundant connections to structures of the limbic system, mainly the *medial* and *central nucleus* of the amygdala (in the monkey[19,116]). There are extensive connections between the olfactory bulbs of the two sides through the *anterior commissure*, which often is regarded as a part of the corpus callosum.

It has been questioned whether the olfactory pathways are interrupted by synaptic transmission in the thalamus as are the other sensory systems but some investigations have shown that olfactory fibers indeed make connections with cells in the thalamus.[202] The connections from the olfactory bulb reach many parts of the CNS (Fig. 3.3E) but it has no direct projections to neocortex. The olfactory pathway project to the *uncus* and in humans probably also to the *entorhinal cortex*, which belongs to the allocortex and is located deep inside the temporal lobe (Fig. 3.4).

5. Anatomical Location of Components of the Ascending Sensory Pathways

The anatomical locations of the structures of the ascending sensory pathways include the medulla (DCN), pons (cochlear nucleus), midbrain (IC), and thalamus (MGB, LGN, VPL, VPM). The cortical regions to which the four sensory systems project are all located on the surface of the brain and are known as the neocortex. The primary sensory cortical regions (Fig. 3.4) are located posterior to the *central sulcus* for somesthesia, in *Heschel's gyrus* for audition (Fig. 3.4C), and in the occipital region for vision (Fig. 3.4A,B). The cerebral cortex for vision is known as the *striate** cortex.

6. Organization of the Ascending Nonclassical Pathways

The nonclassical ascending sensory pathways for touch, hearing, vision, and taste receive input from the classical pathways, and these connections branch off the classical pathways at various levels. The *anteriorlateral tracts*** of the somatosensory system and the nonclassical auditory pathways are the best known of the nonclassical pathways.

The pathways for taste have great species differences. While some species have pathways that may be regarded as nonclassical it is questionable whether primates, including humans, have a separate nonclassical pathway for taste. The pathways for olfaction are different from those of the other four senses but connections from the vomeronasal organ (pheromone) pathway to the accessory olfactory bulb may be regarded as the nonclassical pathway for olfaction. The vomeronasal organ projects to the accessory olfactory bulb, which in turn projects directly to the limbic structure (the *amygdaloid nuclear complex*, the *bed nucleus of the accessory olfactory tract*, and the *posteriomedial cortical nucleus of the amygdala.*[71,116]).

The anteriorlateral system of the somatosensory system receives its input from fibers of dorsal roots of the spinal cord, which give off collaterals that terminate on nerve cells in the dorsal horn of the spinal cord, ascending in the spinal cord as the tract of the dorsal column.[19] Each dorsal root fiber connects to cells in the dorsal horn of several segments (7 to 10), and the output of these dorsal horn

*The *striate* cortex is the primary visual cortex and is so named because of its appearance in histological preparations.

**The anteriorlateral system consists of the spinothalamic tract, the spinomesencephalic tract, and the *spinoreticular tract*. The spinothalamic tract is the best known of these and has been treated as a separate sensory system for pain and temperature in some textbooks.

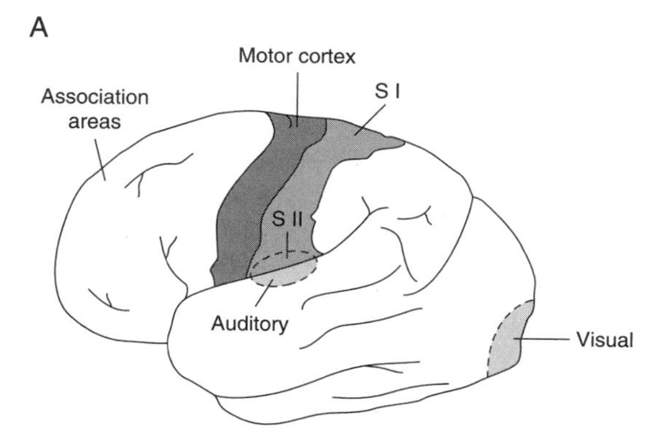

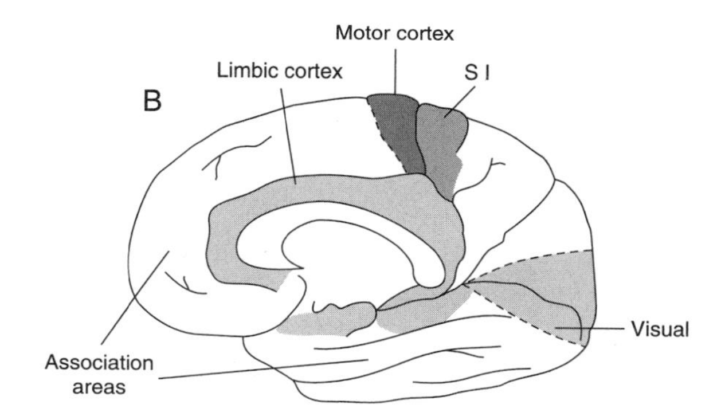

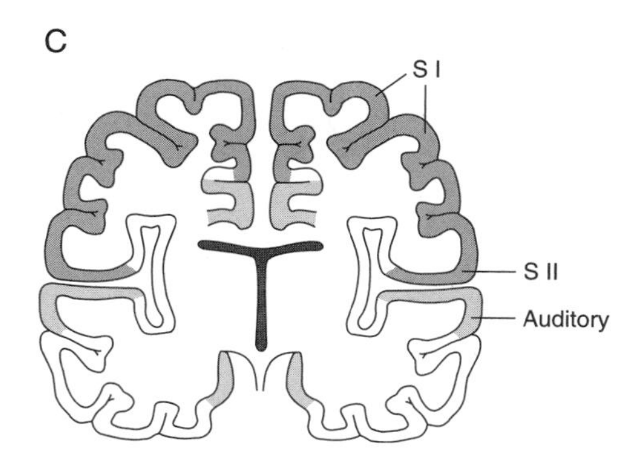

FIGURE 3.4 Anatomical location of primary sensory cortices in a surface view. (A) Somesthesia (SI), visual and auditory. Secondary somatosensory area (SII) and frontal association areas are also shown. (B) The same as (A) but in a midsagittal cut showing the location of the limbic cortex. (C) Coronal cut through the auditory cortex showing the location of SI, and SII.

cells crosses the midline and ascends as the anteriorlateral tract of the spinal cord. The anteriorlateral tract consists of several tracts[19,157] (see Chapter 4), of which the *spinothalamic tract* is the best known. Spinothalamic fibers terminate in the posterior parts of the thalamus known as the *posterior complex* (PO) and some *intralaminar nuclei*.[19] The fibers of the spinothalamic tract also reach the same regions of the thalamus (VPL) as that of the dorsal column system, but not exactly the same neuron groups. Some cells in the *reticular formation* receive input from the spinothalamic and spinoreticular tracts, and these cells also project to the hypothalamus and the amygdala nuclei, providing the basis for affective reactions to pain.

The nonclassical auditory pathways branch off the classical ascending pathways in the IC, where connections from neurons in the ICC reach neurons in the dorsal nucleus (DC) and the *external nucleus of the IC* (ICX).[5,7] Projections from these nuclei reach neurons in the dorsal and medial MGB of the thalamus.[7,100] These neurons project to secondary auditory cortices and association cortices.[7,188] While the classical ascending pathways are specific to each sense, some neurons in the nonclassical pathways of a specific sense also receive input from other senses (Fig. 3.2B).[6,7,198,199,221] These parts of the non-classical pathways are known as *multimodal* or *polysensory* pathways.[188] Connections from the somatosensory system can mediate information that reaches the ascending auditory pathways at several levels. The most peripheral level where such connections have been demonstrated is the cochlear nucleus, which has been shown to receive fibers from the DCN.[81,228] More centrally, the ICX, which receives its auditory input from the ICC, receives somatosensory input from the DCN.[7] The ICX is regarded as a part of the nonclassical auditory pathways. The trigeminal sensory nucleus in the guinea pig sends connections to cochlear nucleus.[199,221] The neurons in the auditory pathways that receive input from other sensory systems may be regarded as *polysensory neurons*. Some investigators,[188] therefore, distinguish between two different nonclassical auditory pathways, namely the *diffuse (nontonotopic)* and the *polysensory* systems.

In the visual system, a small part of the optic tract (approximately 10%) belongs to the nonclassical visual system.[19] These fibers project to the SC,[203] the *pretectal nucleus*, and the pulvinar division of the posterior thalamus.[19,34,163] The projection to the SC and pretectal nucleus mainly mediates visual reflexes, whereas the connections to the pulvinar division of the thalamus provide some visual sensation because these pulvinar neurons project to the cerebral cortex (extrastriate visual cortical areas).[19]

Also, the taste pathways have two parallel pathways, one of which is specific and the other less specific.[19,196] These two pathways may correspond to the classical and nonclassical pathways of hearing, vision, and somatosensory systems. The nonclassical pathways take off from the NST and project to the lateral hypothalamus and the *stria terminalis*, which connects the amygdala and the hypothalamus. In the monkey, both pathways use the NST as relay nucleus.[19,196]

Neurons of the subcortical portions of nonclassical pathways project to nonsensory structures. The neurons in medial and dorsal thalamus project to nuclei of the limbic system, such as the lateral nuclei of the amygdala,[104,116] and ascending fiber tracts of nonclassical pathways give off ample collaterals that connect to neurons in the *brainstem reticular formation*.[19]

The basic organization of nonclassical pathways (Fig. 3.2) can be summarized for the four sensory systems of hearing, somesthesia, taste, and vision:

1. The pathways receive input from sensory receptors and share the sensory nerves and the first or second relay nuclei with the classical ascending pathways.
2. The nonclassical pathways for somesthesia, audition, and vision are interrupted by synaptic transmission in the dorsal and medial portions of the thalamus.
3. The nonclassical system of hearing receives input from other sensory modalities.
4. Fibers of nonclassical sensory systems send collaterals to the brainstem reticular formation, providing *arousal*, before the information reaches the thalamus.
5. Neurons in the medial and dorsal thalamus project to secondary and association cortices and to structures of the limbic system, providing a subcortical route to the limbic system (mainly the amygdala). These connections are mostly reciprocal.

B. Descending Systems

Traditionally, descending pathways are treated separately from ascending pathways, and it is indeed true that there are separate descending systems; however, most of the ascending projections have corresponding descending connections. Whether this is to be regarded as individual ascending and descending systems or whether it is to be treated together as *reciprocal innervation* is a matter of choice.

1. Descending Pathways of the Classical Sensory Systems

The classical sensory pathways include large descending pathways that extend from the cerebral cortex to the peripheral parts of the respective descending sensory pathways, and in some cases the pathways even reach the sensory receptors such as those in the cochlea.

Descending pathways extend from the cerebral cortices to peripheral structures of sensory systems, making connections to neurons in the nuclei of sensory systems. These descending pathways, some of which are more extensive than the ascending pathways, essentially travel in parallel to the ascending pathways, and the descending systems have often been referred to as reciprocal pathways to the ascending pathways. Little is known about the function of descending pathways despite the great number of descending fibers, particularly between the cerebral cortices and the thalamus where the descending (reciprocal) connections are more numerous than the ascending connections.[9,235] There are also abundant descending connections from cortical areas to brainstem structures such as the IC.[10,234] It is generally assumed that the descending systems are important for controlling the transformation and flow of information in the ascending pathways before the information reaches the cerebral cortices, and it has been shown that activity in the descending pathways may modify the signal processing that occurs in the ascending pathways.

2. Descending Nonclassical Pathways

The descending nonclassical pathways are best known for the somatosensory system where there are extensive descending connections from cortical and subcortical structures such as the *periaqueductal gray* (PAG) to neurons along the neural axis reaching as far peripheral as the cells of the dorsal horn of the spinal cord. The neurons that receive nociceptive somatosensory input and are involved in mediating pain receive efferent (inhibitory) input from the *paraventricular hypothalamic nucleus* and the *nucleus raphe magnus*.[55] These descending pathways are important for control of pain, and they have received interests from clinicians because they offer a possibility to treat severe pain conditions. Some of these connections were studied functionally in humans with pain.[17] It has been known for a long time that efferents from the neocortex connect to thalamic neurons that are involved in pain sensation. Recently, these systems have been studied in awake humans through microelectrode recordings from the thalamus.[107,108]

Descending nonclassical auditory pathways are less well known than those of the classical pathways. Ehret and Romand[48] describe two systems of descending pathways of the auditory nervous system, one associated with the diffuse nontonotopic system and the other associated with the polysensory system. The former extends from the *secondary auditory cortex* (AII) to the DC of the IC and also connects to neurons in the dorsal thalamus. The polysensory descending system is more extensive and connects from several auditory cortical areas to the *magnocellular division* of the MGB. Less is known about descending nonclassical systems for the other sensory modalities.

C. The Thalamus is the Gateway to the Cerebral Cortex

The thalamus (Fig. 3.5) plays an important role in both the classical and the nonclassical sensory pathways. We have already mentioned sensory connections that involved the thalamus, but the importance of the thalamus in sensory information processing justifies a separate discussion of the various thalamic sensory nuclei.

The thalamus is the first *common pathway* to the cerebral cortex for sensory information of hearing, vision, somesthesia, and taste that elicits conscious sensations. All incoming fibers of the classical ascending sensory pathways are

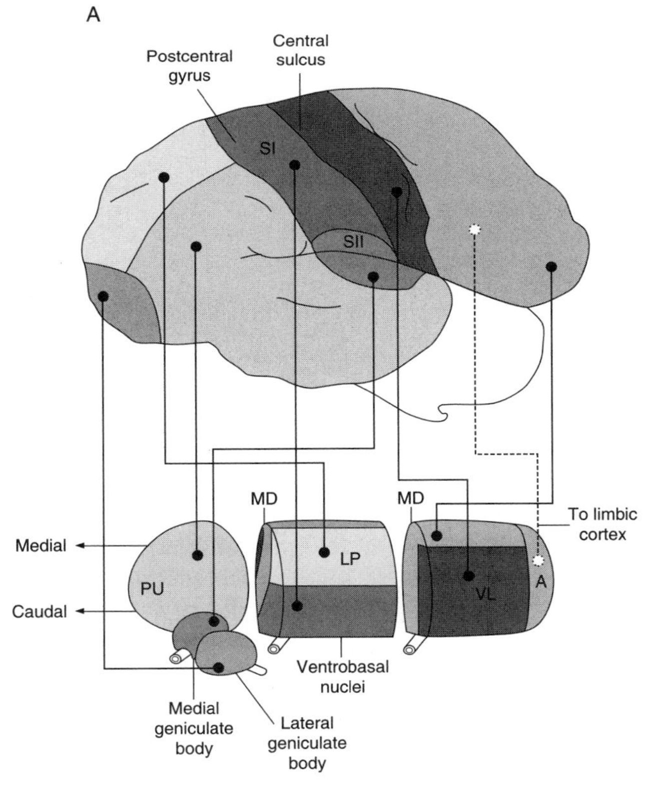

FIGURE 3.5 (A) Connections of the different parts of the thalamus to the cerebral cortex. Abbreviations: LP, lateral posterior nucleus; MD, mediodorsal nucleus; PU, pulvinar. (Adapted from Schmidt, R.F. and Thews, G., *Human Physiology*, Berlin: Springer-Verlag, 1983.) (B) Schematic drawing of thalamus (frontal section) showing anatomical location of the different nuclei of the thalamus. (Adapted from Brodal, P., *The Central Nervous System*, New York: Oxford Press, 1998.)

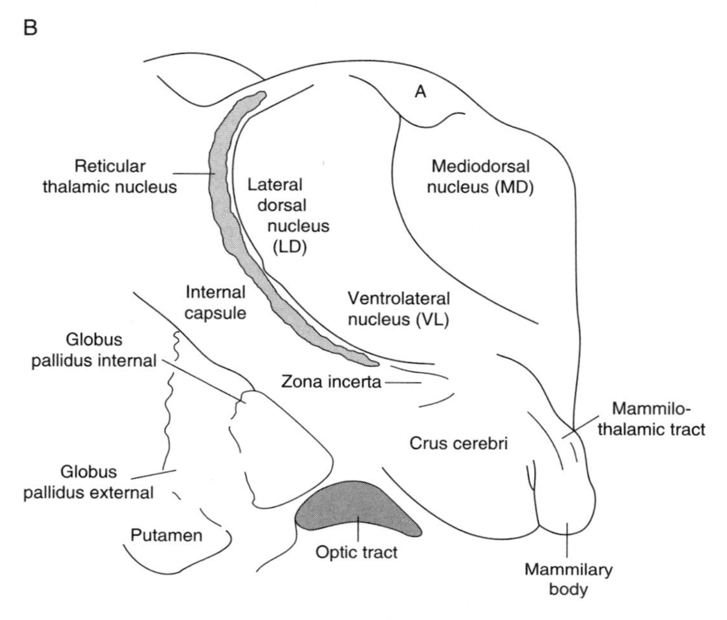

FIGURE 3.5 (*continued*)

interrupted by synaptic transmission in modality-specific nuclei. Extensive processing of information occurs in these different sensory nuclei before the information travels further on to the primary cerebral cortices for the classical sensory pathways. The thalamic nuclei of the nonclassical pathways project to several different regions of cerebral cortices and other brain structures. Information that travels in nonclassical pathways is processed in other thalamic nuclei and directed to many different parts of the CNS. Unlike many other sensory relay nuclei, thalamic nuclei do not seem to have any connections between the two sides.

All fibers of the classical pathways of hearing, somesthesia, vision, and taste are interrupted by synaptic transmission in the *ventral portion of the thalamus* (Fig. 3.5A). These parts of the thalamus project to respective primary sensory cortical areas (Fig. 3.5A). The thalamic nucleus of the classical auditory pathway is the MGB, and the LGN is the thalamic nucleus of the visual pathway. The thalamic nucleus of the somatosensory system is the VPL nucleus for the body and the VPM for the head. The taste pathways are interrupted by synaptic transmission in the VPM portion of the thalamus. These nuclei are modality specific and receive input only from one sensory system.

The nonclassical pathways of these four sensory systems are interrupted by synaptic transmission in the medial and dorsal divisions of the thalamus

(Fig. 3.5B). Many of the neurons in these parts of the thalamus receive input from more than one sensory modality such as somatic and auditory.[7,16,100] While the neurons of the nuclei of the ventral portion of the thalamus mainly project to primary sensory regions of the cerebral cortex, the neurons of the medial and lateral portion of the thalamic sensory nuclei mainly project to secondary and association cortices,[188] as well as to other parts of the CNS such as structures of the limbic system.[102] The nuclei of the dorsal thalamus are phylogenetically older than those of the ventral nucleus.

Some fibers of the olfactory system make connections to thalamic neurons, and afferent fibers from the structures of the thalamus that are involved in olfaction reach the frontal association cortices through the *mediodorsal thalamus.*[202] This would indicate that the entire olfactory pathway may be more similar to nonclassical pathways than to classical pathways.

Neurons of the dorsal nuclei of the thalamus also receive ascending input from nonsensory parts of the CNS such as from the *mesencephalic reticular formation*. Some thalamic nuclei receive information from structures that belong to the limbic system. The anterior thalamic nucleus receives afferents from the *mammillary nucleus* (MN), and the *mediodorsal nucleus* (MD) relays signals from the amygdala to the *frontal lobes*[19] (see pages 158 and 349).

All sensory nuclei of the thalamus have abundant interconnections, and they receive a large amount of descending input from the cerebral cortex. The descending input to thalamic nuclei may be regarded as reciprocal to the afferent innervation, or it may be regarded as a part of a larger descending system. In fact, the efferent input from the cerebral cortex to the ventral sensory nuclei of the thalamus is much larger than the afferent sensory input from the periphery, and the largest number of fibers entering the thalamus for the auditory, visual, and somatosensory pathways originates in the cerebral cortex. This means that a considerable amount of feedback from more central portions of the ascending sensory pathways are available in the thalamus for modifying the processing of signals on the basis of what occurs in the cerebral cortex. This information is based on anatomical data, and it is not known how many of the synapses that connect these fibers to thalamic neurons are open or under which circumstances these synapses may be able to affect the target cells in the thalamus.

1. The Thalamic Reticular Nuclei

All fibers that enter the thalamus — from the periphery as well as from the cerebral cortex — pass through a region of the thalamus known as the *reticular nucleus* (RE), while giving off numerous collaterals. The RE is a thin layer of neurons (Fig. 3.5B) that are mostly inhibitory (GABAergic).[19] The RE projects to all main sensory nuclei of the thalamus and may therefore be regarded as a modulator of both the afferent sensory input to the thalamus and the efferent input from the cerebral cortex to the thalamic sensory nuclei. The RE is divided into regions that respond to somatosensory, visual, and auditory input.

The RE neurons also receive input from other thalamic nuclei, mainly the lateral part of the ventral division of the thalamus (LV) and the ovoidea division of the ventral thalamus (OV) neurons (Fig. 3.5). Because the RE neurons project back to thalamic neurons, they become part of a short, closed feedback loop. The RE neurons also receive input from the *mesencephalic reticular formation*, the activity of which is related to wakefulness. This allows the activity of the RE to be affected by wakefulness and by arousal.[241] The RE neurons also receive input from several other nonsensory parts of the brain such as the PAG which is the termination of one part of the anteriolateral system, namely the *spino-mesencephalic tract*, which is involved in the sensation of pain.

2. The Pulvinar Nuclei

The pulvinar of the thalamus is a large nucleus located in the posterior part of the thalamus in close proximity to the LGN and MGB (Fig. 3.5). The pulvinar is involved in many functions, particularly visually related functions. It receives input from the SC and sends output to extrastriatal (visual) cortices as well as other cortical regions. There are direct connections from the optic nerve to the pulvinar.[19,34,163]

The circuitry of thalamic sensory nuclei thus offers ample possibilities of connecting sensory information to parts of the brain other than those that primarily process sensory information. The nuclei of the thalamus also provides the basis for modulating ascending sensory information by other brain regions, such as the reticular formation, before the sensory information reaches the cerebral cortices.

D. Anatomy of Sensory Cerebral Cortices

The sensory cerebral cortex consists of the neocortex* and the *allocortex*.** The neocortex is a 2- to 4-mm- thick layer that appears as a folded sheet on the external surface of the brain, and the allocortex lies deep within the temporal lobe. The allocortex is phylogenetically older than the neocortex (or isocortex).

The sensory neocortex contains complex networks of neurons that receive input from the thalamus and from other cortical areas. Neurons in the neocortex connect to other cortical areas as well as the thalamus and other subcortical

*The neocortex contains six layers of cells.

**Several regions of the cerebral cortex, including the olfactory cortex and the hippocampus, comprise the allocortex. It has fewer cell layers than the neocortex.

structures. Sensory cortices have been divided into primary sensory cortices and association cortices. Neurons in the primary sensory cortices respond only to one sensory modality, whereas many neurons in the association cortices respond to more than one sensory modality. The connections between primary cortices and secondary cortices are reciprocal, enabling two-way communication. Primary visual, somatosensory, and auditory cortices also send back information to the thalamus. The primary sensory cerebral cortices are often regarded as the end stations of sensory ascending pathways. However, it might be more appropriate to regard the primary sensory cortices as the second common pathway for sensory information that arouses conscious sensations (the first common pathway is the thalamus).

The different regions of the neocortex have many similarities. The sensory cortices have six layers (Fig. 3.6) and are organized in *columns*,[78,155,171] where

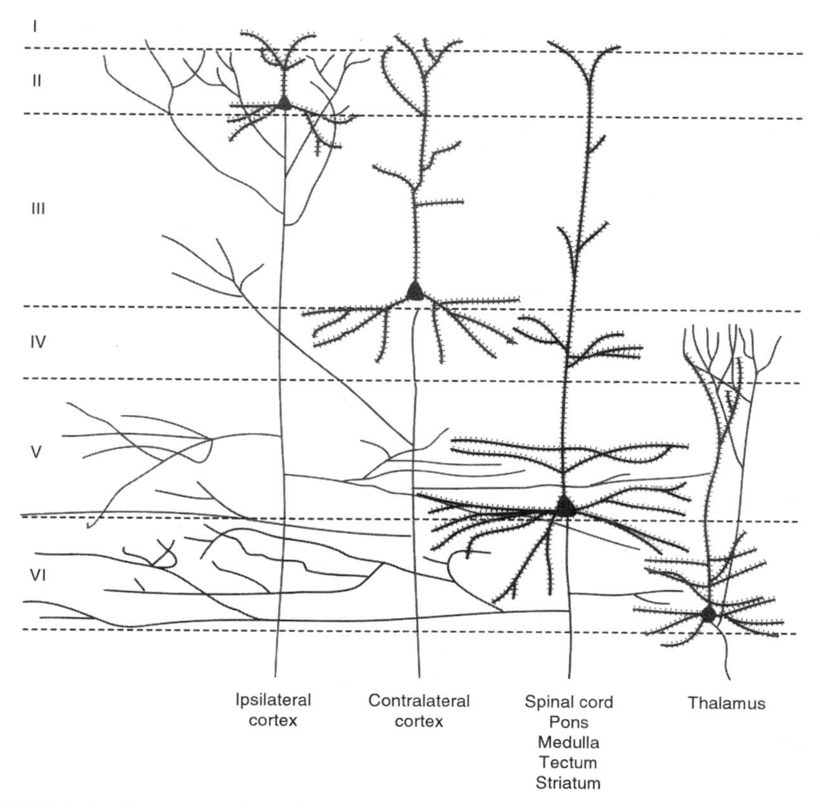

FIGURE 3.6 Cross-section through the sensory cortex to show its layered organization. (From Zigmond, M.J. *et al.*, *Fundamental Neuroscience*, San Diego: Academic Press, 1999. With permission.)

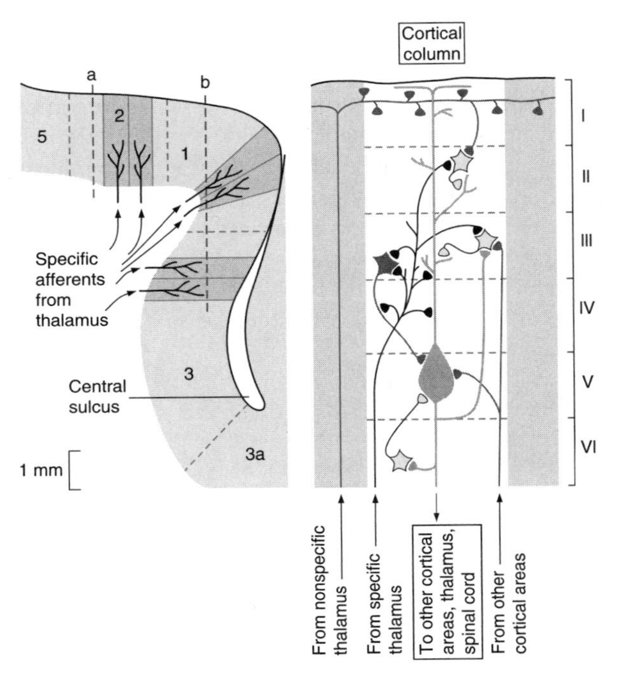

FIGURE 3.7 Schematic illustration of the columnar organization of the sensory cerebral cortex. (Adapted from Schmidt, R.F. and Thews, G., *Human Physiology*, Berlin: Springer-Verlag, 1983.)

neurons respond to the same sensory input (Fig. 3.7). The columnar organization has been demonstrated in various sensory systems, and its discovery was a major breakthrough in studies of the function of sensory systems.[78,171] The thickness of the different layers varies in different regions of the neocortex, but it is always possible to identify six layers. Extensive connections exist between neurons in the different layers of the cortex.[127]

The neocortex has a cell density of 40,000 cells/mm^3. It is composed of mainly three cell types — pyramidal cells (75%), smooth stellate cells (15%), and spiny stellate cells (10%) — axons and dendrites. The density of synapses is estimated to 8×10^8/mm^3, with approximately 20,000 synapses per neuron, approximately 90% of which are excitatory synapses and approximately 10% inhibitory synapses. The average dendritic length is 10 mm. Of the excitatory synapses, approximately 50% make connections from remote sources, and 50% are from local (cortical) sources.[1]

1. Cell Types

Specific types of cells are located in the different layers of the cortex. Layer I (the molecular layer) is composed mostly of *glial* cells and axons traveling laterally (horizontally), and layer II (external granular layer) has small granular cells. Layer III (external pyramidal cells) has small pyramidal cells, and layer IV has small *stellate* cells, which has given it its name "granular regions." Giant pyramidal cells are found in layer V (internal pyramidal layer), and layer VI (multi form layer) contains spindle or *fusiform* cells.

2. Primary Sensory Cerebral Cortices

The primary sensory cortices are the first cortical receiving stations for sensory information that ascends in the classical sensory pathways. The input from the visual, auditory, and somatosensory (and, to some extent, taste) systems originates in the modality-specific sensory nuclei of the thalamus; therefore, these primary sensory cortices are modality specific. The excitability of cortical neurons is modulated by input from other parts of the CNS, most notably from the reticular formation.[153,200]

3. Connections

Neurons in the six layers of the primary cortices have specific connections:

1. Layer I mainly contains connections between local cortical areas. This layer contains few cell bodies. Nonspecific input from the thalamus reaches neurons in layer I.
2. Neurons in layer II receive input from layer I and send connections to neurons in other layers and to other cortical areas on the same side.
3. Layer III provides the main output to other cortical areas. Neurons in layer III send connections to neurons of layer IV of cortical regions on the opposite side.
4. Layer IV is the main receiving area where thalamic fibers from its ventral portion terminate in the internal granular layer. Subdivisions (IVa, IVb, and IVc) of layer IV have been identified in some sensory cortices.
5. The very large cells (pyramidal cells) of layer V have long axons that descend to subcortical structures.
6. Neurons of layer VI receive input from other layers and project back to the thalamus and other more peripheral nuclei of ascending sensory pathways.

The connections of neurons of the different layers of the auditory cortex (Fig. 3.8) can serve as a model of connections in sensory cortices.[127] As seen in the figure, the main input from the thalamic nuclei (MGB) enters layer IV, and the large output back to the thalamus (MGB) exits the cortex from layer V and VI, where layer V also contributes descending connections that reach further peripheral nuclei (the midbrain nuclei, such as the IC).

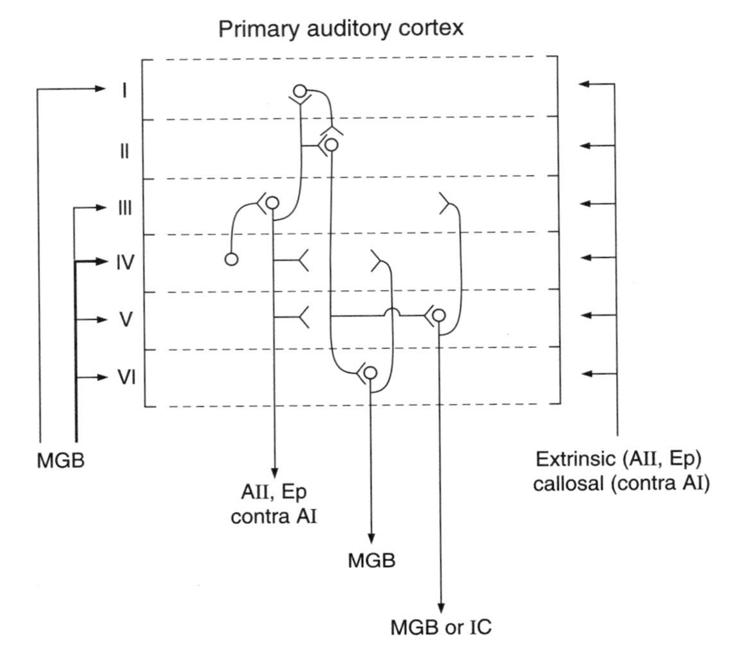

FIGURE 3.8 Schematic drawing of the connections in the primary auditory cortex, as an example of connectivity in primary sensory cortices.[127]

Afferent fibers to the cortex make connections with a large number of neurons. It has been estimated that one afferent fiber from the thalamus can make contact with 5000 cortical neurons. The horizontal axons make both inhibitory and excitatory connections, and laterally and functionally related neurons are interconnected. A single pyramidal nerve cell may integrate information from as many as 600 nearby cortical cells and may receive approximately 60,000 synapses (see above). These estimates are based on studies in monkeys.[19]

These extensive connections from subcortical structures to the cerebral cortex and back again allow for very complex signal processing in the cerebral cortex (see page 350). When connections of the sensory cortices are discussed it is important to keep in mind that these connections are not "hard wired" but are subject to change, as are connections in more peripheral parts of sensory systems (neural plasticity, see pages 165 and 235). Because new connections may be established and existing connections may be eliminated and the synaptic efficacy may vary through neural plasticity, attempts to understand signal processing in sensory systems are complicated.

4. Secondary and Association Cortices

Sensory information from the visual, somatosensory, and auditory systems is processed in the neocortex, but the primary sensory cortices occupy only a small fraction of the neocortex. Information from olfaction and taste is mainly processed in allocortical regions of the cerebral cortex. While primary cortical areas of vision, hearing, and somesthesia are modality specific and receive input from the classical ascending sensory pathways, secondary cortices receive input from several different sources; although mainly modality specific, some neurons may receive input from other modalities. The largest portion of the neocortex is the association cortices. Association cortices represent an even higher order of processing of sensory information, and these cortical areas receive and integrate information from different extrinsic sources such as multiple sensory systems as well as input from intrinsic sources in many parts of the CNS.

5. Pathways to Polysensory Cortical Areas

The three primary routes through which information can be relayed from sensory organs to such *polysensory* (association) cortices are:

1. *Cortico-cortical* connections between primary secondary sensory areas of the cerebral cortex; the input from somatosensory, visual, and auditory cortices ultimately projects onto the *frontal pole*, *orbitofrontal cortex*, and parts of the temporal lobe.
2. Connections from the *lateral posterior* areas of the thalamus and the pulvinar, which in turn receive their input from nuclei of the classical pathways of sensory systems.
3. Projections from the classical ascending pathways of each major afferent system through the reticular formation.
4. Projections from the medial and dorsal thalamic neurons that are parts of the nonclassical sensory pathway.

We will discuss the organization of association cortices later in connection with *stream segregation* and *parallel processing* (pages 110 and 111).

E. Corpus Callosum

The primary cerebral cortical areas for vision, somatic senses, and hearing have ample connections between the two sides through the *corpus callosum*. The corpus callosum is the largest fiber system of the brain. It connects the two hemispheres of the brain[19,50] and ensures that information that enters one hemisphere — such as sensory information from one ear or one half of the visual field — also reaches the other half of the cerebral cortex. Also, tactile information that reaches only one hemisphere from the thalamus is communicated to the other half of the brain by the corpus callosum. The fibers of the corpus callosum can modulate (suppress) activity in the opposite side or transfer information from one side to the other.

The posterior portion of the corpus callosum is devoted to the visual system, while the anterior and middle parts convey information from the somatosensory system. In the cat, connections between the *Primary Auditory Cortical Fields* (AIs) of the two sides as well as secondary auditory cortices (AII) and *anterior* and *posterior auditory fields* (AAF, PAF) are diffuse and traveling in the posterior two thirds of the corpus callosum.[30] Axons from the AAF cross more rostrally than those from the AI and axons from the PAF and AII cross more caudally.

The *anterior commissure* is a part of the corpus callosum that connects the olfactory bulbs on the two sides.[50] The anterior commissure is a round bundle of nerve fibers that crosses the midline of the brain near the anterior limit of the third ventricle. It consists of a smaller anterior part, the fibers of which pass in part to the olfactory bulbs, and a larger posterior part, which interconnects the left and right temporal lobes.

III. INFORMATION PROCESSING IN THE SENSORY NERVOUS SYSTEM

We have seen in the previous part of this chapter that the different sensory systems (except olfaction) have many similarities in anatomical organization. Also, the ascending sensory pathways have many similarities in information processing, particularly regarding vision, hearing, and somesthesia. This part of the chapter will discuss processing of the neural code that the sense organs deliver to the CNS through the respective sensory nerves. General principles for neural processing of sensory information and the control of flow of information in the ascending sensory pathways will be described. Hypotheses about the neural representation of sensory information and its interpretation will be discussed, emphasizing similarities and differences. Later parts of this chapter will be devoted to information processing in the chemical senses and in nonclassical pathways of hearing, vision, and somesthesia. Involvement of the limbic system and the basis for emotional reactions to sensory stimuli will also be discussed in this part of the chapter. Specific detailed features of sensory processing will be discussed in the chapters that cover individual senses (Chapters 4 to 7).

A. Information Processing in the Classical Ascending System

All the properties of physical stimuli are coded in the discharge pattern of the nerve fibers that innervate receptors. That neural code is subjected to extensive processing and reorganization in the ascending sensory pathways as it passes the several nuclei that are located between the receptors and the primary cortices. The primary sensory cortices further process sensory information and convey the information to other cortical areas where processing and integration with other information occur. This means that sensory systems perform a complex hierarchical serial and parallel processing.[88,206]

The classical ascending pathways for the different senses may be viewed as giant signal processors that process the neural code generated by the sense organs in response to physical stimuli. Even though the ascending sensory pathways are less complex than many other parts of the CNS, their complexity is nonetheless far beyond what can be comprehended and it is not possible to fathom in detail what signal processing may take place in these enormously complex networks of neurons and how it is accomplished. Yet, research has brought insight into many important aspects of neural processing of sensory information.

Extracting useful information from a stimulus is accomplished mainly by the action of the CNS, but the properties of the receptors and the media that transmit the physical stimuli to the receptors contribute to the feature extraction that occurs in the CNS (Chapter 2). For example, photoreceptors separate light that reaches the eye according to its wavelength, which is the basis for discrimination of color. Olfactory and taste cells have different sensitivities to odors and tastes and separate such stimuli before they are processed in the nervous system. The cochlea functions as a spectrum analyzer that separates sounds according to their frequency (or spectrum), and the sensory cells (hair cells) are activated accordingly, establishing the separation of sounds on the basis of the frequency.* This means that the neural code in the afferent nerve fibers that innervate the receptors is not an exact image of the physical stimulus that reaches the sense organ. Instead, the neural code reflects the properties of a physical stimulus after it has been modified by the transduction process in the sensory receptors and by the medium that transmits the stimulus to the receptors (see Chapter 2). The basis for our ability to discriminate the frequency (or spectrum*) of a sound, the color of an image, and various odors or tastes is thus laid out in the receptor organs, and the analysis of sensory information continues as the information

*The words *frequency* and *spectrum* are often used synonymously when aspects of hearing are discussed. Frequency (measured in cycles per second, or Hz) can only totally characterize sounds with a single frequency (pure tones). Spectrum is the distribution of energy as a function of the frequency and it can be used to characterize sounds of all kinds.

ascends through the nuclei of the ascending sensory pathways. The processing continues in the primary cerebral cortices and higher order cortices, including the association cortices, where different forms of sensory information are separated according to their properties and information from different sensory modalities is integrated.

1. Separation of Information in the Nervous System

Sensory information is separated in different ways while being processed in the nervous system. *Parallel processing* means that the same information is processed in different populations of neurons, while *stream segregation* means that different kinds of information are processed in different populations of neurons. Stream segregation was first studied in the visual systems, but it has later been shown to occur in other sensory systems.

a. Parallel Processing

The basis for parallel processing is branching of sensory pathways. Parallel processing is prominent in all sensory systems, but perhaps to the greatest extent in hearing, where there is anatomical evidence that parallel processing begins at a peripheral anatomical level, namely where each auditory nerve fiber connects to neurons in each of the three main divisions of the first relay nucleus (the cochlear nucleus) [151,226,227](page 310). The fiber tract that connects the IC with the MGB (the BIC) represents a great degree of parallel processing in that it has approximately 10 times as many nerve fibers as the auditory nerve. This means that that the information that is represented in the neural code in the auditory nerve is divided into many separate channels before it reaches the cerebral cortex.

It has been demonstrated in several sensory systems that cells that process different kinds of information are anatomically segregated and that cells with common properties are anatomically grouped together.[66,155,244] An example of the effect of parallel processing is related to recognizing a speaker and recognizing speech (speech intelligibility), which are different tasks of the auditory system based on different processing of the same information.[22,41]

Also, ascending visual pathways have anatomical signs of parallel pathways.[19,125] Parallel processing is less prominent in the somatosensory system. However, the fact that nociceptive stimulation (painful stimulation) is processed in (at least) two different parts of the CNS[20,21] is an indication of parallel processing. Studies using functional imaging and recordings of evoked potentials have shown that pain activates two anatomically different regions of the CNS and that the activation of these two regions results in different qualities of sensation, which means that the same information is processed in two anatomically different locations of the CNS.

The conscious awareness of pain and its location seem to be a result of processing in the *secondary somatosensory cortices* while the emotional and aversive component of pain is a result of processing in limbic structures, specifically in the posterior cingulum.[35] These assumptions are supported by the finding that lesions of the *primary somatosensory cortices* do not cause analgesia.[35,210] Pain pathways project to nuclei in both the lateral and the dorsal thalamus[119] emphasizing the role of the thalamus in pain and the importance of spinoreticulothalamic pathways.[35,107,108] The dorsal thalamus is mainly involved in the emotional components of pain while the lateral thalamus is involved in sensory discrimination of pain. These pathways are not static and the processing and connectivity in the thalamus can change as a result of neural plasticity,[109] as is the case for processing of pain at more peripheral (spinal cord) levels.[38,173]

b. Stream Segregation

The classical and best known examples of stream segregation is the separate processing of *spatial* and *object* information that occurs in the visual *extrastriatal cortices,* but similar separation of processing of different kinds of information has been demonstrated to occur in other sensory systems, such as the auditory system. Stream segregation has been less studied in somesthesia and the chemical senses, but future research may find that similar forms of sensory processing occur in these senses. The term *steam segregation* has specifically been used to describe a form of specialization that occurs in large groups of neurons in association cortices.

There is evidence that specific regions of the association cortices* perform processing of sensory stimuli of specific kinds. Such separation, now known as stream segregation, was first shown in the visual system.[126,218,219] The separation of spatial and object information is an example of stream segregation[218] and has been extensively studied in the visual system in animals.[67,125,170,225] In humans, separate regions for spatial and temporal (motion) characteristics have been identified.[36] Stream segregation has more recently been shown to occur also in other sensory systems, most clearly perhaps in the auditory system.[90,177–179,186,214]

The segregation of different kinds of information (for example, spatial and object information) that has been shown to occur in the visual and the auditory systems is an important feature of information processing in sensory systems that recently has attracted the attention of many

*Association cortices receive input from primary sensory receiving areas from more than one sensory system and integrate information from different sensory systems.

scientists.[2,37,51,56,67,90,125,179,186,209,216,225] The coding of the properties of the physical stimulation in the neural discharges of nerve fibers and nerve cells may be regarded as the "what" aspect of sensory stimulation, whereas spatial features and movement properties that are represented in the nerve system by activation of specific populations of neurons may be regarded as the "where" aspect of sensory stimulation.

In the visual system, the concept that spatial and object ("what" and "where") visual information is processed in different extrastriatal cortical regions has been supported by many investigators. While Ungeleider and Mishkin[218] were the first to publish results showing that spatial visual information ("where") is processed in a dorsal stream while object information ("what") is processed in a ventral stream, other investigators[67,125] have suggested that the neural substrate for visual perception is different from that involved in visual control of action such as visual control of movement. These investigators also reported evidence indicating that the ventral stream of visual projections from the striate cortex is involved in identification of objects by visual inspection, while the dorsal stream is involved in providing visual guidance of movements such as that of the hand. The two streams of visual information, the dorsal and the ventral, seem to become integrated as the information propagates toward the prefrontal cortex.[184,185] In one study, it was shown that forms that were defined by motion activated both dorsal and ventral extrastriatal cortical regions.[225]

Before the study by Ungeleider and Mishkin[218] was published, the concept that object and spatial information is processed in anatomically segregated populations of neurons was described for the visual system by Schneider.[193] He regarded the phylogenetically old (*retinotectal*) pathway (which in this book we call the nonclassical pathway) as a processor of information related to localization while the phylogenetically younger *geniculostriate* pathway (which we call the classical pathway) processed information that was related to identification of the stimulus. Schneider found that ablation of the SC and the visual cerebral cortex (Brodmann's areas 17 and 18) (see Fig. 4.12), which interrupt the *geniculocortical* pathway, had two different effects on vision: The lesion of the SC caused an inability to *locate* an object, while ablation of the visual cortex caused an inability to *identify* an object. Schneider[193] was probably the first to use the notions of "where is it" and "what is it" to describe the differences between visual localization and identification of an object.

The concept that Schneider developed is no longer accepted, but the principle that different kinds of information such as spatial and object information are processed in anatomically different populations of nerve cells is now known as stream segregation. What Schneider[193] described was the difference between the classical and the nonclassical visual systems, and what Ungeleider and Mishkin[218] described was a separation of streams in the same system, namely the classical visual system. (The anatomy of the nonclassical visual system was discussed in the first part of this chapter and will be discussed in more detail in Chapter 6, where the differences between the function of the classical and the nonclassical visual systems will also be described).

It is not only spatial and object information that are separated and processed in different populations of neurons. Identification of *faces* is an example of information that is separated into different streams. Many mammals use olfaction for identification of members of their own species and use scents for communication of sexual signals. Humans instead use other senses for such purposes. We use face recognition for identifying members of our own species, and we also use facial expressions for communicating emotion and sexual signals. In fact, facial expressions are important for communicating, such as in conversation where such feedback to the person speaking is essential for carrying on a conversation.

Evidence from studies of individuals with prosopagnosia (difficulty in recognizing familiar faces) due to defined lesions in the brain have shown that the recognition of familiar faces and distinguishing facial expressions are based on processing in two anatomically different parts of the brain.[2,37,216] Destruction of the amygdala eliminates the perception of fear in facial expressions but the ability to recognize faces is preserved after destruction of the amygdala,[2,37] which indicates that facial expressions of emotion are directed to the amygdala nuclei, whereas face information that relates to the identity of the person depends on bilateral structures. Recognition of (familiar) faces, on the other hand is based on the *mesial occipitotemporal* visual region.[37] Bilateral (but not unilateral) damage to the amygdala affects the recognition of fear in facial expressions while leaving the ability to recognize familiar faces intact.[2] This was taken as an indication that these two qualities, perceiving fear and identification of faces, were carried in two different streams of information. One stream would carry information about the expression of fear to the amygdala, while the other stream would carry information about the identity of the face to the visual nervous system. This is another manifestation of stream segregation in the visual system that concerns processing images of the face.

Other studies have shown that viewing the face of a beautiful woman activates regions of the brains of men that are usually activated by addictive drugs. This means that female faces for men have reward values comparable to addictive substances.[3] Different kinds of information gained from observation of the face of another person are processed in different extrastriatal regions of the visual system, but it is not known to which extent this separation of information is completed at a peripheral level and whether the nonclassical visual pathways may be involved in separating these different forms of visual information.

Separation of spatial and temporal components of visual information is another example of stream segregation in the visual system that has been studied by several investigators.[8,11,36] The results showed that information about motion bypassed the primary (V1) visual cortex and reached the MT/V5 cortex before visual information reached V1.[11] These investigators used transcranial magnetic stimulation* to block different parts of the MT/V5 visual cortex in awake humans (either by depolarizing cortical neurons or by eliciting *inhibitory postsynaptic potentials,* IPSPs)[8] and observed the resulting deficits in perception of movement of a visual pattern.

Some investigators[244] have pointed out that the concept of "what" and "where" information being processed in two separate pathways originating in the primary visual cortex is an oversimplification, that the separation in these two major streams is not total, that spatial information may also be processed in ventral cortical pathways and that there are more than two different pathways emanating from the primary visual cortex (V1). Zeki[244] has suggested that the difference between the types of information processed in such different streams is not as distinct as other investigators[218] have indicated. This means that the ventral pathway, normally associated with perception of form may also convey spatial information and that the dorsal pathway may not exclusively communicate spatial information. The concept of stream segregation may therefore be regarded as an oversimplification.

Although object ("what") and spatial ("where") information are directed to different parts of extrastriatal cortices, such separation of information does not mean that information is interpreted in these parts of the extrastriatal cortices. In fact, some studies have shown that the two streams of "what" and "where" information converge on to the same prefrontal cortical areas. Anyway, the concept seems to be useful for understanding the basic functions of sensory systems, and there is ample evidence both physiological and behavioral, of the existence of separation of distinctly different types of sensory information.

Studies of the auditory system have shown that directional information is processed in anatomically separate locations in the auditory nervous system.[72] Neurons in the lateral belt of the rhesus monkey auditory cortex respond to different types of auditory information.[178,214] Neurons in the anterior portion of this belt prefer complex sounds such as species-specific communication sounds ("what"), whereas neurons in the caudal portion of the belt region show the greatest spatial specificity ("where"). The *superior temporal gyrus* of the monkey (macaques) is organized in two areas with different functions. One, the most rostral streams seems to be involved in processing of object information such as that carried out by complex sounds (for instance, vocalization). The other one of the two streams, located caudally, is involved in processing of spatial information. Auditory spatial information that concerns directional information is not related to the location on a receptor surface as is the case for visual and somatosensory information but rather is based on the relationship before input from the two ears. This means that auditory directional information is not related

*Transcranial magnetic stimulation consists of applying a strong impulse of a magnetic field to the head using a coil through which a large electrical current flows. The magnetic field induces an electrical current in tissue under the coil. The induced current can be used to stimulate selected areas of the brain, or, if strong enough, it will block neural conduction to the selected areas.

to a specific sensory surface such as the retina or the skin but instead is derived from manipulation of information from the two ears.

While speech perception is better when listening with the right ear (right ear advantage),[76] there is no hemispheric difference with regard to identification of a speaker.[99] This is another indication that these two aspects of speech perception and speech recognition are processed in different parts of the brain. This process may be similar to those of face recognition to determine the identity of a person and discrimination of facial expressions, both of which are based on different kinds of information that may be regarded as spatial and object information, respectively.

In the auditory system, spectral information may be regarded as spatial information that is likely to be processed in different populations of neurons than those that process object information such as the amplitude and the time patterns of sounds. Changes in the spectrum of sounds change the location of the maximal excitation of the basilar membrane, a form of movement information that is spatial in nature. Spectral information may therefore be similar to spatial information of an image in vision. The cerebral cortex and neurons of the nuclei of the ascending auditory pathways are anatomically organized according to the frequency of sounds to which they respond best. The strength and the time pattern of sound may be regarded as object properties. However, such spatial coding is different from the three-dimensional auditory space that is created by comparison of the input to the two ears.

If this hypothesis is correct it would mean that the frequency or spectrum of sounds may be processed in similar ways as other spatial information, such as directional information, and the spectrum of sounds may be processed by populations of neurons different than those that process the temporal pattern of sounds and the intensity of sounds. (It is interesting that frequency is also coded as a temporal pattern, which may be regarded as an object coding.) This hypothesis has so far not been experimentally verified.

In the flying bat, the cortical representation of distance to an object is the interval between the emission of a high-frequency sound and receipt of the echo of that sound; that time difference is coded in the discharge pattern of individual neurons. Other studies have shown specific coding of sound intervals (duration of silence) in neurons in the auditory pathways.[129,240] Such information may be regarded as spatial information because it refers to a location. Information such as that of low-frequency communication sounds that the bat uses while flying, may be regarded as object information, and there is evidence that these two kinds of information are separated at the midbrain level (IC); however, these two streams of information are joined again in the cortex, where the same neurons process both kinds of information.[180] Sound duration may also be coded specifically in the auditory system.[29]

When a point on the skin is stimulated, two kinds of information are sent to the CNS, namely information about the nature and strength of the stimulus (object information, or "what") and information about where on the skin the stimulation has occurred (spatial information, or "where"). These two kinds of information are likely to be processed in different populations of nerve cells but that has not been confirmed experimentally. The distinction between forms of objects (for instance, cubes vs. spheres) represents objective information[4] that is also likely to be processed in populations of neurons that are different from those processing spatial information.

The chemical senses seem to have less separation of stimulus types, as there are fewer patterns of stimulation of the chemical senses. Stimulation of the chemical senses may be regarded to be entirely object information.

B. Processing of Object and Spatial Properties of Sensory Stimuli

Because there is evidence that object and spatial information is processed by different populations of neurons in general, we will discuss the processing of these two properties of sensory stimulation separately when discussing processing of sensory information in the following section of this chapter. We will first discuss how the object properties of stimuli are processed in the nervous system and later we will discuss how spatial information is processed.

1. Processing of Object Properties of Stimuli

The characteristics of the discharge pattern of neurons in the sensory pathways, including the primary cortices, have been studied in great detail in animal experiments using recordings from single nerve cells and fibers using microelectrodes. The time pattern of discharges in response to various kinds of stimuli for hearing, somesthesia, and vision has been studied. The relationship between the time pattern of discharges of single neurons and that of the sensory stimuli that affect the discharge pattern has been studied extensively.

2. Coding of Stimulus Strength

In general, the intensity (strength) of sensory stimulation can be coded in two ways in the nerve fibers that innervate sensory receptors — namely, by the discharge rate in individual nerve fibers and by the number of nerve fibers that respond with an increased firing rate. That means that the more intense the stimulation, the higher the discharge rate of individual neurons and the greater the number of neurons activated. The *temporal coherence* of firing of populations of neurons may be important in coding of the intensity of sensory stimuli[44] and may also play a role in determining the threshold of sensory stimulation such as has been hypothesized for the auditory system.[45,143]

The discharge rate of nerve fibers that innervate *rapid adapting* (RA) mechanoreceptors (velocity detectors) increases as a power function with increasing stimulus intensity.[158] (When the discharge rate is plotted as a function of the stimulus intensity on log–log coordinates, the discharge rate appears as a straight line;[158] see Chapter 4). This is similar to the growth of the perceived intensity of a stimulus (see Chapter 1). There is thus a clear and direct relationship between the discharge rate of the fibers that innervate these RA mechanoreceptors and perception of the strength of the stimulation.

This dependence of the discharge rate on stimulus intensity is, however, not as clear in afferent nerve fibers that innervate other receptors in the skin, such as the low-threshold receptors in the *glabrous* (hair-free) skin. The afferent nerve fibers of these receptors do not communicate the intensity of a stimulus as precisely[86] because the discharge rate reaches saturation just above threshold, and perception of stimulus strength must therefore be based on the response of other types of receptors or other forms of neural coding.

The discharge rate of many of the nerve fibers that innervate receptors of other sensory systems is less directly related to the physical strength of sensory stimuli. For example, most individual nerve fibers in the auditory nerve reach saturation about 25 to 35 dB above the threshold of hearing[92,159] (Fig. 3.9) despite the fact

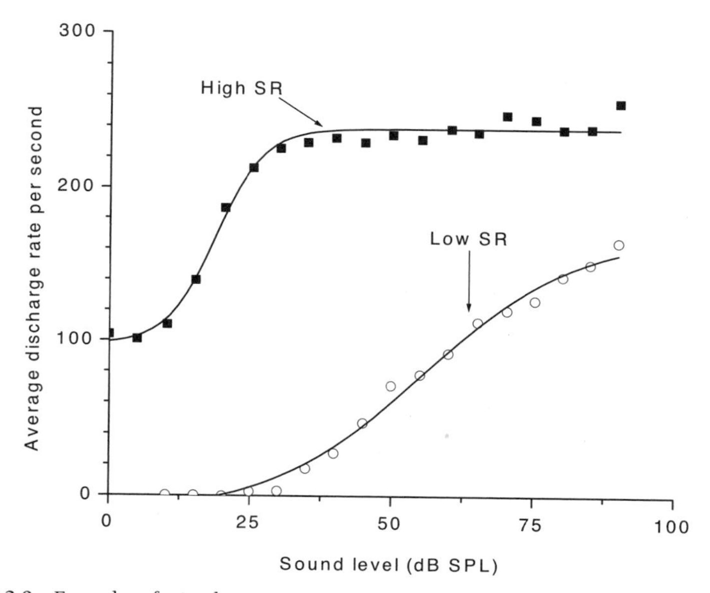

FIGURE 3.9 Examples of stimulus-intensity curves of auditory nerve fibers showing input–output functions from a fiber with a high spontaneous rate (SR) over a small range of sound intensities a fiber with a low SR and large range of sound intensities. (Data from Müller *et al.*[159])

that hearing covers a very large range of sound intensities extending to at least 90 dB above the hearing threshold (see Chapter 5). The psychoacoustic finding that sound intensity can be discriminated over a very large range may be explained by the response characteristics of a few auditory nerve fibers that have a large dynamic range (Fig. 3.9). The discharge rate of these nerve fibers increases monotonically with increasing stimulus intensity over a large range of stimulus intensities,[159] and these nerve fibers are capable of communicating information about the intensity of sounds over the range of normal hearing. It is possible that these few nerve fibers may supply all the information about the loudness of a sound.

Coding the intensity of sounds in the nervous system by the number of nerve fibers activated also reflects other features of stimulation. In the auditory system, the number of nerve fibers that are activated depends on the spectrum of sounds, and in the skin the number of nerve fibers that are activated depends on the size of the area of the skin that is stimulated. This means that the number of nerve fibers that are activated by a stimulus is an ambiguous measure of stimulus strength, but the CNS may have ways to discern different causes of activation of many nerve fibers.

3. Coding of the Time Pattern of Physical Stimuli

The time pattern of physical stimuli plays an important role in conveying sensory information, in particular in the auditory system. Timing also plays some role for somatosensory perception, but the visual system is relatively insensitive to the timing of visual stimuli. Timing of stimulation is of little importance in the chemical senses.

In the auditory system the temporal pattern of sounds is important for discrimination of the frequency of sounds and for discrimination of natural sounds such as speech sounds (timing is also important for directional hearing, thus for processing of spatial information, where the ability to determine the direction to a sound source is based on the difference in the arrival time of a sound at the two ears, see pages 287 and 341).

All use of timing information depends on the ability of the nervous system to preserve timing information as the information ascends in the sensory nervous systems to the level where the timing is interpreted or converted to a different code. In the auditory system, accurate preservation of timing is especially important for discrimination of complex sounds such as speech sounds (and for directional hearing). Timing on the order of a few microseconds can be resolved by the auditory system.

Several aspects of neural transmission appear as obstacles for preserving the high degree of precision of timing. One is related to the probabilistic nature of synaptic transmission. The fact that the delay in synaptic transmission depends on the strength of activation also appears to hinder accurate preservation of timing.

Psychophysical studies show that the required accuracy of timing in the auditory system must be in the order of 10 to 20 μS to explain the ability to determine the direction to a sound source on the basis of the difference in the time of arrival of sounds at the two ears.[215] This means that the timing must be preserved to a high degree of precision in the discharge pattern in sensory nerve fibers and in cells in the ascending sensory pathways until the information from the two ears can be compared and the directional information can be converted into another code. Decoding of timing of information related to directional hearing has been shown to occur in neurons in the *superior olivary complex* (SOC), and it has also been assumed that decoding of timing information of sounds that arrive at one ear occurs in these nuclei (see Chapter 5). This means that the information must pass three synapses before it is interpreted: cochlear hair cells, the cochlear nucleus, and neurons of the SOC. The documented ability to discriminate very small time differences therefore depends on preservation of timing of neural transmission in at least two synapses in addition to the neural transduction in hair cells.

Synaptic transmission involves discrete quanta of a chemical substance (transmitter substance) to be passed over the synaptic cleft to the cell that receives the input. This means that the firing of a nerve cell that is activated through synaptic transmission will not occur at exactly the same time when identical stimulations are repeated. Because of that, synaptic transmission is assumed to cause blurring of the timing of nerve impulses (often referred to as *synaptic jitter*).

In brief, synaptic transmission implies that an action potential in a presynaptic cell causes a release of a neural transmitter which in turn leads to the generation of a synaptic potential in the postsynaptic cell. The action potential causes transmitter substances to be released in fixed quanta. Each quantum of a transmitter released by an excitatory synapse produces an excitatory postsynaptic potential (EPSP) of a certain size. The EPSP is a positive potential that reduces the normal negative resting potential of a cell (depolarizes the cell). The EPSP evoked by the release of a sequence of transmitter substance quanta will therefore result in a stepwise increase in the membrane potential for each quantum that arrives over the synaptic cleft. The quantum nature of synaptic transmission provides a (random) uncertainty in the size of the EPSP. Because the axon that leads away from the cell fires when the EPSP reaches a certain threshold level, the firing will have a temporal uncertainty (jitter) when the same presynaptic stimulation is repeated.

While it is true that neural timing is impaired in synaptic transmission, as demonstrated by synaptic jitter when one or a few synapses form the input to a nerve cell, the situation is different in most nerve cells because they receive many synaptic inputs. Nerve cells that receive many inputs function as signal averagers, and that action decreases the temporal variability. Each such quantum potential gives only a small contribution to the resulting EPSP which is why the EPSP appears as a smooth decrease in the negative membrane potential of the cell.

The jitter of the firing of a nerve cell has been studied using the "integrate and fire" model of neurons. In this model, the sum of the incoming postsynaptic potentials

(PSPs) generates an action potential in the axon of the cell when the PSP reaches the threshold of the impulse generation in the axon. (This model dates back almost 100 years[101] and has been successful in explaining many phenomena related to the behavior of neurons in the nervous system.)

Temporal precision of synaptic transmission revealed from studying an integrate and fire model with a leaky integrator[201] has shown that the jitter of the discharges in the output (temporal variability of the discharges) of a neuron was less in neurons with many synapses compared to neurons with few synapses.[24] The decrease in the output jitter compared with the input jitter is a linear function of the logarithm to the number of synaptic inputs (Fig. 3.10). The variability of the firing of a neuron may become smaller than that of the individual inputs in cells with many inputs. The effect of this *spatial averaging* is similar to the (temporal) averaging used to recover evoked potentials that appear in noise. The decrease in the output jitter with increasing number of inputs is in good agreement with the theory that the signal-to-noise ratio increases by a factor that is equal to the square root of N, where N is the number of inputs, provided that the noise is random noise. When many inputs converge on to a neuron, it therefore functions as a spatial integrator of its input that counteracts the blurring of the temporal pattern of the discharges.

The fact that a neuron with many inputs acts as a spatial integrator not only will cause it to preserve the temporal code through synaptic transmission but can also cause the temporal variability of the discharge pattern to become less than that of the

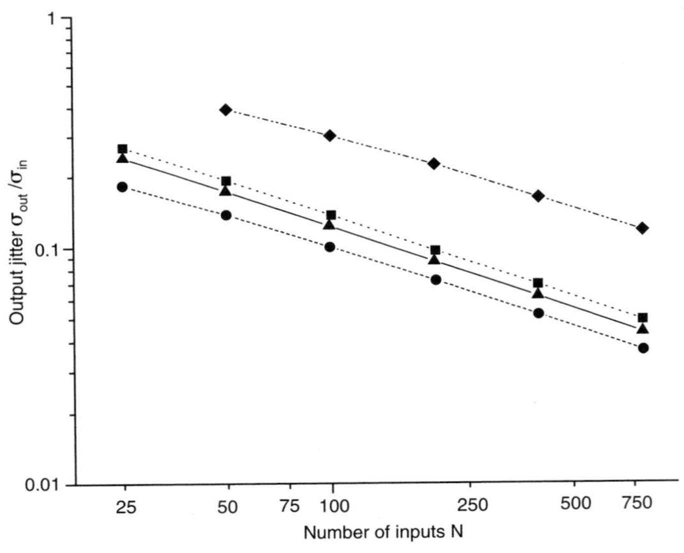

FIGURE 3.10 Improvement of temporal precision as a function of the number of inputs to a neuron. Log–log plot of output jitter as a function of the number of synaptic inputs to a neuron. Solid line with triangles; perfect integrator; dotted line with squares, Stein model; dashed line with circles, a general integrate-and-fire model; dash and dot line with diamonds, Stein model with inhibition with a ratio between inhibition and excitation of 0.5. (Data from Burkitt and Clark.[24])

inputs to the neuron. In other words, synaptic transmission in neurons that have a sufficient number of inputs, may, in fact, *decrease* the variability in timing of nerve impulses rather than cause it to increase (Fig. 3.10). The gain in the *signal-to-noise ratio* (SNR), or temporal precision of firing, with increasing number of excitatory synaptic terminals is, however, counteracted by the number of active *inhibitory* synapses that provide inputs to a nerve cell. The gain in SNR is also (slightly) dependent on the threshold of the neuron.[24]

It is the number of *active* synapses conveying input to the neuron in question that determines the temporal precision of neural firing, and that number can change as a result of external factors that affect synaptic efficacy, such as through expression of neural plasticity (see page 235). Psychoactive drugs may also affect synaptic efficacy.

Different nerve cells in the cochlear nucleus receive different numbers of auditory nerve fibers, from a few to several hundred, perhaps thousands.[93,151] Experimental studies of the discharge patterns of cells in the cochlear nucleus in response to repetitive click sounds have shown that some cells respond in a similar way as auditory nerve fibers, revealing a considerable temporal variation (jitter) in their response (Fig. 3.11, left), indicating that these cells receive only a few synaptic inputs. The timing of the responses of other cells is much more precise[129] (Fig. 3.11, right), indicating that such cells receive many auditory nerve fibers. Other studies have revealed other forms of diversity in the response pattern of nerve cells in the cochlear nucleus to tone bursts,[169] and these differences have been related to the different numbers of synaptic inputs to the cells from the auditory nerve.[93]

Another source of uncertainty in timing that would impair the preservation of timing of information is related to the fact that the latency of the response of neurons in sensory systems depends on the stimulus strength. Generally, the latency of neural activity decreases with increasing stimulus intensity because the EPSP climbs to reach the firing threshold at a faster rate in response to a strong stimulus than in response to a weaker stimulus. The stronger the stimulus, the faster the EPSP reaches the firing threshold. This is the reason why the latency of the response from sensory neurons decreases with increasing intensity. The influence of stimulus intensity on timing (latency of response) has been taken as a common characteristic of neural transmission, but it depends on the way neural transmission is usually studied. For instance, the timing of the response to small changes in the amplitude of a sound is preserved accurately, at least in the peripheral portion of the ascending auditory pathways (Fig. 3.12).[136,141] This means that the timing of sounds that occur in a background of sounds, which is the way natural sounds normally occur, is almost independent of the stimulus intensity. Studies of the auditory system are nearly always done by presenting stimuli in a silent background, while sounds normally occur in a background of other sounds. This unnatural way of presenting sound stimuli in experimental studies has exaggerated the dependence of the latency of the response on the intensity of the stimulus.

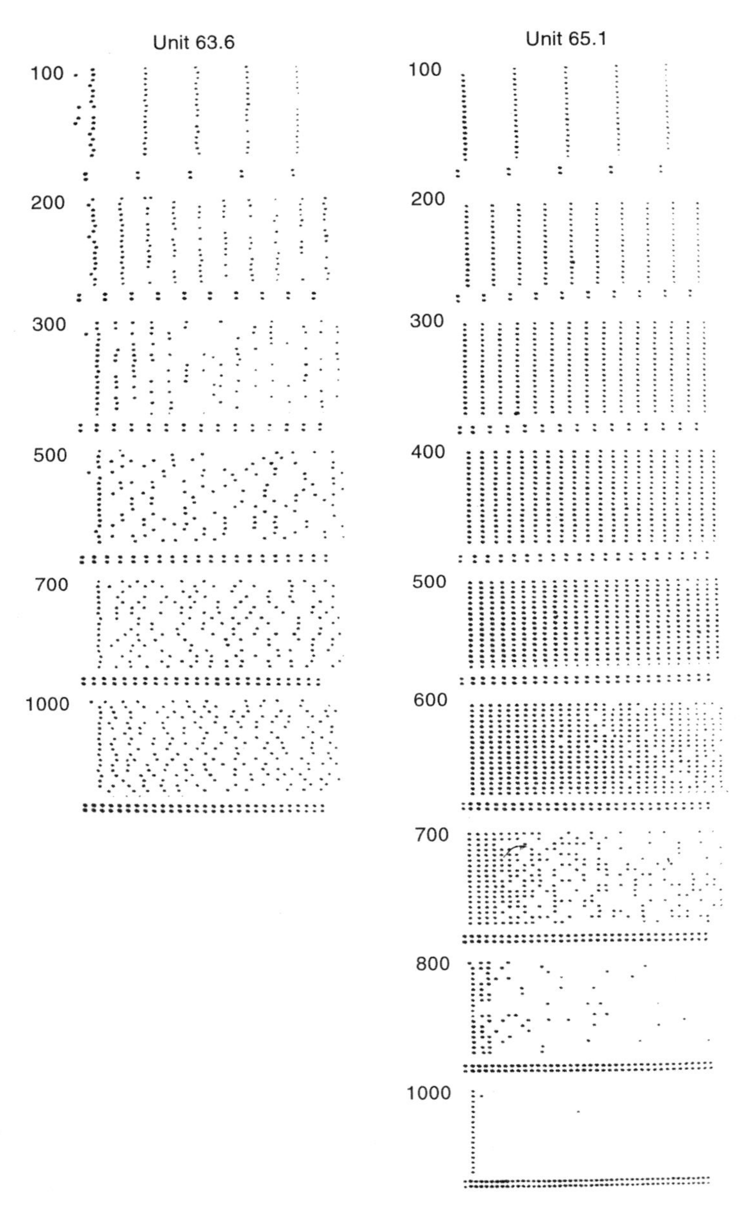

FIGURE 3.11 Response from two different types of neurons in the CN to repetitive click stimulation. The one to the left shows similar temporal jitter similar to that seen in the auditory nerve; the one to the right shows very little jitter, presumed to be due to the neurons receiving many auditory nerves as input. (Adapted from Møller.[129])

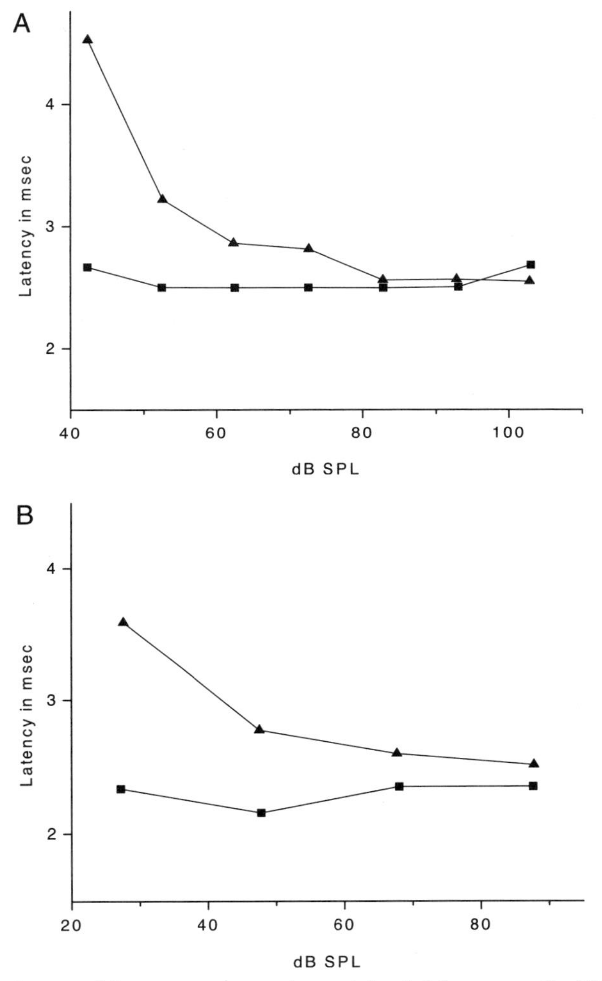

FIGURE 3.12 Latency of the response to tone bursts (triangles) from two cells, (A) and (B), in the cochlear nucleus compared with the latency of the response to small changes in the amplitude of tones (amplitude modulation; squares) shown as a function of the sound intensity.[136]

4. Coding of Small Changes in the Stimulus Strength

Neural coding and processing of small changes in the stimulus strength such as the intensity of sounds, the intensity of illumination, or the strength of skin stimulation are important for discrimination of many natural sensory stimuli. In fact, the intensity of practically all normal sensory stimuli changes more or less rapidly, and these variations in the intensity of sensory stimuli often represent important sensory information. We have already seen how the timing of small changes in amplitude is preserved more accurately in the nervous system than that of sounds that are presented in a silent background. Small changes in the intensity of stimuli are known as *amplitude modulation*. It is therefore not surprising that sensory systems include mechanisms that enhance the coding of small changes in the intensity of stimuli.

One of the mechanisms that can enhance the response to small changes in the intensity of a physical stimulus is *adaptation*. Adaptation can reduce unimportant information by emphasizing changes. Adaptation occurs at many stages of processing of sensory stimuli. It occurs generally in the medium that conducts the physical stimulus to the receptors such as those of the skin and it occurs in receptor cells (see Chapter 2). Adaptation is also prominent in the ascending sensory pathways of the nervous system, where it is an important part of neural processing of sensory information.

The response of nerve cells to amplitude modulation has been studied most extensively in the auditory system, where it has been shown that the coding of small rapid changes in the amplitude of sounds is enhanced in the response from nerve cells in the ascending auditory pathways, and many neurons attenuate slow changes in the intensity of stimuli. The neural response to amplitude-modulated stimuli can be illustrated by observing period histograms* of the discharges of neurons in the auditory nervous system in response to sounds the amplitudes of which are modulated[58,59,130,135,137] and using that information to obtain *modulation transfer functions*.

The shape of the modulation transfer functions depends on the methods used to obtain such functions. Some investigators have determined the ratio between the modulation of period histograms of the discharges and the modulation of the amplitude of the stimulus and plotted these values as a function of the modulation frequency.[58,59,130,135,137] This method has been commonly used in connection with sinusoidally amplitude-modulated sounds or light. Other investigators have displayed the discharge rate as a function of the modulation rate. This method has been used to describe the response to impulses presented at different rates or bursts of sounds presented at different rates.

The modulation of discharges of cochlear nucleus cells in response to amplitude-modulated sounds can be greater than that of the modulation of the discharges of auditory nerve fibers (Fig. 3.13).[59,138] This means that the *gain* of the modulation (namely, the ratio between the modulation of the sound and that of the histograms of the nerve impulses) is greater for cells in the CN than for auditory nerve fibers, and it is a sign of neural processing to enhance the

*A period histogram of the response to modulated stimuli shows the distribution of nerve impulses over one period of the modulation.

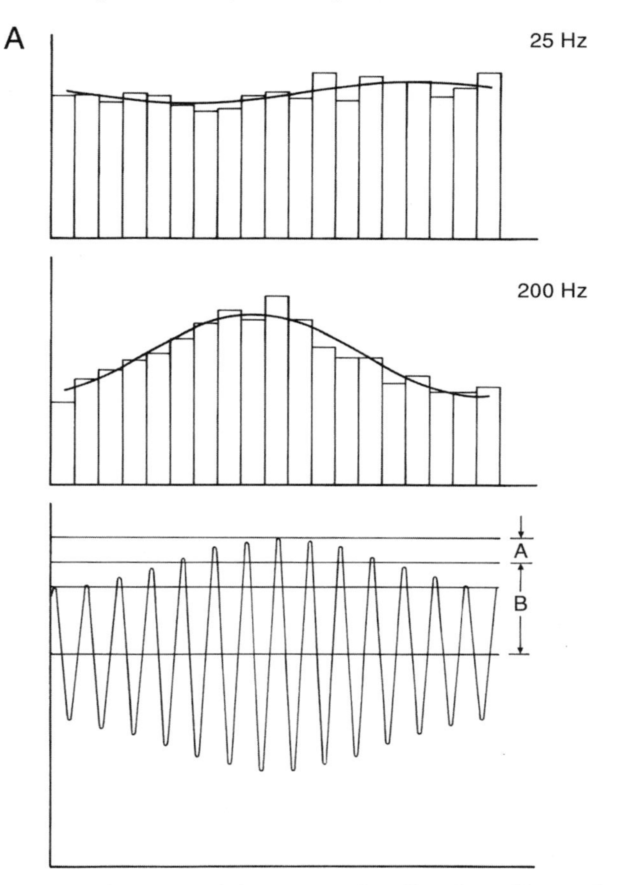

FIGURE 3.13 (A) Period histograms of the response of a cell in the cochlear nucleus of a rat to amplitude modulated tones at two modulation frequencies (25 and 200 Hz). Below: One period of modulated sound.[148] (B) Comparison of the modulation transfer function for an auditory nerve fiber (thin line) and a cell in the CN (thick line).[137]

representation of small changes in the amplitude of sounds. It has been shown that the gain of modulation is related to the number of auditory nerve fibers that converge upon a cell in the CN.[59–61]

The response to amplitude modulation of light stimuli (*grating patterns*) has been studied in the retina in animal experiments and used for characterizing retinal ganglion cells by Enroth-Cugell and colleagues.[49] Other investigators have used similar techniques to study the responses from cells in the LGN.[122]

Some studies of the neural coding of small changes in stimulus intensities have used stimuli that were amplitude modulated by sinusoidal

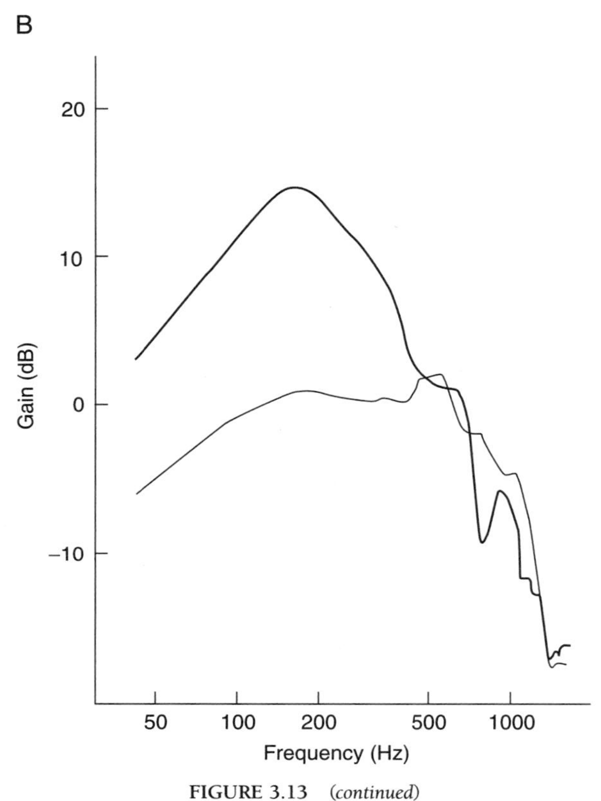

FIGURE 3.13 *(continued)*

waveforms,[49,58,59,61,122,130,132] whereas other investigators have used stimuli that were amplitude modulated by random waveforms (noise).[131,133,135-138]

5. Improving the Signal-to-Noise Ratio of Neural Discharges

The input to the CNS from sensory systems such as the auditory, somatosensory, or visual system may appear to be very noisy if viewed in the recordings from single fibers of the respective sensory nerves. This has been claimed to be an obstacle for preserving many features of sensory information such as that of coding of small changes in the intensity of sensory stimuli. Practically all studies of coding of sensory stimuli in the discharge pattern of single nerve fibers or nerve cells must use some form of averaging of the response to many repetitions of the same stimuli in order to show how sensory stimuli are coded, another sign of the effect of neural noise.

The level of noise present in the discharge pattern of neurons in the sensory pathways limits coding of small changes because the change in the neural discharge rate induced by the stimulus in question must be greater than the random changes in the discharge rate in order to be detected in the CNS. This means that the level of random changes in the discharge rate sets the threshold for how small changes in the sensory stimulation can be transmitted to higher CNS centers where it can be detected (decoded).

The neural noise is mainly caused by similar processes as the temporal uncertainty (namely, by synaptic transmission), and the noise can be reduced by spatial averaging, which occurs when many inputs converge onto the same neuron, increasing the signal-to-noise ratio effectively.[24] Spatial averaging is similar to the signal averaging commonly used for enhancement of evoked potentials in a background of noise. This means that the spatial averaging that occurs because of convergence of many inputs to a nerve cell will not only increase the precision of timing of the coding of a physical stimulus, as was discussed earlier, but it will also increase the SNR of the coding of changes in a stimulus such as changes of the amplitude of sounds (modulation). Enhancement of amplitude modulation in auditory system has been demonstrated in the studies of neurons in the dorsal cochlear nucleus.[61] This means that such spatial neural integration is essential to several aspects of the normal function of sensory systems in discriminating sensory stimuli.

C. Processing of Spatial Information

Information about the localization of stimulation of the skin, information about visual objects, and localization of sound sources are usually regarded as spatial information. We will also regard coding of specific sensory information such as color, taste, or smell as spatial information, as well as representation of the frequency of sounds according to the place principle of frequency discrimination.

1. Coding of Specific Properties of Sensory Stimuli

Specific qualities of sensory stimulation may be coded in the nervous system in two different ways. One way consists of each quality activating distinctly different neural elements (the *labeled line* hypothesis). The other way consists of activation of overlapping populations of neurons (the population hypothesis).

a. Labeled Lines

The labeled line hypothesis assumes that different physical properties of sensory stimuli activate different neurons and implies that specific qualities of stimuli are conveyed in separate nerve fibers and nerve cells. This hypothesis consequently links distinct perceptual qualities such as the frequency of a tone or the color of an object to selective activation of specific neural elements in the CNS. For example, the theory about labeled lines would state that a specific shade of green light would evoke responses in specific neurons, or that a 1000-Hz tone would activate specific auditory nerve fibers and specific neurons in the ascending auditory pathway. The labeled line hypothesis also states that activity in the 1000-Hz labeled lines will always be interpreted as a sound of the frequency of 1000 Hz independent of its strength, duration, or how the sound is applied, to one ear or both ears.

The labeled line hypothesis is an intuitively attractive hypothesis; however, it has not received experimental support except, perhaps in one specific sense — namely, for the vomeronasal system, which detects pheromones (the vomeronasal system is a part of the olfactory system and resembles nonclassical sensory systems). The sensory cells of the vomeronasal organ seem to respond selectively to specific pheromones, and these cells have been named *odor specialists.*[196] Most sensory systems, however, seem to code the physical qualities of sensory stimuli according to the population hypothesis, which is based on overlapping responses to different physical qualities of sensory stimuli. (Sensory cells in the olfactory system that respond to several qualities of a sensory modality are known as *odor generalist.*[196])

b. Population Hypothesis

It has been shown in many studies that different specific sensory qualities elicit neural activity in overlapping populations of many neurons, where certain neurons are activated to a greater extent than others. One of the first investigators to publish evidence of the population hypothesis was Pfaffman et al.,[168] who showed that individual single nerve fibers in the chorda tympani could respond to several taste qualities but each nerve fiber responded best to one specific taste quality. This must mean that each nerve fiber innervates more than one taste receptor or that each taste receptor responds to more than one taste quality.

This way of coding different sensory qualities is prominent for all sensory systems (except perhaps the vomeronasal system), and it can be demonstrated throughout the sensory nervous system. It is assumed that the population hypothesis is the neurophysiologic basis for perception of specific sensory qualities such as color, vision, taste, odors, the type of stimulation of the skin, and, in hearing, perhaps, the frequency of sounds. This means that, for example, the populations of neurons that respond to two tones clearly perceived as being different are distinguished from each other on the basis of the responses of overlapping populations of neurons activated by two tones. It also means that, for instance, different shades of green light and even red light and blue light may activate the same neurons when the light intensity is sufficient, but the degree of overlap defines the exact color of the light.

In the visual system, it is possible to obtain precise information about a specific quality of sensory stimulation by comparing the response of (three) different populations of neurons, and the color of light can be judged on the basis of information from these three types of receptors despite the fact that all three types of receptors may respond to the light in question. Similar reasoning can be applied to taste and olfaction.

The basis for discrimination of specific sensory qualities thus seems to involve the relationship between the response of members of a large population of nerve cells rather than specificity of the response of individual nerve cells. This principle of coding of stimulus properties requires fewer neurons than needed for the labeled line hypothesis. Coding of sensory information according to the labeled line hypothesis would require a large number of neurons in order to represent every physical property, and coding according to the population hypothesis is more economic regarding the use of neural elements.

2. Neural Representation of Spatial Information

Discrimination of spatial and movement information is based on differences in the *receptive fields* of neurons in the sensory nervous system that are stimulated. The size of receptive fields of neurons is important for the ability to discriminate the location of stimulation because it affects the ability to locate the site of stimulation and discriminate among different spatial patterns of stimulation of arrays of sensory receptors.

The receptive fields of receptors were discussed in Chapter 2, as were the receptive fields of the nerve fibers that innervate sensory receptor cells. In the somatosensory system, the receptive fields are the patches of skin in which stimulation of cutaneous (skin) receptors activate specific cells in the CNS. In the visual system, the receptive fields are regions of the visual field where light causes a cell in the CNS to increase its discharge rate. In the auditory system, the receptive field of auditory nerve fibers is the range of frequencies to which a nerve fiber responds (frequency tuning curves). These receptive fields are established in the sense organs and further refined by the neural processing that occurs in the nuclei of the ascending sensory pathways and in the cerebral sensory cortices.

3. Transformation of Receptive Fields

Most nerve cells in the ascending sensory pathways, including those of the cerebral cortices, have distinct receptive fields. The receptive field of a neuron

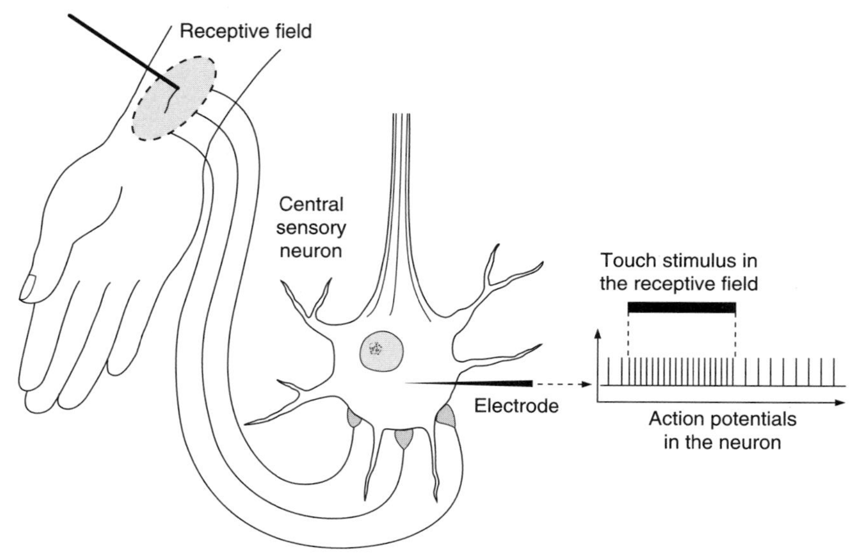

FIGURE 3.14 Mapping of the receptive field of a nerve cell that receives input from nerve fibers that innervate several skin receptors.

in the sensory pathways represents the input from all receptors that can influence its activity (Fig. 3.14).

The receptive field of nerve cells in the nuclei of the ascending sensory pathways is related to the receptive fields of receptors but they differ from the receptive fields of nerve fibers that innervate the receptors because most neurons in the CNS receive input from more than one receptor, resulting in *convergence* of input. Convergence of excitatory input makes the receptive fields of nerve cells in the ascending sensory pathways become broader than those of the nerve fibers that innervate receptors (Fig. 3.15). The transformation of the receptive fields of receptors that occurs in the ascending is an important aspect of neural processing of sensory information.

Another difference between the receptive fields of cells in the nuclei of the ascending sensory pathways and those of receptors or their afferent nerve fibers is that nerve cells in the nuclei of the ascending pathway receive *inhibitory* input in addition to excitatory input. While the effect of convergence of excitatory input is a broadening of receptive fields, the result of inhibition is a narrowing of receptive field (Fig. 3.16). The receptive fields of neurons in the ascending sensory pathways are more complex than those of the nerve fibers that innervate the sensory organs because of multiple inputs and may have complex shapes because of input from different populations of receptor cells.

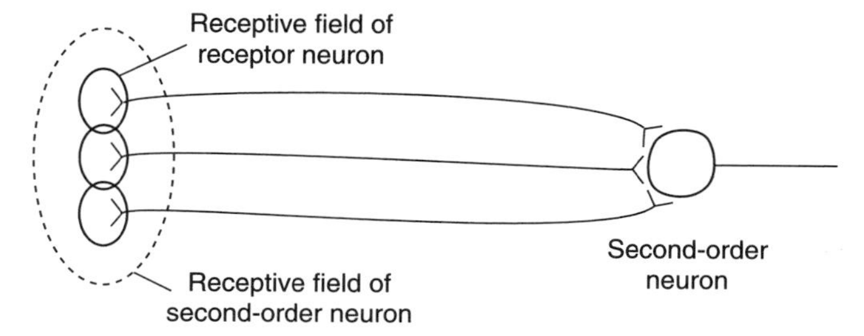

FIGURE 3.15 Illustration of how the outputs of several receptors converge onto a neuron in the dorsal column nucleus; the resulting receptive field becomes larger than that of the individual neurons.

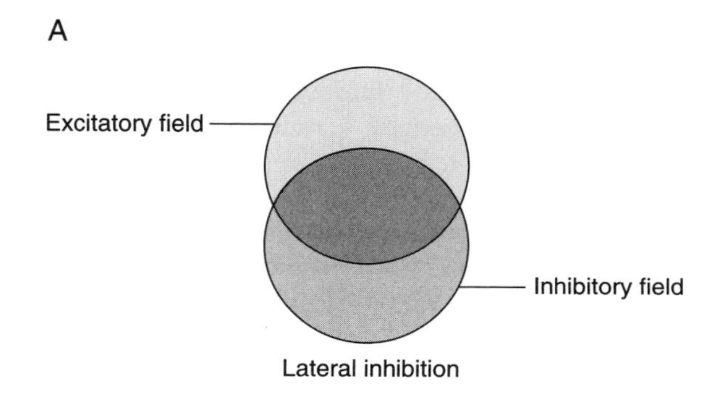

FIGURE 3.16 Lateral inhibition, (A) Overlapping excitatory and inhibitory fields, and (B) simple neural circuitry that can perform lateral inhibition.

a. Interaction Between Excitatory and Inhibitory Receptive Fields

Neurons with both excitatory and inhibitory input have distinct inhibitory and excitatory receptive fields. Stimulation within the excitatory receptive field increases the discharge rate of the nerve fiber from which recording is made, while stimulation within the inhibitory receptive fields causes a decrease of the discharge rate that is evoked by stimulation within the excitatory response area. That means that stimulation within inhibitory fields counteracts stimulation within the excitatory fields. Inhibition may be caused by activation of a specific population of receptors that connect to inhibitory synapses on cells in nuclei of the ascending sensory pathways or it may be caused by internal circuits in nuclei of ascending sensory pathways. Inhibitory response areas may be located on each side of excitatory response areas (known as *lateral inhibition*) (Fig. 3.16) or they may surround excitatory areas (*surround inhibition*) (Fig. 3.17). An excitatory response area may also surround an inhibitory response area (*center inhibition*). Auditory nerve fibers have inhibitory response areas within which a tone inhibits the response evoked by a tone located within the excitatory response area (known as *two-tone inhibition*) (see Chapter 5, page 322). Interplay between excitation and inhibition in hearing affects the representation of frequency in the auditory nervous system, and neurons in the nuclei of the ascending auditory pathways have similar, although more complex, response areas.

This pattern of interlaced inhibition and excitation is accomplished in different ways in the three sensory systems of vision, somesthesia, and hearing. In the eye, it is a result of interaction between the neurons in the network

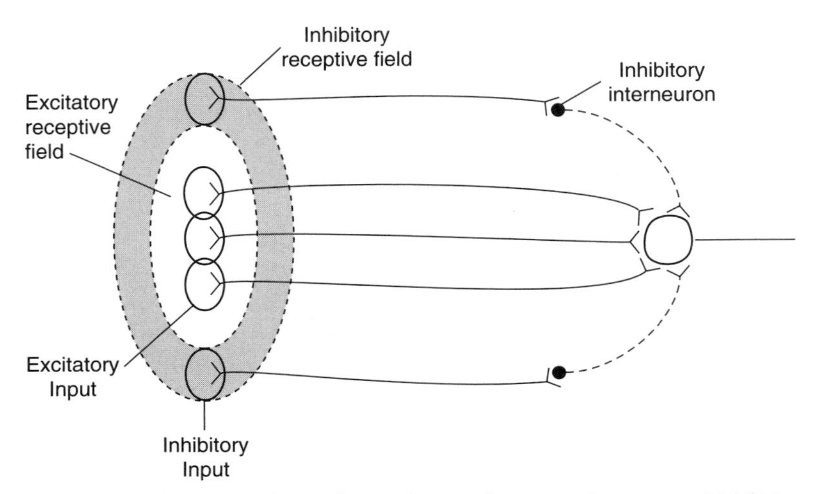

FIGURE 3.17 Illustration of a simple neural circuit that can perform surround inhibition.

of neurons in the retina and can be observed in the responses of the ganglion cells of the retina. In the skin, the basis for lateral inhibition is the neuronal network in the DCN. In the cochlea, the lateral inhibition (two-tone inhibition) is a result of mechanical interaction between the outer hair cells and the basilar membrane, which is a property of the medium conducing the stimulus to the receptors.

The arrangement of excitatory and inhibitory response areas of cells in nuclei of sensory pathways can be as simple as shown in Figs. 3.16 and 3.17, where inhibition is added to a neural network that causes spatial integration, but the convergence of many nerve fibers onto the same neuron increases the complexity of the arrangements of inhibitory and excitatory response areas. Interaction between inhibition and excitation continues to modify the response areas of neurons as information moves along the ascending sensory pathways, including the primary sensory cortex. Recurrent feedback from cells of the nuclei of the classical ascending sensory pathways including those of the primary cerebral cortex adds complexity to neural processing in sensory pathways.

Experimental determination of receptive fields is usually done by using steady-state stimuli of certain intensity. Receptive fields of neurons, however, depend on the characteristics of the stimuli (intensity, rate of change, etc.) but that is rarely studied. For example, the receptive fields (response areas) of auditory nerve fibers are highly dependent on the stimulus intensity,[139] but most studies have used the threshold for determining auditory receptive fields. Other factors, such as the rate with which the frequency of sounds (such as a tone) is changed, affect the receptive fields and the width of the response areas of auditory neurons.[134] If the frequency of a tone is changed at a high rate, the width of the response area of neurons in the CN decreases but the response areas of auditory nerve fibers depend little on the rate of change of the frequency of sounds.[134,140] (see page 337). This is an example of transformation that occurs in the ascending sensory pathways. Such matters have not been studied in detail for other sensory systems.

Receptive fields can also change as result of external circumstance and through the expression of neural plasticity. Stimulation (experience) can cause receptive fields to contract or expand, more so at central levels of sensory systems such as the cerebral cortex than at lower levels (see Chapters 4 to 6).

4. Basis for Spatial Discrimination

Spatial discrimination of sensory stimuli depends on the separation of the population of neurons that are activated by stimulation at two locations. For example, the ability to discriminate the stimulation of the skin at two nearby points as two separate stimuli (two-point discrimination) depends on the distance between the two points that are stimulated; when that become smaller than a certain distance, the perception of such two-point stimulation fuses into a perception of a single stimulation (Chapter 1). Stimulation of two

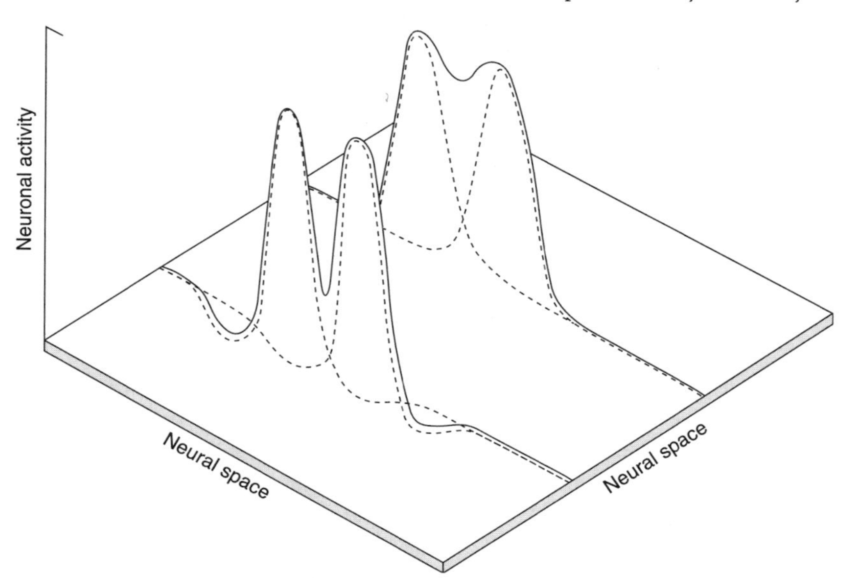

FIGURE 3.18 Schematic illustration of the response of a population of neurons to stimulation of two points of the skin showing how the response depends on the width of the two populations of neurons that are activated. (Adapted from Kandel *et al.*[91])

points on the skin activates two overlapping populations of nerve cells in the nuclei of the ascending pathways as well as in the cerebral cortex as has been illustrated in two-dimensional graphs such as that shown in Fig. 3.18. In this graph, simulated neural activity is displayed for an array of neurons that are located in a "neural space" that refers to a plane sheet of neurons, for instance in a nucleus such as those of sensory pathways. The graph in Fig. 3.18 is a theoretical illustration of how neurons in such a sheet of the DCN may respond to stimulation of a point on the skin. The degree of overlap depends on the distance between the two points that are stimulated and on the width of the two populations of neurons activated by stimulation at the two sites on the skin. Activation of neurons in a nucleus can be described as a stack of such planes to form a three-dimensional array of nuclei (Fig. 3.19).

The neurophysiologic basis for two-point discrimination is the difference between the two populations of neurons that are activated by two stimulations but it is not necessary for totally separate populations of neurons to be activated by each of the two stimuli in order to perceive stimulation at two distinctly different locations.

The example from skin stimulation given above serves as a model for other forms of spatial discrimination in other sensory systems. In vision, a light spot

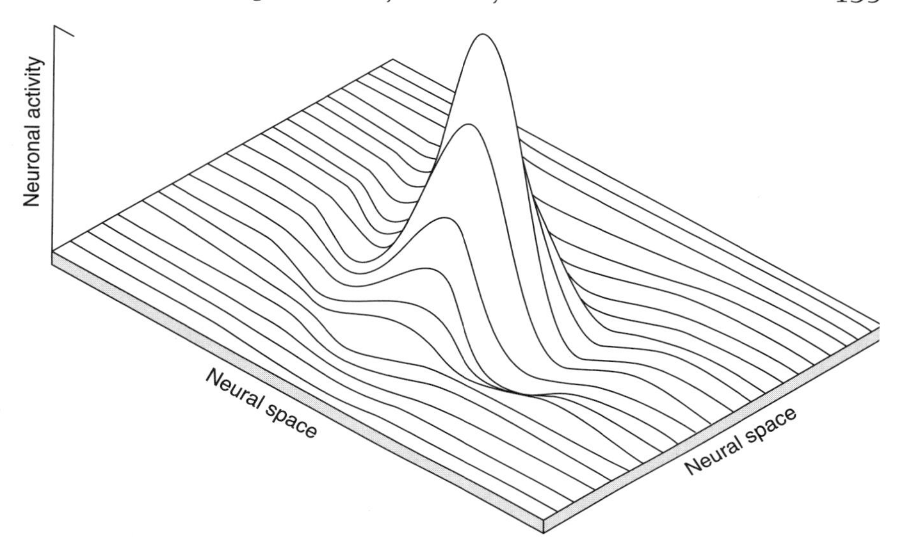

FIGURE 3.19 Schematic illustration of the response of a population of neurons to stimulation of one point of the skin. (Adapted from Kandel *et al.*[91])

that is projected on the retina elicits responses from specific populations of photoreceptors, and that causes activation of a specific population of ganglion cells and subsequently optic nerve fibers. Another spot of light that is projected onto the retina at a different location activates responses in another population of optic nerve fibers. The difference between the two populations of nerve fibers activated by these two different light spots provides information about the separation of the light spots which reaches the retina. When the separation is sufficiently large, the stimulation is perceived as two separate spots of light. However, the ability to discern stripes of black and white is more commonly used to determine the spatial resolution in vision. When the number of stripes is increased beyond a certain number per unit angle of view, individual stripes can no longer be discerned and that is the limit for spatial discrimination.

In the auditory system, frequency or spectrum of sounds may also be regarded as spatial information that activates different populations of the receptors that are located along the basilar membrane. Two tones of slightly different frequency activate two different but overlapping populations of auditory nerve fibers. However, discrimination of sounds on the basis of the difference in their frequency is not only a result of separation of the receptive fields of individual auditory nerve fibers; the coding of the temporal pattern of sound is also important for discrimination of frequency and the difference in the frequency of two sounds. This feature, known as *phase locking*, may be regarded as an object type of information (see Chapter 5).

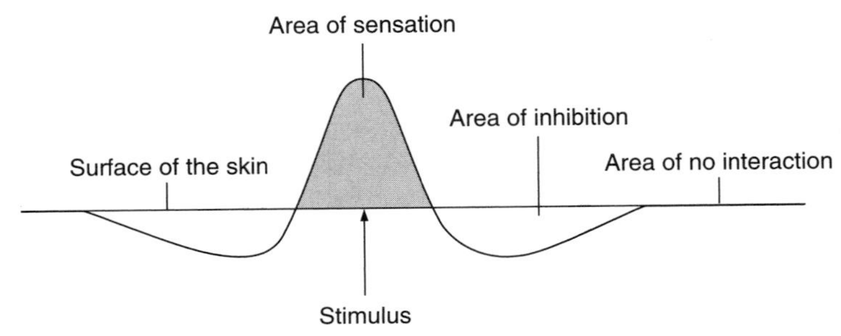

FIGURE 3.20 Schematic illustration of how lateral inhibition can sharpen the receptive field of receptors.[12]

5. Inhibition Can Enhance Spatial Contrast

Various forms of inhibition can enhance contrast by making receptive fields narrower, thus increasing the separation of populations of neurons that are activated by stimuli with slightly different spatial properties.

a. Lateral Inhibition

One such form of inhibition, known as lateral inhibition,[74,176] consists of inhibitory receptive fields that are located adjacent to excitatory fields. Lateral inhibition decreases the width of the receptive field of a neuron and thereby improves spatial discrimination (Fig. 3.20), by increasing the separation between two populations of neurons such as those that would respond to stimulation of two nearby points on the skin (Fig. 3.18).

Lateral inhibition affects the information that travels from the receptors toward the cerebral cortex in vision, hearing, and somesthesia. The interaction between the excitation and inhibition is the basis for some of the complex signal processing that occurs in these sensory systems.

Lateral inhibition was originally studied in the facet eyes of horseshoe crabs (*Limulus*). It was shown that the response of one of the many individual eyes of the horseshoe crab could be inhibited by illuminating an adjacent eye.[74] Lateral inhibition is also prominent in the vertebrate eye, where the result can be demonstrated by observing a series of stripes with different grayness. When the grayness within each stripe is the same, the region adjacent to a darker band is perceived as being lighter than areas that are farther away from the transitions and areas adjacent to a lighter band are perceived to be darker at the transition than away from the transition. This is a result of *contrast enhancement* by lateral inhibition. The area that is perceived lighter than the middle of the bands is known as a *Mach band* (see Chapter 1).[176] Lateral inhibition as it occurs in the periphery of sensory systems enhances contrast between spatial stimulation at different points such as of the skin[12] and in a similar way along the basilar membrane of the cochlea or between small spots of light shining at different points on the retina.[74,176]

Not only is the effect of lateral inhibition on the response of retinal ganglion cells an enhancement of contrast and increased spatial resolution, but the enhancement of contrast is also a form of suppression of unimportant information. When a region of the retina is evenly illuminated, excitatory and inhibitory response areas of retinal ganglion cells become evenly activated, and the resulting activation is minimized because the net excitation is the difference between excitation and inhibition. This means that activation of the visual nervous system in response to an even illumination is minimal. This is advantageous because it suppresses unimportant information. The interaction between inhibition and excitation that occurs in the skin has a similar effect if a large area is stimulated evenly.

The spatial discrimination depends on the temporal pattern of the stimulation. Thus, if the pattern of black and white stripes changes with time such as in a rotating drum with stripes, detection of the stripes becomes a function of how fast the drum is rotated. This is probably the result of the temporal characteristics for inhibition and excitation being different, which means that it may take a longer time to activate an inhibitory response than an excitatory response. In a similar way, representation of the frequency of steady sounds in the response of neurons in the auditory system is different from that of sounds for which the frequency changes rapidly and may be a result of such differences in the temporal properties of inhibition and excitation. However, the effect of lateral inhibition on discrimination of sounds on the basis of their frequency is not as obvious as it is in somesthesia and vision because the representation of frequency in the auditory nervous system not only depends on the receptive fields of auditory nerve fibers but also on the sound intensity[139,142] and the rate with which the frequency of a sound is changed.[134,148] Spatial representation of sensory stimuli also depends on previous exposure, or learning and is affected by neural plasticity, which is a general feature of the CNS.

6. Detection of Movement

A shift in the population of nerve fibers that are activated by a sensory stimulus is the basis for detection of movement. When the location of stimulation of the skin changes rapidly from one location to another, it typically generates neural activity that is interpreted as a movement of the site of stimulation. This can be demonstrated by stimulating two slightly different places on the skin and varying the interval between the stimulation which creates a sense of motion of the stimulus from one point to the other. When the stimulations occur at the same time, the perception is that the

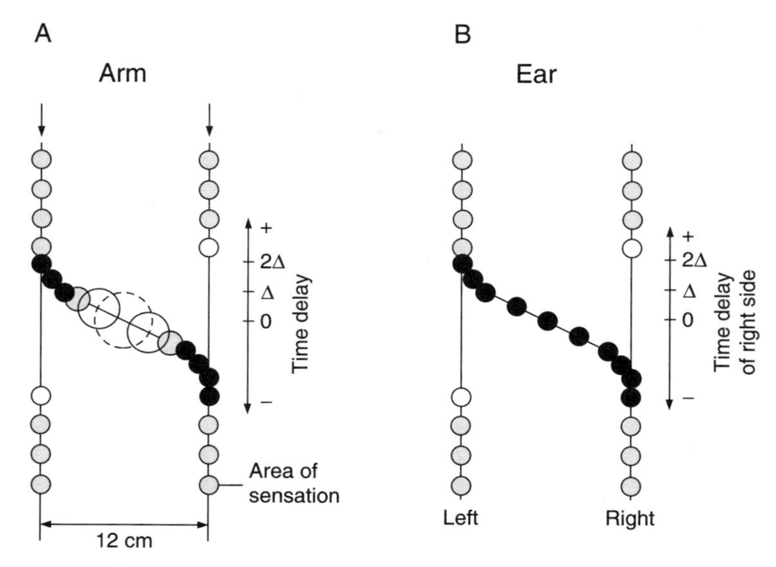

FIGURE 3.21 The perceived localization of two stimuli (vibrations) applied to the forearm (A) or two sounds applied to each ear (B) depends on the time interval between the two stimuli. (Adapted from Békésy.[12])

location of the stimulation is in between the two locations that are physically stimulated (Fig. 3.21).[12] The impression of movement caused by stimulation of two points on the skin with a small time interval in between (Fig. 3.21A) can be explained by considering the complex arrangement of receptive fields and the interaction between inhibition and excitation (Fig. 3.22).

a. Directional Information from Paired Sense Organs

The directional information from paired sense organs (eye and ear) can provide a three-dimensional representation of sensory input based on the differences in the physical stimuli that reach the receptors of the paired organs. The sense of hearing makes use of the difference in arrival time of the sound at the two ears, together with the difference in the sound intensity at the two ears, to determine the direction to a sound source. This is possible because of the unique relationship between the differences in the sounds that reaches the two ears and the direction to a sound source in the horizontal plane. These differences in the sound reaching the two ears create neural maps in the auditory nervous system, but these maps are not related to the receptor surfaces as the visual and somatosensory maps are; rather, the auditory maps are derived from

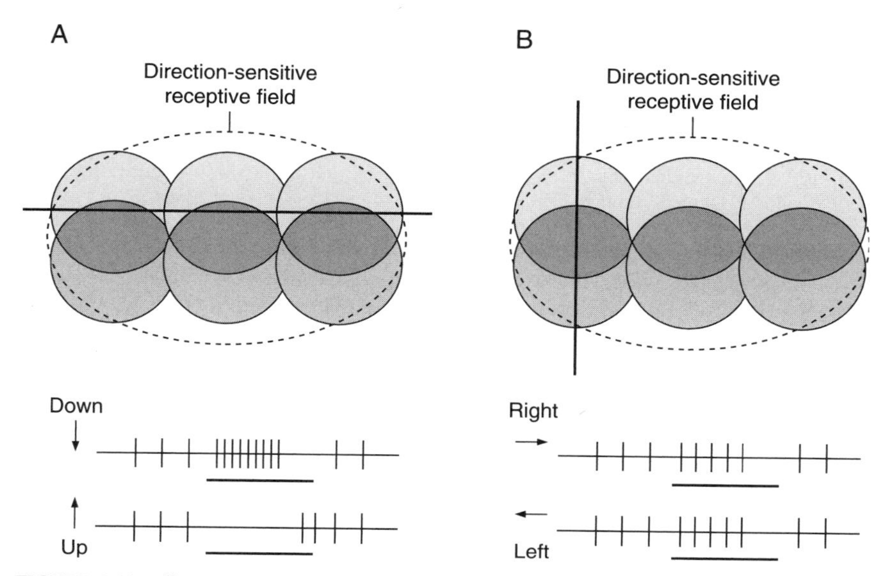

FIGURE 3.22 Illustration of how the arrangement of inhibition and excitation in the skin can produce a response that is sensitive to the direction of stimulation. (Adapted from Kandel *et al.*[91])

manipulation (comparison) of the input to the two ears. Such maps are known as *computational maps.*

The difference in the representation of an object on the retinas of the two eyes provides information about the three-dimensional location of the object in space and that information is also represented in the visual nervous system as spatial information. The combination of input from the two eyes creates a three dimensional description of the environment (stereoscopic vision) in the CNS. Information from eye muscles and neck muscles aids in determining the direction to an object. The *stereoscopic vision* that is based on the angle between the optical axes of the two eyes that provides fusion of the images from the two retinas functions for short distances only. Judgment of the distance to an object at a long distance is based on other factors such as knowledge of the size of the object in question. Depth perception for objects at long distances can therefore be performed without the aid of stereoscopic vision and can be achieved with one eye (monocular).

A sense of a moving sound source can be created by presenting sounds to the two ears with a small time difference (Fig. 3.21B).[12] Because the difference in time of arrival of sound at the two ears determines the direction to a sound source, changing this time interval can create the sense of a moving sound source (see Chapter 1). The same is the case for the perception of movement of a visual image caused by presentation of two light spots in the visual field of one eye with a small interval.

The creation of a three-dimensional sensation of sound (*stereophonic sound*) not only is a result of the difference in the sound that reaches the two ears but also depends on the ability to move the head. The latter is one of several reasons that music and other sounds presented through earphones do not sound exactly as they do in the environment where the sounds were recorded. Another reason is that the head and the outer ear modify the spectrum of sounds and that modification is lost when listening through earphones. The quality of stereophonic sound has been improved very much since electronic devices were introduced to modify the spectrum of sounds to mimic the spectral alterations that normally result from the effect of the acoustic properties of the head and ear canal.[15]

The dual representation of sensory stimuli from paired sense organs has implications other than providing the ability to determine the direction to the source or for stereoscopic vision and stereophonic hearing. It also improves the selection of specific stimuli in a background of other stimuli.

The "cocktail party problem," where a person can select to listen to a specific speaker in an environment where many people are talking, is an example of directional information being used for selecting a speaker. A person who has even slight hearing loss in one ear is handicapped in such a situation, because this loss impairs directional hearing. (A person who has a hearing loss in both ears for certain frequencies is also handicapped because all information of speech is important for distinguishing one speaker from another.)

The superior colliculus (SC) integrates auditory and visual information, and this nucleus is involved in creating a perception of auditory space.[236] Perception of auditory space is aided by input from the eyes, and the directional information from these two senses is integrated in the SC.[111] Factors such as the size of a visual object and the spectrum of sounds are important factors in creating the perception of space.

IV. NEURAL CONTROL OF SENSORY PROCESSING IN ASCENDING SENSORY PATHWAYS

The flow of information in the sensory pathways and processing in the sensory cerebral cortices are affected by neural activity of both *extrinsic** and *intrinsic*** origin. The ascending flow of information in the sensory pathways and, in particular, processing in the cerebral cortices are affected by arousal.

*Extrinsic means generated by sources outside the brain, thus by sensory input.
**Intrinsic means generated in the brain itself.

The flow of information in the ascending sensory pathways is also under the control of descending activity that is a result of the sensory stimulation of the same modality and other sensory modalities (extrinsic activity). The processing in the ascending sensory pathways and in sensory cortices is also under the influence of activity generated in the CNS (intrinsic activity) such as attention.

The influence of arousal and attention on neural processing of sensory information explains why sensory input alone may not give rise to a conscious perception. The other requirement for a conscious sensory response is sufficient arousal and, to some extent, attention.

A. Arousal

Transmission of sensory information to higher centers of the CNS requires a certain level of arousal. A low level of arousal is associated with drowsiness and high arousal is associated with agitation. Wakefulness, or vigilance, is an expression of arousal. Arousal mainly asserts its influence on the excitability of cortical neurons but it also influences the flow of information in the ascending sensory pathways. Arousal is mediated by the *reticular formation* of the brainstem.

The reticular formation is a diffuse group of neurons that are located in the brainstem core and project to neurons of the cerebral cortex. The function of the reticular formation in activating the cerebral cortex was discovered by Moruzzi and Magoun,[57,153] who in the late 1940s showed that high-frequency stimulation of the brainstem reticular formation could increase the excitability of the cerebral cortex (arousal).[57,200]

The reticular formation gets most of its input from collaterals from ascending fibers of sensory systems[200] and from the dorsal and medial thalamus. While the reticular formation mainly acts on the cerebral cortex,[57,153] it also acts on the reticular nucleus (RE) neurons in the thalamus, where it modulates sensory information such as visual.[242] There are also descending connections from the reticular formation, which through the *reticulospinal tract* can affect (inhibit)[82,157] the responsiveness of spinal motoneurons.

The central nucleus of the amygdala is another important source of arousal. This nucleus receives input from sensory systems via the lateral nucleus and the *basolateral nucleus of the amygdala,* and its effect on arousal is mediated through the *nucleus basalis* in the forebrain.[104,116,229,230] Sensory input to the *lateral nucleus of the amygdala* originates from the dorsal and medial thalamus (the subcortical or *low route*)[104] and from association cortices via primary and secondarv sensory cortices[116] (the *high route*).[104] The low route is fast and carries information that is little processed, while the high route carries highly processed information and is subject to being affected by conscious input from other parts of the CNS.

Arousal evoked by activation of one sensory modality can therefore affect processing in the ascending pathways of other sensory modalities. This means that perception of a sensory stimulus depends not only on the properties of the stimulus of that modality that reaches the sensory receptors but is also affected by sensory input of other modalities, as well as by factors that are internal to the CNS.

In general, input from the nonclassical pathways are assumed to be more important for arousal than input from the classical pathways both because of the input from the dorsal thalamus to the amygdala and because of the abundant collaterals from ascending sensory pathways that project to neurons of the brainstem reticular formation. The fact that neurons of the central nucleus of the amygdala project to the subthalamic nuclei and intralaminar nuclei of the thalamus makes it possible for many other forms of input to the CNS to affect sensory processing.

B. Attention

Attention controls the awareness of sensory stimuli. Attention is the difference between looking and seeing or the difference between hearing and listening. While arousal increases the excitability of neurons in the sensory system in general, attention directs focus on specific sensory modalities and makes it possible to focus on specific sensory stimuli. Arousal facilitates transmission of information in general, whereas attention can facilitate or suppress specific information. Attention may be regarded as focused arousal. The mother who wakes up to the cry of her child while sleeping through much louder noises is an often-used example of attention.

Specific information may also be suppressed as an effect of attention and may be referred to as *inattention* which implies that certain information is ignored despite activation of the sensory organs. Suppression of unwanted information is important for reducing the amount of information that reaches higher CNS regions. Suppression of certain specific aspects of sensory stimuli is different from selection of specific stimuli. Suppression of sensory input can be unintentional or intentional. We all know situations of unintentional suppression of sensory input. For example, the search for an object may fail even though the object is perfectly visible. Also, people report their cars being hit by other cars that they never saw despite the fact that they must have appeared clearly in the drivers' visible field. The ability to ignore more complex visual input such as commercials while watching television is an example of ignoring information based on its content, which is controlled from high CNS centers (intrinsic effect).

Attention and inattention are examples of how high CNS functions can rule which parts of the information reaching our sense organs will reach awareness. The neurophysiologic basis for attention (and inattention) is poorly understood.

C. Control of Ascending Neural Activity by Descending Pathways

The extensive systems of descending pathways, which may be regarded as reciprocal to ascending pathways, can influence processing of sensory information that ascends toward the primary sensory cortices. Efferent activity exerts its effect on the cells of the nuclei of the ascending sensory pathways, where it can act as presynaptic inhibition or postsynaptic inhibition of the neural activity that ascends in the sensory pathways.

The descending (efferent) activity can modify processing at several levels in the ascending sensory pathways on the basis of the activity at more central levels. It can reduce the information that reaches more central structures of the ascending sensory pathways by suppressing unimportant information and thereby help to extract useful information from noise. Descending activity can also modify the way sensory information is processed.

Efferent innervation can adjust processing in more peripheral nuclei on the basis of what arrives at a more central location. In the auditory system, efferent activity can also modulate the sensitivity of sensory receptors and even affect the mechanical properties of the medium that conducts the physical stimulus (sound) to the receptor (the cochlea) because the mechanical properties of the cochlea are affected by the outer hair cells, which receive efferent innervation from more central parts of the auditory pathways.

The abundant innervation of the thalamus from the same area of the cerebral cortex that receives input from the thalamus can adjust processing in the thalamus on the basis of information that reaches the cortex. For example, it has been demonstrated in the auditory system that inactivation of cortical cells can cause the frequency tuning of neurons in the thalamus to change (Fig. 3.23).[245] (Recall that tuning of cells in the ascending auditory pathways is based on the tuning of the basilar membrane in the cochlea, but convergence of input to nerve cells makes the tuning of nerve cells in the ascending auditory pathways different from that of the basilar membrane.) The observed shift in tuning may be explained by abolishment of efferent input to the thalamus from the cortex, which affects the array of neurons that converge on the cells from which these recordings were made. If the efferent activity from the cortex blocks some of these converging inputs, a release of such blocks can shift the array of inputs to cells in the MGB or IC, which can explain the shift in the frequency to which the cells are tuned. Another explanation would be that the descending activity reached all the way peripherally to the cochlea, where it affected the mechanical properties of the outer hair cells through activation of the *olivocochlear bundle* (OCB)

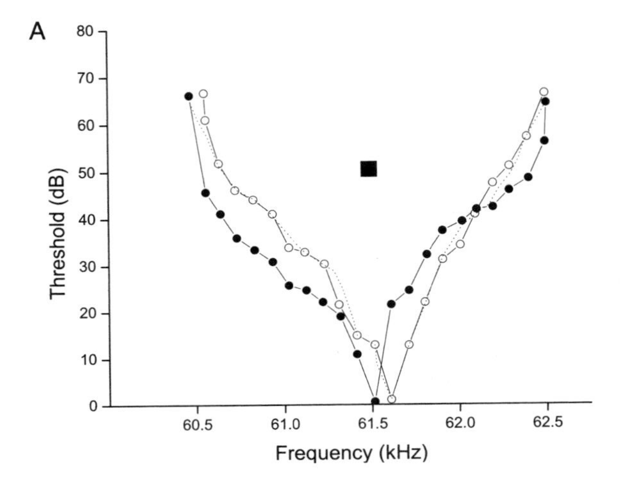

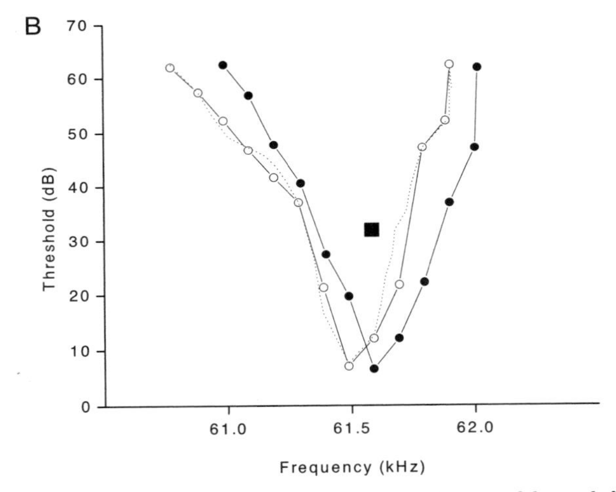

FIGURE 3.23 Changes in the tuning of neurons in the ventral division of the medial geniculate body of the thalamus (A) and in the posterior dorsal position of the inferior colliculus (B) before (open circles) and after (closed circles) inactivation of neurons in the primary auditory cerebral cortex using local anesthetics.[245]

(see pages 284 and 315). That could change the tuning of the basilar membrane and the tuning of all neurons in the ascending pathways.

Other investigators have shown that the frequency tuning of auditory cortical cells can change as a result of electrical stimulation of the cerebral cortex, indicating that internal processing in the cerebral cortex can be modified by local neural activity.[189] (see page 358).

Descending pathways are less prominent in the visual system and in the classical somatosensory system but are abundant in the nonclassical somatosensory system, where they have an effect on transmission of pain signals. The effect of the abundance of reciprocal connections in the ascending classical sensory pathways indicates that the picture of the environment that we finally achieve on the basis of sensory input may be obtained by adjusting the processing at lower levels of the sensory nervous system on the basis of the neural activity that arrives at higher levels (such as the cerebral cortex). This means that the final interpretation of sensory information may be arrived at by successive iterations of different solutions to the analysis that occurs in the ascending sensory pathways and the cerebral cortices.

V. PROCESSING OF INFORMATION IN THE NONCLASSICAL PATHWAYS

The function of the nonclassical sensory pathways is less wellknown than that of the classical sensory pathways but it is assumed that the analysis that takes place in these pathways is less extensive than is the case for the classical pathways. Neurons of nonclassical sensory pathways respond less distinctly and less specifically to sensory stimuli than those of the classical pathway. In the somatosensory system, the classical pathway (dorsal column pathways) is involved in fine touch, whereas the nonclassical pathways (*anteriorlateral pathways*, see page 210) communicate pain, temperature, and some deep touch. While neurons in the classical sensory pathways (up to the primary sensory cortices) respond only to one sensory modality, some neurons in the subcortical portion of nonclassical pathways respond to more than one sensory modality. Some parts of the nonclassical pathways are therefore *polysensory* in nature.

Little is known about the normal function of the nonclassical pathways of hearing and vision, but the anterior lateral system is known to be functional under normal conditions, where it mediates the sensation of temperature, pain, and some deep touch. Some studies show involvement of the nonclassical auditory system in young individuals[150] and in some patients with tinnitus.[145]

The input to the anteriorlateral system comes from many different types of receptors. These inputs converge onto the *wide dynamic range* (WDR) neurons[19,172] in the spinal cord and the brainstem, where a complex interaction occurs between excitatory and inhibitory input. As the information ascends in the three or four different pathways that constitute the anterior lateral system it is modified and subjected to modifications from other sensory inputs and from input from higher CNS centers (See chapter 4). The neural processing in pain pathways has been studied extensively because of its great clinical importance.

These different pathways have slightly different targets of subcortical structures and reach different parts of the cerebral cortex (see page 210).

Neurons in the nonclassical auditory pathways typically respond to a broad range of sounds and the response of many neurons is less specific than neurons in the classical pathways. The responses from neurons in the nonclassical auditory system have been studied most extensively in recordings from the parts of the IC that are believed to belong to the nonclassical system (the ICX and DC).[7,152] Many neurons in the nonclassical auditory system receive input from more than one sensory modality and respond to more than one sensory modality while neurons in the classical pathways respond only to one modality of sensory input. Studies in animals have shown interaction with the response to sound stimulation from stimulation of the somatosensory system at the levels of the IC[7,211,212] and at the cochlear nucleus.[81,198] Sound-evoked activity in neurons in the thalamus can be affected by stimulation of the visual system and from stimulation of the somatosensory system in animals.[77]

In the visual system, the *superior colliculus pathway* can be regarded as the nonclassical pathway that mediates visual reflexes and some crude visual perception. It was mentioned above that the earliest published studies that showed that different pathways process different types of information involved the classical and nonclassical visual pathways.[193] These studies showed that the nonclassical pathways process spatial information while the classical pathways mainly process objective information.[193]

A. Developmental Aspects of Classical and Nonclassical Pathways

The nonclassical sensory systems served well when it was mainly important to be able to detect sensory input as a warning signal and specificity of sensory discrimination was less important. This would explain the lack of specificity in the responses of neurons in the nonclassical systems. The connections from the nonclassical system to allocortical regions*[,222] and structures of the limbic system fit the hypothesis that sensory input mainly served as warning signals. The classical ascending systems that developed as species became more sophisticated improved the ability to discriminate sensory stimuli for identification of prey or foe and for intraspecies communication.

The decrease in emphasis on the nonclassical pathways that may have occurred during ontogeny concomitant with development of classical sensory

*Allocortical region is Vogt's term denoting several regions of the cerebral cortex, in particular the olfactory cortex and the hippocampus, characterized by fewer cell layers than the *isocortex*.

systems could have been caused by pruning of axons, *apoptosis*, or by adjustment (decrease) of the efficacy of the synapses that provide sensory input to the nonclassical systems. The change in synaptic efficacy may be completed after birth during childhood and adolescence.[150] The hypothesis about change in synaptic efficacy is supported by the finding that the nonclassical auditory pathway is active in individuals with some forms of tinnitus, as indicated by the demonstration that the tinnitus in some patients changes as a result of activation of the somatosensory system,[25–27,145] indicating involvement of nonclassical auditory pathways. This means that pathological conditions may restore the connections from the ear to the nonclassical pathway that was interrupted during childhood development.

Some individuals with tinnitus experience sensations of sound in response to skin stimulation, such as rubbing with a towel, and muscle activity (such as eye muscles and jaw muscles) can alter the sensation of tinnitus in some individuals,[26,27] yet other signs of interaction between the somatosensory and the auditory systems that can only occur through the nonclassical auditory system because the classical auditory system does not receive input from the somatosensory system.

Involvement of the nonclassical auditory pathways means that a subcortical route of auditory information to the limbic system has been established in these patients. This has been confirmed by studies using *functional magnetic resonance imaging* (fMRI)[110] that have shown that limbic structures are indeed activated in some patients with tinnitus. The fact than affective disorders (such as depression) are common in patients with severe tinnitus further supports the hypothesis that the limbic system is activated in some patients with tinnitus.

B. Processing of Information in Descending Nonclassical Systems

The nonclassical sensory systems have descending (reciprocal) pathways like the classical sensory pathways, but little is known about the function of such descending pathways with one exception, namely pain systems. It has been known for many years that pain impulses that ascend in the anterior lateral pathways of the somatosensory system can be modulated by descending impulses,[55] as well as by activity in large-diameter spinal afferents.[118]

Descending activity of supraspinal origin (mainly from the PAG[233] but also from the thalamus) has been shown to modulate ascending impulse traffic in pain pathways.[55] The descending influence on pain is not simply inhibitory as has been assumed earlier; recent studies indicate that the supraspinal influence can facilitate or disinhibit spinal transmission of pain impulses (for a recent review, see Urban and Gebhart[220]).

VI. HOW IS THE NEURAL CODE OF SENSORY INFORMATION INTERPRETED?

We have concentrated on how sensory information is processed in the receptor organs and subsequently in the CNS. In this section, we will discuss how the neural code is interpreted to build the basis for our perception of sensory stimuli. While we must agree that our knowledge about how information is processed in sensory systems is incomplete, our knowledge about how sensory information is interpreted is even more sparse.

It is known from many studies that physical stimuli that are different are represented in several ways in the nervous system, but our knowledge is limited regarding which of such representations are important for discrimination of sensory stimuli. Neural activity evoked by sensory stimulation can be described by the characteristics of the firing pattern of single individual neurons (discharge rate and temporal pattern of the evoked discharges), or it can be described by the degree of coherence in the firing of an array of neurons. We do not know which one of these two descriptions is the most relevant with regard to the interpretation done in the CNS and which is the basis for perception of sensory stimuli.

Lack of understanding of how the nervous system interprets the information contained in a neural code has hampered progress in understanding neural processing of sensory information. It seems obvious to assume that features of sensory stimuli that can be distinguished psychophysically are also represented in distinctly different ways in the nervous system.

A. Coding in Single Nerve Cells

Because most studies have used recordings from single nerve fibers and nerve cells, the interpretation of the neural message has focused on the discharges in single neural elements. Such studies have provided much information about how sensory stimuli are coded in the discharge pattern of individual neurons and how this pattern is transformed as it ascends in the neural pathways of the sensory systems. This is known as the *rate hypothesis,* and the prevailing hypothesis of interpretation of the neural code has for a long time been that the rate of neural firing was the basis for interpretation. It was assumed that sensory stimulation that evoked more discharges per unit time in a neuron was more important than stimulation that evoked fewer discharges. That approach has been successful in many instances in providing insight into some fundamental processes in the nervous system, such as in vision. However, the discharge rate of a single neuron is rather irregular, and stable records of firing rate, latency, and other properties have only been obtained by averaging many responses to identical stimuli.

As our knowledge about the function of the nervous system increases, it has become more and more apparent that recording from single nerve cells of nerve fibers is an exceedingly unphysiologic way of studying the function of the nervous system. Instead of relying on the response of individual neurons, the coherent firing of large populations of neurons would seem to be more appropriate. The population hypothesis has support from studies done in the somatosensory system[40] as well as in other sensory systems.[44,47]

The nervous system consists of a very large number of neurons, and the activity in many neurons and the correlation between the firing of large populations of neurons is likely to be a much more important basis for interpretation of sensory information than the discharge rate or even the pattern of discharges in a single neuron or a few neurons. Difficulties in studying experimentally the simultaneous activity in many neurons have hampered studies of the correlation between neural activity in large populations of neurons.

B. Coherent Firing of Groups of Neurons

The hypothesis that the coherence of firing of large population of neurons is important for interpretation of sensory information has gained increasing support and inspired many different kinds of experiments,[44,47,54,190] and it seems likely that the interrelationship between the timing of neurons in large populations of neurons plays an important role for processing and interpretation of information. For example, it has been shown that the internal synchronization of the activity in many neurons is affected by internal cognitive factors.[181] Coherent firing of many neurons can only be studied by simultaneous recordings from many neural units, but such recording are technically demanding and have only been done in a few studies.

The hypothesis that neural messages are interpreted by the degree of temporally coherent firing of many neurons within a specific anatomical location is attractive, but the experimental support is yet insufficient to draw definite conclusions regarding the importance of that type of coding.

C. Maps

Neural maps show the anatomical organization of neurons that innervate receptor surfaces such as the basilar membrane of the cochlea, the skin, or the retina of the eye. The existence of such maps has been demonstrated in many studies. The image that is projected onto the retina is represented on the

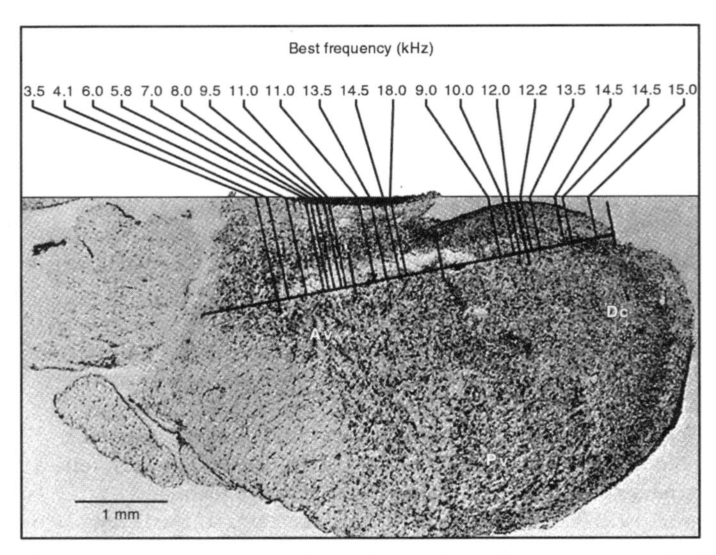

FIGURE 3.24 Tonotopic organization in the cochlear nucleus of a cat, as an example of tonotopic organization in the auditory system. (From Rose, J.E., Galambos, R., and Hughes, J.R.., *Bull. Johns Hopkins Hosp.*, 104: 211–251, 1959. [Copyright Holder]. Reprinted with permission of the Johns Hopkins University Press.)

surface of the primary visual cortex and the body surface (skin) is represented as a two-dimensional map on the somatosensory cortex.[167] Also, spectral information about sounds (frequency distribution) is represented as maps throughout the auditory nervous system but these maps are one dimensional (Fig. 3.24). Such maps reflect spatial properties of sensory stimulation, but neural maps may also reflect different qualities of stimuli, such as the temporal pattern of a sound[240] or the color of light, which activate different populations of nerve fibers.[184,185]

A few studies have demonstrated maps that reflect features of specific sensory qualities that are known to be important for perception. Suga and his collaborators[204] have shown that the characteristics of sounds that echolocating bats receive when flying create maps on the surface of their auditory cortex that reflect features directly related to important information gained by sound echolocation, such as the distance to a target, target velocity, and velocity of the animal in flight. These maps do not represent the projection of a receptor surface but instead are created by processing information from sensory systems. Such maps are therefore known as computational maps.

The flying bat navigates during flying by using echolocation, relying upon sound instead of vision. This means that these animals emit short impulses of sound and listen to the echoes. The time between emission of a sound and the arrival of the echo is a measure of the distance to an object (range). The bats can calculate the velocity of a target in relation to the bat, and the result is also represented in cortical maps (Fig. 3.25).[204,205] Cortical maps of echolocating sounds also represent features such as

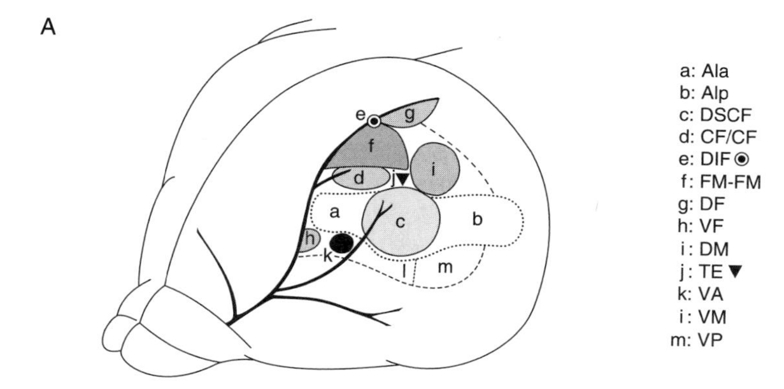

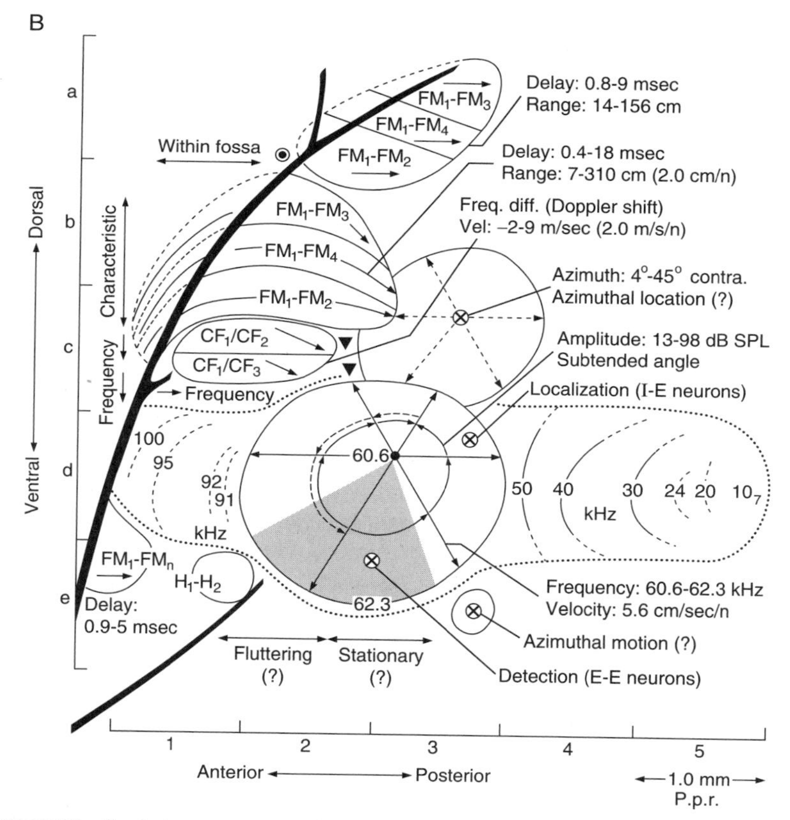

FIGURE 3.25 Cortical representation of different aspects of sound emitted by flying bats. (Adapted from Suga.[204,205])

the harmonic content of the sounds, which may be important for an individual bat to identify its own echo. Bats fly in large groups and therefore must have a giant "cocktail party" problem (see Chapter 5, page 341) in regard to distinguishing the echoes originating from their own emitted sounds from the sounds made by their fellow bats.

Computational maps have also been demonstrated to project the space around an observer onto the SC.[111,121,236] These maps represent input from both the ears and the eyes. Although the SC is primarily involved in eye movements, its neurons also receive input from the auditory system,[165] and visual and auditory information are known to interact in neurons in the SC,[121,165] which may be the basis for perception of space, although the output of that nucleus is not known to reach the conscious part of the brain.

D. Where is the Anatomical Location for Interpretation of Sensory Information?

We do not know exactly in which anatomical regions of the CNS that sensory information is interpreted. Knowledge about where interpretation of the sensory information occurs is perhaps less important than information about which features are important for the proper interpretation of sensory information. However, the question about which parts of the CNS are activated by specific sensory stimuli has recently been studied extensively using many different methods, such as recordings of *evoked potentials*, *magneto encephalography* (recording of *magnetic evoked potentials*, MEPs), and more recently imaging techniques* such as fMRI, *positron emission tomography* (PET), and *single-photon emission computed tomography* (SPECT).*

Studies of changes in sensory perception resulting from known permanent lesions or temporary inactivation are well-established methods for investigating the anatomical locations of neural activity important for perception.[36,112,164] That method has been used in animal experiments, where controlled lesions can be made, and in humans, whose lesions are caused by diseases such as stroke or are the result of surgery.

Studies using different techniques show consistently that wide areas of the association cortices are activated during sensory stimulation. This, however, does not mean that the activated regions of the CNS are those where interpretation of sensory information occurs, but the results can equally well show that the identified structure is involved in transmission of the information to the

*The so-called functional imaging techniques measure changes in blood flow using MRI or PET techniques. Their use in studies of localization of regions with high neural activity is based on the assumption that increased blood flow means increased neural activity. That assumption has been challenged, however, and even if the assumption is correct the method is rather crude and requires considerable signal averaging.

anatomical location where it is being interpreted. The same ambiguity exists regarding interpretation of the results of studies of the effect of lesions of the CNS that produce loss of specific abilities, leaving others intact. Such studies have, however, provided important information about the anatomical location of structures involved in transmission and processing of sensory messages.

Studies of the effect of localized lesions in sensory (visual) association cortices have indicated that specific forms of *visual agnosia** such as *prosopagnosia*[37,216] (inability to recognize familiar faces), *autotopagnosia* (inability to recognize one's own body parts), and *simultagnosia* (inability to recognize more than one object at the same time) are associated with specific regions of the extrastriate association cortices. Tactile *agnosia* (inability to recognize objects by touch) seems to be associated with lesions in the parietal–temporal junction.

The large individual variation in the anatomy of the human brain is a noticeable obstacle in studies of the location of the anatomical location where sensory information is interpreted,[128,164] but this factor is often ignored.

VII. SENSORY INFORMATION CAN REACH NONSENSORY REGIONS OF THE CNS

Sensory information can reach many different regions of the CNS that are not usually regarded as sensory regions. We have already mentioned that some parts of sensory pathways, in particular nonclassical pathways, project to areas of the CNS where neural activity does not give rise to direct awareness in the same way as information that reaches the primary sensory cortices. Nonsensory regions that are activated by sensory stimuli include motor systems and structures that are involved in learning and emotional and autonomic reactions, such as the various components of the limbic system. Sensory input to these structures can elicit reactions such as pleasure or suffering or autonomic reactions such as a change in blood pressure and heart rate, vomiting, and perhaps more complex reactions such as attraction between sexes.

A. Motor Systems

Examples of sensory reflexes that are mediated by subcortical connections are spinal reflexes elicited by cutaneous stimulation and the acoustic middle ear

**Agnosia* means "impairment of ability to recognize, or comprehend the meaning of, various sensory stimuli, not attributable to disorders of the primary receptors or general intellect; agnosias are receptive defects caused by lesions in various portions of the cerebrum" (Stedman's Medical Dictionary).

reflex, where the muscles of the middle ear contract due to a loud sound through connections from pontine auditory structures and the facial motonucleus. The pupil reflex causes the pupil to constrict in strong light. The blink reflex, consisting of closing the eye as a result of stimulating the skin around the eye, is a trigeminal reflex that is protective of nature. Of the several spinal reflexes, the withdrawal reflex is probably the most prominent. Eye movements are partly controlled by visual and auditory sensory input through the SC. Visual input is also important for such activities as maintaining posture and walking, particularly in individuals with impaired vestibular function.

The *startle response* is a general contraction of many muscles most readily elicited by strong and unexpected sounds. It is mediated through the brainstem reticular formation but can also be elicited by other sensory stimuli such as from stimulation of skin receptors or by strong light.

> The threshold and the strength of spinal reflexes depend on alertness and on the facilitory input to motoneurons from higher CNS centers. The threshold of the startle response is especially variable and depends primarily on the state of alertness, which is an indication of the involvement of the brainstem reticular formation. The startle response to one sensory modality is affected by activation of other sensory modalities. A technique where the startle response is modulated (decreased) by presenting a short weak sound before the (strong) sound that is used to elicit the startle response (pre-pulse inhibition)[191] has been used extensively in studies of many brain functions.[18]

Reflex responses are different from coordinated motor movements in that the former is elicited only by sensory input, and the coordinated muscle responses are elicited by a combination of sensory input and central input that relates to conscious decisions to make movements. Conscious motor activity receives complex input from sensory systems such as skin receptors and, in particular, vision. Proprioceptive input from sensory organs in muscles, tendons, and joints are important for the proper function of motor systems. These inputs serve as feedback from motor activity to the CNS. Other hypotheses postulate that interpretation of sensory input, such as speech sounds, is accomplished by mimicking motor activity.

B. Autonomic Reactions to Sensory Stimulation

The chemical senses particularly can activate autonomic systems. Vomiting from taste and smell are examples of activations of the autonomic systems by sensory systems. Other autonomic reactions such as sweating may also arise from activation of taste and smell. The autonomic reactions that are mediated by hearing, somesthesia, and vision are mostly related to the message that such stimulation contains.

Pain is often associated with activation of autonomic systems, and certain forms of pain can cause secretion of norepinephrine by the autonomic (sympathetic) nerve endings that are located close to somatosensory receptors in the skin. This secretion increases the sensitivity of somatosensory receptors, including nociceptors. If sufficient, norepinephrine secreted close to the nociceptors may cause them to become activated without any other stimulation. Activation of nociceptors may initiate a vicious circle that activates the sympathetic nervous system, which causes increased secretion of norepinephrine, which in turn increases activation of nociceptors, resulting in increased pain. This condition, known as *reflex sympathetic dystrophy** (RSD)[39,172] (see Chapter 4), is a severe pain disorder that is directly related to the autonomic system and which must be treated with little delay because it can become permanent and untreatable, probably through permanently establishing new neural connections (neural plasticity).

C. Emotional Reactions to Sensory Stimuli

We have already discussed the abundant connections from sensory systems to the amygdala (pages 83 and 90) but other structures of the limbic system also receive input from sensory systems. This section will discuss the implications of sensory activation of limbic structures

1. The Limbic System

Many different structures of the limbic system receive input from sensory systems, and that input is integrated with input from many other parts of the brain. The most important part of the limbic structures with regard to processing of sensory input is the amygdala. The limbic structures relay information to many parts of the CNS.[19] The limbic system is regarded as a part of the old brain and many of its components receive input from the chemical senses. Neurons of the olfactory bulb project to limbic structures such as the *periform cortex* in the *uncus* and the *hippocampal gyrus*. The pathways of taste also have strong projections to limbic structures. Pain pathways (nonclassical somatosensory pathways) have connections to limbic structures such as the *anterior cingulate gyrus*, and pain impulses that travel in the spinomesencephalic tract can therefore elicit affective components of pain (see pages 216 and 240).

Olfaction plays an important role in sexual signals for many animals, while similar communications in humans rely upon vision and hearing. The

*Other names and classifications have evolved, and these disorders are now often known as *sympathetic maintained pain* (SMP).[33]

vomeronasal pathways project to the *corticomedial amygdaloid nucleus* of the amygdala. The input to the amygdala from the accessory olfactory bulb is large, and its targets include the *bed nucleus of the accessory olfactory bulb*. The vomeronasal organ provides information to the amygdala that is less processed than the information reaching the amygdala through the cortical pathways (the "high route"). This means that the vomeronasal pathway may be similar to nonclassical pathways in other sensory systems and may represent the equivalent of the low route to the amygdala that has been described for other sensory systems.[104] The corticomedial nuclei of the amygdala are believed to relay pheromoneal input from the accessory olfactory bulb to the medial forebrain and the hypothalamus.

2. The Amygdala

The amygdala has many different functions, only a few of which will be discussed here. The amygdala receives input from all sensory systems[64,116] and is involved in affective (emotional) reactions to sensory stimuli[105] and memory, especially of emotional events.[53] The amygdala is also involved in neural plasticity through its input to the nucleus basalis and is a mediator of arousal.

a. Anatomy of the Amygdala

The amygdala is a cluster of three main groups of nuclei, namely the basolateral, medial, and central nuclei.[64,116] The basolateral group of nuclei are especially of interest in connection with sensory systems.[19,116] The amygdala has connection to and receives input from many parts of the CNS.[19,64,116] The central and medial nuclei (the corticomedial group of amygdaloid nuclei) receive extensive input from the olfactory system.[116]

The amygdala nuclei receive considerable input from the olfactory bulb that projects to the medial nucleus of the amygdala. The accessory olfactory bulb receives input from the vomeronasal organ and projects exclusively to the corticomedial nucleus of the amygdala.

The basolateral nucleus also receives input from all other sensory systems via the lateral nucleus of the amygdala and connects to the central nucleus of the amygdala. The auditory, somatosensory, and visual systems provide input to the basolateral nuclei and central nuclei through the lateral nucleus of the amygdala (Fig. 3.26) via mainly two routes: a direct (subcortical, short) route from the dorsal and medial thalamus (low route) (Fig. 3.26B)[104,105] and an indirect (cortical, long) route ("high route") via primary sensory cortices, secondary cortices, and association cortices (Fig. 3.26A).[19,42,104] The classical sensory systems supply information to the amygdala only through the high (cortical) route. The low route to the amygdala carries raw information that is little

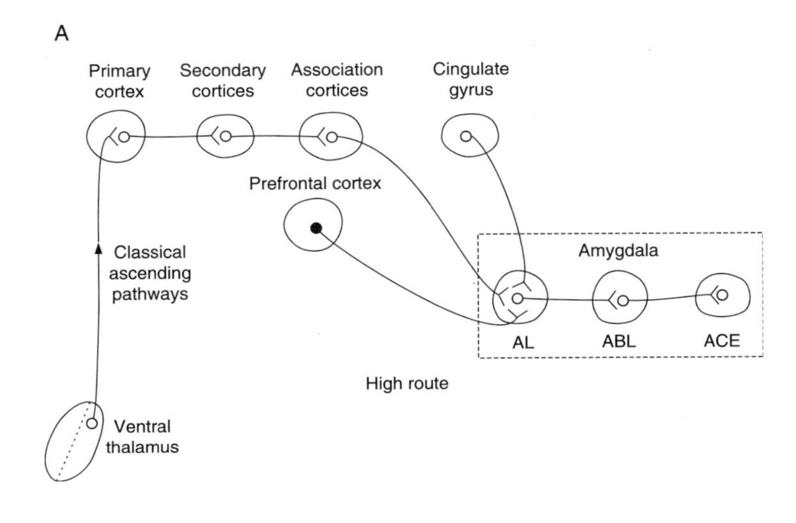

FIGURE 3.26 Schematic drawings of the connections between the auditory system and the lateral and basal nuclei of the amygdala. (A) Cortical connections from classical ascending pathways (high route); (B) Subcortical connections from nonclassical pathways (low route).

processed. The cortical route (high route) is much longer and carries highly processed information subject to interaction from intrinsic brain activity as well as all modalities of sensory information (Fig. 3.26A). (The low route has also been labeled "quick and dirty" and the high route "slow but accurate."[106])

Connections from taste and the spinothalamic neurons in the thalamus project to the several nuclei of the amygdala.[217] Thalamic neurons that receive input from sensory receptors in the viscera in the thorax and the abdomen have abundant projections to the amygdala as have other thalamic sensory nuclei.[217]

The dorsal and medial auditory thalamic nuclei that are part of the low route to the amygdala receive input mainly from the midbrain auditory nuclei (IC) but it has recently been shown that axons from cells in the dorsal cochlear nucleus travel uninterrupted to the thalamus and make synaptic connections with neurons in the medial MGB.[113] This direct auditory route from the dorsal cochlear nucleus to the medial MGB provides an even shorter and faster route for auditory information to reach the amygdala than the low route through the IC.

The connections from the amygdala are abundant and mostly reciprocal to the afferent connections (Fig. 3.27). The central nucleus connects to cortical regions such as the prefrontal cortex brainstem structures and the hypothalamus.[19,91] The hypothalamus and visceral sensory structures such as the NST (Fig. 3.3D) are also important targets for these nuclei. The brainstem

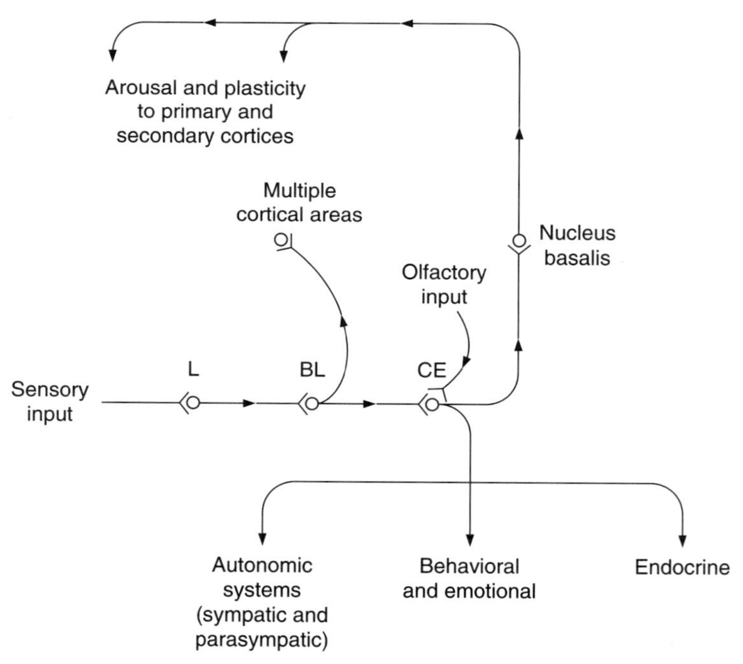

FIGURE 3.27 Important connections from the basolateral and central nuclei of the amygdala. Abbreviations: L, lateral nucleus of the amygdala; BL, basolateral nuclei; CE, central nuclei of the amygdala.

structures to which the central nucleus projects are involved in emotional, endocrine, and autonomic reactions. The connections to paraventricular hypothalamic nucleus can cause increases in secretion of *adrenocorticotropin* (ACTH) with increased secretion of corticosteroids, which are typical components of stress responses. Any of these reactions may be reduced by injuries to the central nucleus of amygdala. The central nucleus also projects to the nucleus basalis, which is important in arousal and for promoting neural plasticity (Fig. 3.27). Arousal and increased vigilance are therefore typical reactions to activation of the central nucleus of the amygdala. Projections from the amygdala to brainstem nuclei can potentiate reflexes such as the startle response and also mediate arousal by connections to the reticular formation. Another target is the lateral hypothalamus, through which activity from the amygdala can cause sympathetic activation (elevated blood pressure, tachycardia, pupil dilatation, etc.).

Yet another important target of connections from the basolateral and central nuclei of the amygdala is the nucleus of the vagus nerve (cranial nerve X), causing parasympathetic activation resulting in bradycardia, increased intestinal activity, etc. Other projections from nuclei of the amygdala including the PAG, and the stria terminalis that covers the thalamus have also been identified using anatomical techniques.[19,116] The connections to the PAG may mediate pain sensations related to fear. Structures belonging to the allocortex, especially the hippocampus, septal nuclei, nucleus accumbens, and the nucleus basalis, also receive input from the amygdala (Fig. 3.27).

There are also ample reciprocal connections between the nuclei of the amygdala and the medial and dorsal thalamic nuclei through which the output from the amygdala may reach prefrontal cortices. Recent studies of the auditory system have shown connections from the amygdala to the ICC,[115] thus a structure associated with classical sensory (auditory) pathways. The projection from the amygdala to the ICC may close a loop consisting of the DC and the ICX (of the IC) connecting to the medial and dorsal MGB (of the thalamus) and the basolateral nuclei (via the lateral nucleus of the amygdala) to the central nucleus of the amygdala.

Most of the connections to and from the amygdala are identified using anatomical methods and the functions of only a few of these connections are known in detail. The synapses that establish the functional connections may be closed (dormant) with the possibility to open under certain circumstances such as through the expression of neural plasticity.

b. Function of the Amygdala

The central and medial nuclei are involved in autonomic functions, whereas the basolateral nuclei are involved in conscious functions that are related to the frontal and temporal lobes[19,64] and numerous other functions such as providing arousal and promoting neural plasticity (through the nucleus basalis).[104,229,230]

The basolateral and central nuclei of the amygdala are involved in memory consolidation. McGaugh and his colleagues[52,53] have shown that learning of emotional events depends on adrenergic substances and that learning is impaired after administering β-adrenergic blocker such as propanolol. Stimulation that reaches the basolateral nucleus of the amygdala either from the thalamus (the subcortical or low route) or from the cerebral cortex (the cortical or high route) can change synaptic efficacy as an expression of learning. Such learning may also be regarded as neural plasticity and it can switch on neural circuits that are not normally activated.

Because the dorsal MGB supplies important input to the amygdala (low route) it can be assumed that the nonclassical sensory pathways that project to the dorsal and medial MGBs may play an important role as mediators of emotional sensory input through their direct (subcortical) connections to the amygdala.

The basolateral nucleus of the amygdala through its projections to the central nucleus of the amygdala and the nucleus basalis of the forebrain provides arousal to both primary and association cortices. Nucleus basalis also facilitates neural plasticity,[94,95,229,230] which may change the connectivity of the cerebral cortex.

In general, the amygdala processes sensory stimuli with regard to their emotional value.[19] The amygdala is involved in many basic functions such as learned fear, aggression,[43] and social behaviors and emotions. The amygdala seems to be important for unconscious processing of visual cues of fear such as from facial expressions, as concluded from studies in monkeys where the amygdala nuclei were ablated. Fear and fear conditioning are important for survival because they make it possible to store information about events and objects that represent severe risks to the animal in question.

For example, the visual information about emotional facial expressions activates structures of the limbic systems while information related to identity is processed in prefrontal cortex. While some mammals use facial expressions to show anger, hostility, etc., the perception of facial expressions in humans are more detailed. We are capable of making and identifying many more different forms of smiles and other expressions of emotions than, for instance, cats and dogs can. It is not known if that means that vision and hearing have stronger connections to limbic structures in humans than in animals that rely on olfaction for such communication.

Different forms of facial expressions seem to be processed by different populations of neurons. Adolphs et al.[2] have shown evidence that facial expressions of fear are processed in populations of neurons that are different from those processing information regarding the identity of faces (see page 415). This is a form of stream segregation. Bilateral but not unilateral destruction of the amygdala removes recognition of fear from the expression of faces but leaves the ability to recognize faces unimpaired,[2,37] indicating that one stream carrying fear information uses the amygdala while the other stream carrying innocuous information about the identity of a person uses the visual nervous system.

The amygdala is important for olfaction and especially for processing of information from the vomeronasal organ that responds to pheromones. While pheromones are known to play important roles in sexual behavior in many animals (especially insects, but also many nonprimate mammals), it is not known whether pheromones have any noticeable role in humans, but some studies indicate that pheromones may control some reproductive functions even in humans[68] (see page 448).

The basolateral group of nuclei of the amygdala is especially well developed in primates including humans. This may have to do with the lesser importance of olfaction in humans compared with that of many mammals, including the rat.

The amygdala is also important for decision making. Its nuclei have extensive connections with many regions of the CNS that are involved in basic functions such as the prefrontal cortex, but its nuclei also connect to systems that control basic body functions such as the endocrine and visceral systems.

The Amygdala and Fear Reactions The connections from sensory systems to the limbic system are the basis for evoking emotional responses to sensory stimuli. The lateral nucleus of the amygdala receives input from all sensory systems from thalamic sensory nuclei (Fig. 3.26B) and from sensory cortices (Fig. 3.26A), and this nucleus may be regarded as the gateway between the *dorsal* and *medial* division of the MGB (dMGB and mMGB), the *suprageniculate nucleus* (SG), and the *posterior intralaminar nucleus* (PIN).[103] These polysensory pathways seem to be more important for learned fear than the modality-specific projections to the amygdala from the neocortex, as ablation of cortical sensory areas does not prevent conditioning of fear responses to auditory stimuli.[102,103] Instead, acoustically mediated conditioned fear responses are controlled by the input to the lateral nucleus of the amygdala from the medial and dorsal thalamus (mMGB and dMGB).

In addition to learned fear reactions, the amygdala is also involved in many other behaviorally related functions. For example, the amygdala in human has been shown to be involved in memory, dreams, fear, sexuality, and anxiety and probably also the ability to make social contacts. The amygdala has been associated with developmental disorders such as autism.

Abnormal access to the amygdala through nonclassical auditory pathways may occur in diseases such as severe tinnitus, which has been associated with abnormal activation of nonclassical pathways[110,145] (see page 355). This activation may then open subcortical connections to limbic structures, which may be the reason why some individuals with severe tinnitus experience fear of sound (phonophobia).[145] Other studies have shown indications that the nonclassical auditory pathways are active in young children but rarely are in adults[150] (see page 356).

The function of the amygdala has been studied using many different methods, one of which is ablation of the amygdala in animals. The best known of such studies is that reported by Klüver and Bucy.[96] These investigators showed dramatic changes in the behavior of monkeys after ablation of the amygdala; an often-cited syndrome known as the Klüver–Bucy syndrome includes lack of social interaction and lack of fear and lowered aggression. However, some of these changes in behavior were probably caused by the ablation of extensive parts of the temporal lobe[64] by these investigators[97] when they ablated the amygdala. More recent studies in which the amygdala was ablated bilaterally show less pronounced effects on social behavior but such ablation did make the animals less aggressive (tame).

The ample connections from the amygdala to the autonomic nervous system can explain why, for instance, fear causes autonomic reactions such as palpation, sweating, and redirection of blood flow from viscera to muscle (fight or flight reaction). Fear also causes increased muscle tonus, and it may, paradoxically, cause increased visceral motility. That may be mediated by the connections to nuclei of the vagus nerve. Other (strong) emotional reactions to sensory input may cause fainting spells because of a paradoxical (parasympathetic) response from the cardiovascular system. Seeing blood, particularly one's own, is a strong stimulus for such reactions. Medical students' first visits to the operating room can cause fainting spells.

c. The Perception of Pleasure from Sensory Stimuli

The feeling of pleasure from sensory stimulation is also an emotional reaction, but little is known about its mechanisms. The neural mechanisms that are the basis for the perception of pleasure from sensory stimulation have similarities to those causing other forms of emotional reactions to sensory stimuli, and the limbic structures are undoubtedly involved in one way or another.

It seems that only very few kinds of sensory stimulation provide a sensation of pleasure the first time they are experienced. One kind of stimulation that stands out as being unconditionally pleasant at the first exposure is the taste of sweet (sucrose). In addition to sugar, some animals such as cattle seem to be attracted by the taste of salt after being on a diet of low salt. Some sexually related somatosensory stimulations may fall in the same category as may the gentle stroking of the skin of a child.

Some of the sensory input seems to require extensive processing to elicit the perception of pleasure, whereas other sensory input seems to elicit such feeling with little processing. Sexually related feelings of pleasure from somatosensory stimulation are only partly learned, but the sensory information seems to be subjected to a high degree of processing. The pleasure of eating in addition to serving to satisfy hunger is learned. Many forms of food elicit pleasure and induce a desire to eat more than necessary for satisfying hunger. Pleasure from most taste and smells (food, etc.) are learned and require neural processing, but with one clear exception: sugar (sweet). Visually induced pleasure such as that from art is learned, and the sensory information requires a high degree of processing. The appreciation of (visual) art, music, and many forms of food is thus based on extensive learning. Some authors, however, believe that enjoyment from music is not learned.[243]

The myriad of sensory stimulations that provide a sensation of pleasure to the adult person are thus learned pleasures. Addictions to hypnotics and narcotics of various kinds including alcohol and nicotine are learned. The pleasure from indulging alcohol and drugs that are abused for the purpose of evoking euphoria is related to their effect rather than their sensory stimulation. The reactions to these substances are therefore controlled by their effect or anticipated effect (reward effect) rather than the sensory perception.

The first response to such substances does not evoke a feeling of pleasure, and the feeling of pleasure is not evoked until after multiple exposures. However, not all individuals who are exposed to addictive substances become addicted even after repeated exposures.

Many sensory stimuli are perceived to be unpleasant without involving learning. Painful stimuli are probably the most obvious of such stimuli but bitter taste and also the taste of sour are probably tastes that evoke an unpleasant sensation that is not learned. The smell of many substances also has that property. Bromide is perhaps the clearest example of a scent that evokes an unpleasant perception by most individuals and indeed most mammals.

3. Species Differences

Considering species differences is important when research results on the function of the amygdala are evaluated. Most of the published studies of the low route to the amygdala have been made in the rat, and little is known about species differences. It is, however, known that such subcortical connections to the amygdala also exist in other sensory systems.[217] Some behavioral studies related to the amygdala were done in primates, and there are some studies that relate to humans. There are some known anatomical differences such as the basolateral nuclei being relatively larger in primates, including humans, than in rodents, for example.

VIII. PROCESSING OF INFORMATION IN THE SENSORY NERVOUS SYSTEM IS DYNAMIC

Connectivity in the CNS depends on two factors, namely the morphologic basis for connections (axon, synapses, and nerve cells) and the ability of synapses to activate the target cell. In sensory systems, the morphological basis for connections has been studied to a much greater extent than the functional aspects of connections between nerve cells. The pathways of sensory systems as we know them now are mainly determined in morphologic studies using tracer methods. Little is known about the normal efficacy of synaptic transmission, which is equally important as the morphology in understanding the function of sensory systems. Studies show that normal neural activity in the axons that connect cells in sensory systems can cause the cells to which they connect to fire, but they are few. Neural transmission in nerve cells may be blocked by ineffective synapses (see page 235). That may occur if the normal firing of such axons cannot produce a sufficiently large EPSP to raise the membrane potential of the target nerve cell so that it exceeds the threshold of firing. Such synapses are regarded as dormant.[114,208,224,246] This means that not all anatomically verified connections conduct nerve impulses and it is therefore important to verify which anatomical connections are functional. That can be done using physiological methods but only a few of the connections in sensory pathways that have been verified using anatomical methods have in fact been studied with respect to their functionality. The fact that the number of connections in the CNS seems to increase steadily as more anatomical studies are done may not have enormous implications, because probably only a few of these connections are functional under normal conditions.

Furthermore, morphologists seem to focus on specific parts of neural systems, ignoring others, which gives a skewed impression of which parts of the CNS are connected together. For example, the classical ascending pathways and the brainstem reticular formation have been studied to a greater extent than the nonclassical pathways. The descending pathways, despite their abundance, have attracted even less attention from morphologists, and physiologic studies on the nonclassical pathways and descending pathways are few. Many parts of the CNS have not been studied in detail anatomically so we can expect that the number of known connections in the CNS will increase with further study. Future physiological studies should also increase our knowledge about functional connections.

A. Neural Plasticity

The ability of the nervous systems to change its organization in response to external or internal factors is known as *neural plasticity*. The role of neural plasticity in the function of the sensory system has gained increased interest during recent years. There are two ways that neuroplasticity can occur. One way involves morphologic changes, such as eliminating existing connections or establishing new connections (sprouting). The other way is functional and involves altering the threshold of synapses and their efficacy. Neural plasticity can change functions according to changing demands or because of injuries. Neural plasticity can also cause symptoms of disease such as some forms of neuropathic pain, phantom limb syndrome, and hyperactive disorders such as muscle spasm and tinnitus, and paresthesia.[149] We will call the first kind of neural plasticity *good plasticity* and the other form *bad plasticity*.

It has emerged from studies of the past few decades that it is not only the developing brain that is plastic but also the adult brain can change its function. Studies have shown that it is also possible to induce substantial changes in the organization of the adult CNS. This means that connections between neurons and synaptic efficacy are not permanently established and are subject to change during life. Neural plasticity can be evoked by external and internal events such as deprivation of input, both normal and abnormal. Injury to neural tissue can also induce neural plasticity.

During childhood development (ontogeny), the CNS changes, apoptosis reduces the number of nerve cells, and pruning of axons and dendrites reduces the number of connections between nerve cells. During that period the nervous system is most sensitive to external and internal events that may invoke neural plasticity in the wiring of the nervous system.

One of the first experimental demonstrations of change in the organization of a sensory system as a result of sensory deprivation showed that the representation of the two eyes on the striate cortex changed in kittens when one eyelid was sutured just after birth.[231] Normally the two eyes are represented on the primary visual (striate) cortex in alternating bands from the two eyes. When one eye is deprived of input at early stage of life the cortical region of the other eye expands and that normally representing the blinded eye shrinks.[79,231]

Reorganization of the somatosensory cortex has been shown to occur after deprivation of sensory input (such as by amputation of a finger).[123,124] However, the amount of anatomical shift of cortical areas shown by these investigators[124] was small (2 to 4 mm). Subsequent studies showed that much more dramatic changes in neural organization of sensory cortical areas could occur.[89] For example, reorganization of the somatosensory cortex after limb amputation has been reported to result in light touch to the face producing a sensation not only in the face but also in the amputated arm.[175]

Other forms of change, in response to cortical cells due to neural plasticity, have been demonstrated by Robertson and coworkers[183] and other investigators[80] who have shown that the tuning of cells in the auditory cortex can be affected by sound deprivation. Synaptic activity in AI cells in congenital deaf animals is altered[46] and has shown that hearing loss from acoustic trauma alters the cortical map of tone frequency and the spontaneous activity of neurons.

Kilgard and Merzenich [94,95] have shown that novel auditory stimulation can cause considerable changes in the response area (tuning) of neurons in the auditory cortex. These investigators studied the influence of nucleus basalis in facilitating the expression of neural plasticity. Specifically, they showed that both cortical receptive field sizes and maximum temporal following rates changed when specific sound stimulation was paired with electrical stimulation of the nucleus basalis.[*,94] Most of these studies concerned primary cortical receiving areas but most likely reorganization of secondary cortices and association cortices was also involved.

The function of auditory midbrain structures of the ascending auditory pathways also changes after sound deprivation,[62,174] particularly in developing animals.[73] Expression of neural plasticity in sensory systems typically leads to hyperactivity and change of temporal summation in the nuclei of the ascending auditory pathways.[62,63]

In a more peripheral structure of a sensory system, Wall[224] showed that stimulation of the skin could not fire dorsal horn cells in spinal segments that were distant from the entry of the dorsal root innervating the skin that was stimulated. Wall coined the term *dormant synapses* to describe synapses that exist morphologically but are not functional. He showed that such synapses could be brought to function (*be unmasked*) by depriving input. This was one of the first demonstrations of how synaptic efficacy can change function.

> Dorsal root fibers connect (morphologically) to dorsal horn cells at several levels up and down the spinal cord from the segment where they enter the spinal cord (Fig. 3.28), but only the synapses that are closest to the segment where the dorsal root enters the spinal cord conduct nerve impulses under normal circumstances. Under abnormal circumstances, however, synapses in segments that are farther away form the root's entry can also be made to conduct (unmasking of dormant synapses)[224] and stimulation of the skin can activate dorsal horn cells that are located several segments from where the activated dorsal roots enter the spinal cord. The abnormal situation induced in these experiments consisted of severing dorsal roots that entered the spinal cord at the segment where recordings were made and several segments from those that were stimulated. These connections opened with very short delay,[224] which means that they could not have been the result of new anatomical connections (sprouting) but

*The nucleus basalis of the forebrain is known to promote neural plasticity.[104,229,230]

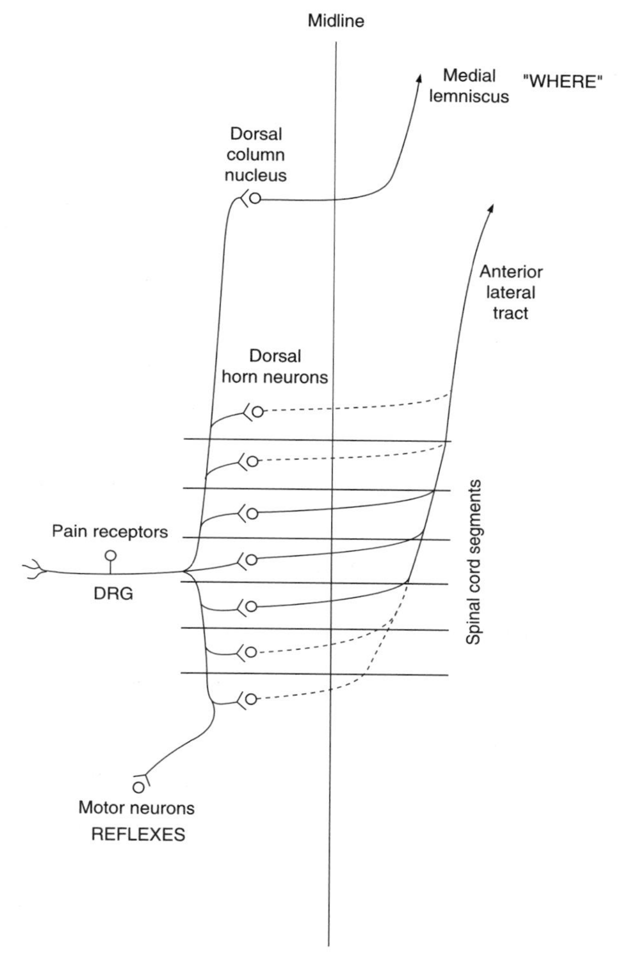

FIGURE 3.28 Illustration of connections that are important for pain sensation. Dorsal root fibers ascend and descend in the spinal cord while giving off collaterals to dorsal horn cells of several segments. The ascending limb reaches the dorsal column nucleus (DCN) at the dorsal column system. Cells of the DCN cross the midline at the medial lemniscus (ML). The descending branch can even reach the motoneurons in the ventral horn. Axons of dorsal horn cells cross the midline at the segmental level and join in the ascending anterior lateral tract.

must have been caused by changes in the threshold (or efficacy) of the synapses on dorsal horn cells.

Dormant synapses that become unmasked by abnormal events (such as severing of a dorsal root)[224] are examples of how processing of sensory information can change as a result of neural plasticity. More recently the matter of silent synapses has been studied in detail by Zhou and co-workers,[246] who especially studied the role of the neural transmitter glutamate. It is possible that unmasking of dormant synapses plays a role in many forms of neural plasticity, particularly such changes that occur with little delay after inducing the stimulation (or deprivation of input) that induces the plasticity.

The change in the receptive fields of neurons in the ascending pathways that occurs as a result of neural plasticity may be explained by changes in synaptic threshold and synaptic efficacy, similar to that observed by Wall.[224] While the basis for the receptive fields of neurons in the ascending pathways, including the primary sensory cortices, is established in the periphery by the organization of the receptors, the size and shape of the receptive fields of neurons in the ascending pathways depend on convergence of cells in the ascending pathways and the interplay between excitation and inhibition. Because the number of active inputs depends on the synaptic efficacy, changes in synaptic efficacy can alter the receptive fields of neurons without morphological changes such as elimination or sprouting of axons or dendrites. Changes in synaptic threshold and synaptic efficacy can cause receptive fields to contract, expand, or shift in location. Changes in the interplay between inhibition and excitation can also change receptive fields. Another explanation for the observed changes in the response of cells in the ascending sensory pathways would be that the receptive field of sensory receptors changed. For example, changes in the tuning of cochlear nerve fibers could occur as a result of efferent input to outer hair cells. It seems more likely, however, that the observed changes in the receptive fields of neurons in sensory pathways as a result of neural plasticity are caused by changes in synaptic efficacy.

When dealing with neural plasticity and rewiring of the CNS by induced neural activity, Hebb's principle[75] is often mentioned. Hebb's principle proposes that neurons that are active together may also become morphologically attached to each other. The observation that neural connection can be established by specific stimulation supports this hypothesis. Hebb's hypothesis has been expanded upon in recent years by the discovery of nerve growth factors necessary for the growth and maintenance of synapses.[14] These growth factors reside within vesicles at the synapse, and release into the synapse affects both the presynaptic and postsynaptic cell every time the cell fires, making the connection at an active synapse stronger than that of an inactive (dormant) synapse.[246] This concept has recently been expanded to include the effect of timing of pre- and postsynaptic activity.[114,208]

A slightly different form of change in the function of the CNS, known as *kindling*,* was first described by Goddard,[65] who showed that repeated

*The term *kindling* was introduced by Goddard[65] to describe the epileptic seizures in rats that resulted from repeated electrical stimulation of the amygdala nuclei. It has later been shown that many other nuclei can be kindled to generate repetitive activity.[223] Typically, novel stimulation such as electric stimulation done for a short period (10 to 20 minutes) and repeated every day causes the development of self-sustained activity after 3 to 4 weeks of such stimulation. Kindling has been associated with several disorders including muscle spasm.

electrical stimulation of the amygdala in rats could lead to epileptic seizures. Subsequent studies have shown that other nuclei can be kindled, including the cerebral cortex,[46] with the result of hyperactivity. Even motonuclei such as the facial motonucleus can be kindled (by antidromic stimulation of a motor nerve).[144]

> Kindling is a form of induction of neural plasticity that is believed to occur because the repetitive stimulation increases the synaptic strength of the excitatory synapses beyond the conductive capabilities of the inhibitory synapses, thereby overwhelming the control mechanism for the system. Kindling is an extreme case of neural plasticity that may affect all classical and nonclassical sensory systems.

Neural plasticity in one form or another thus seems to be associated with most senses; however, the olfactory nervous system seems to be hardwired (genetically programmed) to a greater extent than what is the case for the other four senses.[248]

Neural plasticity and unmasking of dormant synapses (and masking of normally functioning synapses) may open connections to parts of the brain that are normally not activated by a specific sensory stimulation or may close connections that are normally open. The abundance of morphologically verified connections would indicate that most parts of the brain may be connected to most other parts. However, it seems equally clear that many of these connections do not normally conduct neural activity because of ineffective synapses, but the existence of morphological connections opens the possibility of functional connections when specific synapses become unmasked. It is possible that adjustment of synaptic efficacy is a process equally important as the much better known apoptosis and severance of (morphological) connections that occur during ontogenetic development.

For example, evidence has been presented that information that enters through one sense is processed in neurons that are dedicated to that sensory modality,[197] and neurons in parts of the CNS that normally respond to stimulation of other sensory modalities may be activated. This means that auditory stimulation could under certain circumstances activate neurons that belong to the visual system. The absence of function of one sensory modality from birth seems to make neural circuitry that normally processes one modality available for other modalities.[166,207] For instance, the auditory cortex can be rewired* to receive visual input when an animal is deprived of auditory input early in life.[166] One consequence of such early deprivation of auditory input is that visual input activates neural structures that normally are activated by sound.

*When the term rewiring is used here, not only does it include establishment of new connections by outgrowth of dendrites and axons (sprouting) but the term is also used to describe establishment of new functional connections by change (increase) in synaptic efficacy without morphological changes.

However, despite the fact that visual stimuli activate neurons that normally (only) respond to auditory input, the visual stimulation is interpreted as being visual when such rewiring has occurred.[117] This process of redirecting visual information to auditory areas changes the function of these rewired neurons and similar signs of switching of function of cortical neurons have also been observed in studies in humans where it was shown that visual evoked potentials are larger in individuals born deaf.[160,161]

Other investigators, however, have found little evidence of such switching on the responses of cells in the AI. This means that the question of switching of input from one group of neurons to neurons of another sense that have been deprived of normal input is probably more complex than originally assumed.

Interaction between sensory modalities seems to occur normally in children where the loudness of a loud sound is affected by somatosensory stimulation (electrical stimulation of the median nerve). This effect decreases with age and is very small above the age of 18 years in a population of normal individuals.[150] It is also possible that the specialization regarding sensory modality which is included in Müller's hypothesis of the "specific nerve energies" in fact develops during ontogeny rather than being inherited.

The interaction between the somatosensory system and the auditory system has also been shown to occur in some patients with severe tinnitus.[145] This means that the neural circuitry for such interaction must be commonly present automatically but only activated under specific (abnormal) circumstances such as disease processes.

B. Plastic Changes in the Nervous System May Cause Symptoms and Signs of Disease

Neural plasticity may be regarded as a reorganization that is beneficial after injuries because it makes it possible to shift functions to undamaged parts of the CNS. However, neural plasticity can also cause symptoms and signs of disease (bad plasticity) such as pain, hyperactive disorders such as muscle spasm, severe tinnitus, and paresthesia. The changes in the nervous system that cause such effects do not appear to be of any advantage to the organism. *Phantom pain* and other abnormal sensations that often occur after limb amputations[84,120] are probably the strongest indication that symptoms and signs can be caused by reorganization of the CNS induced by external or internal events, causing functional reorganization of the specific parts of the nervous system.

Disorders that result from unmasking dormant synapses are often hyperactive by nature. Hyperactivity may be induced by reduction in normally occurring central inhibition that may occur when inhibitory input from the periphery is reduced. Examples of that have been demonstrated in research on pain where it is believed that reduction of input to the somatosensory system from large fibers (A-fibers) can cause normally innocuous stimulation of the skin to become painful, a condition known as *allodynia*,*,[38,213] *hyperalgesia*,** and *hyperpathia*.*** These symptoms often accompany chronic pain conditions.[31,32,38,172] Some forms of neuropathic pain that result from compression of peripheral nerves and spinal nerve roots[28,38] are caused by plastic changes in the nervous system.

Establishment of abnormal functional connections from low-threshold mechanoreceptors to pain circuits in the CNS is another example of morphological connections that normally are not potent but may become conducting under abnormal circumstances causing symptoms.

These abnormal sensations are assumed to be caused by new functional connections that are established by unmasking of dormant synapses that connect existing axons to target cells where these synapses have normally been closed (ineffective). Some studies indicate that some of such new connections are caused by new axons (sprouting),[98,238,239] but it seems likely that at least part of the reorganization that cause such symptoms is the result of unmasking of dormant synapses.[31,237,238] A measurable sign of reorganization of the CNS in patients with chronic neuropathic pain is an increased sensitivity to pain elicited by electrical stimulation and a change in temporal integration of pain stimuli.[146]

It is also possible that the symptoms and signs of some developmental disorders are caused by an overemphasis on nonclassical sensory pathways. There are signs that the nonclassical pathways are not normally intact at adult life but that sensory input to the nonclassical sensory systems is diminished during (normal) ontogenetic development. The nonclassical pathways may become active during adult life through reestablishment of connections that were interrupted during ontogeny. It is also possible that nonclassical pathways may become active because of incomplete pruning and apoptosis during ontogeny. Abnormal activation of the nonclassical auditory pathways may be the cause of *hyperacusis***** and *phonophobia*,***** which have been observed in connection with conditions such as severe tinnitus.[147] Insufficient pruning and apoptosis during ontogeny may cause some of the symptoms seen in

**Allodynia* is a painful sensation from stimulation (of the skin) that is normally innocuous.

***Hyperalgesia* is an extreme sensitivity to painful stimuli.

****Hyperpathia* is an exaggerated reaction to painful stimuli, often with a continuing sensation of pain after that the painful stimulation has ended.

****Hyperacusis is sound that appears stronger than normal.

*****Phonophobia is fear of sound.

developmental disorders such as autism. All these symptoms and signs may be caused by incorrect synaptic threshold and efficacy.

While novel stimulation can evoke neural plasticity that causes hyperactivity and hypersensitivity, appropriately designed stimulation can also have the opposite effect and even reverse hyperactivity and hypersensitivity caused by neural plasticity. Many of the adverse conditions that are caused by neural plasticity can be treated successfully by novel sensory stimulation, which means methods that are different from conventional surgical or pharmacological treatments. Examples are treatment of pain by *transdermal electrical nerve stimulation* (TENS)[232] and sound exposure in treatment of patients with tinnitus (tinnitus retraining therapy, TRT).[83,85] Successful treatment requires a thorough knowledge about the normal function of sensory systems as well as an understanding of the pathologies that can affect the function of these systems.

REFERENCES

1. Abeles, M., *Corticonics*, Cambridge, U.K.: Cambridge University Press, 1991.
2. Adolphs, R.D., Tranel, H., Damasio, H., and Damasio, A.R., Fear and the Human Amygdala. *J. Neurosci.*, 15: 5879–5891, 1995.
3. Aharon, I., Etcoff, N., Ariely, D., Chabris, C.F., O'Connor, E., and Breiter, H.C., Beautiful Faces Have Variable Reward Value; fMRI and Behavioral Evidence. *Neuron,* 32: 537–551, 2001.
4. Ahissar, E. and Zacksenhouse, M., Temporal and Spatial Coding in the Rat Vibrissal System. *Progr. Brain Res.,* 130: 75–87, 2001.
5. Aitkin, L.M., Dickhaus, H., Schult, W., and Zimmermann, M., External Nucleus of Inferior Colliculus: Auditory and Spinal Somatosensory Afferents and Their Interactions. *J. Neurophysiol.,* 41: 837–847, 1978.
6. Aitkin, L.M., Kenyon, C.E., and Philpott, P., The Representation of Auditory and Somatosensory Systems in the External Nucleus of the Cat Inferior Colliculus. *J. Comp. Neurol.,* 196: 25–40, 1981.
7. Aitkin, L.M., *The Auditory Midbrain, Structure and Function in the Central Auditory Pathway*. Clifton, NJ: Humana Press, 1986.
8. Amassian, V.E., Cracco, R.Q., Maccabee, P.J., Cracco, J.B., Rudell, A., and Eberle, L., Suppression of Visual Perception by Magnetic Coil Stimulation of Human Occipital Cortex. *Electroencephalogr. Clin. Neurophysiol.,* 74: 458–462, 1989.
9. Andersen, P., Knight, P.L., and Merzenich, M.M., The Thalamocortical and Corticothalamic Connections of AI, AII, and the Anterior Field (AAF) in the Cat: Evidence for Two Largely Segregated Systems of Connections. *J. Comp. Neurol.,* 194: 663–701, 1980.
10. Andersen, P., Snyder, R.L., and Merzenich, M.M., The Topographic Organization of Corticocollicular Projections from Physiologically Identified Loci in the AI, AII, and Anterior Auditory Cortical Fields of the Cat. *J. Comp. Neurol.,* 191: 479–494, 1980.
11. Beckers, G. and Zeki, S.T., The Consequences of Inactivating Areas V1 and V5 on Visual Motion Perception. *Brain,* 118: 49–60, 1995.
12. Békésy, G. von, *Sensory Inhibition*. Princeton: Princeton University Press, 1967.
13. Berson, D.M., Dunn, F.A., and Takao, M., Phototransduction by Retinal Ganglion Cells That Set the Circadian Clock. *Science*, 295: 1070–1073, 2002.

14. Bi, G. and Poo, M., Synaptic Modifications in Cultured Hippocampal Neurons: Dependence on Spike Timing, Synaptic Strength, and Postsynaptic Cell Type. *J. Neurosci.*, 18: 10464–10472, 1998.
15. Blauert, J., Binaural Localization. *Scand. Audiol. (Stockholm) Suppl.*, 57: 7–26, 1982.
16. Blum, P.S., Abraham, L.D., and Gilman, S., Vestibular, Auditory, and Somatic Input to the Posterior Thalamus of the Cat. *Exp. Brain Res.*, 34: 1–9, 1979.
17. Boivie, J. and Meyerson, B.A., A Correlative Anatomical and Clinical Study of Pain Suppression by Deep Brain Stimulation. *Pain*, 13: 113–126, 1982.
18. Breier, A., Cognitive Deficit in Schizophrenia and Its Neurochemical Basis. *Br. J. Psychol. Suppl.*, 37: 16–18, 1999.
19. Brodal, P., *The Central Nervous System*. New York: Oxford Press, 1998.
20. Bromm, B. and Desmedt, J.E., *Pain and the Brain: From Nociception to Cognition*. New York: Raven Press, 1995.
21. Bromm, B., Brain Images of Pain. *News Physiol. Sci.*, 16: 244-249, 2001.
22. Brown, R., An Experimental Study of the Relative Importance of Acoustic Parameters for Auditory Speaker Recognition. *Language Speech*, 24: 295–310, 1981.
23. Brownell, W.E., Observation on the Motile Response in Isolated Hair Cells. In: *Mechanisms of Hearing*, edited by W.R. Webster and L. M. Aitken. Melbourne: Monash University Press, 1983, pp. 5–10.
24. Burkitt, A.N. and Clark, G.M., Analysis of Integrate-and-Fire Neurons: Synchronization of Synaptic Input and Spike Output. *Neural Computation*, 11: 871–901, 1999.
25. Cacace, A.T., Lovely, T.J., McFarland, D.J., Parnes, S.M., and Winter, D.F., Anomalous Cross-Modal Plasticity Following Posterior Fossa Surgery: Some Speculations on Gaze-Evoked Tinnitus. *Hear. Res.*, 81: 22–32, 1994.
26. Cacace, A.T., Cousins, J.P., Parnes, S.M., McFarland, D.J., Semenoff, D., Holmes, T., Davenport, C., Stegbauer, K., and Lovely, T.J., Cutaneous-Evoked Tinnitus. II. Review of Neuroanatomical, Physiological and Functional Imaging Studies. *Audiol. Neuro-Otol.*, 4: 258–268, 1999.
27. Cacace, A.T., Cousins, J.P., Parnes, S.M., Semenoff, D., Holmes, T., McFarland, D.J., Davenport, C., Stegbauer, K., and Lovely, T.J., Cutaneous-Evoked Tinnitus. I. Phenomenology, Psychophysics and Functional Imaging. *Audiol. Neuro-Otol.*, 4: 247–257, 1999.
28. Carstens, E., Altered Spinal Processing in Animal Models of Radicular and Neuropathic Pain. In: *Nervous System Plasticity and Chronic Pain*, edited by J. Sandkühler, B. Bromm, and G.F. Gebhart. Amsterdam: Elsevier, 2000.
29. Casseday, J.H., Ehrlich, D., and Covey, E., Neural Tuning for Sound Duration: Role of Inhibitory Mechanisms in the Inferior Colliculus. *Science*, 264: 847–850, 1994.
30. Clarke, S.F., Ribaupierre, F. de, Bajo, V.M., Rouiller, E.M., and Kraftsik, R., The Auditory Pathway in Cat Corpus Callosum. *Exp. Brain Res.*, 104: 534–540, 1995.
31. Coderre, T.J., Katz, J., Vaccarino, A.L., and Melzack, R., Contribution of Central Neuroplasticity to Pathological Pain: Review of Clinical and Experimental Evidence. *Pain*, 52: 259–285, 1993.
32. Coggeshall, R.E., Dougherty, P.M., Pover, C.M., and Carlton, S.M., Is Large Myelinated Fiber Loss Associated with Hyperalgesia in a Model of Experimental Peripheral Neuropathy in the Rat? *Pain*, 52: 233–242, 1993.
33. Cousins, M. and Power, I., Acute and Postoperative Pain. In: *Textbook of Pain, 4th ed.*, edited by P.D. Wall and R. Melzack. Edinburgh: Churchill-Livingstone, 1999, pp. 447–491.
34. Cowey, A., Stoerig, P., and Bannister, M., Retinal Ganglion Cells Labeled from the Pulvinar Nucleus in Macaque Monkeys. *Neuroscience*, 61: 691–705, 1994.
35. Craig, A.D. and Dostrovsky, J.O., Medulla to Thalamus. In: *Textbook of Pain*, 4th ed., edited by P.D. Wall and R. Melzack. Edinburgh: Churchill-Livingstone, 1999, pp. 183–214.

36. d'Alfonso, A.A.L., Honk, J. van, Schutter, D.J.L.G., Caffé, A.R., Postama, A., and Haan, E.H.F. de, Spatial and Temporal Characteristics of Visual Motion Perception Involving V5 Visual Cortex. *Neurol Res.*, 24: 266–270, 2002.
37. Damasio, A.R., Damasio, H., and Van Hoesen, G.W., Prosopagnosia: Anatomic Basis and Behavioral Mechanisms. *Neurology*, 32: 331–341, 1982.
38. Devor, M., Central Changes Mediating Neuropathic Pain. In: *Proceedings of the Fifth World Congress on Pain*, edited by R. Dubner, G. Gebhart, and M. Bond. Amsterdam: Elsevier, 1988, pp. 114–128.
39. Devor, M. and Seltzer, Z., Pathophysiology of Damaged Nerves in Relation to Chronic Pain. In: *Textbook of Pain*, 4th ed., edited by P.D. Wall and R. Melzack. Edinburgh: Churchill-Livingstone, 1999, pp. 129–164.
40. Doetsch, G.S., Patterns in the Brain. Neuronal Population Coding in the Somatosensory System. *Physiol. Behav.*, 69: 187–201, 2000.
41. Dommelen, W.A. van, Acoustic Parameters in Human Speaker Recognition. *Language Speech* 33: 259–272, 1990.
42. Doron, N.N. and LeDoux, J.E., Cells in the Posterior Thalamus Project to Both Amygdala and Temporal Cortex: A Quantitative Retrograde Double-Labeling Study in the Rat. *J. Comp. Neurol.*, 425: 257–274, 2000.
43. Downer, J.L.C., Changes in Visual Agnostic Functions and Emotional Behaviors Following Unilateral Temporal Pole Damage in the "Split-Brain" Monkey. *Nature*, 191: 50–51, 1961.
44. Eggermont, J.J., *The Correlative Brain. Theory and Experiment in Neural Interaction*. Berlin: Springer-Verlag, 1990.
45. Eggermont, J.J., On the Pathophysiology of Tinnitus: A Review and a Peripheral Model. *Hear. Res.*, 48: 111–124, 1990.
46. Eggermont, J.J. and Komiya, H., Moderate Noise Trauma in Juvenile Cats Results in Profound Cortical Topographic Map Changes in Adulthood. *Hear. Res.*, 142: 89–101, 2000.
47. Eggermont, J.J., Between Sound and Perception: Reviewing the Search for a Neural Code. *Hear. Res.*, 157: 1–42, 2001.
48. Ehret, G. and Romand, R., *The Central Auditory Pathway*. New York: Oxford University Press, 1997.
49. Enroth-Cugell, C. and Robson, J.G., The Contrast Sensitivity of Retinal Ganglion Cells of the Cat. *J. Physiol. (London)*, 187: 517–552, 1966.
50. Everett, N.B., *Functional Neuroanatomy*. Philadelphia: Lea & Feibiger, 1971.
51. Fay, R.R., Auditory Stream Segregation in Goldfish (Carassius auratus). *Hear. Res.*, 120: 69–76, 1998.
52. Ferry, B., Roozendaal, B., and McGaugh, J.L., Role of Norepinephrine in Mediating Stress Hormone Regulation of Long Term Memory Storage: A Critical Involvement of the Amygdala. *Biol. Psychiatry*, 46: 1142–1152, 1999.
53. Ferry, B. and McGaugh, J.L., Role of Amygdala Norepinephrine in Mediating Stress Hormone Regulation of Memory Storage. *Acta Pharmacol. Sinica*, 21: 481–493, 2000.
54. Fetz, E.E., Temporal Coding in Neural Populations?, *Science*, 278: 1901–1902, 1997.
55. Fields, H.L. and Basbaum, A.I., Central Nervous System Mechanism of Pain Modulation. In: *Textbook of Pain*, 4th ed., edited by P.D. Wall and R. Melzack. Edinburgh: Churchill-Livingstone, 1999, pp. 309–329.
56. Fishman, Y.I., Reser, D.H., Arezzo, J.C., and Steinschneider, M., Neural Correlates of Auditory Stream Segregation in Primary Auditory Cortex of the Awake Monkey. *Hear. Res.*, 151: 167–187, 2001.
57. French, J.D., Amerongen, F.K. von, and Magoun, H.W., An Activating System in Brain Stem of Monkey. *AMA Arch. Neurol. Psychol.*, 68: 577–590, 1952.

58. Frisina, R.D., Smith, R.L., and Chamberlain, S.C., Differential Encoding of Rapid Changes in Sound Amplitude by Second Order Auditory Neurons. *Exp Brain Res.*, 60: 417–422, 1985.
59. Frisina, R.D., Smith, R.L., and Chamberlain, S.C., Encoding of Amplitude Modulation in the Gerbil Cochlear Nucleus. I. A Hierarchy of Enhancement. *Hear. Res.*, 44: 99–122, 1990.
60. Frisina, R.D., Smith, R.L., and Chamberlain, S.C., Encoding of Amplitude Modulation in the Gerbil Cochlear Nucleus. II. Possible Neural Mechanisms. *Hear. Res.*, 44: 123–142, 1990.
61. Frisina, R.D., Walton, J.P., and Karcich, K.J., Dorsal Cochlear Nucleus Single Neurons Can Enhance Temporal Processing Capabilities in Background Noise. *Exp. Brain Res.*, 102: 160–164, 1994.
62. Gerken, G.M., Saunders, S.S., and Paul, R.E., Hypersensitivity to Electrical Stimulation of Auditory Nuclei Follows Hearing Loss in Cats. *Hear. Res.*, 13: 249–260, 1984.
63. Gerken, G.M., Solecki, J.M., and Boettcher, F.A., Temporal Integration of Electrical Stimulation of Auditory Nuclei in Normal Hearing and Hearing-Impaired Cat. *Hear. Res.*, 53: 101–112, 1991.
64. Gloor P., *The Temporal Lobe and Limbic System*. New York: Oxford Press, 1997.
65. Goddard, G.V., Amygdaloid Stimulation and Learning in the Rat. *J. Comp. Physiol. Psychol.*, 58: 23–30, 1964.
66. Goldman-Rakic, P.S., Modular Organization of Prefrontal Cortex. *Trends Neurosci.*, 7: 419–429, 1984.
67. Goodale, M.A. and Milner, A.D., Separate Pathways for Perception and Action. *Trends Neurosci.*, 15: 20–25, 1992.
68. Graham, C.A. and McGrew, W.C., Menstrual Synchrony in Female Undergraduates Living on a Coeducational Campus. *Psychoneuroendocrinology*, 5: 245–252, 1980.
69. Graybiel, A.M., Some Fiber Pathways Related to the Posterior Thalamic Region in the Cat. *Brain Behav. Evol.*, 6: 363–393, 1972.
70. Guinan, Jr., J.J. and Gifford, M.L., Effects of Electrical Stimulation of Efferent Olivocochlear Neurons on Cat Auditory-Nerve Fibers: II. Spontaneous Rate. *Hear. Res.*, 33: 115–128, 1988.
71. Halpern, M., The Organization and Function of the Vomeronasal System. *Ann. Rev. Neurosci.*, 10: 325–362, 1987.
72. Harrington, I.A. and Heffner, H.E., A Behavioral Investigation of "Separate Processing Streams" within Macaque Auditory Cortex. *Assoc. Res. Otolaryngol. Abstr.*, 25: 120, 2002.
73. Harrison, J.M., Ibrahim, D., and Mount, R.J., Plasticity of Tonotopic Maps in Auditory Midbrain Following Partial Cochlear Damage in the Developing Chinchilla. *Exp. Brain Res.*, 123: 449–460, 1998.
74. Hartline, H.K., Response of Single Optic Nerve Fibers of the Vertebrate Eye to Illumination of the Retina. *Am. J. Physiol.*, 121: 400–415, 1938.
75. Hebb, D.O., *The Organization of Behavior*. New York: Wiley, 1949.
76. Hellige, J., *Hemispheric Asymmetry: What's Right and What's Left*. Cambridge, MA: Harvard University Press, 1993.
77. Hotta, T. and Kameda, K., Interactions Between Somatic and Visual or Auditory Responses in the Thalamus of the Cat. *Exp. Neurol.*, 8: 1–13, 1963.
78. Hubel, D.H. and Wiesel, T.N., Receptive Fields, Binocular Interaction and Functional Architecture in the Cat's Visual Cortex. *J. Physiol. (London)*, 160: 106–154, 1962.
79. Hubel, D.H. and Wiesel, T.N., The Period of Susceptibility to Physiological Effects of Unilateral Eye Closure in Kittens. *J. Physiol. (London)*, 206: 419–436, 1970.
80. Irvine, D.R. and Rajan, R., Injury-Induced Reorganization of Frequency Maps in Adult Auditory Cortex: The Role of Unmasking of Normally Inhibited Inputs. *Acta Otolaryng. (Stockholm)*, 532: 39–45, 1997.
81. Itoh, K., Kamiya, H., Mitani, A., Yasui, Y., Takada, M., and Mizuno, N., Direct Projections from Dorsal Column Nuclei and the Spinal Trigeminal Nuclei to the Cochlear Nuclei in the Cat. *Brain Res.*, 400: 145–150, 1987.

82. Jankowska, E., Lund, S., and Lundberg, A.P.O., Inhibitory Effects Evoked through Ventral Reticulospinal Pathways. *Arch. Ital. Biol.*, 106: 124–140, 1968.
83. Jastreboff, P.J., Gray, W.C., and Gold, S.L., Neurophysiological Approach to Tinnitus Patients. *Am. J. Otol.*, 17: 236–240, 1966.
84. Jastreboff, P.J., Phantom Auditory Perception (Tinnitus): Mechanisms of Generation and Perception. *Neurosci. Res.*, 8: 221–254, 1990.
85. Jastreboff, P.J., Tinnitus as a Phantom Perception: Theories and Clinical Implications. In: *Mechanisms of Tinnitus*, edited by J.A. Vernon and A.R. Møller, Boston: Allyn & Bacon, 1995.
86. Johansson, R.S., Tactile Afferent Units with Small and Well Demarcated Receptive Fields in Glabrous Skin Area of the Human Hand. In: *Sensory Functions of the Skin of Humans*, edited by D. R. Kenshalo. New York: Plenum, 1979, pp. 129–145.
87. Johnson, L.R., Westrum, L.E., and Henry, M.A., Anatomic Organization of the Trigeminal System and the Effects of Deafferentation. In: *Trigeminal Neuralgia*, edited by G.H. Fromm and B.J. Sessle. Boston: Butterworth-Heinemann, 1991, pp. 27–69.
88. Kaas, J.H., and Garraghty, P.E., Hierarchical, Parallel, and Serial Arrangements of Sensory Cortical Areas: Connection Patterns and Functional Aspects. *Neurobiology,* 1: 248–251, 1991.
89. Kaas, J.H., Florence, S.L., and Jain, N., Subcortical Contributions of Massive Cortical Reorganization. *Neuron,* 22: 657–660, 1999.
90. Kaas, J.H. and Hackett, T.A., Subdivisions of Auditory Cortex and Processing Streams in Primates. *Proc. Nat. Acad. Sci. USA,* 97: 11793–11799, 2000.
91. Kandel, E.R., Schwartz, J.H., and Jessel, T.M., *Principles of Neural Sciences*. New York: McGraw-Hill, 2000.
92. Kiang, N.Y.S., Watanabe, T., Thomas, E.C., and Clark, L., *Discharge Patterns of Single Fibers in the Cat's Auditory Nerve*. Cambridge, MA: MIT Press, 1965.
93. Kiang, N.Y.S., Stimulus Representation in the Discharge Patterns of Auditory Neurons. In: *The Nervous System*, edited by E.L. Eagles. New York: Raven Press, 1975, pp. 81–96.
94. Kilgard, M.P. and Merzenich, M.M., Plasticity of Temporal Information Processing in the Primary Auditory Cortex. *Nature Neurosci.,* 1: 727–731, 1998.
95. Kilgard, M.P. and Merzenich, M.M., Cortical Map Reorganization Enabled by Nucleus Basalis Activity. *Science,* 279: 1714–1718, 1998.
96. Klüver, H. and Bucy, P.C., "Psychic Blindness" and Other Symptoms Following Bilateral Temporal Lobectomy in Rhesus Monkeys. *Am. J. Physiol.*, 119: 352–353, 1937.
97. Klüver, H. and Bucy, P.C., Preliminary Analysis of Functions of the Temporal Lobes in Monkeys. *Arch. Neurol. Psychiatry,* 42: 979–1000, 1939.
98. Kohama, I., Ishikawa, K., and Kocsis, J.D., Synaptic Reorganization in the Substantia Gelatinosa after Peripheral Nerve Neuroma Formation: Aberrant Innervation of Lamina II Neurons by Beta Afferents. *J. Neurosci.*, 20: 1538–1549, 2000.
99. Kreiman, J. and Van Lancker, D., Hemispheric Specialization for Voice Recognition: Evidence from Dichotic Listening. *Brain Language,* 34: 246–252, 1988.
100. Kvasnak, E., Suta, D., Popelar, J., and J.S., Neuronal Connections in the Medial Geniculate Body of the Guinea-Pig. *Exp. Brain Res.*, 132: 87–102, 2000.
101. Lapicque, L., Recherches Quantitatives Sur L'excitation Electrique Des Nerfs Traitee Comme Une Polarization. *J. Physiol. (Paris),* 9: 620–635, 1907.
102. LeDoux, J.E., Sakaguchi, A., and Reis, D.J., Subcortical Efferent Projections of the Medial Geniculate Mediate Emotional Responses Conditioned by Acoustic Stimuli. *J. Neurosci.,* 4: 683–698, 1984.
103. LeDoux, J.E., Cicchetti, P., Xagoraris, A., and Romanski, L.M., The Lateral Amygdaloid Nucleus: Sensory Interface of the Amygdala in Fear Conditioning. *J. Neurosci.*, 10: 1062–1069, 1990.
104. LeDoux, J.E., Brain Mechanisms of Emotion and Emotional Learning. *Curr. Opin. Neurobiol.* 2: 191–197, 1992.

105. LeDoux, J.E., *The Emotional Brain*. New York: Touchstone, 1996.
106. LeDoux, J.E., *Synaptic Self*. New York: Viking, 2002.
107. Lenz, F.A., Dostrovsky, J.O., Tasker, R.R., Yamashito, K., Kwan, H.C., and Murphy, J.T., Single-Unit Analysis of the Human Ventral Thalamic Nuclear Group: Somatosensory Responses. *J. Neurophysiol.,* 59: 299–316, 1988.
108. Lenz, F.A. and Dougherty, P.M., Pain Processing in the Ventrocaudal Nucleus of the Human Thalamus. In: *Pain and the Brain*, edited by B. Bromm and J.E. Desmedt. New York: Raven Press, 1995, pp. 175–185.
109. Lenz, F.A., Lee, J.I., Garonzik, I.M., Rowland, L.H., Dougherty, P.M., and Hua, S.E., Plasticity of Pain-Related Neuronal Activity in the Human Thalamus. *Progr. Brain Res.,* 129: 253–273, 2000.
110. Lockwood, A., Salvi, R., Coad, M., Towsley, M., Wack, D., and Murphy, B., The Functional Neuroanatomy of Tinnitus. Evidence for Limbic System Links and Neural Plasticity. *Neurology,* 50: 114–120, 1998.
111. Lomber, S.G., Payne, B.R., and Cornwell, P., Role of Superior Colliculus in Analyses of Space: Superficial and Intermediate Layer Contributions to Visual Orienting, Auditory Orienting, and Visuospatial Discriminations During Unilateral and Bilateral Deactivations. *J. Comp. Neurol.,* 441: 44–57, 2001.
112. Lomber, S.G., Learning to See the Trees Before the Forest: Reversible Deactivation of the Superior Colliculus During Learning of Local and Global Features. *Proc. Nat. Acad. Sci.,* 99: 4049–4054, 2002.
113. Malmierca, M.S., Oliver, D.L., Henkel, C.K., and Merchan, M.A., A Novel Projection from Dorsal Cochlear Nucleus to the Medial Division of the Medial Geniculate Body of the Rat. *Assoc. Res. Otolaryngol. Abstr.,* 25: 176, 2002.
114. Markram, H., Lubke, J., Frotscher, M., and Sakmann, B., Regulation of Synaptic Efficacy by Coincidence of Postsynaptic APs and EPSPs. *Science,* 275: 213–215, 1997.
115. Marsh, R.A., Grose, C.D., Wenstrup, J.J., and Fuzessery, Z.M., A Novel Projection from the Basolateral Nucleus of the Amygdala to the Inferior Colliculus in Bats. *Soc. Neurosci. Abstr.,* 25: 1417, 1999.
116. McDonald, A.J., Cortical Pathways to the Mammalian Amygdala. *Progr. Neurobiol.,* 55: 257–332, 1998.
117. Melchner, L. von, Pallas, S.L., and Sur, M., Visual Behavior Mediated by Retinal Projections Directed to the Auditory Pathways. *Nature,* 404: 871–876, 2000.
118. Melzack, R. and Wall, P.D., Pain Mechanisms: A New Theory. *Science,* 150: 971–979, 1965.
119. Melzack, R. and Casey, K.I., Sensory, Motivational, and Central Control Determinants of Pain. In: *The Skin Senses*, edited by D.R. Kenshalo DR. Springfield, IL:Charles C Thomas, 1968, pp. 423–443.
120. Melzack, R., Phantom Limbs. *Sci. Am.* 266: 120–126, 1992.
121. Meridith, M.A., Nemitz, J.W., and Stein, B.E., Determinants of Multisensory Integration in Superior Colliculus Neurons. I. Temporal Factors. *J. Neurosci.,* 10: 3215–3229, 1987.
122. Merigan, W.H. and Maunsell, J.H.R., Macaque Vision after Magnocellular Lateral Geniculate Lesions. *Vis. Neurosci.,* 5: 347–352, 1993.
123. Merzenich, M.M., Kaas, J.H., Wall, J., Nelson, R.J., Sur, M., and Felleman, D., Topographic Reorganization of Somatosensory Cortical Areas 3b and 1 in Adult Monkeys Following Restricted Deafferentation. *Neuroscience,* 8: 3–55, 1983.
124. Merzenich, M.M., Nelson, R.J., Stryker, M.P., Cynader, M.S., Schoppmann, A., and Zook, J.M., Somatosensory Cortical Map Changes Following Digit Amputation in Adult Monkeys. *J. Comp. Neurol.,* 224: 591–605, 1984.
125. Milner, A.D. and Goodale, M.A., Visual Pathways to Perception and Action. *Progr. Brain Res.,* 95: 317–337, 1993.

126. Mishkin, M., Ungeleider, L.G., and Macko, K.A., Object Vision and Spatial Vision: Two Cortical Pathways. *Trends Neurosci.,* 6: 415–417, 1983.
127. Mitani, A. and Shimokouchi, M., Neural Connections in the Primary Auditory Cortex: An Electrophysiologic Study in the Cat. *J. Comp. Neurol.,* 235: 417–429, 1985.
128. Modayur, B., Prothero, J., Ojemann, G., Maravilla, K., and Brinkley, J., Visualization-Based Mapping of Language Function in the Brain. *Neuroimage,* 6: 245–258, 1997.
129. Møller, A.R., Unit Responses in the Rat Cochlear Nucleus to Repetitive Transient Sounds. *Acta Physiol. Scand.,* 75: 542–551, 1969.
130. Møller, A.R., Coding of Amplitude and Frequency Modulated Sounds in the Cochlear Nucleus of the Rat. *Acta Physiol. Scand.,* 86: 223–238, 1972.
131. Møller, A.R., Statistical Evaluation of the Dynamic Properties of Cochlear Nucleus Units Using Stimuli Modulated with Pseudorandom Noise. *Brain Res.,* 57: 443–456, 1973.
132. Møller, A.R., Responses of Units in the Cochlear Nucleus to Sinusoidally Amplitude Modulated Tones. *Exp. Neurol.,* 45: 104–117, 1974.
133. Møller, A.R., Use of Stochastic Signals in Evaluation of the Dynamic Properties of a Neuronal System. *Scand. J. Rehab. Med.,* 3: 37–44, 1974.
134. Møller, A.R., Coding of Sounds with Rapidly Varying Spectrum in the Cochlear Nucleus. *J. Acoust. Soc. Am.,* 55: 631–640, 1974.
135. Møller, A.R., Dynamic Properties of Excitation and Inhibition in the Cochlear Nucleus. *Acta Physiol. Scand.,* 93: 442–454, 1975.
136. Møller, A.R., Latency of Unit Responses in the Cochlear Nucleus Determined in Two Different Ways. *J. Neurophysiol.,* 38: 812–821, 1975.
137. Møller, A.R., Dynamic Properties of the Responses of Single Neurons in the Cochlear Nucleus. *J. Physiol.,* 259: 63–82, 1976.
138. Møller, A.R., Dynamic Properties of Primary Auditory Fibers Compared with Cells in the Cochlear Nucleus. *Acta Physiol. Scand.,* 98: 157–167, 1976.
139. Møller, A.R., Frequency Selectivity of Single Auditory Nerve Fibers in Response to Broadband Noise Stimuli. *J. Acoust. Soc. Am.,* 62: 135–142, 1977.
140. Møller, A.R., Coding of Time-Varying Sounds in the Cochlear Nucleus. *Int. Audiol.,* 17: 446–468, 1978.
141. Møller, A.R., Neural Delay in the Ascending Auditory Pathway. *Exp. Brain Res.,* 43: 93–100, 1981.
142. Møller, A.R., Frequency Selectivity of Phase-Locking of Complex Sounds in the Auditory Nerve of the Rat. *Hear. Res.,* 11: 267–284, 1983.
143. Møller, A.R., Pathophysiology of Tinnitus. *Ann. Otol. Rhinol. Laryngol.,* 93: 39–44, 1984.
144. Møller, A.R. and Jannetta, P.J., On the Origin of Synkinesis in Hemifacial Spasm: Results of Intracranial Recordings. *J. Neurosurg.,* 61: 569–576, 1984.
145. Møller, A.R., Møller, M.B., and Yokota, M., Some Forms of Tinnitus May Involve the Extralemniscal Auditory Pathway. *Laryngoscope,* 102: 1165–1171, 1992.
146. Møller, A.R. and Pinkerton, T., Temporal Integration of Pain from Electrical Stimulation of the Skin. *Neurol. Res.,* 19: 481–488, 1997.
147. Møller, A.R., Similarities between Severe Tinnitus and Chronic Pain. *J. Am. Acad. Audiol.,* 11: 115–124, 2000.
148. Møller, A.R., *Hearing: Its Physiology and Pathophysiology*. San Diego: Academic Press, 2000.
149. Møller, A.R., Symptoms and Signs Caused by Neural Plasticity. *Neurol. Res.,* 23: 565–572, 2001.
150. Møller, A.R. and Rollins, P., The Non-Classical Auditory System Is Active in Children but Not in Adults. *Neurosci. Lett.,* 319: 41–44, 2002.
151. Morest, D.K., Structural Organization of Auditory Pathways. In: *The Nervous System*, edited by E.L. Eagles. New York: Raven Press, 1975, pp. 19–29.

152. Morest, D.K. and Oliver, D.L., The Neuronal Architecture of the Inferior Colliculus in the Cat. *J. Comp. Neurol.,* 222: 209–236, 1984.
153. Moruzzi, G. and Magoun, H.W., Brain Stem Reticular Formation and Activation of the EEG. *Electroenceph. Clin. Neurophysiol.,* 1: 455–473, 1949.
154. Mountain, D.C., Geisler, C.D., and Hubbard, A.E., Stimulation of Efferents Alters the Cochlear Microphonic and Sound-Induced Resistance Changes Measured in the Scala Media of the Guinea Pig. *Hear. Res.,* 3: 231–240, 1980.
155. Mountcastle, V.B., Modality and Topographic Properties of Single Neurons of Cat's Somatic Cortex. *J. Neurophysiol.,* 20: 408–434, 1957.
156. Mountcastle, V.B. and Darian-Smith, I., Neural Mechanisms in Somesthesia, In: *Medical Physiology*, edited by V.B. Mountcastle, St. Louis: C.V. Mosby, 1968.
157. Mountcastle, V.B., ed., *Medical Physiology*. St. Louis: C.V. Mosby, 1974.
158. Mountcastle, V.B., Neural Mechanisms in Somesthesia. In: *Medical Physiology*, edited by V.B. Mountcastle, St Louis: C.V Mosby, 1974.
159. Müller, M., Robertson, D., and Yates, G.K., Rate-Versus-Level Functions of Primary Auditory Nerve Fibres: Evidence of Square Law Behavior of All Fibre Categories in the Guinea Pig. *Hear. Res.,* 55: 50–56, 1991.
160. Neville, H.J., Schmidt, A., and Kutas, M., Altered Visual-Evoked Potentials in Congenitally Deaf Adults. *Brain Res.,* 266, 1983.
161. Neville, H.J. and Lawson, D., Attention to Central and Peripheral Visual Space in Movement Detection Task: An Event-Related Potential and Behavioral Study. II. Congenitally Deaf Adults. *Brain Res.,* 405: 268–283, 1987.
162. Norgren, R., Neuroanatomy of Gustatory and Visceral Afferent Systems in Rat and Monkey. In: *Olfaction and Taste*, edited by H. van der Starre. London: IRL Press, 1980, pp. 288.
163. O'Brien, B.J., Abel, P.L., and Olavarria, J.F., The Retinal Input to Calbindin-D28k-Defined Subdivisions in Macaque Inferior Pulvinar. *Neurosci. Lett.,* 312: 145–148, 2001.
164. Ojemann, G.A., Effect of Cortical and Subcortical Stimulation on Human Language and Verbal Memory. *Res. Publ. Assoc. Res. Nervous Mental Dis.,* 66: 101–115, 1988.
165. Oliver, D.L., and Huerta, M.F., Inferior and Superior Colliculi. In: *The Mammalian Auditory Pathway: Neuroanatomy*, edited by D.B. Webster, A.N. Popper, and R.R. Fay. New York: Springer-Verlag, 1992.
166. Pallas, S.L., Roe, A.W., and Sur, M., Visual Projections Induced into the Auditory Pathway of Ferrets. I. Novel Inputs to Primary Auditory Cortex. (AI) from the LP/Pulvinar Complex and the Topography of the MGN-AI Projection. *J. Comp. Neurol.,* 298: 50–68, 1990.
167. Penfield, W. and Boldrey, E., Somatic Motor and Sensory Representation in the Cerebral Cortex of Man as Studied by Electrical Stimulation. *Brain,* 60: 389–443, 1937.
168. Pfaffman, C., Frank, M., Bartoshuk, L.M., and Snell, T.C., Coding Gustatory Information in the Squirrel Monkey Chorda Tympani. In: *Progress in Psychobiology and Physiological Psychology*, edited by J.M. Sprague, and A.N. Epstein. New York: Academic Press, 1976, pp. 1–27.
169. Pfeiffer, R.R., Classification of Response Patterns of Spike Discharges for Units in the Cochlear Nucleus: Tone Burst Stimulation. *Exp. Brain. Res.,* 1: 220–235, 1966.
170. Postle, B.R. and D'Esposito, M., "What"–Then–"Where" in Visual Working Memory: An Event-Related FMRI Study. *J. Cog. Neurosci.,* 11: 585–597, 1999.
171. Powell, T.P.S. and Mountcastle, V.B., Some Aspects on the Functional Organization of the Cortex of the Postcentral Gyrus Obtained in a Single Unit Analysis with Cytoarchitecture. *Bull. Johns Hopkins Hosp.,* 105: 133–162, 1959.
172. Price, D.D., Long, S., and Huitt, C., Sensory Testing of Pathophysiological Mechanisms of Pain in Patients with Reflex Sympathetic Dystrophy. *Pain,* 63–173, 1992.
173. Price, D.D., Psychological and Neural Mechanisms of the Affective Dimension of Pain. *Science,* 288: 1769–1772, 2000.

174. Rajan, R. and Irvine, D.R., Neuronal Responses across Cortical Field A1 in Plasticity Induced by Peripheral Auditory Organ Damage. *Audiol. Neurootol.*, 3: 123–144, 1998.
175. Ramachandran, V.S., Rogers-Ramachandran, D., and Stewart, M., Perceptual Correlates of Massive Cortical Reorganization [letter; comment]. *Science,* 258: 1159–1160, 1992.
176. Ratliff, F., *Mach Bands. Quantitative Studies on Neural Networks in the Retina.* San Francisco: Holden-Day, 1965.
177. Rauschecker, J.P., Tian, B., and Hauser, M., Processing of Complex Sounds in the Macaque Nonprimary Auditory Cortex. *Science,* 268: 111–114, 1995.
178. Rauschecker, J.P., Cortical Processing of Complex Sounds. *Curr. Opin. Neurobiol.*, 8: 516–521, 1998.
179. Rauschecker, J.P. and Tian, B., Mechanisms and Streams for Processing of "What" and "Where" in Auditory Cortex. *Proc. Nat. Acad. Sci.USA,* 97: 11800–11806, 2000.
180. Razak, K.A., Fuzessery, Z.M., and Lohuis, T.D., Single Cortical Neurons Serve Both Echolocation and Passive Sound Localization. *J. Neurophysiol.,* 81: 1438–1442, 1999.
181. Riehle, A., Grun, S., Diesmann, M., and Aertsen, A., Spike Synchronization and Rate Modualtion Differently Involved in Motor Cortical Function. *Science,* 278: 1950–1953, 1997.
182. Rimmer, D.W., Boivin, D.B., Shanahan, T.L., Kronauer, R.E., Duffy, J.F., and Czeisler, C.A., Dynamic Resetting of the Human Circadian Pacemaker by Intermittent Bright Light. *Am. J. Physiol. Regul. Integr. Comp. Physiol.,* 279: R1574–1579, 2000.
183. Robertson, D. and Irvine, D.R., Plasticity of Frequency Organization in Auditory Cortex of Guinea Pigs with Partial Unilateral Deafness. *J. Comp. Neurol.,* 282: 456–471, 1989.
184. Roe, A.W. and Ts'o, D.Y., Visual Topography in Primate V2: Multiple Representation across Functional Stripes. *J. Neurosci.,* 15: 3689–3715, 1995.
185. Roe, A.W. and Ts'O, D.Y., Specificity of Color Connectivity between Primate V1 and V2. *J. Neurophysiol.,* 82: 2719–2730, 1999.
186. Romanski, L.M., Tian, B., Fritz, J., Mishkin, M., Goldman-Rakic, P.S., and Rauschecker, J.P., Dual Streams of Auditory Afferents Target Multiple Domains in the Primate Prefrontal Cortex. *Nature Neurosci.,* 2: 1131–1136, 1999.
187. Rose, J.E., Galambos, R., and Hughes, J.R., Microelectrode Studies of the Cochlear Nuclei in the Cat. *Bull. Johns Hopkins Hosp.,* 104: 211–251, 1959.
188. Rouiller, E.M., Functional Organization of the Auditory System. In: *The Central Auditory System*, edited by G. Ehret and R. Romand. New York: Oxford University Press, 1997, pp. 3–96.
189. Sakai, M.D. and Suga, N., Plasticity of the Cochleotopic (Frequency) Map in Specialized and Nonspecialized Cortices. *Proc. Nat. Acad. Sci. USA,* 98: 3507–3512, 2001.
190. Sakurai, Y., Population Coding by Cell Assemblies—What It Really Is in the Brain. *Neurosci. Res.* 26: 1–16, 1996.
191. Schicatano, E.J., Peshori, K.R., Gopalaswamy, R., Sahay, E., and Evinger, C., Reflex Excitability Regulates Prepulse Inhibition. *J. Neurosci.,* 20: 4240–4247, 2000.
192. Schmidt, R.F. and Thews, G., *Human Physiology*. Berlin: Springer-Verlag, 1983.
193. Schneider, G.E., Two Visual Systems. *Science*, 163: 895–902, 1969.
194. Shepherd, G.M., Microcircuits in the Nervous System. *Sci. Am.*, 238: 92–103, 1978.
195. Shepherd, G.M., *The Synaptic Organization of the Brain*. New York: Oxford University Press, 1990.
196. Shepherd, G.M., *Neurobiology*. New York: Oxford University Press, 1994.
197. Shinsuke, S. and Shams, L., Sensory Modalities Are Not Separate Modalities: Plasticity and Interactions. *Curr. Opin. Neurobiol.*, 11: 505–509, 2001.
198. Shore, S.E., Godfrey, D.A., Helfert, R.H., Altschuler, R.A., and Bledsoe, S.C., Connections between the Cochlear Nuclei in Guinea Pig. *Hear. Res.,* 62: 16–26, 1992.
199. Shore, S.E., Vass, Z., Wys, N.L., and Altschuler, R.A., Trigeminal Ganglion Innervates the Auditory Brainstem. *J. Comp. Neurol.,* 419: 271–285, 2000.

200. Starzl, T.E. and Magoun, H.W., Organization of the Diffuse Thalamic Projection System. *J. Neurophysiol.*, 14: 133–146, 1951.
201. Stein, R.B., A Theoretical Analysis of Neuronal Variability. *Biophys. J.*, 5: 173–194, 1965.
202. Stepniewska, I. and Rajkowska, G., The Sensory Projections to the Frontal Association Cortex in the Dog. *Acta Neurobiol. Exp.*, 49: 299–310, 1989.
203. Stepniewska, I., Ql, H.X., and Kaas, J.H., Projections of the Superior Colliculus to Subdivisions of the Inferior Pulvinar in New World and Old World Monkeys. *Visual Neurosci.*, 17: 529–549, 2000.
204. Suga, N., Auditory Neuroetology and Speech Processing: Complex Sound Processing by Combination-Sensitive Neurons. In: *Auditory Function*, edited by G.M. Edelman, W.E. Gall and W.M. Cowan. New York: Wiley, 1988, pp. 679–720.
205. Suga, N., Principles of Auditory Information Processing Derived from Neuroethology. *J. Exp. Biol.*, 146: 277–286, 1989.
206. Suga, N., Parallel-Hierarchical Processing of Complex Sounds for Specialized Auditory Function. In: *Encyclopedia of Acoustics*, edited by M.J. Crocker, New York: Wiley, 1997.
207. Sur, M., Pallas, S.L., and Roe, A.W., Cross-Modal Plasticity in Cortical Development: Differentiation and Specification of Sensory Neocortex. *Trends Neurosci.*, 13, 227–233, 1990.
208. Sur, M., Schummers, J., and Dragoi, V., Cortical Plasticity: Time for a Change. *Curr. Biol.*, 12: 168–170, 2002.
209. Sussman, E., Ceponiene, R., Shestakova, A., Naatanen, R., and Winkler, I., Auditory Stream Segregation Processes Operate Similarly in School-Aged Children and Adults. *Hear. Res.*, 153: 108–114, 2001.
210. Sweet, W.H., Cerebral Localization of Pain. In: *New Perspectives in Cerebral Localization*, edited by R.A. Thompson, and J.R. Green. New York: Raven Press, 1982, p. 205–242.
211. Syka, J., Popelar, J., and Kvasnak, E., Response Properties of Neurons in the Central Nucleus and External and Dorsal Cortices of the Inferior Colliculus in Guinea Pig. *Exp. Brain Res.*, 133: 254–266, 2000.
212. Szczepaniak, W.S. and Møller, A.R., Interaction between Auditory and Somatosensory Systems: A Study of Evoked Potentials in the Inferior Colliculus. *Electroenceph. Clin. Neurophysiol.*, 88: 508–515, 1993.
213. Tabo, E., Jinks, S.L., Eisle, J.H., and Carstens, E., Behavioral Manifestations of Neuropathic Pain and Mechanical Allodynia, and Changes in Spinal Dorsal Horn Neurons, Following L4-L6 Root Constriction. *Pain,* 66: 503–520, 1999.
214. Tian, B., Reser, D., Durham, A., Kustov, A. and Rauschecker, J.P., Functional Specialization in Rhesus Monkey Auditory Cortex. *Science,* 292: 290–293, 2001.
215. Tobias, J.V., and Zerlin.S., Lateralization Threshold as a Function of Stimulus Duration. *J. Acoust. Soc. Am.*, 31: 1591–1594, 1959.
216. Tranel, D., Damasio, A.R., and Damasio, H., Intact Recognition of Facial Expression, Gender, and Age in Patients with Impaired Recognition of Face Identity. *Neurology,* 38: 690–696, 1988.
217. Turner, B.H. and Herkenham, M., Thalamoamydaloid Projections in the Rat: A Test of the Amygdala's Role in Sensory Processing. *J. Comp. Neurol.*, 313: 295–325, 1991.
218. Ungeleider, L.G. and Mishkin, M., Analysis of Visual Behavior. In: *Analysis of Visual Behavior*, edited by D.J. Ingle, M.A. Goodale, and R.J.W. Mansfield. Cambridge, MA: MIT Press, 1982.
219. Ungeleider, L.G. and Haxby, J.V., "What" and "Where" in the Human Brain. *Curr. Opin. Neurobiol.*, 4: 157–165, 1994.
220. Urban, M.O. and Gebhart, G.F., Supraspinal Contribution to Hyperalgesia. *Proc. Nat. Acad. Sci. USA,* 96: 7687–7692, 1999.
221. Vass, Z., Shore, S.E., Nuttall, A.L., Jabncso, G., Brechtelsbauer, P.B., and Miller, J.M., Trigeminal Ganglion Innervation of the Cochlea—A Retrograde Transport Study. *Neuroscience,* 79: 605–615, 1997.

222. Vogt, B.A., Sikes, R.W., and Vogt, L.J., Anterior Cingulate Cortex and the Medial Pain System. In: *The Neurobiology of the Cingulate Cortex and Limbic Thalamus*, edited by B.A. Vogt and M. Gabriel. Boston: Birkhauser, 1993.
223. Wada, J.A., *Kindling* 2. New York: Raven Press, 1981.
224. Wall, P.D., The Presence of Ineffective Synapses and Circumstances Which Unmask Them. *Phil. Trans. Royal Soc. (London),* 278: 361–372, 1977.
225. Wang, J., Zhou, T., Qiu, M., Du, A., Cai, K., Wang, Z., Zhou, C., Meng, M., Zhou, Y., Fan, S., and Chen, L., Relationship between Ventral Stream for Object Vision and Dorsal Stream for Spatial Vision: An FMRI+ERP Study. *Human Brain Mapping,* 8: 170–181, 1999.
226. Webster, D.B., An Overview of Mammalian Auditory Pathways with an Emphasis on Humans. In: *The Mammalian Auditory Pathway: Neuroanatomy*, edited by D.B. Webster, A.N. Popper, and R.R. Fay. New York: Springer-Verlag, 1992, pp. 1–22.
227. Webster, D.B., Popper, A.N., and Fay, R.R., *The Mammalian Auditory Pathway: Neuroanatomy*. New York: Springer-Verlag, 1992.
228. Weinberg, R.J. and Rustioni, A., A Cuneocochlear Pathway in the Rat. *Neuroscience,* 20: 209–219, 1987.
229. Weinberger, N.M., Javid, R., and Lepan, B., Long-Term Retention of Learning-Induced Receptive-Field Plasticity. *Proc. Nat. Acad. Sci USA.*,90: 2394–2398, 1993.
230. Weinberger, N.M., Learning-Induced Physiological Memory in Adult Primary Auditory Cortex: Receptive Field Plasticity, Model, and Mechanisms. *Audiol. Neuro-Otol.* 3: 145–167, 1998.
231. Wiesel, T.N. and Hubel, D.H., Extent of Recovery from the Effects of Visual Deprivation in Kittens. *J. Neurophysiol.,* 28: 1060–1072, 1965.
232. Willer, J.C., Relieving Effect of Tens on Painful Muscle Contraction Produced by an Impairment of Reciprocal Innervation: An Electrophysiological Analysis. *Pain,* 32: 271–274, 1988.
233. Willis, W.D., From Nociceptor to Cortical Activity. In: *Pain and the Brain*, edited by B. Bromm and J.E. Desmedt. New York: Raven Press, 1995, pp. 1–19.
234. Winer, J.A., Larue, D.T., Diehl, J.J., and Hefti, B.J., Auditory Cortical Projections to the Cat Inferior Colliculus. *J. Comp. Neurol.,* 400: 147–174, 1998.
235. Winer, J.A., Diehl, J.J., and Larue, D.T., Projections of Auditory Cortex to the Medial Geniculate Body of the Cat. *J. Comp. Neurol.,* 430: 27–55, 2001.
236. Wise, L.Z. and Irvine, D.R.F., Topographic Organization of Interaural Intensity Difference Sensitivity in Deep Layers of Cat Superior Colliculus: Implications for Auditory Spatial Representation. *J. Neurophysiol.,* 54: 185–211, 1985.
237. Woolf, C.J., Evidence of a Central Component of Postinjury Pain Hypersensitivity. *Nature,* 308: 686–688, 1983.
238. Woolf, C.J. and Thompson, S.W.N., The Induction and Maintenance of Central Sensitization Is Dependent on *N*-Methyl-D-Aspartic Acid Receptor Activation: Implications for the Treatment of Post-Injury Pain Hypersensitivity States. *Pain,* 44: 293–299, 1991.
239. Woolf, C.J., Shortland, P., and Cogershall, R.E., Peripheral Nerve Injury Triggers Central Sprouting of Myelinated Afferents. *Nature,* 355: 75–78, 1992.
240. Yan, J. and Suga, N., The Midbrain Creates and the Thalamus Sharpens Echo-Delay Tuning for Cortical Representation of Target-Distance Information in the Mustached Bat. *Hear. Res.,* 93: 102–110, 1996.
241. Yingling, C.D. and Skinner, J.E., Regulation of Unit Activity in Nucleus Reticularis Thalami by the Mesencephalic Reticular Formation and the Frontal Granular Cortex. *Electroenceph. Clin. Neurophysiol.,* 39: 635–642, 1975.
242. Yingling, C.D. and Skinner, J.E., Selective Regulation of Thalamic Sensory Relay Nuclei by Nucleus Reticularis Thalami. *Electroenceph. Clin. Neurophysiol.,* 41: 476–482, 1976.
243. Zatorre, R.J., *The Biological Foundations of Music*. New York: New York Academy of Sciences, 2001.

244. Zeki, S., *A Vision of the Brain*. London: Blackwell Scientific, 1993.
245. Zhang, M., Suga, N., and Yan, J., Corticofugal Modulation of Frequency Processing in Bat Auditory System. *Nature,* 387: 900–903, 1997.
246. Zhou, M., Silent Glutamatergic Synapses and Long-Term Facilitation in Spinal Dorsal Horn Neurons. In: *Nervous System Plasticity and Chronic Pain*, edited by J. Sandkühler, B. Bromm, and G.H. Gebhart. Amsterdam: Elsevier, 2000.
247. Zigmond, M.J., Bloom, F.E., Landis, S.C., Roberts, J.L., and Squire, L.R., *Fundamental Neuroscience*. San Diego: Academic Press, 1999.
248. Zou, Z., Horowitz, L.F., Montmayeur, J.-P., Snapper, S., and Buck, L.B., Genetic Tracing Reveals a Stereotyped Sensory Map in the Olfactory Cortex. *Nature,* 414: 173–179, 2001.

CHAPTER 4

Somatosensory System

ABBREVIATIONS

CNS : Central nervous system
CRPS : Complex regional pain syndrome
fMRI : Functional magnetic resonance imaging
HTM : High-threshold mechanoreceptors
LTM : Low-threshold mechanoreceptors
ML : Medial lemniscus
MRI : Magnetic resonance imaging
PAG : Periaqueductal gray
PET : Positron emission tomography
PO : Posterior part of the thalamus
RA : Rapid-adapting receptors
rCBF : Regional cerebral blood flow
RF : Reticular formation
RSD : Reflex sympathetic dystrophy
SA : Slow-adapting receptors
SI : Primary somatosensory cortex

SII : Secondary somatosensory cortex
SMP : Sympathetically maintained pain
TENS : Transdermal electric nerve stimulation
TGN : Trigeminal neuralgia
VB : Ventrobasal (nucleus of the thalamus)
VI : Ventralis intermedius (nucleus of thalamus)
VPL : Ventral–posterior–lateral (nucleus of the thalamus)
VPM : Ventral–posterior–medial (nucleus of the thalamus)
WDR : Wide dynamic range (neurons)

ABSTRACT

1. The somatosensory system provides information about touch, vibration, temperature of the skin, and pain.
2. The classical somatosensory system is the dorsal column pathway, the rostral trigeminal sensory nucleus, and the ventral thalamus projecting to the primary somatosensory cortex and it provides the sensation of fine touch and vibration senses.
3. The nonclassical somatosensory system is the anteriorlateral system, the caudal trigeminal nucleus, and the dorsal, medial thalamus, projecting to several regions of the CNS including limbic structures, and it mediates the sensation of pain, temperature, and some deep touch.

Classical Somatosensory System

4. The classical somatosensory system receives input from five main types of cutaneous mechanoreceptors in the skin that respond to innocuous stimulation:
   a. Meissner's corpuscles
   b. Merkel's discs
   c. Pacinian corpuscles
   d. Ruffini endings
   e. Hair follicle receptors
5. In addition, cool and warmth receptors in the skin supply input to the classical somatosensory system.
6. Mechanoreceptors are classified according to their type of adaptation:
   a. Slow adapting: Merkel's discs and Ruffini endings
   b. Rapid adapting: Meissner's corpuscles
   c. Very rapid adapting: Pacinian corpuscles

7. The cutaneous mechanoreceptors are innervated by Aβ nerve fibers (large myelinated fibers) that enter the spinal cord through the dorsal roots, and from the face the afferent fibers of mechanoreceptors enter the brainstem through the trigeminal nerve.
8. Each dorsal root of the spinal cord innervates patches of skin known as dermatomes. The dermatomes of the face are innervated by the three branches of the trigeminal nerve (V1, V2, and V3).
9. The large dorsal root fibers that innervate mechanoreceptors of the skin ascend uninterrupted in the dorsal column of the spinal cord on the same side and synapse with cells in the dorsal column nuclei.
10. The fibers of each dorsal root give off collaterals to cells in the dorsal horn of the spinal cord at several segments above and below their entry into the spinal cord. Some collaterals reach motoneurons in the ventral horn.
11. Axons from cells in the dorsal column nuclei cross the midline and form the medial lemniscus.
12. The fibers of the medial lemniscus proceed uninterrupted to nuclei in the ventral posterior lateral nucleus of the thalamus which project to the primary somatosensory cortex.
13. The fibers of the medial lemniscus give off collaterals to the reticular formation of the brainstem, thereby contributing to arousal.
14. Fibers from cutaneous receptors in the face and receptors of the mouthtravel in the three branches of the sensory part of the trigeminal nerve (cranial nerve V), and some cutaneous receptors of the face are innervated by fibers of cranial nerves IX and X.
15. The fibers of the trigeminal nerve that innervate cutaneous receptors responding to innocuous stimulation synapse with cells in the rostral trigeminal nucleus.
16. The axons of the cells in the rostral sensory trigeminal nucleus cross the midline, join the medial lemniscus, and then reach nuclei in the ventral posterior medial nucleus of the thalamus. The fibers of the thalamic nuclei project from the ventral thalamus to the primary somatosensory cortex.
17. Four different regions (1, 2, 3a, and 3b) of the primary somatosensory cortex have been identified. The surface of the body is represented on the surface of each of these regions.
18. Neurons of the primary somatosensory cortex project to the secondary somatosensory cortex and association cortices as well as other regions of the brain.
19. Interplay between inhibition and excitation (lateral inhibition) modifies spatial resolution (sharper two-point discrimination)

20. Activity in descending pathways can modify the ascending neural activity as far peripherally as the dorsal horn.

Nonclassical Somatosensory System

21. The input to the nonclassical somatosensory system originates in nociceptors (pain, cold, and heat receptors) that are located in the skin, joints, tendons, and viscera.
22. Nociceptors respond to noxious stimulation and are innervated by small myelinated fibers (Aδ; diameter, 1–5 μm; conduction velocity, 5–30 m/sec) and unmyelinated C-fibers (diameter, 0.2–2 μm; conduction velocity, 0.5–1 m/sec), which travel in the dorsal roots. Nociceptor fibers for the head travel in the trigeminal, glossopharyngeal, and vagus nerves.
23. Dorsal root fibers from nociceptors travel several (5 to 7) segments up and down the spinal cord, and their collaterals synapse with cells in the dorsal horn of the spinal cord.
24. Only dorsal root fibers that terminate on neurons in spinal segments that are close to the entry of the dorsal root can normally activate dorsal horn cells, and the distant synapses are normally nonconducting (dormant) but may become unmasked by deprivation of input from nearby dorsal roots.
25. Nociceptor afferents from the face terminate in cells in the caudal trigeminal nucleus.
26. The axons of the dorsal horn cells cross the midline at the segmental level and ascend as the anteriolateral tract consisting mainly of the spinothalamic, mesencephalic, and spinoreticular tracts.
27. The anteriorlateral tracts proceed to several nuclei in the medial and dorsal thalamus after giving off collaterals to the reticular formation of the brainstem, providing a strong driving force on large portions of the forebrain (arousal).
28. Axons of cells in the caudal part of the trigeminal nucleus cross the midline and ascend in the trigonothalamic tract to reach the ventral posterior medial and the intralaminar nuclei of the thalamus.
29. The anteriorlateral pathways project to the primary and secondary somatosensory cortices and several noncortical regions such as the periaquaductal gray, limbic structures, and the brainstem reticular formation.
30. Wide dynamic range neurons in the spinal cord, the brainstem, and the medulla integrate the output of many types of skin receptors, high-threshold and low-threshold mechanoreceptors, and cool, warmth, heat, cold, and pain receptors. The function of these neurons can change through neural plasticity.

Pain

31. Body pain can be caused by:
    a. Stimulation of pain receptors
    b. Stimulation of heat and cold receptors
    c. Injury, asphyxia, and inflammation to body tissue
    d. Overstimulation of mechanoreceptors
32. Neuropathic pain can be caused by:
    a. Injury to the peripheral nerves
    b. Injury, necrosis, and asphyxia of central nervous system tissue
    c. Change in synaptic efficacy and neural connectivity mediated through expression of neural plasticity
33. Acute body pain sensation is caused by stimulation of nociceptors and by overstimulation of other receptors.
34. Acute body pain has two phases: fast and slow.
    a. The fast phase of pain is mediated by myelinated fibers (A).
    b. The slow and delayed pain is mediated by unmyelinated fibers (C-fibers).
35. Chronic body pain may be caused by:
    a. Chronic inflammation
    b. Sensitization of skin receptors
36. Chronic body pain and neuropathic pain may induce changes in the function of the somatosensory system that cause:
    a. Normally innocuous stimulation to become painful (allodynia)
    b. Stimuli that normally cause mild pain now causing an exaggerated reaction (hyperpathia)
37. Chronic pain may involve the sympathetic nervous system (sympathetically maintained pain and reflex sympathetic dystrophy).
38. Itching has similarities with pain, but little is known about the mechanisms that cause itching.

I. INTRODUCTION

The mechanoreceptors in the skin sense touch, pressure, and vibration that are applied to the skin. In reality, the sense of touch is more complex. Not only do we feel vibration and pressure but we can also feel the difference between wet or dry, warm or cool, and pain. We can feel the size and the form of an object and the texture of its surface. The separation between stimulations at two locations on the skin and the difference in time between two stimulations provide distinct sensations.

The output of the receptors in the skin communicates several forms of fine touch and vibration through the *dorsal column system* and the *rostral trigeminal system* to higher centers of the central nervous system. The dorsal column system mainly communicates innocuous stimulations such as touch, vibration of the skin, and cool or warmth. We will regard that as the *classical* part of the somatosensory system. Noxious stimulation, such as that causing the sensation of pain, heat, and cold, are communicated by the *anteriorlateral system* and the *caudal trigeminal system*. These two systems also communicate some deep touch. We will regard them as being the *nonclassical* part of the somatosensory system (Chapter 3). Receptors in muscles, tendons, and joints (proprioceptors) also send information through the dorsal column system to higher centers. Such proprioceptive information does not normally reach consciousness; therefore, proprioception will not be discussed in this book.

In this chapter, we discuss specific features of the somatosensory system. The general features of the somatosensory system were discussed in Chapter 2 and 3 together with that of other sensory systems. First, we discuss the anatomical organization of the classical and the nonclassical somatosensory systems, followed by discussions of the physiology of these systems. A separate section on the perception and physiology of pain concludes this chapter.

II. ANATOMY OF THE SOMATOSENSORY SYSTEM

This section describes the anatomy of the classical and the nonclassical somatosensory system, including the receptors, the ascending and descending neural pathways, and their nuclei. The anatomical organization of the cerebral cortices and their projections to higher CNS centers are also discussed.

A. Classical Somatosensory System

The classical somatosensory system consists of receptors that respond to innocuous stimulation and produce a conscious perception. The ascending sensory pathways are the dorsal column system of the spinal cord and the trigeminal sensory system. These pathways proceed to the thalamic relay nuclei from which connections project to the somatosensory cerebral cortices.

1. Receptors

The receptors for the somatosensory system are endings of afferent nerve fibers. Specialized regions of the membranes of these nerve endings are sensitive to specific modalities of (physical) stimulation.[42] We have called such receptors Type I receptors (Chapter 2). Mechanoreceptors are prevalent in the skin, where thermoreceptors are also located. Mechanoreceptors are also found in muscles, joints, tendons, and viscera, but these receptors belong to the proprioceptive system that is outside the scope of this book. Receptors that respond to noxious and injurious stimulation and produce the sensation of pain (nociceptors) are also abundant in the skin, muscles, tendons, and joints, as well as in viscera. These receptors together with *polymodal* receptors provide the input to the nonclassical somatosensory system.

a. Mechanoreceptors of the Skin

Mechanoreceptors in the skin can be divided into five different types according to their morphology. There are two types of receptors in the superficial skin: *Meissner's corpuscles* and *Merkel's discs*. The three other types of receptors, the *Pacinian corpuscles*, *Ruffini endings*, and *hair-follicle receptors*, are deep in the skin. Ruffini endings may also be in joint capsules. All these receptors are nerve endings with specialized regions of the cell membrane that are sensitive to mechanical deformation. These nerve endings are surrounded by different kinds of tissue. In the Pacinian corpuscles, the nerve endings are surrounded by an onion-shaped structure of alternating concentric layers of cellular membranes and fluid-filled spaces.

The receptors in the *glabrous* (hairless) skin are different from those found in hairy skin. Glabrous skin has three different types of receptors: Meissner's corpuscles, Merkel's discs (Ruffini endings), and Pacinian corpuscles (Fig. 4.1). Hairy skin has hair-follicle receptors, tactile disks (*pincus domes*) and Ruffini endings (Fig. 4.2). Pacinian corpuscles are also found deep in hairy skin.[79] In addition to these mechanoreceptors there are warmth and cool receptors and nociceptors (heat and cold receptors and pain receptors) in the skin.[38]

b. Thermoreceptors

Four kinds of receptors are temperature sensitive. Two of these, known as cool and warmth receptors, are regarded as sensory receptors that respond to innocuous stimulation. The two other kinds of thermoreceptors, known as heat and cold receptors, are nociceptors,[40,50]and they will be discussed later. Little is known about the morphology of thermoreceptors, but it is generally assumed that they are specialized membrane patches on the distal portion of the axons of afferent sensory nerve fibers. The function of thermal (cool, warmth, cold, and heat) receptors is poorly understood, but recent studies using molecular biology methods of cloning receptor channels have increased our understanding of the function of these receptors.[63] Studies of the molecular

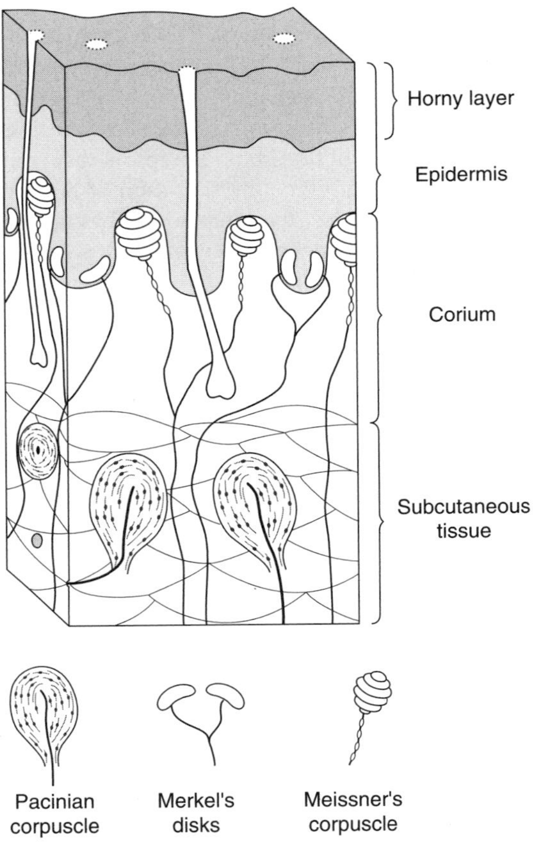

FIGURE 4.1 Schematic drawings of mechanoreceptors in hairless (glabrous) skin showing three types of mechanoreceptors. (Adapted from Schmidt, R.F., *Fundamentals of Sensory Physiology*, New York: Springer-Verlag, 1981.)

mechanisms of nociceptors have revealed mechanisms involved in sensitization of nociceptors that has importance for explaining chronic pain.[48]

2. Media That Conduct the Stimulus to the Receptors

The physical stimulus for the mechanoreceptors in the skin is deformation of the skin, which is conveyed to the receptor cells through various kinds of tissue. The tissue conducts the stimulus to the sensitive part of the receptors and transforms the stimulus in various ways before it reaches the receptors, depending on the properties of the tissue that the physical stimuli must pass through before it activates the receptors.

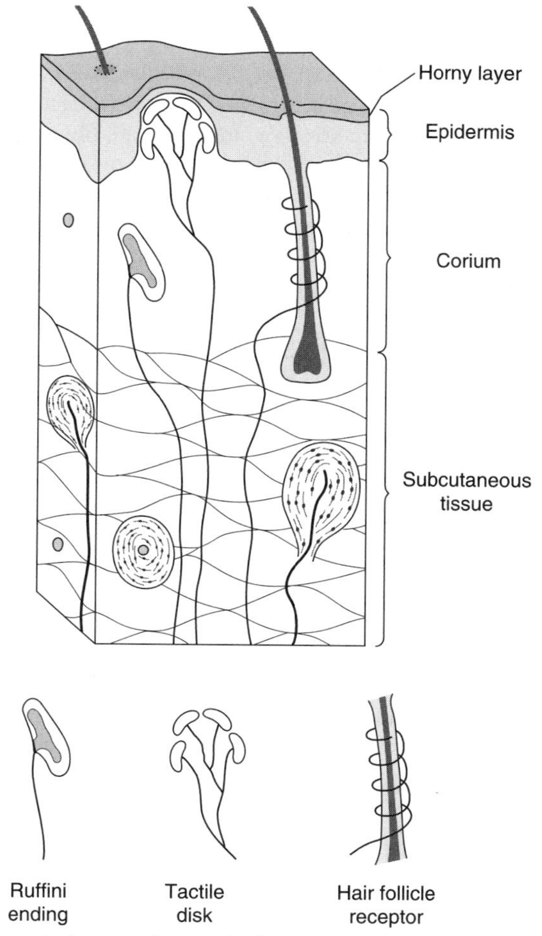

FIGURE 4.2 Schematic drawings showing the location of three types of mechanoreceptors in hairy skin. (Adapted from Schmidt, R.F., *Fundamentals of Sensory Physiology*, New York: Springer-Verlag, 1981.)

Differences in the properties of the media that conduct the physical stimulus to the receptors contribute to the different way the receptors respond to deformation of the skin. The adaptation of receptors in the skin is to a great extent caused by the media that conduct the stimulus to the receptors. On the basis of their adaptation (see page 222), receptors in the skin have been divided into *rapid adapting* (RA) and *slow adapting* (SA) receptors (see Fig. 4.21). The properties of the receptors also contribute to adaptation of receptors (see page 58).

3. Innervation of Mechanoreceptors

Meissner's corpuscles, Merkel's discs, Ruffini endings, and Pacinian corpuscles are innervated by large myelinated fibers (Aβ-fibers), while some receptors on hair follicles that respond to light stroking are innervated by smaller fibers (Aδ). Thermal receptors (cool and warmth) are innervated by Aδ-fibers (see page 52 and Fig. 4.34). It has also been shown that some low-threshold mechanoreceptors in humans that respond to innocuous stimulation are innervated by unmyelinated C-fibers.[110]

The axons of mechanoreceptors of the skin of the body travel in peripheral nerves together with those of other receptors and motor nerve fibers. These sensory fibers (from the surface of the body, excluding the head) enter the spinal cord together with proprioceptor fibers through the dorsal roots of spinal nerves. (Motor fibers exit from the spinal cord as ventral roots.) The dorsal root fibers have their cell bodies in the dorsal root ganglia.

Most of the receptors of the skin of the face are innervated by the sensory part (portio major) of the *trigeminal nerve* (CN V) together with receptors of the mouth.[47] (Portio minor of the trigeminal nerve is a motor nerve that innervates muscles of mastication, some muscles in the pharynx, and the tensor tympani in the middle ear). The cell bodies of the sensory portion of the trigeminal nerve are located in the trigeminal ganglion (Gasserian ganglion). Some skin receptors around the ear and in the ear canal are innervated by fibers of the *glossopharyngeal nerve* (CN IX) and the *vagus nerve* (CN X). Receptors in the pharynx, the upper throat, and the larynx are innervated by the glossopharyngeal nerve (CN IX).

a. Dermatomes

The patches of skin that are innervated by an individual dorsal root of the spinal cord are known as spinal *dermatomes*. In a similar way, the patches of skin of the face that are innervated by each of the three branches of the trigeminal nerve are known as dermatomes of the face.

Spinal Dermatomes The dermatomes of the body are labeled according to the vertebra (C_2–C_8; T_1–T_{12}; L_1–L_5; S_1–S_4) from which the dorsal root that innervates the dermatome originates (Fig. 4.3). The dermatome boundaries are not as distinct as they often appear on drawings of dermatomes, such as those of Fig. 4.3; there is a considerable overlap of dermatomes for light touch, and any point on the skin is innervated by at least two dorsal roots. (Dermatomes for pain have less overlap, see page 208.) The dermatomes have a certain individual variation in size and location, which is evident from the differences that are present in different published maps of dermatomes.

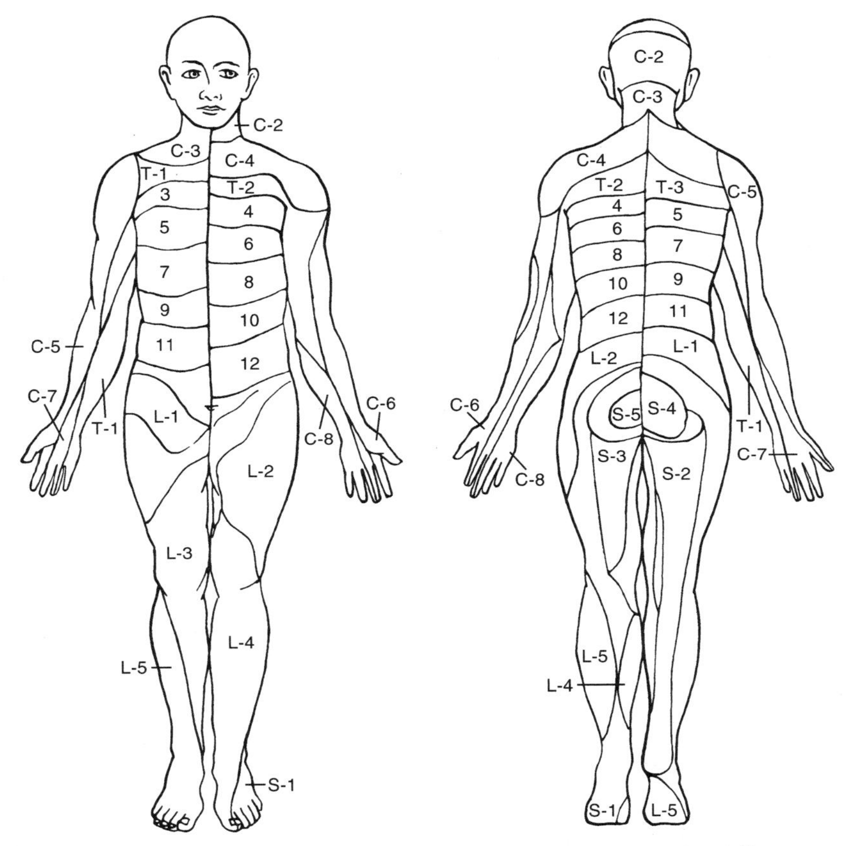

FIGURE 4.3 Regions of the skin that are innervated by dorsal roots of the spinal cord (dermatomes).

Dermatomes of the Face The trigeminal nerve has three branches (thus the name), usually labeled by the Roman numeral V followed by Arabic (or sometimes Roman) numbers for the different branches. The first branch (V1) innervates dermatomes of the upper part of the face, the second branch (V2) the middle part, and the third branch (V3) innervates the lower face, including the mouth (Fig. 4.4). Some axons from sensory receptors of the face, the mouth, and upper pharynx are innervated by cranial nerves IX and X, which contain fibers that innervate skin receptors, particularly in the region around the ear and of the ear canal. The skin of the upper part of the head, forehead, and scalp, in addition to being innervated by the V1 branch of the trigeminal nerve, is also innervated by the C2 and C3 cervical spinal roots (the C1 vertebra has no dorsal root).

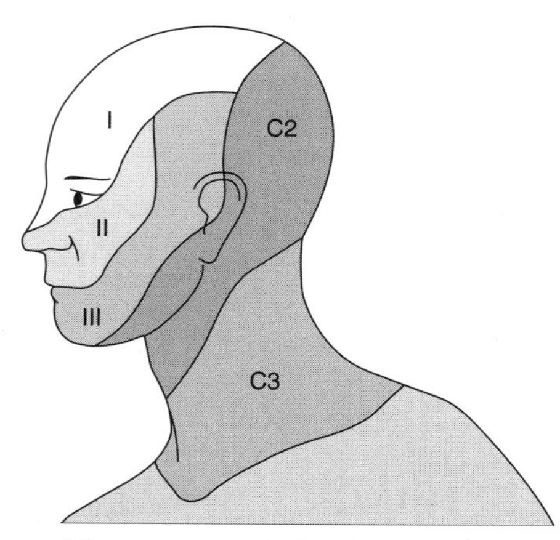

FIGURE 4.4 Innervation of skin receptors in the face (dermatomes) by the three different branches of the trigeminal nerve (I, II, III) and innervation by two cervical dorsal roots (C2 and C3; C1 has no dorsal root).

Dermatome maps are important for diagnostic purposes. Probing the skin in individuals with spinal cord injuries can help determine the level of the spinal cord where the injury is located. Thus, an individual with a spinal cord injury will respond to light pinpricks of the skin from areas that are innervated by dorsal roots that enter the spinal cord above the level of injury but not below. Because of the considerable overlap of dermatomes, injury to a single dorsal root results in only a partial sensory deficit that may extend over several dermatomes. In the face, probing dermatomes provide information about which branch of the trigeminal nerve may be affected by disease processes.

4. Ascending Classical Somatosensory Pathways

The ascending classical somatosensory pathways consist of the spinal and trigeminal pathways. The main spinal ascending pathway of the classical somatosensory system is the *dorsal column system*, which communicates several forms of fine touch and vibration but does not seem to be necessary for the perception of temperature and light touch. The input to that system comes through dorsal root fibers that carry responses to innocuous stimulation. The trigeminal system conveys information from receptors in the face and mouth to higher centers in a similar way as the dorsal column system. The somatosensory system is the least complex of all the ascending sensory pathways, and the pathway from the receptors to the primary sensory cortex has fewer relay nuclei than any of the other sensory systems; there are few or no uncrossed fibers and no significant connections between the two sides.

After entering the spinal cord, dorsal root fibers ascend as the *dorsal column tract*, the fibers of which terminate in the *dorsal column nuclei*.[6,79] The dorsal root fibers also give off collaterals that make synaptic contact with cells of the dorsal horn.

The classical ascending somatosensory pathway from the head is the trigeminal sensory system.[102] The fibers of the trigeminal (sensory) nerve terminate in the *trigeminal nucleus*, the axons of which join the axons of the cell in the dorsal column nucleus to form the *medial lemniscus* (ML), which crosses the midline before reaching the thalamic relay nuclei, which projects to the primary sensory cortices.

a. Dorsal Column Tract

The uninterrupted ascending branch of dorsal root fibers ascends on the same side of the spinal cord as the dorsal column tracts, consisting of the *cuneate funiculus*, which carries information from the upper body, and the *gracilis funiculus*, which carries some of the information of the lower body (mainly information from skin receptors) (Fig. 4.5).[6] The cuneate funiculus travels laterally to the gracilis funiculus in the cervical spinal cord. The first synapses of each of these two dorsal column tracts are located in the *nucleus cuneatus* and *nucleus gracilis*, respectively. These nuclei together are known as the dorsal column nuclei.

The dorsal column system of the upper body (cuneate funiculus) also carries proprioceptive information from muscles and joints to the dorsal column nuclei in addition to information from skin receptors.[79] The cuneate funiculus contains fibers that innervate receptors in the skin, muscle, and joints (proprioception) while the dorsal column tract from the lower body (fascicules gracilis) mainly carries information from skin receptors and mostly SA receptors.[31,55,56] The fibers from the lower body that innervate fast adapting muscle and joint receptors travel in a *lateral funiculus*.

The ascending dorsal root fibers have collaterals that make synaptic contact with cells of the dorsal horn, not only at the spinal level where the root enters the spinal cord, but also at 5 to 7 segments up the spinal cord. A collateral fiber of the descending branch gives off collaterals to dorsal horn cells at several levels down from the point where the dorsal root enters the spinal cord (Fig. 4.6). Some collaterals reach vertically to motor neurons in the dorsal horn, thus serving as the pathway for spinal reflexes. The ascending branch of these fibers curve dorsally when they enter the spinal cord and ascend uninterrupted as the dorsal column on the ipsilateral side of the spinal cord and ascend as the dorsal column tract. Collaterals of these dorsal root fibers terminate on cells in Rexed's[93] layer III, IV, V, and VI of the dorsal horn (Fig. 4.6).[6] The descending branch of dorsal root fibers gives off collaterals, which make synaptic contact with dorsal horn cells several (7 to 10) segments below (caudally to) the entry of the dorsal roots. These descending fibers also give off collaterals that terminate on dorsal horn cells. Some of the collaterals extend as far ventrally as cells in the ventral horn, where they terminate on motor neurons and thus mediate fast reflexes.

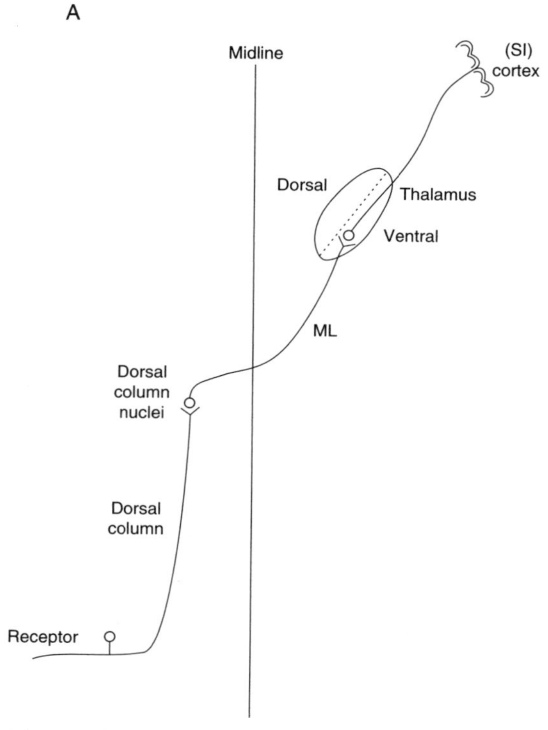

FIGURE 4.5 (A) Schematic diagram of the classical ascending spinal somatosensory pathway (dorsal column system). Fibers of dorsal roots ascend as the dorsal column and reach the dorsal column nuclei on the same side. Axons of cells in the dorsal column nuclei cross the midline as the medial lemniscus (ML) and reach cells in the ventral thalamus, from which fibers ascend to the primary somatosensory cortex (SI). (B) Anatomical locations of the main components of the ascending spinal somatosensory pathways.

The dorsal root fibers that innervate receptors that respond to noxious stimulation (Aδ- and C-fibers) do not ascend in the dorsal column, but these fibers and their collaterals terminate in layers I and II (*substancia gelantinosa*), and some Aδ-fibers terminate in layer V of the dorsal horn (Fig. 4.6).[6] These parts of the dorsal horn belong to the nonclassical somatosensory system that will be discussed later. (see page 206). Many dorsal horn cells belong to an internal network of neurons (interneurons) that do not send ascending fibers.

Dorsal Column Nuclei All fibers in the dorsal column are interrupted by synaptic transmission in the dorsal column nuclei (the gracilis and cuneatus nuclei), which are located in the lower medulla. The cuneate nucleus is located more lateral than the gracilis nucleus, which is located close to the midline.

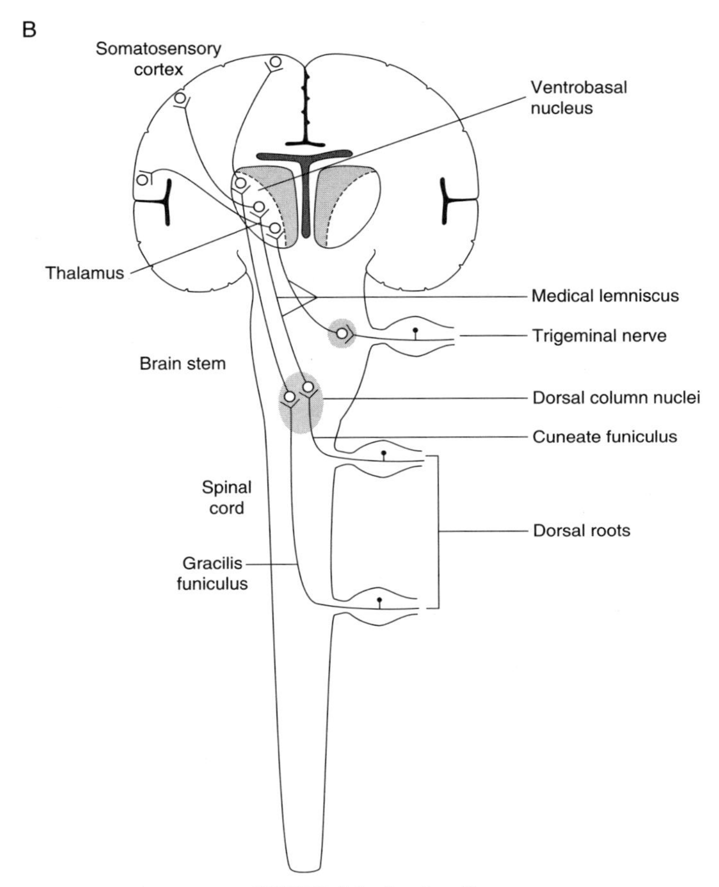

FIGURE 4.5 *(continued)*

The fibers that carry proprioceptive input from the lower body as well as some afferents of low threshold cutaneous receptors travel in a dorsolateral fiber tract in the spinal cord are interrupted by synaptic contacts with cells in the *nucleus Z* instead of the gracilis nucleus.[31,55,67] The nucleus Z is located slightly more rostral and medial than the gracilis nucleus.[55] The axons from the cells in the dorsal column nuclei and the nucleus Z cross to the opposite side and ascend together as the ML.[4,6]

b. Trigeminal Pathways

The trigeminal pathway communicates sensory tactile and temperature information, including pain, from the face and mouth. The trigeminal nerve

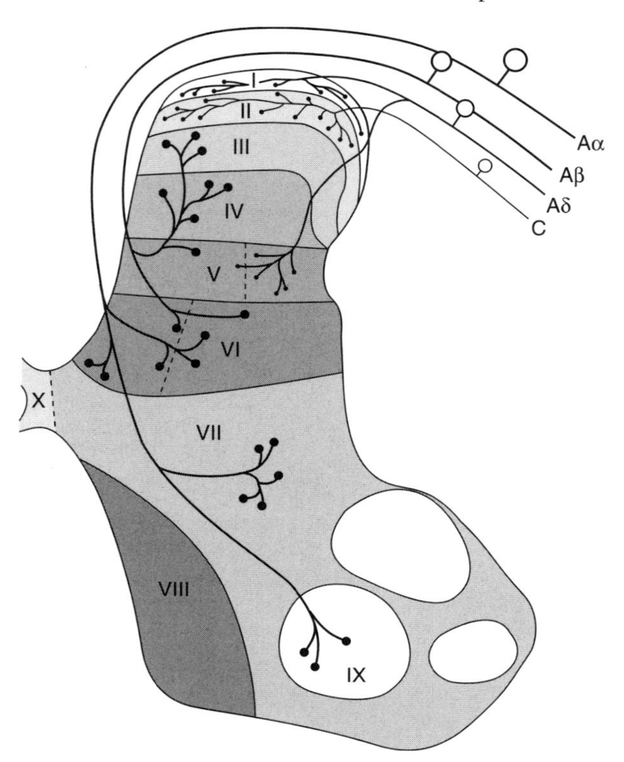

FIGURE 4.6 Schematic illustration of how different kinds of dorsal root fibers terminate on cells in different layers of the dorsal horn of the spinal cord at the location of the entry of the dorsal root (Adapted from Brodal, P., *The Central Nervous System*. New York: Oxford Press, 1998.)

(CN V) has two parts, a sensory and a motor part (portio major and portio minor, respectively). The motor part innervates muscles of mastication and muscles in the pharynx and the tensor tympani muscle. The focus of this book is sensory systems; therefore, we limit the discussion of the trigeminal nerve to the portio major.

The Trigeminal Nucleus The fibers of the three branches of the sensory part of the trigeminal nerve reach the different parts of the *trigeminal sensory nucleus* (Fig. 4.7) on the same side.[47,102] The trigeminal (sensory) nucleus has many similarities with both the dorsal horn of the spinal cord and the dorsal column nuclei. Anatomically it is an elongated nucleus that has several parts extending caudally from the rostral midbrain to the upper spinal cord. The rostral part of the trigeminal nucleus corresponds to the dorsal column nuclei, and that part of the nucleus can be regarded as belonging to the classical

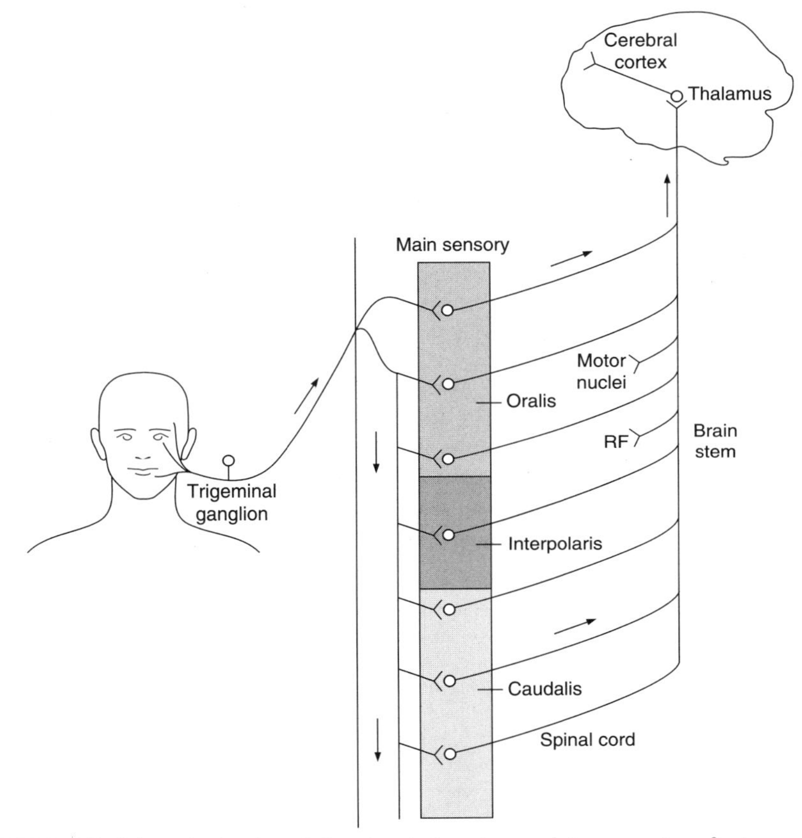

FIGURE 4.7 Schematic drawing of the trigeminal nucleus and its connections for transmitting sensory information from the mouth and the face to central structures. (Adapted from Sessle, B.J., *J. Endon*. 12: 435–444, 1986.)

somatosensory system. The caudal (spinal) portion may be regarded as a part of the nonclassical somatosensory system that is involved in pain (see page 205) and it has similarities with the dorsal horn of the spinal cord. Fibers that innervate tactile receptors in the face and the mouth and which are innervated by myelinated fibers terminate in the rostral part of the trigeminal nucleus. The trigeminal nucleus also receives input from fibers of the glossopharyngeal and vagus nerves (CN IX and X), which innervate receptors in the mouth, pharynx, and the ear canal. Fibers from the trigeminal nucleus cross the midline to reach the ML and join fibers that originate in the dorsal column nuclei (Fig. 4.5B).

5. Medial Lemniscus

After crossing the midline, the ML continues uninterrupted to the somatosensory nuclei in the ventral part of the thalamus. During their course, the fibers of the ML give off collaterals to the *brainstem reticular formation*. The brainstem reticular formation influences the excitability of large portions of the cerebral cortex (*arousal*). The reticular formation also exerts control over the autonomic nervous systems eliciting mainly sympathetic activity. A high degree of arousal is associated with high sympathetic activation.

6. Somatosensory Thalamic Nuclei

The main somatosensory thalamic nuclei of the classical somatosensory system are the *ventrobasal nuclei* (VB). The fibers of the ML terminate in the VB and make synaptic contact with cells in the *ventroposterior lateral* (VPL) and the *ventroposterior medial* (VPM) nuclei of the VB. The somatic afferents from the body reach the VPL, and those from the face (the trigeminal nerve) reach the VPM from the rostral trigeminal nucleus (Fig. 4.8).[6]

Studies of these pathways have mainly been performed in animals. The thalamus in humans is different from that of animals including monkeys. Studies in humans are few, some of which[59,61] have shown that neurons in the core area of the ventrocaudal nucleus of the thalamus respond to innocuous somatosensory stimulation. It has been concluded that this nucleus is the main somatosensory nucleus where all somatosensory information is interrupted by synaptic transmission.

7. Somatosensory Cortices

The two main *somatosensory cortical areas* are the *SI* and the *SII*. The SI is the *primary somatosensory cortex* and is located posterior to the *central sulcus* (Fig. 4.9). The much smaller SII is the secondary receiving area, which is located closer to the central sulcus on its upper wall. The monkey has been the most studied species, in that respect, and much more is known about the somatosensory cortex of the monkey than of any other species including humans.

a. Primary Somatosensory Cortex (SI)

The neurons of the SI receive input from the VB of the thalamus on the same side as SI thus information only from the opposite side of the body. Layers V and VI of the SI cortex send fibers back to the thalamus as a part of the reciprocal innervation that is abundant in sensory pathways.[72] There are also connections from the SI to other cortical areas. Four distinctly different regions of SI in the postcentral gyrus have been identified (3a, 3b, 1, 2) (Fig. 4.9).

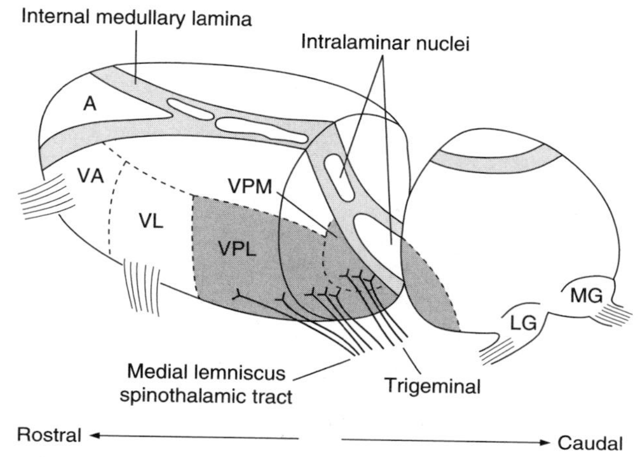

FIGURE 4.8 Schematic drawing of the thalamus showing the anatomical locations of the ventral posterior medial (VPM) and ventral posterior lateral (VPL) nuclei indicating their input from the trigeminal nerve and the medial lemniscus. VL, ventral lateral nucleus; VA, ventral anterior nucleus; A, anterior thalamic nucleus; LG, lateral geniculate body, MG, medial geniculate body.[6]

Each individual part of the SI region of the somatosensory cortex receives its input from different parts of the thalamus (Fig. 4.10).

Area 3b receives approximately 70% of the axons from the thalamus (VPL and VPM), thus it receives more input than the three other regions together. The input to area 3b terminates in layers III and IV, from which neural activity spreads to other cortical layers. Layer IV of the three other regions of SI receives input from layers II and III of area 3b. Area 3b is the first step in the *serial processing* that occurs in the cerebral cortex, and the three other regions are regarded to perform higher order processing. Extensive processing also occurs within each cortical region through connections between neurons in the different layers of the cortex. Area 3a receives input from deep body receptors, especially muscle afferents from the *ventralis intermedius* (VI) nucleus of the thalamus. Area 1 and 2 receive their thalamic input from the VPL and VPM nuclei and *ventralis intermedius* (VI) nucleus. Area 2 receives input from deep receptors, while area 1 and 3b receive input from cutaneous receptors. These two cortical areas are the second step in the serial processing that occurs in the primary somatosensory cortex. The neurons of areas 3a, 1, and 2 of the primary somatosensory cortex send connections to the secondary somatosensory cortex (SII). These connections are also reciprocal, which means that there are connections between these parietal regions and the primary somatosensory cortex.

Connections between neurons in the different layers of the cortices are extensive and important for information processing in the different parts of the primary somatosensory cortices.[72] In general, information entering layer IV connect to layer III and from there to layer I, II, and back to layer IV. Neurons in layer II connect to neurons in layer V and VI.[72]

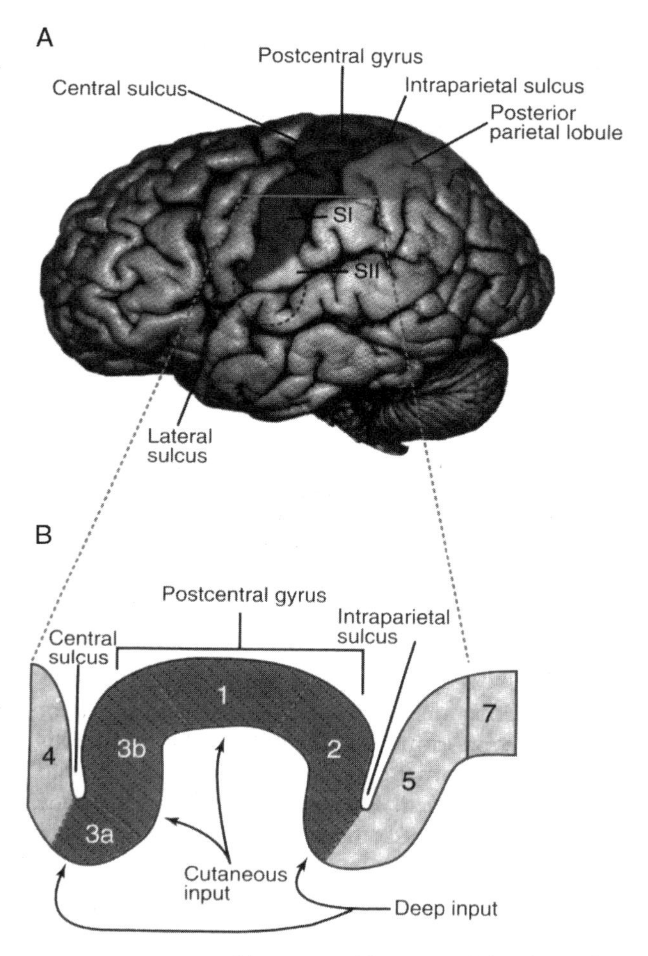

FIGURE 4.9 Somatosensory cortices: (A) Location of the primary (SI) and secondary somatosensory cortex (SII). (B) The primary cortex and its divisions in four different regions. (From Zigmond, *et al.*, *Fundamental Neuroscience*. San Diego: Academic Press, 1999.[133] With permission.)

b. Secondary Somatosensory Cortex

The secondary cortical area is located on the lateral end of the *postcentral gyrus* near the upper wall of the *sylvian fissure* of the temporal lobe (Fig. 4.11). The SII area is smaller than the SI, approximately one fourth the size of the SI. The SII area is phylogenetically older than the SI. Neurons of the SII area receive input from neurons in the SI and also directly from the somatosensory ventral

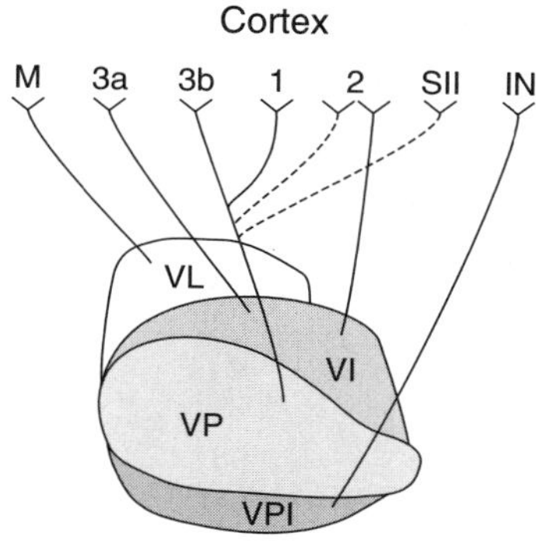

FIGURE 4.10 Connections between the thalamus and the different regions of the primary (SI) and secondary somatosensory cortex (SII). VI, ventralis intermedius nucleus; VL, ventralis lateralis nucleus; VP, ventroposterior nucleus; VPI, ventroposterior inferior nucleus; M, motor-sensory area; 3a, 3b, 1, 2, divisions of the primary somatosensory cortex (see Fig. 4.9); IN, insular cortex. (Adapted from Merzenich and Kaas.[68])

thalamus (VPL and VPM). The neurons of the SII cortex send extensive connections back to these thalamic nuclei (reciprocal innervation). The input from SI neurons reach layer IV of the SII cortex. The neurons of the SII also receive input directly from the VB thalamic complex, but different from the SI, the SII cortical neurons receive input from *both sides of the body*. Input from the ipsilateral body also arrives at the SII area through fibers of the *corpus callosum*. Neurons in SII send axons to the structures of the limbic system and the insular cortex and from there to the amygdala and the hippocampus (see Chapter 3).

c. *Parietal (Secondary and Association) Cortical Areas*

The two cortical areas that are located posterior to Brodman's areas area 2 of the SI, namely Brodman's areas 5 and 7 (Fig. 4.12), also receive somatosensory input and consequently respond to stimulation of somatosensory receptors. The rostral portion, areas 5a and 7b, are innervated by area 2 (in monkeys) and respond to low-intensity somatosensory stimulation.

B. Nonclassical System

In this book, the anteriolateral system is regarded as being the nonclassical ascending somatosensory system together with the caudal trigeminal system.

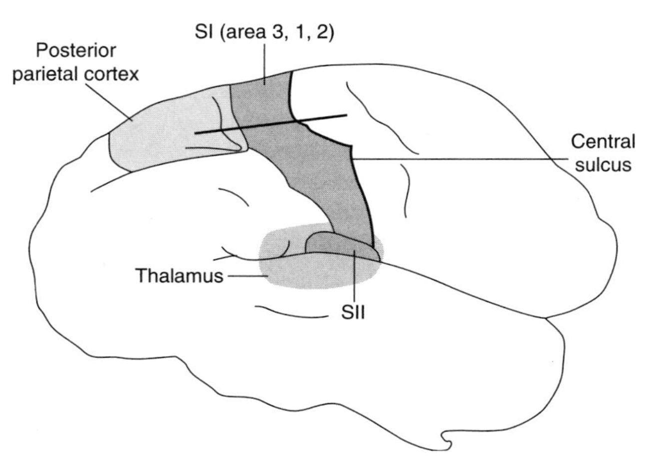

FIGURE 4.11 Locations of the somatosensory cortical areas.

The receptors and pathways of the nonclassical system differ from those of the classical system in that they respond primarily to noxious stimulation.

1. Receptors

The input of the nonclassical pathways mostly comes from nociceptors located in the skin, tendons, joints, and viscera. Mechanoreceptors located in deep tissue also serve as input to the nonclassical somatosensory pathways.

Three main classes of nociceptors have been identified: mechanoreceptors, heat and cold receptors, and polymodal receptors. These three types of nociceptors are distributed widely in the skin and deep tissue and can be activated by mechanical, chemical, or thermal stimulation. Polymodal nociceptors are high threshold receptors that can be activated by noxious stimuli of several modalities. In addition, *silent nociceptors* are located in viscera. These receptors are not normally activated by noxious stimulation; however, their threshold can be lowered by inflammation and certain chemical insults. Receptors that mediate pain are primarily bare axons. Itch has many similarities to pain, and it is believed that bare axons different from pain receptors mediate itch.

It was shown many years ago that menthol can potentiate the response from trigeminal fibers that respond to cold by shifting their threshold for activation by cold toward warmer temperatures.[39] More recently the molecular basis for such activation has been studied by cloning the receptor channels.[63] Capsaicin activates a receptor that is also activated by heat (above 43°C).[106] Specific membrane channels have been identified that respond to noxious stimuli such as capsaicin.[11] This means that thermoreceptors can respond to specific substances such as capsaicin, which activates heat receptors, and menthol, which activates cool receptors.

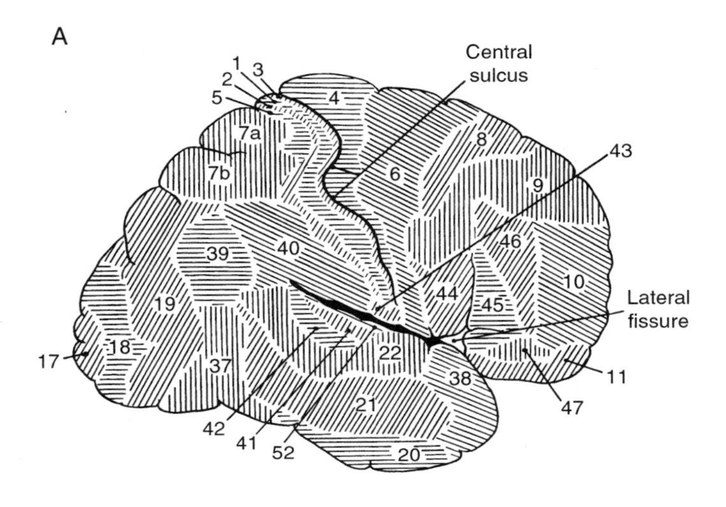

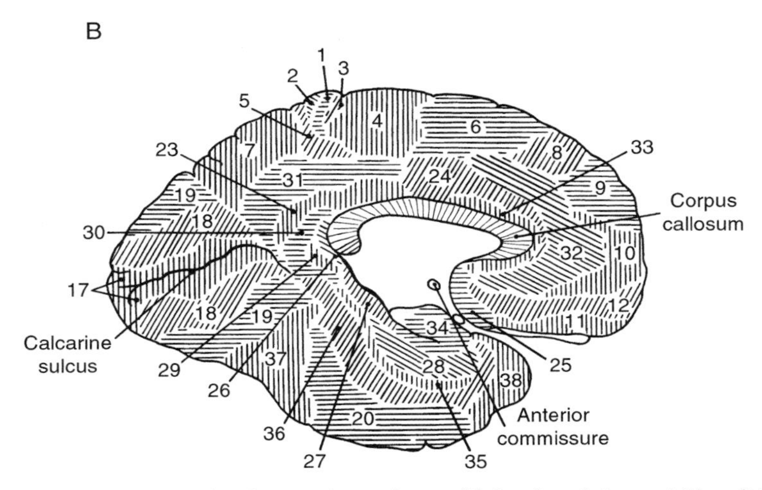

FIGURE 4.12 Brodmann's classification of cortical areas: (A) dorsolateral view, and (B) medial view. (From Everett, N.B., *Functional Neuroanatomy*. Philadelphia: Lea & Fibiger, 1971. With permission.)

a. Innervation of Nociceptors

Nociceptors (pain and heat and cold receptors) are innervated by small-diameter myelinated fibers (Aδ) and unmyelinated fibers (C1), (see page 51). These fibers also enter the spinal cord in the dorsal roots together with other sensory fibers. Similar to sensory fibers, the fibers that innervate nociceptors innervate patches of skin. The area of skin that they innervate is related to where they enter the spinal cord. This means that nociceptors also have dermatomes for pain. These dermatomes are similar to the dermatomes for innocuous stimulation, but the dermatomes for pain tend to overlap less than the sensory dermatomes.

Nerve fibers that innervate nociceptors in the face and the mouth follow the paths of the sensory fibers in the three branches of the trigeminal nerve, and dermatomes similar to those for somatic stimulation can be identified for pain.

Two types of receptors that are innervated by Aδ-fibers mediate mechanical and heat nociceptive stimulation.[38] One type that has a high heat threshold (approximately 53°C) is found in glabrous skin. The other type of thermoreceptor has a lower threshold (approximately 43°C) and are located in hairy skin. These receptors are innervated by Aδ-fibers that conduct at a rate of 8 to 15 m/sec.

2. Dorsal Horn

Much can be explained about pain from observing the organization and function of the dorsal horn. The afferent nerve fibers that innervate nociceptors (Aδ- and C-fibers) enter the spinal cord in the dorsal roots, together with fibers that carry sensory information from receptors that respond to innocuous stimulation (Fig. 4.6). Dorsal root fibers bifurcate when they enter the spinal cord, and one branch ascends and the other descends several (7 to 10) segments. The dorsal root fibers that innervate nociceptors and their collaterals make synaptic contacts with dorsal horn cells at each of these segments (Fig. 4.13). The axons of these dorsal horn neurons cross the midline at segmental levels and ascend as the anterior lateral system. The fact that some of the dorsal root fibers travel ventrally in the spinal cord and make synaptic contact with motoneurons is the basis for spinal reflexes, such as the withdrawal reflex.

Different layers of dorsal horn cells receive fibers of different types (Fig. 4.6).[6] The fibers that innervate nociceptors terminate in cells in the superficial layer of the dorsal horn (lamina I and II of Rexed),[6,93] also known as *substantia gelatinosa*. Large-diameter dorsal root fibers (Aα and Aβ) mediate innocuous sensory information and terminate in cells in deeper (more ventral) layers of the dorsal horn (layers III to VI). Some Aδ-fibers reach lamina V.

The axons of the cells in lamina I and II of the dorsal horn (on which thin fibers terminate) cross at the segmental level after which they ascend as the *anteriorlateral tracts*, which here are regarded as part of the nonclassical spinal somatosensory system. It is important to note that the dorsal horns contain complex networks of interneurons that connect neurons in different layers and which reach several segments of the spinal cord (Fig. 4.13).

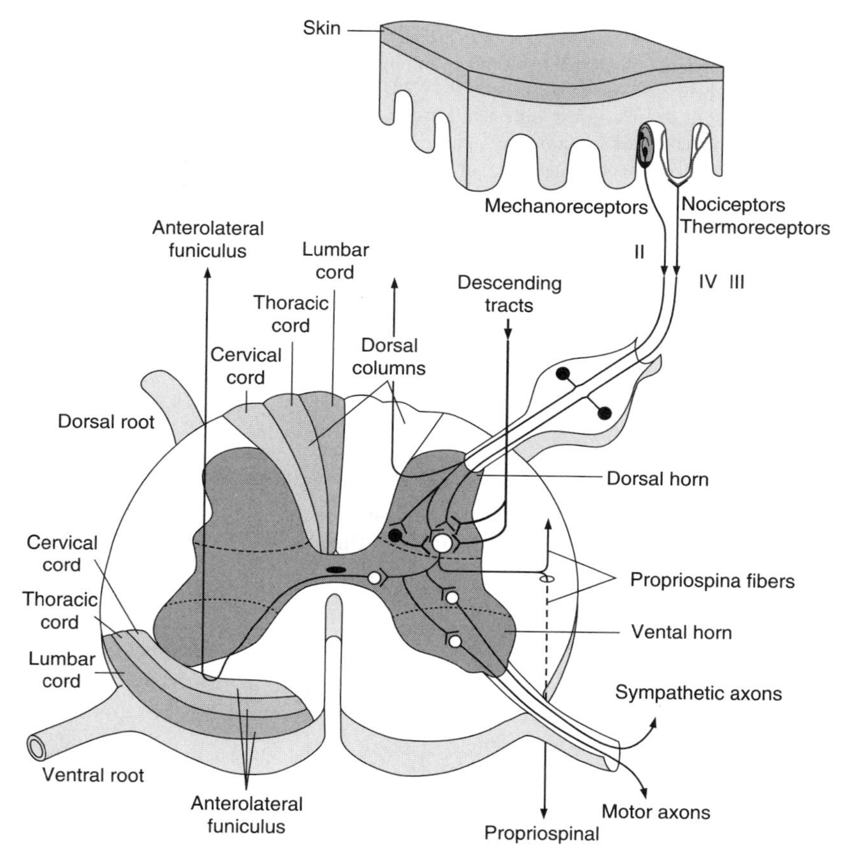

FIGURE 4.13 Termination of dorsal and ventral root fibers in cells of the dorsal horn of the spinal cord at the segment where the roots enter the spinal cord. (Adapted from Schmidt, R.F., *Fundamentals of Sensory Physiology*. New York: Springer-Verlag, 1981.)

Four different groups of cells of the dorsal horn give rise to the anteriorlateral tracts after crossing the midline at the segmental level:[6]

1. Cells that are innervated by low-threshold mechanoreceptors that respond to light touch
2. Cells that are innervated by high-threshold nociceptors
3. Cells that are innervated by heat-sensitive skin receptors
4. *Wide dynamic range* (WDR) neurons

Wide dynamic range neurons are located throughout the dorsal horn but are found most densely in deeper layers. They receive input from several different pain receptors. The input to WDR neurons is mediated through nerve fibers (Aβ, Aδ, and C) with varying conduction velocities (Fig. 4.14): high-threshold mechanoreceptors and heat receptors that are innervated by Aδ-fibers and by several kinds of nociceptors that are innervated by unmyelinated C-fibers.[6,86] WDR neurons are also innervated by collaterals of Aβ-fibers that innervate low-threshold mechanoreceptors and which ascend in the dorsal column tract, which is a part of the classical somatosensory pathways that normally respond to innocuous stimulation of the skin. Some of the input from these large myelinated fibers that make synaptic contact with the WDR neurons is inhibitory. The WDR neurons also receive input from high-threshold mechanoreceptors, and that input is excitatory.

The WDR neurons send axons ascending mainly in the *spinothalamic tract* (a part of the anterior lateral system), after crossing the midline. The WDR neurons can respond within a large range of stimulus intensities and are involved in pain that is caused by neural plasticity (see page 253).

3. Anteriorlateral Pathways

The anteriorlateral system consists of several separate systems that receive their input from dorsal root fibers that innervate nociceptors and by mechanoreceptors that are located deep in the body. These fibers, including

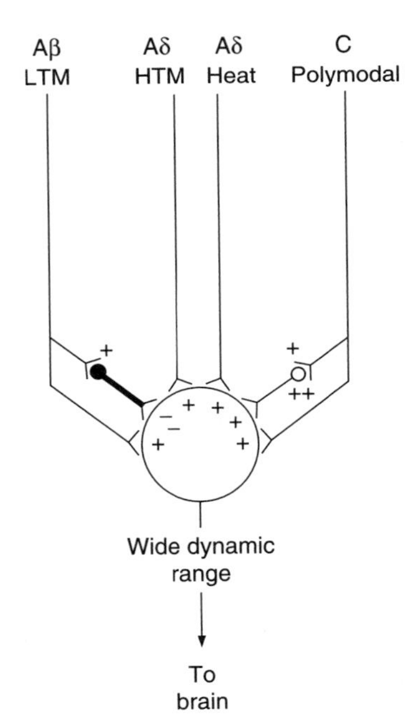

FIGURE 4.14 Input to WDR neurons. LTM, low-threshold mechanoreceptors. HTM, high-threshold mechanoreceptors. (Adapted from Price, D.D., Long, S., and Huitt, C., *Pain*. 49: 63–173, 1992.)

their collaterals, are interrupted in the dorsal horn of the spinal cord, mainly in laminae I and II (substantia gelatinosa) where they make synaptic contacts with dorsal horn cells. Some Aδ-fibers also terminate in neurons in layer V (see page 200). Fibers from receptors in viscera terminate in layers I and V.

The axons of these dorsal horn cells cross the midline at segmental levels and ascend and form the anterior lateral tracts. Some fibers of the anterior lateral tract also receive input from other cells in the dorsal horn.

The main anatomical difference between the anteriorlateral system and the dorsal column system can be found where these tracts cross the midline. The anteriorlateral pathways are interrupted by synapses at cells in the dorsal horn and of the spinal cord, and the axons of these cells cross the midline at the segmental level to form the ascending (anteriorlateral) tracts. The dorsal column system crosses the midline first after making synaptic contact with cells in the dorsal column nuclei, which are located in the lower medulla.

Another important difference between the anterior lateral and dorsal column system is that the anteriorlateral system uses the medial and dorsal thalamic relay nuclei, while the dorsal column system uses the ventral thalamic nuclei (VB) as their obligatory relay nuclei. The anteriorlateral system has more direct connections to many different parts of the CNS than the classical somatosensory system.[6,19] For the head, several cranial nerves carry information similar to that of the anteriorlateral system, and that information is also interrupted by synaptic communication in the dorsal thalamus.

The anteriorlateral system consists of at least four separate systems that all are important for perception of pain resulting from activation of nociceptors. (Pain will be discussed in a separate section, page 241).

a. Spinothalamic Tract

The *spinothalamic* system is the most important of the four main ascending pathways that make up the anteriorlateral pathways (at least it is the best known of these systems) (Fig. 4.15). This tract is regarded as the main mediator of pain. Because of its great clinical importance, the connections of the spinothalamic tract with cells in the various nuclei of the thalamus have been studied extensively (see Wall and Melzack,[114] Craig and Dostrovsky,[19] and Simpson[103]).

The spinothalamic tract is mainly crossed, and it projects to the ventral and dorsal thalamus without interruptions. The fibers of the spinothalamic tract that originate in the dorsal portion of the dorsal horn (laminae I and II) give off many collaterals to the reticular formation in the brainstem and to the periaqueductal gray (PAG) and terminate in various nuclei of the thalamus. The axons from cells in deeper layers of the dorsal horn (laminae IV and V, which contain WDR

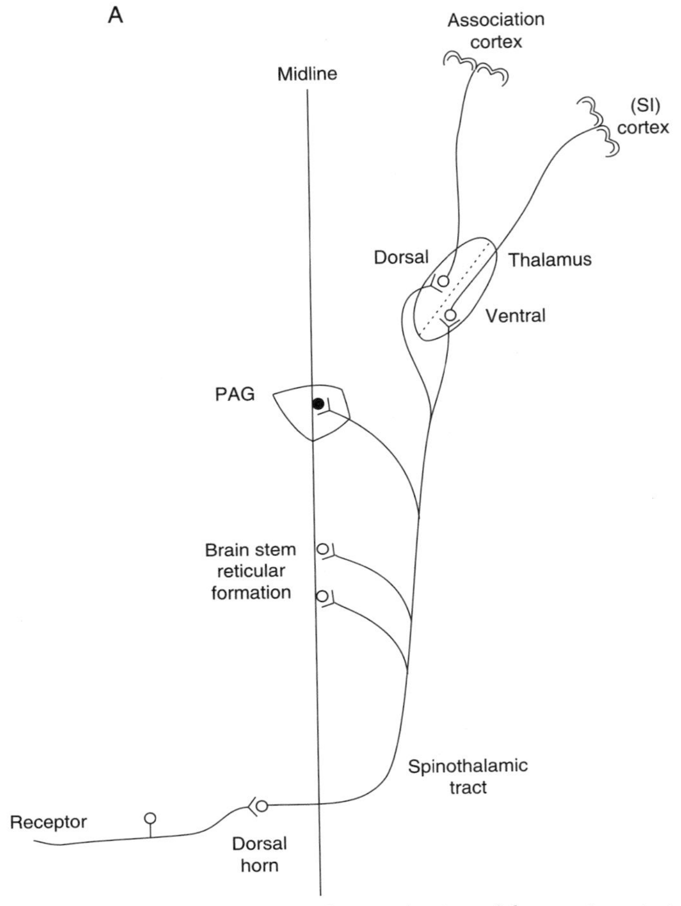

FIGURE 4.15 The spinothalamic tract. (A) Schematic drawings of the most important parts of the spinothalamic pathways. (B) Anatomical locations of the components of the spinothalamic pathways.

neurons) also ascend in the spinothalamic tract and send collaterals to the brainstem reticular formation.[6,19]

Most fibers of the spinothalamic tract terminate in the dorsal and medial thalamus. Some fibers also terminate in the VPL of the thalamus but in a slightly different region than those receiving input from the classical pathways through the medial lemniscus (see Fig. 4.8). In the monkey, the spinothalamic tract fibers that originate in laminae I and II of the dorsal horn and which innervate cold and polymodal nociceptive receptors terminate in the ventromedial nucleus of the posterior portion of the thalamus, the ventrolateral and

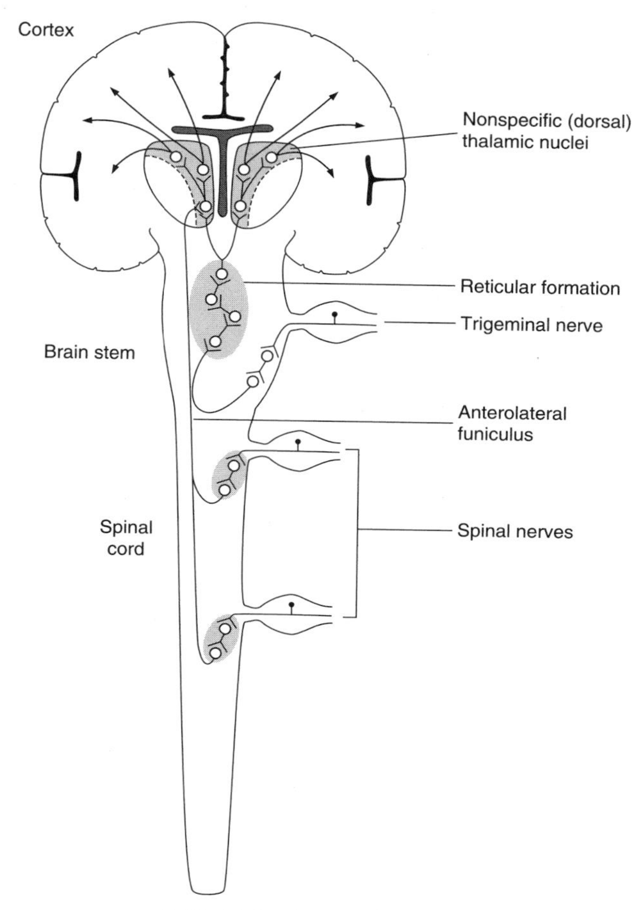

FIGURE 4.15 (*continued*)

central lateral nuclei, and the ventral caudal portion of the medial dorsal nucleus of the thalamus. The part of the spinothalamic tract that originates in layers IV and V of the dorsal horn terminates in the ventroposterior inferior nucleus, the ventroposterior nucleus, and the central nucleus of the thalamus. The fact that thalamic neurons share input from the skin and viscera is probably responsible for *referred* pain, where activation of pain receptors in the viscera are felt as pain at specific locations of the skin. Fibers of the spinothalamic tract also enter the *intralaminar nuclei* and the *posterior complex* (PO) of the thalamus (Fig. 4.8).[6]

The VPM and PO nuclei of the thalamus receive input both from the dorsal column system and from the anteriorlateral tract, mainly from the WDR nuclei of the dorsal horn of the spinal cord. It is believed that these neurons mediate information about the location of injury and pain. Input from WDR neurons reaches cells in the central-lateral portion of the thalamus (PO, VPL, and intralaminar nuclei) from where connections proceed to regions of the association cortex embracing the caudal parts of the secondary somatosensory cortex[32] and association cortices.[6,19]

b. Spinoreticular Pathway

The *spinoreticular tract* (Figure 4.16) has a peripheral course similar to that of the spinothalamic tract, with the exception that it ascends both on the ipsilateral and contralateral side of the spinal cord. This tract thus has bilateral central paths, while the other components of the anteriorlateral system (and the dorsal column system) are mainly unilateral (crossed) pathways. The fiber tract of the spinoreticular pathway is located in the anteriorlateral quadrant of the spinal cord. The fibers of the spinoreticular tract also give off collaterals to the

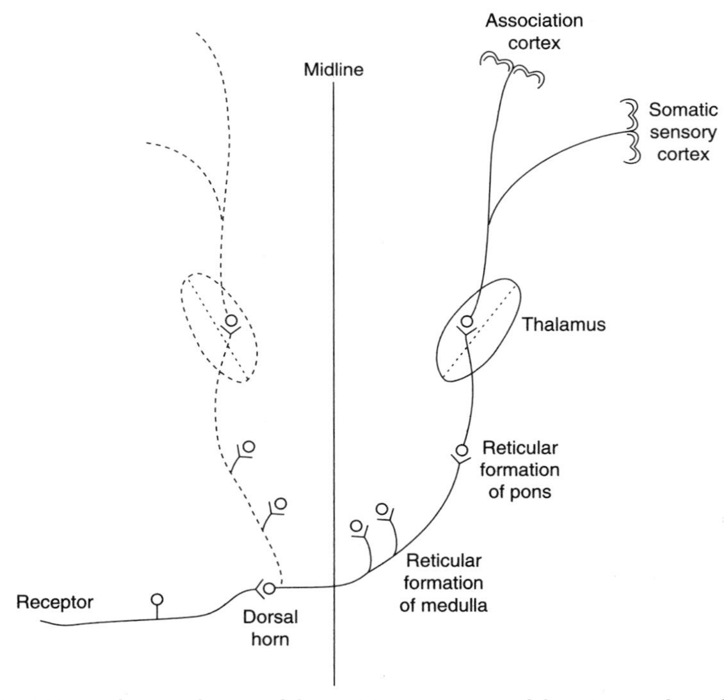

FIGURE 4.16 Schematic drawing of the most important parts of the spinoreticular pathways.

reticular formation of the medulla, and some fibers are interrupted in the reticular formation of the pons.[119]

The spinoreticular tract is important for mediating emotional (affective) components of pain. Many of the fibers of the spinoreticular tract terminate in cells in the intralaminar thalamic nuclei, which have been associated with relaying emotional (affective) components of pain. Input to the intralaminar nuclei is assumed to provide the discomfort and affective components of pain.

c. *Spinomesencephalic Tract*

The third main ascending tract for pain, the *spinomesencephalic tract* does not reach the cerebral cortex but terminates in the periaqueductal gray. The PAG is especially involved in pain (Fig. 4.17) and the spinomesencephalic tract is therefore important for pain.

d. *Other Spinothalamic Tracts*

The *spinocervicothalamic tract* is the fourth component of the anteriorlateral pathways. This tract originates in the lateral cervical nucleus, which is located

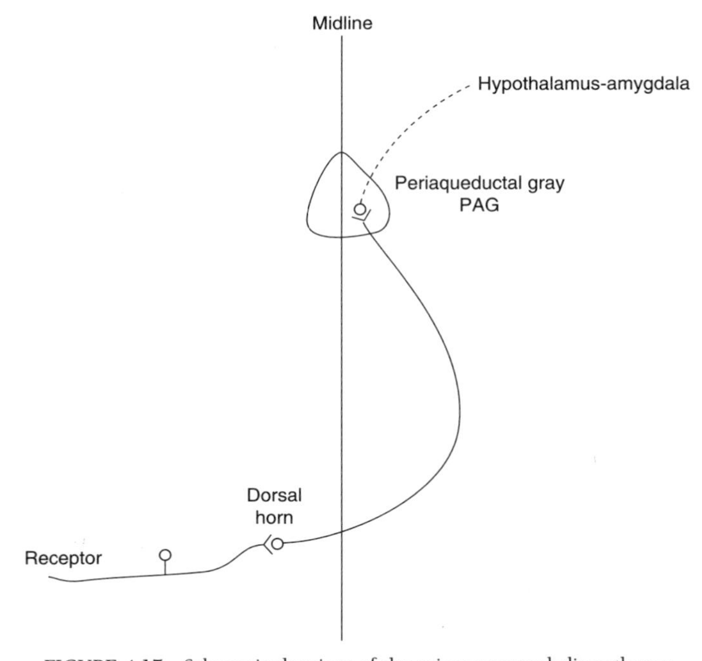

FIGURE 4.17 Schematic drawings of the spinomesencephalic pathways.

in the upper two cervical segments of the spinal cord, and it connects to the thalamus. In cats and monkeys, cells in the lateral funiculus at the cervical level (lateral cervical nucleus) give rise to the spinocervicothalamic tract, which conveys information from nociceptors and low-threshold mechanoreceptors to the VPL of the thalamus. This pathway may not exist in humans. A little-known tract, the *paleospinal tract* is regarded to be the earliest ascending somatosensory pathway to occur in the evolution of mammals. The medial nuclear group of the thalamus receives input from neurons in the dorsal horn of the spinal cord via the paleospinal tract.

4. Trigeminal Pathway

The sensory portion of the trigeminal nerve (the portio major of the fifth cranial nerve) communicates noxious stimulation of the face and the mouth. The first relay nucleus is the caudal trigeminal nucleus (Fig. 4.15B).[30,102] Some noxious stimulation is also mediated by other cranial nerves such as the glossopharyngeal (CN IX) and the vagus nerve (CN X).

The trigeminal nerve carries fibers that innervate nociceptors in the face and the mouth together with fibers that carry innocuous stimulation. The teeth are particularly well represented by pain fibers and these have often been used in studies of the physiology of pain.

a. Trigeminal Nucleus

Pain fibers from the face and the mouth, particularly the teeth, terminate in the caudal (spinal) portion of the trigeminal nuclei (see page 201). Some fibers of CN IX and X, which mediate painful stimulation from receptors in the back of the mouth and throat, also terminate in cells in the caudal trigeminal nucleus. The caudal trigeminal nuclei may therefore be regarded as corresponding to the nuclei of the dorsal horn of the spinal cord.

Fibers from the caudal nucleus of the trigeminal nerve (CN V) (Fig. 4.7) reach two thalamic nuclei that are located in the lateral part of the thalamus, namely the VPM nucleus and the posterior nucleus (PO) (see Fig. 4.8).[6,102] These nuclei are important for processing nociceptive information from the face. The caudal part of the trigeminal nucleus and its ascending tract are known for their role in a form of face pain (trigeminal neuralgia, TGN).[30]

5. Projections from the Dorsal and Medial Thalamus to the Cerebral Cortex

The regions of the thalamus that receive input from the anteriorlateral tract project to both SI and SII cortical regions and also to such areas as the *anterior cingulate gyrus*, the *insula*, and a small area between Brodmann's areas 1 and 3 (see Fig. 4.12). In addition some cell groups of the thalamus project to the hypothalamus and the amygdala.

While neurons of the VPL nucleus project only to somatic sensory cortical areas, neurons of the intralaminar nuclei project to larger areas of the cerebral cortex and to other parts of the brain such as the basal ganglia. Neurons of the posterior nucleus project to parietal cortical regions. The connections from the thalamus to the cerebral cortex have descending counterparts, and these connections may be regarded as being reciprocal.

The cells in the ventral thalamus that receive input from nociceptors do not project to primary somatosensory cortical areas as do those that receive input from the dorsal column system; instead, the cells that receive input from the anteriorlateral system project to the dorsal and anterior insula and to SII.[18,19,119]

Cells in the dorsomedial thalamus make abundant connections to limbic structures, particularly the hyppocampus and the basolateral amygdala nuclei. The neurons in the medial thalamus that receive input from that part of the spinothalamic tract project to area 24C which lies within the anterior cingulate sulcus.[18,19,119]

The structures of the limbic system that are reached by input from the anteriorlateral system are heavily involved in many functions including affective reactions to pain. The subcortical connection from the dorsal thalamus to the amygdala and other limbic structures may be responsible for such affective reactions to pain as depression and fear (see Chapter 3). Somatosensory information also reaches limbic structures other than the basolateral amygdala nuclei and hypothalamus.

Many of the neurons in the medial thalamus project to the basal ganglia and many different areas of the cerebral cortex that are not specifically involved in processing of somatosensory information. Some of these connections are assumed to mediate non-specific arousal reactions.

C. Descending Somatosensory Pathways

There are extensive descending pathways from the cerebral cortex to the thalamus, and some connections reach as far caudally as the dorsal column nuclei and dorsal horn of the spinal cord and the trigeminal sensory nucleus. Many of these descending pathways are reciprocal to the ascending pathways.

1. Classical Descending System

Abundant descending connections from the SI to more peripheral structures (Fig. 4.18) have been identified. Some of these tracts reach neurons in the VB nuclei of the thalamus and the dorsal column nuclei as well as cells in the

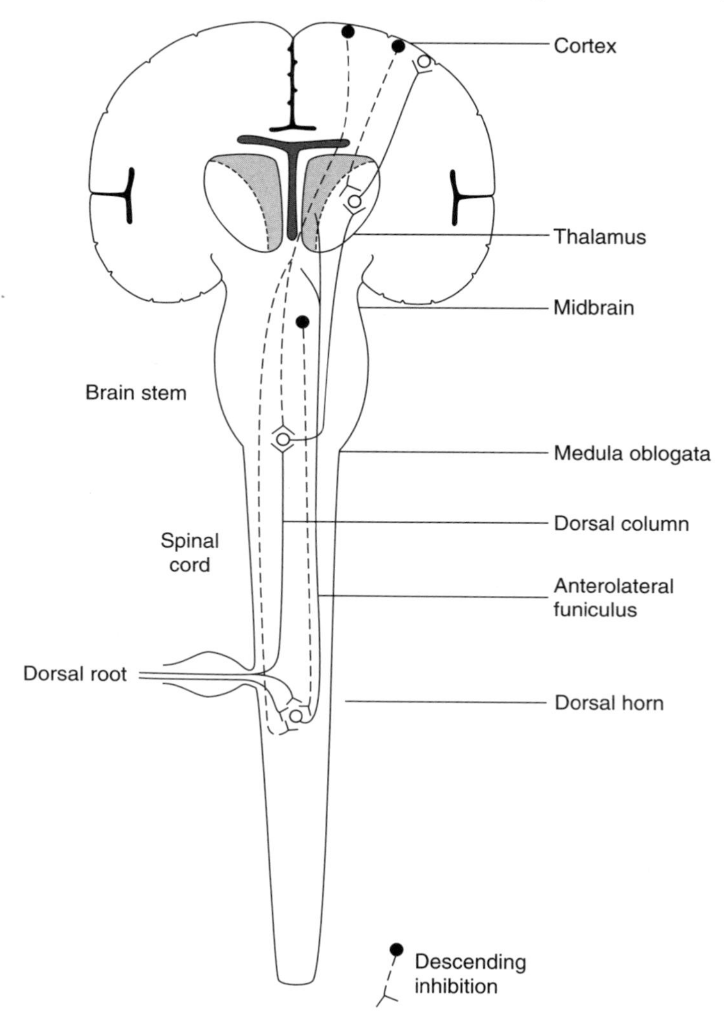

FIGURE 4.18 Descending connections from the cerebral cortex that reach the thalamus, the dorsal column nuclei, and the dorsal horns.

dorsal horns. Only some of the many efferent connections to nuclei in the ascending somatosensory pathway are illustrated in Fig. 4.18. Inhibition from descending connections that reach the dorsal column nuclei has the potential to change the receptive fields of neurons in the CNS because of the way they are organized (Fig. 4.19). Both the surrounding inhibitory area and the center excitatory areas can be affected independently by descending activity.

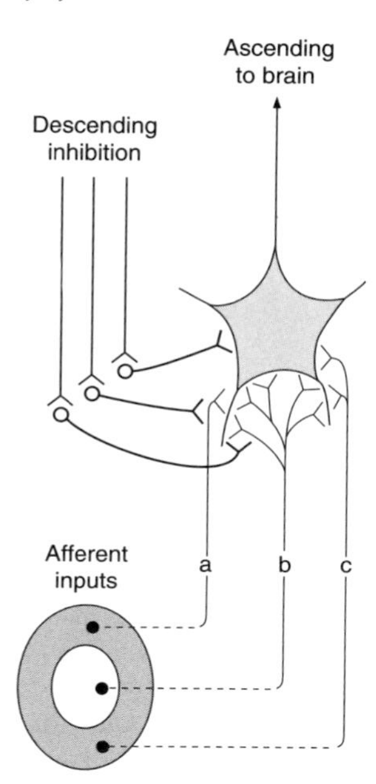

FIGURE 4.19 Illustration of how descending activity can affect the receptive field of somatosensory neurons in the CNS.

2. Nonclassical Descending System

The nonclassical pathways also have extensive descending pathways, most of which are reciprocal to the ascending pathways. The descending connections from the PAG to neurons in the dorsal horn (Fig. 4.20) are especially important because these connections can mediate control of pain.[29] These extensive descending pathways from the PAG terminate in cells in the dorsal horn of the spinal cord where they can modulate (block) ascending pain impulses by regulating excitability of dorsal horn neurons that are involved in pain. The connections from the thalamus to limbic structures are to a great extent reciprocal, and limbic structures make ample reciprocal connections with many parts of the forebrain.

It must be remembered that illustrations such as those showing sensory pathways are greatly simplified. In reality, the connections between the receptors that mediate somatosensory information to the cerebral cortex in the preceeding illustrations are much more complex than what can appear in any illustration. First, the number of nerve cells is enormous and each nerve cell can receive many hundreds and indeed thousands of afferent and efferent synapses, which means that an individual nerve cell can receive input from a very large number of nerve cells. Second, the importance of the anatomical connections depends on the efficacy of the synapses involved. The synaptic

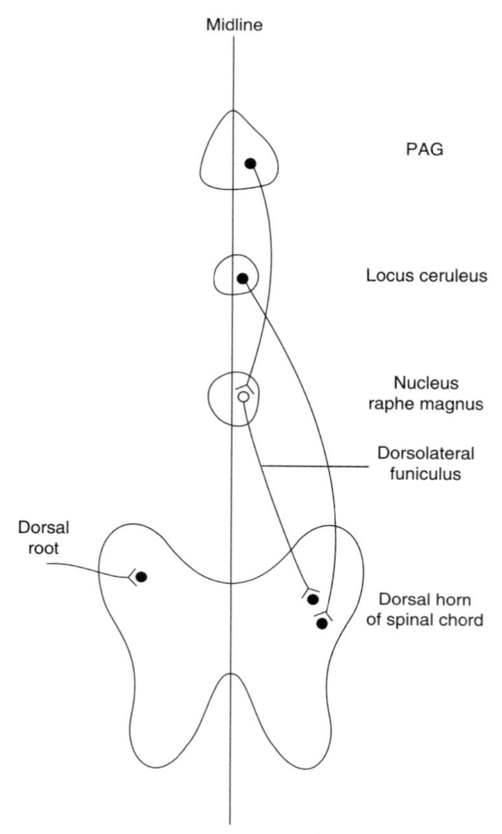

FIGURE 4.20 Descending pathways from the periagueductal gray (PAG) that reach the dorsal horn in the spinal cord.

efficacy can change over time or through expression of neural plasticity. This means that not all anatomically verified connections are functional, and the connections that are functional can change over time as a result of exogenous as well as endogenous factors.

III. PHYSIOLOGY OF THE SOMATOSENSORY SYSTEM

In this section, the physiology of the somatosensory system is discussed on the basis of studies in animals as well as some studies in humans. We begin with the classical pathways, the physiology of which is better known than that of the nonclassical pathways. The general physiology of the nonclassical pathways is then discussed. Pain is discussed in a separate section.

A. The Classical Somatosensory System

The physiology of the classical somatosensory system (dorsal column and trigeminal systems) has been studied extensively. The focus has been on mechanoreceptors and the transformation and restructuring of information that begins at the receptor levels and continues on to the nuclei of the ascending pathways, including the somatosensory cerebral cortices. The separations of different types of information that occur along the ascending pathways and the influence on information processing of descending activity from higher centers have also been studied. In this section, the function of receptors and conduction of the stimulus to the receptors are discussed first, and the processing of information in the somatosensory nervous system will be discussed after that.

1. Receptors

The main input to the classical somatosensory system that causes conscious awareness comes from mechanoreceptors located in the skin; some additional input comes from thermoreceptors. Although this system also receives proprioceptive input from muscles, tendons and joints, such input does not usually reach conscious awareness.

a. Mechanoreceptors

The sensitivity of mechanoreceptors that mediate the sensation of touch is determined by the property of the medium that conducts the stimulus to the receptors, the property of the region of the cell membrane of the receptors that respond to mechanical deformation, and the threshold of the neural transduction that occurs in the receptors (see Chapter 2, page 36). The dynamic response of mechanoreceptors is determined by the properties of the tissue located between the receptor and the location of the mechanical stimulation and the properties of neural transduction occurring in the receptors. The properties of the medium that conducts the stimulus to the receptors mainly provide adaptation, whereas the properties of the receptors themselves provide both adaptation and temporal summation. These factors are important because they modify the response to time-varying stimuli (see Chapter 2) and therefore affect neural coding of natural stimuli that are characterized by their temporal pattern.

Conduction of the Stimulus to the Receptors Adaptation is the most important property of the medium that conducts the stimulus to the receptors. Adaptation suppresses continuous transmission of steady and slow deformations of the skin to the receptors; therefore, receptors respond best to changes in stimulation (see Fig. 2.11, Chapter 2). The medium that conducts the stimulus to the receptors may also attenuate the stimulus in general and thereby lower the sensitivity of receptors.

Classification of Mechanoreceptors According to Their Adaptation Mechanoreceptors are often classified according to their modes of adaptation. Mechanoreceptors in the skin are divided into *slow adapting* (SA) and *rapid adapting* (RA) receptors (Fig. 4.21). The RA receptors respond preferentially to stimuli that change rapidly (Fig. 4.21A), whereas the slowly adapting receptors respond well to steady and slowly varying stimuli (Fig. 4.21B). Pacinian corpuscles belong to a group of their own because of their very rapid adaptation, which means that they can best be described as responding to *acceleration* of the skin caused by a force applied to the skin (Fig. 4.21C). The RA receptors are known as *velocity* detectors, whereas the SA receptors are *displacement* detectors. Slow adapting receptors are also known as "tonic" receptors, whereas the rapid adapting receptors are known as "phasic" receptors because of the way they respond to steady pressure applied to the skin.

As a result of the adaptation, the number of nerve impulses in an afferent nerve fiber that innervates an RA mechanoreceptor decreases rapidly with time after that, a constant force is applied to the skin (Fig. 4.22). Conversely, a fiber that innervates SA receptors maintains its discharge rate after a constant force is applied or the discharge rate decreases only slightly over time.

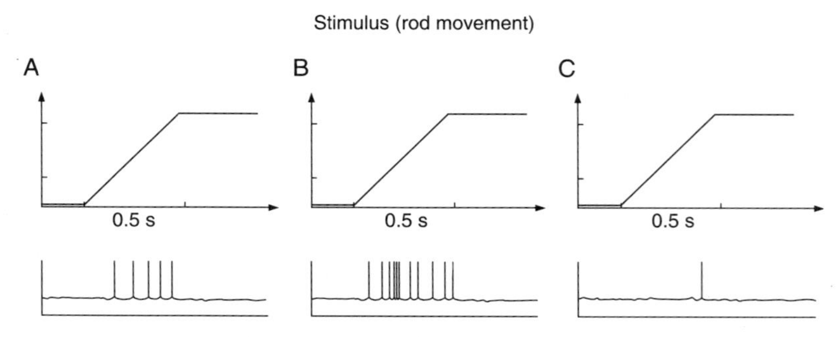

FIGURE 4.21 Examples of discharge pattern of the afferent nerve fibers of mechanoreceptors with different degree of adaptation. (A) Rapid adapting receptor; (B) slow adapting receptor; C: very rapid adapting receptor.

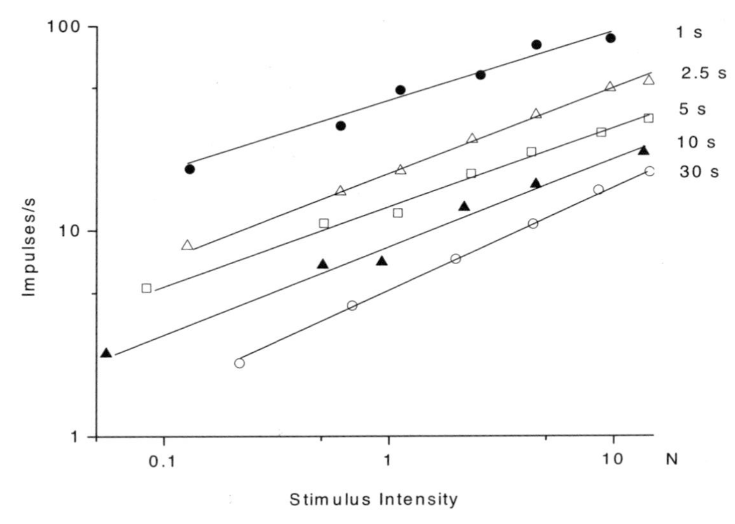

FIGURE 4.22 Discharge rate of an afferent neuron from a mechanoreceptor in response to a constant force as a function of stimulus intensity (in Newton, N) at different times after application of pressure to the skin.

The adaptation of mechanoreceptors also determines their response to vibrations such as (constant) sinusoidal vibrations. RA mechanoreceptors respond best to high-frequency vibrations, whereas SA receptors also respond well to low-frequency vibrations. The Pacinian corpuscles have a distinct frequency selectivity regarding vibration of the skin and may be regarded as being tuned to a specific frequency (see Chapter 2). The characteristics of Pacinian corpuscles are a result of the onion-shaped structures that surround the receptor. Pacinian corpuscles are typically tuned to frequencies of approximately 200 Hz (see Chapter 2, Fig. 2.15).

Displacement, velocity, and acceleration of mechanical deformation of the skin are often referred to as the *adequate* stimuli for the different types of receptors. Of the receptors in glabrous skin, Merkel's discs are slowly adapting while Meissner's corpuscles are moderately rapid adapting. That means that Meissner's corpuscles mostly respond to the velocity of a stimulus while Merkel's discs can maintain their response to steady pressure on the skin. Meissner's corpuscles therefore respond best to rapidly increasing force. Pacinian corpuscles respond best to the acceleration component of mechanical stimulation of the skin. In hairy skin, tactile disks and Ruffini endings are SA receptors. The hair-follicle receptors that sense stimulation of hairs are RA receptors, which are most sensitive to the velocity of mechanical stimulation of the hairs. Pacinian corpuscles are also found in hairy skin.

Sensitivity of Mechanoreceptors The threshold of perception of vibration of the skin in humans is similar to the threshold of individual mechanoreceptors when determined by using electrophysiologic methods (recording from fibers of the afferent nerves) in monkeys.[77] The lowest thresholds for detecting sinusoidal vibration applied to the palms of humans and of monkeys are similar (Fig. 4.23).

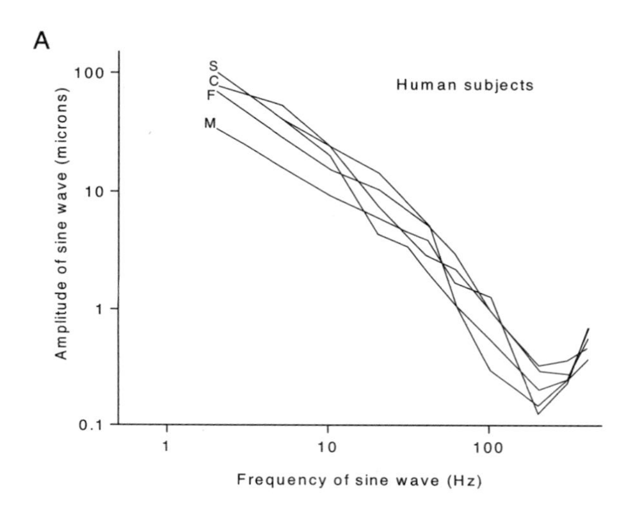

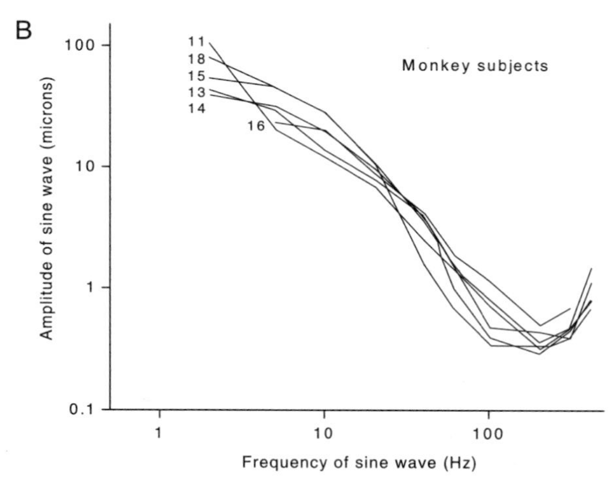

FIGURE 4.23 (A) Frequency threshold functions for five experienced human observers, shown as a function of the frequency of the vibration.[77] (B) Similar results obtained in six monkeys;[77] the threshold was defined as 50% correct response.

For humans, the threshold of detection of sinusoidal vibration of the skin is lowest at a frequency of approximately 200 Hz, where the threshold is approximately 0.2 μm skin displacement. The threshold is slightly higher for the monkey (approximately 0.5 μm). The threshold functions obtained in human and monkeys by stimulation of glabrous skin (Fig. 4.23) resemble the frequency threshold functions to sinusoidal vibration stimuli of Pacinian corpuscles (see Chapter 2, Fig. 2.15). This confirms that Pacinian corpuscles sense vibrations applied to the skin.

Representation of Stimulus Intensity The discharge rate in afferent nerve fibers that innervate most mechanoreceptors increases with the stimulus intensity (Fig. 4.24). When the responses (number of nerve impulses per stimulation) are plotted on double logarithmic coordinates the responses appear as linear functions (Fig. 4.24B). These curves have different slopes for different types of mechanoreceptors, and in some receptors these functions are not straight lines when plotted on logarithmic coordinates.

When assessed using psychophysical methods, the number of correct responses to vibration stimuli applied to the skin increases with the intensity of the stimulation as a sigmoidally shaped function (Fig. 4.25).

Representation of the Stimulus Time Pattern The illustrations of the responses of mechanoreceptors shown above also indicate that the waveforms of vibrations of the skin are coded in the time pattern of the discharges of the nerve fibers that innervate receptors. The neural coding of the temporal pattern of mechanical stimulation is important for interpreting several kinds of stimulation of the skin, such as vibrations and the texture of a surface. The sense of movement that is perceived when two points of the skin are stimulated with a short time in between depends on interpretation of the temporal code of the stimulation.

b. Temperature Receptors

The system that signals cool is distinctly different from that which signals warmth. The cool receptors signal skin cooling to approximately 25°C, whereas warmth receptors signal skin warming to approximately 41°C. Thermal receptors sense both static temperatures and transient temperature but are more sensitive to changes in temperature (Fig. 4.26).

Thresholds of warmth and cold depend on the initial (adapting) temperature (Fig. 4.27). Paradoxical sensations sometimes occur, such as a cold sensation in response to a strong warm stimulation.

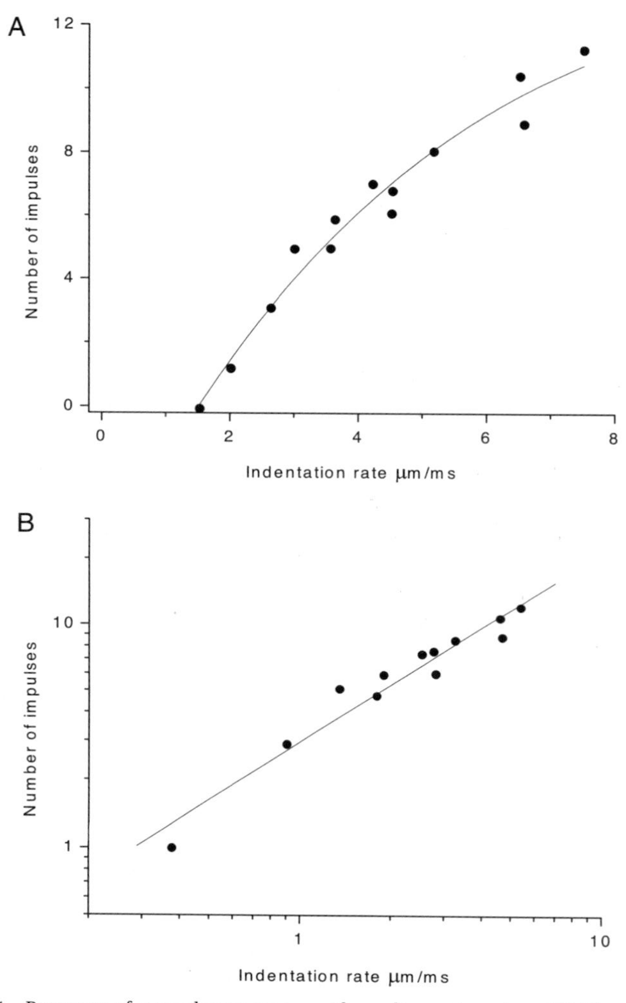

FIGURE 4.24 Response of a mechanoreceptor with medium adaptation. (A) The response as a function of the rate of indentation shown on a linear scale. (B) The same data shown on a log scale. (Adapted from Schmidt.[100])

Both cool and warmth receptor activation can cause autonomic reactions such as decreased or increased peripheral blood circulation. Activation of warmth receptors may cause sweating and may open arteriovenous shunts in the skin to aid in regulating body temperature. Cooling of the skin can cause involuntary muscle contractions (shivering).

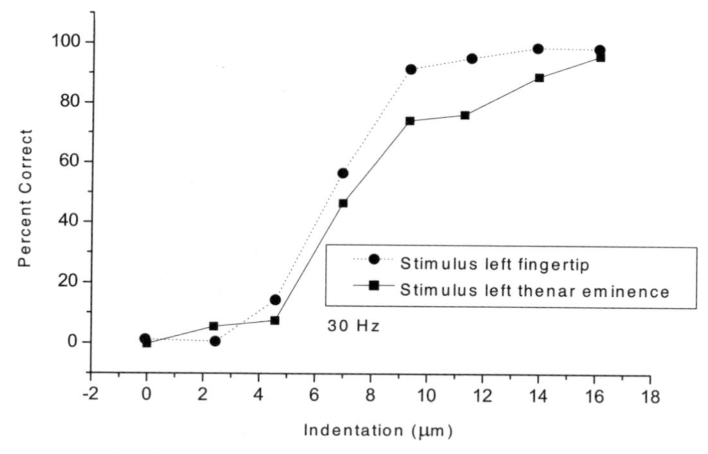

FIGURE 4.25 Number of correct responses to stimulation of the hands with 30-Hz vibration shown as a function of the vibration amplitude. Results were obtained in a human observer and show the average of six separate experiments using eight amplitude classes.[77]

2. Processing of Information in the Classical Ascending Pathways

While the properties of the medium that conducts the stimulus to the receptor perform some transformation of the mechanical stimulation of the skin, the most extensive signal processing of tactile information occurs in the nuclei of

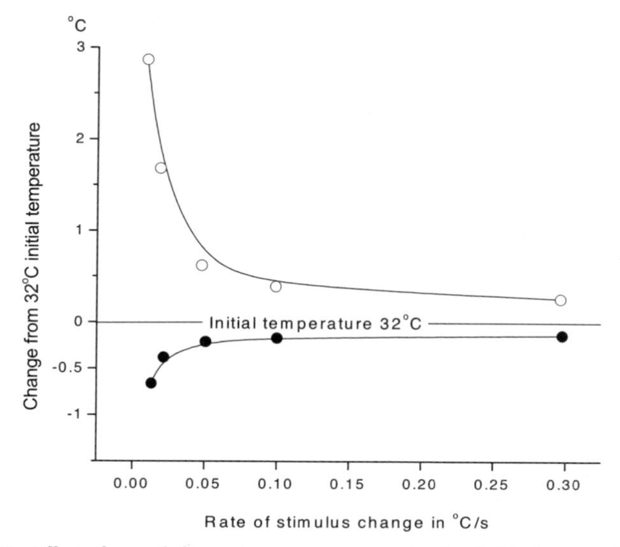

FIGURE 4.26 Effect of rate of change in temperature on the thresholds for warmth and cool.[50]

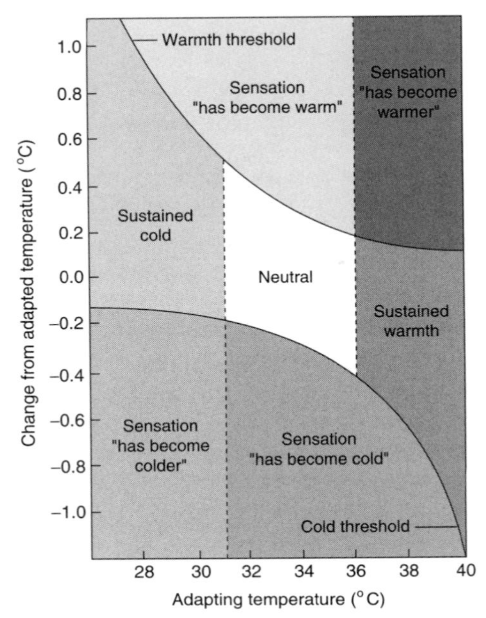

FIGURE 4.27 Thresholds for warmth and cool as functions of the adapting (initial) skin temperature.[50]

the ascending somatosensory pathways and in the cerebral cortex. Feedback mediated by the descending neural pathways can modify the processing in the ascending somatosensory pathway. Neural activity in other parts of the CNS can also affect the flow of information in the ascending somatosensory pathway. The same information may be processed differently (*parallel processing*) or various kinds of information may be processed in different populations of neurons (*stream segregation*).

In this section, we discuss neural coding and processing in the classical ascending somatosensory pathways (the dorsal column system). This pathway is less complex than other sensory systems. The nonclassical somatosensory system, which receives input from nociceptors (pain and temperature receptors), is discussed above (page 206).

a. Somatotopic Organization

Neurons throughout the somatosensory system are organized in a *somatotopic* way. That means that neurons are anatomically organized in accordance with the parts of the skin to which they respond. Most of the neurons in the VB nuclei of the thalamus are organized in a somatotopic fashion. Each neuron is excited by a single type of receptor (slow or rapid adapting receptors or hair follicle receptors in the skin).

Each of the four regions of the primary somatosensory cortex in the postcentral gyrus contains a map of the body surface. The representation of the body surface on the primary (SI) cerebral cortex is often illustrated by a homunculus representing of the opposite side, after the classical work by Penfield and co-workers in the late 1930s (Fig. 4.28).[84] The representation of the body surface is not uniform, and representations of the hands, especially the fingers, and the face are much larger than those of the rest of the body surface. The face is represented by the most lateral area of SI and the legs are represented by the most medial area, including the part of the cortex that is located in the midline.

The somatosensory cortex is organized in columns like other parts of the neocortex. The dendrites extend perpendicular to the surface of the cortex and make connections between cells in the six layers of the cortex (see page 103). Each column connects to receptors in a small patch of skin, and each region of the

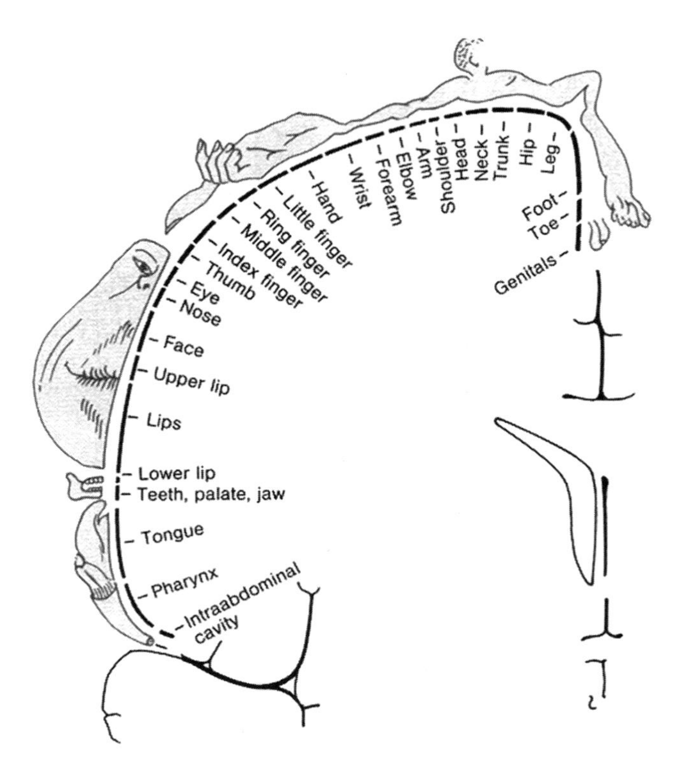

FIGURE 4.28 Representation of the body surface on the primary somatosensory cortex by Penfield and co-workers. (From Everett, N.B., *Functional Neuroanatomy*. Philadelphia: Lea & Fibiger, 1971.)

cortex receives input from one type of receptor (either SA or RA). The different columns of the SI cortex are activated by different types of skin receptors (SA, RA, or hair follicle receptors), thus are similar to the cells in the VB of the thalamus.

All four regions of SI have similar somatotopic representations as an indication of parallel processing of somatosensory information. Each of these four areas of the SI seems to specialize in different specific processing. It was mentioned earlier that areas 3b and 1 receive input from SA and RA skin receptors, whereas areas 3a and 2 receive input from receptors located deep in tissue such as muscle afferents. Area 2 receives both afferents from cutaneous receptors and from proprioceptors such as those in muscles. Area 3b seems to be critical in discrimination of shape and texture of objects. Area 1 and 2 are specifically involved in perception of texture and size, respectively. Area 1 seems to be even more specialized in that lesions of that area disrupt performance based on texture but do not affect discrimination of the size of objects, and lesions of area 2 have the opposite effect.[9,92]

Because the SII cortical area receives input from the ipsilateral SI as well as from the contralateral SI through the corpus callosum, the body is represented *bilaterally* in the response of SII neurons. The SII cortical area is therefore important for creating a fused impression of the body surface, or of body space. The parietal association cortices, especially Brodmann's area 7b, play an important role in such integration. In addition, these cortical areas integrate input from somatosensory receptors with that from other sensory systems, such as vision. For instance, the caudal sensory cortical areas are involved in directing eye movements.

The response from SII (but not SI) neurons is affected by attention, as shown in studies using functional magnetic resonance imaging (fMRI*). The complex circuitry around the SII cortical neurons causes the responses of SII neurons to be more complex than those of SI neurons. SII also furnishes input to motor cortical regions. One of the roles of the SII cortex may therefore be to coordinate sensation and motor activity from the two sides of the body.

b. Receptive Fields

The receptive field of a nerve fiber that innervates a mechanoreceptor of the skin is defined as the area of the skin where mechanical stimulation elicits a response.* The various mechanoreceptors have different receptive

*Results obtained using functional magnetic resonance imaging (fMRI) have been interpreted to show the anatomical localizations of increased neural activity. However, the results of fMRI studies have no simple relationship with neural activity. It is a measured change in blood flow. Recent studies have been increasingly skeptical as to the previous interpretations of fMRI studies. Other techniques in addition to fMRI rely on measurements of changes in blood flow, such as positron emission tomography (PET) and single-photon emission computed tomography (SPECT).

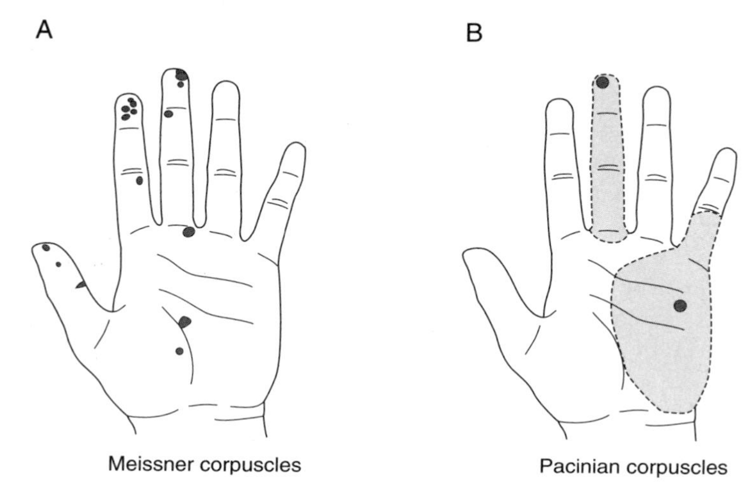

FIGURE 4.29 Receptive field of two different types of types of mechanoreceptors. (A) The (small) receptive field of Meissner's corpuscles. (B) The (much larger) receptive fields of Pacinian corpuscles.[109]

fields.[46,109,134] Merkel's discs have small and localized receptive fields, whereas Ruffini endings have large receptive fields. Pacinian corpuscles may be activated by stimulation of a large area of skin, but their highest sensitivity is localized to a small area directly over the location of the receptor (Fig. 4.29).

The receptive fields of neurons in the dorsal column nuclei (second-order neurons), as well as in neurons of other nuclei of the somatosensory system, including those of the cerebral cortex, are different from the receptive fields of the afferent nerve fibers that innervate receptors. The reason is that nerve cells of the CNS receive input from more than one afferent fiber. The *convergence* of nerve fibers on a nerve cell (which means that a nerve cell has excitatory input from several nerve fibers) causes the receptive field of the nerve cell to become larger than that of the nerve fibers that supply the input (see Chapter 3, Fig. 3.15). Such *spatial integration* is important for increasing the precision of coding of sensory information by enhancing the signal-to-noise ratio, where uncertainty in neural transmission is regarded as "noise."

When the stimulus intensity is increased, the stimulation can exceed the threshold of nearby receptors because the spread of the mechanical stimulus in the skin acts to widen the receptive field due to activation of distant receptors that converge onto the cell from which the response is recorded.

*The original use of the term was for describing the response areas of fibers of the optic nerve. Now it is also used to describe the area of response of nerve cells and fibers in the somatosensory and visual nervous systems. It is sometimes also used in other sensory systems.

The size of the receptive fields of cells in the dorsal column is different for different locations on the body because the density of innervation (number of receptors per unit skin area) differs (see Chapter 1, Fig. 1.23). The density of innervation is greatest for the fingertips and the lips and is least on the belly, legs, and the arms. The two-point discrimination is best in the areas of the body where the density of the innervation is greatest.

Another way to view receptive fields is to observe the response from a population of neurons to stimulation of a single point on the skin (Fig. 3.17), which is probably more similar to the way the nervous system processes somatosensory information than discharges in individual neurons but technically is more difficult to study experimentally because it requires recording from many neurons to the same stimulation. Computer simulations such as those illustrated in Fig. 3.17 provide insight into how populations of neurons may respond to stimulation of the skin of a single point.

c. *Interplay Between Inhibition and Excitation*

In addition to the excitatory response areas that were discussed above, neurons in the ascending somatosensory pathways also have *inhibitory response areas*. Often neurons that respond to stimulation of the skin have an excitatory area of the skin that is located between two inhibitory response areas or an excitatory skin area is surrounded by an inhibitory area. While stimulation of receptors that are within a neuron's excitatory area causes an increase in discharge rate, simultaneous stimulation of receptors within the inhibitory response area of a neuron causes a decrease of the discharge rate. The interaction between inhibition and excitation can sharpen the receptive fields (lateral inhibition, surround inhibition; see Chapter 3, Fig. 3.18). Areas of inhibition that are located adjacent to areas of excitation are known as *lateral inhibition*.

Neurons that are located more centrally than the dorsal column nuclei (thalamus and cortex) may have more complex receptive fields because these neurons may have many excitatory and inhibitory inputs that converge on a neuron, making the interaction between inhibition and excitation more complex.

d. *Basis for Spatial Resolution of Stimulation of the Skin*

When two different points of the skin are stimulated, each stimulation evokes responses in two populations of neurons in the CNS. These two populations may overlap more or less depending on the distance between the points of stimulation and the size of the population evoked by the two stimulations. When the distance between the two points being stimulated is small, the receptors that are stimulated activate populations of neurons that may fuse into a single (broad) representation; the stimulation is perceived as a single stimulation. When the distance between the two points of skin stimulation is sufficiently large, two different populations of neurons will be activated, and two separate stimulations will be perceived.

Stimulation of two points perceived as being separate stimulations does not require activation of separate populations of neurons, but activation of overlapping populations of neurons may be perceived as stimulation at separate locations. Whether such stimulation is perceived as touch at two points or whether the stimulation fuses into perception of stimulation at a single point depends on the degree of overlap of the two populations of neurons that are activated.

The degree of overlap of the population of neurons that are activated depends on several factors. One factor is the density of mechanoreceptors in the skin that are stimulated. Another is the degree of the convergence of nerve fibers onto neurons in the CNS and the interaction between inhibitory and excitatory response areas.

Interplay between excitation and inhibition can increase the separation of the two populations of nerve cells that respond to two-point stimulation, which means that it can increase the two-point discrimination (Fig. 3.18). Lateral inhibition increases the separation between the two populations of neurons that respond to stimulation of two nearby points of the skin and thereby enhancing spatial discrimination. Interplay between inhibition and excitation is abundant in the ascending somatosensory pathways and in the cerebral cortex of the somatosensory system. An asymmetric arrangement of excitatory and inhibitory fields is important for feature detection such as the shape of objects and for detection of movements of the stimulation.[41]

e. Basis for Discrimination of the Shape of Objects

Discrimination of the shape of an object is an example of interpretation of complex stimuli by the somatosensory system. That ability depends on the properties of the mechanoreceptors in the skin and the neural processing that occurs in the nuclei of the ascending somatosensory pathways.[104]

The selectivity of mechanoreceptors with regard to the shape of an object with which the skin is stroked is an example of processing of information at the receptor level as well as in the nervous system. Slow adapting (SA) mechanoreceptors respond in accordance to the displacement of the skin perpendicular to the skin surface and the change of the curvature of the skin at the most sensitive spot of its receptive field. The response of rapid adapting (RA) receptors is a function of the velocity of the displacement of the skin that occurs in a direction that is perpendicular to the skin surface. The response also depends on the rate of change in the curvature of the surface that is probed (and thus that of the skin). The discharge rate of the nerve fibers that innervate RA receptors is highest when the velocity perpendicular to the surface is high or when the rate of change in the curvature of the surface is high. These properties explain the response to stimulation with corrugated surfaces, and these properties are the basis for the ability to discriminate objects with different curvatures, which in turn is important for discrimination of the shape of objects.

The basis for discrimination of the form of objects can be illustrated from a nerve fiber that innervates a Merkel's disc in response to indentations in the skin made by objects of different shapes (Fig. 4.30).[104]

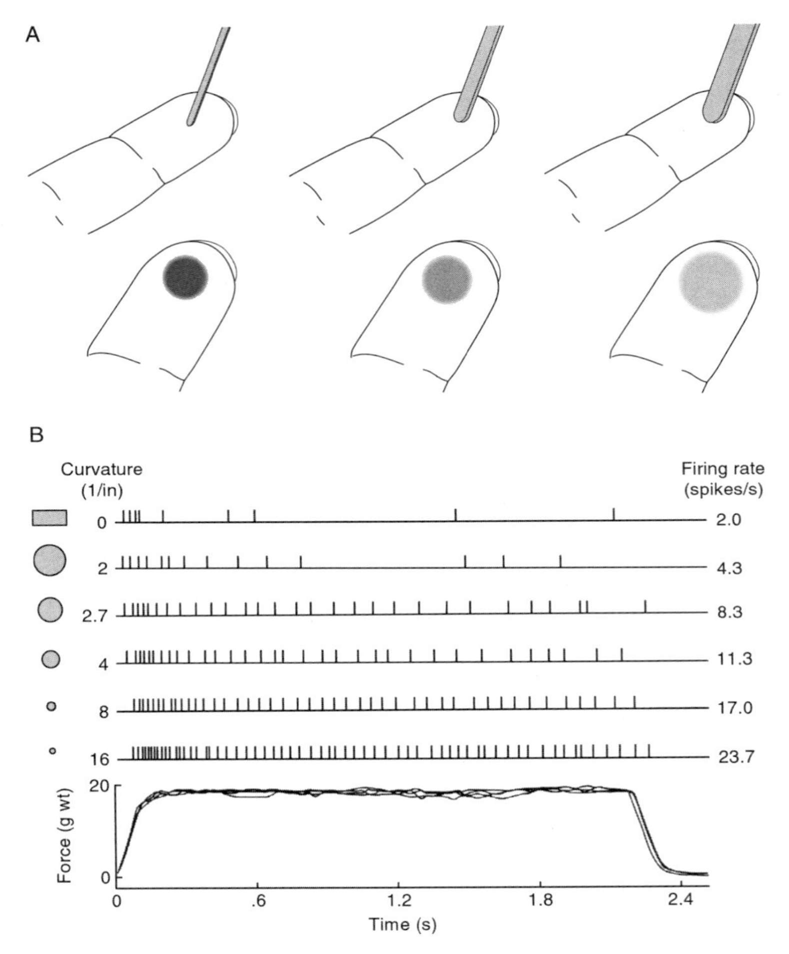

FIGURE 4.30 Responses of a nerve fiber that innervate mechanoreceptors in the tip of a finger in response to objects of different shapes. (A) Different shapes of the tips of the rods used as stimulators. (B) Responses to the same force applied by rods with tips of different shapes. (From Srinivasan, M.A. and LaMotte, R.H., *Encoding of Shape in the Responses to Cutaneous Mechanoreceptors*. New York: Macmillan, 1991. With permission.)

f. Stream Segregation

Discrimination of the location of stimulation on the body surface and two-point discrimination are both forms of spatial information ("where"), and discrimination of shape and vibration is a form of object ("what") information. It is, not known, however, if these kinds of spatial and object information are processed in separate populations of neurons in the somatosensory cerebral cortices (stream segregation) as has been shown for spatial and object information for vision and hearing.

3. Neural Plasticity

The somatosensory system was one of the first sensory systems where neural plasticity was demonstrated experimentally. Several investigators[44,69,70,112] have shown experimental evidence that the function of the somatosensory system can change as a result of external circumstances such as deprivation of input. These changes are especially important for pain and will be discussed later in this chapter.

Wall[112] was one of the first investigators to publish results suggesting that the conduction of synapses is affected by external factors. As mentioned earlier, nerve cells in the dorsal horns receive synaptic input from dorsal root fibers that enter the spinal cord at the same segment as well as from collaterals of dorsal root fibers that enter the spinal cord at several levels below or above that segment. The synapses that connect collaterals of ascending or descending dorsal root fibers far from their entry into the spinal cord do not normally conduct nerve impulses, but severing the dorsal root that enters the segment makes the synapses of distant dorsal root fibers conduct nerve impulses (see Chapter 3 Fig. 3.28).[112] Deprivation of the synaptic input from dorsal root fibers that enter at the segment in question thus results in increased excitability (efficacy) of synapses receiving input from collaterals of dorsal root fibers that entered at distant segments. These synapses conduct nerve impulses only under abnormal conditions, such as after severing some other connections to the nerve cells. Wall coined the term *dormant synapses* to describe synaptic connections that exist morphologically but which are blocked because the connecting synapse is ineffective. (The properties of such "silent synapses" and long-term facilitation have recently been studied extensively; for an overview see Zhou, 2000.[132])

Neural plasticity has also been demonstrated in the somatosensory cortex, and it has been shown that somatotopic maps of the body surface can change as a result of external circumstances such as injuries or deprivation of input. Perhaps the most cited studies on neural plasticity in the somatosensory system are those by Michael Merzenich and his colleagues.[69,70] These investigators showed that the receptive field of somatosensory cortical neurons expand to neighboring regions of the skin when the input is decreased by amputation of a monkey's finger. Amputation of a finger causes deprivation of input from receptors in specific patches of skin, and such deprivation is a strong promoter of neural plasticity. In the cortex, it caused a change in the cortical map of the hand in such a way that the cortical areas that belonged to the skin of the finger responded after amputation to stimulation of the skin of the adjacent fingers. These changes in responsiveness are most likely caused by opening of ineffective, or dormant, synapses that connect input from mechanoreceptors to wide areas of the cerebral cortex.

4. Physiologic Basis for the Normal Function of the Somatosensory System

We have limited physiologic evidence regarding many common functions of the somatosensory system. Studies of neural firing in response to isolated stimulation of a small area of the skin provide only very basic information about the function of the somatosensory system. Under normal circumstances, complex interactions occur between the input from receptors in various parts of the skin, and, indeed, information from other sensory systems. Also, sensory input is compared with stored information and, when cutaneous stimulation is similar to earlier (stored) responses, it may evoke complex reactions that are different from those evoked by stimuli that we have not previously experienced.

The function of the somatosensory system is far more complex than what the results of studies of the responses to simple stimuli reveal. For example, one would expect that probing one's own body surface with one's fingers would provide faithful information about the body part that was probed. A simple experiment reveals that this is far from the truth. When a person's face and mouth are anesthetized during a visit to the dentist, probing the face reveals a size and shape much different from actual. This means that the loss of sensibility in the face has changed the perception obtained from probing despite the fact that the sensibility of the fingers is unchanged. The reason for that is that the perception of the size and form of the face resulting from probing by one's own fingers not only is based on the output of mechanoreceptors in the finger used to probe the skin but also depends on the stimulation of mechanoreceptors in the face from the stimulation by the finger. That is just one small example of how interaction between the output of the receptors in different parts of the body normally interact to provide such basic information as an impression of size and form

Similar complex interactions cause the perception of being probed by another person to be very different from probing one's own body. An example of such difference is the sensitivity to being tickled. One cannot evoke that feeling oneself; only another can cause tickling. It has been hypothesized that the difference between the perception of one's own stimulation of cutaneous receptors and that done by others is related to predictions of the sensory consequences based on motor commands.[3]

5. Nonperceptual Effects of Somatosensory Stimulation

Normal stimulation of the sensory receptors that are part of the proprioceptive system apparently does not cause any conscious sensory response and awareness because such stimulation does not activate any parts of the CNS involved in conscious perception (proprioception is not covered here). Stimulation of somatosensory receptors that normally evoke conscious awareness can, however, also activate parts of the CNS not associated with consciousness and, consequently, it can activate parts of the CNS not involved in conscious perception. Activation of the reticular formation and limbic structures are examples of activation of neural systems that do not result in direct conscious awareness but which may occur simultaneously with activation of populations of neurons that result in conscious awareness.

a. Arousal

Stimulation of receptors in the skin can provide arousal through mainly two routes. One route is through the brainstem reticular formation, which receives input from the somatosensory system by collaterals from the medial lemniscus (ML). The other route is through the cerebral cortex, where activation of nuclei of the amygdala can provide arousal. Activation of the amygdala can occur via the dorsal thalamus or via a much longer route through the primary sensory cerebral cortex and association cortices. This cortical route, in contrast to the subcortical connections to the brainstem reticular formation and from the dorsal thalamus to the amygdala, carries highly processed information. The arousal evoked through this route therefore depends on the message of the stimulation and previous experience.

Stimulation of the skin can also reduce arousal. For example, stroking of the skin can induce sleep in a child or calm down an agitated person. This effect of skin stimulation on arousal caused by stimulation of skin receptors is most likely mediated by cortical connections after processing of the information because it is related not only to the way stimulation occurs but also to other factors such as other sensory input and intrinsic CNS activity. Strong stimulation, however, will always cause an increased arousal probably mediated directly to the reticular formation.

b. The Role of the Amygdala

The two routes to the amygdala mentioned earlier were discussed in Chapter 3,[58] where the subcortical (low route) extends from the dorsal thalamus to the basolateral amygdala nuclei via the lateral nucleus of the amygdala. The other route, the high route, is reached through both the classical and the nonclassical ascending sensory pathways and carries highly processed information. Sensory information that travels in the nonclassical pathways can use the low route but can also use the cortical route (see pages 141 and 155).

Connections from the amygdala extend to many different parts of the brain, including the reticular formation, which mediates arousal. The variety of routes and connections taken by information from skin receptors probably explains why stimulation of the skin can cause so many different reactions.

Because the high route to limbic structures mediates highly processed information, it is likely that this route is responsible for the complex arousal reactions to stimulation of the skin, whereas the increased alertness from skin stimulation, for example upon awakening a sleeping person is probably mediated through the subcortical connections as well as through the connections from the ML to the reticular formation.

The affective and sexual response to cutaneous stimulation that involves the limbic structure is based on highly processed information and therefore must involve a cortical route to limbic structures. Learning is also involved in such reactions, which require involvement of high regions of the CNS.

B. Physiology of the Nonclassical System

The nonclassical somatosensory system mediates the output of nociceptors to several higher brain centers, which are involved in the sensation of pain and in causing other reactions such as motor and autonomic responses (discussed later). These systems also receive neural activity elicited by stimulation of mechanoreceptor located deep in the body, but the functional importance of that is unknown.

1. Receptors

Like other receptors, nociceptors are specialized patches of membranes of axons of the nerve fibers that mediate the sensation, thus Type I receptors (see Chapter 2). It is not known exactly how noxious stimuli activate the free nerve endings, but it is known that the membranes of some of these receptors have proteins that can convert noxious stimuli into a depolarizing electrical potential (receptor potential) that controls the firing of the afferent nerve fiber and is continuous with the receptor region, connecting with the central nervous system. One such protein is the receptor for capsaicin, which is the active ingredient in hot peppers. The same kind of receptor responds to painful heat.[81] These receptors are known as polymodal receptors because they respond to many different modalities of stimuli including noxious stimulation. Ordinary mechanoreceptors of the skin may also produce a sensation of pain when overstimulated.

2 The Dorsal Horn

The dorsal horn plays an important role in the function of the nonclassical somatosensory system. As was mentioned earlier, small fibers of the dorsal roots (Aδ- and C-fibers) terminate in layers I and II of the dorsal horn (substantia gelatinosa), and layer V.[6] It was also mentioned that dorsal root fibers give off collaterals to cells in the dorsal horn over several segments both above and below the entry of the dorsal root fibers into the spinal cord. There are many interneurons and abundant interconnections in the dorsal horn that contribute to the processing of somatosensory information. Processing in the dorsal horn is the beginning of the complex processing of noxious stimulation of receptors in the skin, muscles, joints, and viscera that occurs in the nonclassical somatosensory system. The connections in these circuits are dynamic because the synapses may change their efficacy as a result of external and internal factors (neural plasticity).[112]

Neural plasticity plays an important role in processing of information in the nonclassical somatosensory pathways. Identifying the changes that can occur in the function of the dorsal horn is important for understanding many pain conditions.

Under normal conditions, the efficacy of the synapses with which the collaterals connect to dorsal horn neurons is lower than that of the synapses at the segment where the fibers enter the spinal cord. Synapses that are several segments away from the entry of the dorsal root do not normally conduct activity that is elicited by stimulation of skin receptors. These dormant synapses may be activated (unmasked) through expression of neural plasticity.[112] The discovery of such dormant synapses has explained many features of pain and it deserves to be mentioned that the original observation by Wall[112] represented a fundamental new thinking regarding the mechanisms of pain. His meticulous studies and observations and their interpretations have remained valid over time. The molecular mechanisms by which synapses can become dormant have been studied recently by many investigators, and it has become evident that there are several ways in which synapses can become nonconducting. (For a recent overview, see Zhou.[132])

3. The Anteriorlateral System

The connections of the anteriorlateral system are the basis for many of the functions that are ascribed to the nonclassical somatosensory system, and there is evidence that the nonclassical somatosensory system (anteriorlateral and caudal-trigeminal systems) provides a major driving force to the reticular formation of the brainstem, which is involved in controlling the excitability of the forebrain[78] and controlling wakefulness (arousal). The ample connections from the anteriorlateral system to limbic structures are the basis for emotional reactions of pain and for arousal. Some of these connections are subcortical, as discussed earlier and some are cortical. The information that reaches the amygdala via the subcortical connections (low route) is subjected to little neural processing and is under little influence from higher (conscious) CNS centers, whereas the information that is conveyed through the cortical route to the amygdala (the high route) is highly processed and influenced by conscious processes.[58]

The physiologic implications of the connections from the somatosensory system to limbic structures is not known in detail, but the fact that pain can cause affective reactions indicates that these connections indeed are active. The connections from pain pathways to other parts of the brain, such as the autonomic system, are probably responsible for the effect of pain on such basic functions as blood pressure, heart rate, and other autonomic reactions, such as vomiting. These are signs that noxious stimulation can activate parts of the CNS that normally are not reached by the dorsal column system (classical somatosensory system). Pain is often associated with abnormal sensations such as *allodynia*,* which is assumed to be caused by rerouting of information that normally travels in the dorsal column to the anteriorlateral system. *Hyperpathia*,** another abnormal reaction to pain, is probably also a result of opening of pathways that are normally dormant. Such rerouting can occur as a result of unmasking of dormant synapses or by creating new connections (sprouting), or neural plasticity.

The anteriorlateral system connects to motor systems both at spinal and higher CNS levels, and activation of the anteriorlateral system can evoke motor activity including reflexes. The connections from dorsal root fibers directly to motoneurons are the anatomical basis for the withdrawal reflex (a monosynaptic reflex). These matters will be discussed in more detail in the following section on pain (page 241).

The conscious perception of pain is eliminated by general anesthesia, but there is evidence that the processing of pain stimuli in the cingulum is not affected. Painful stimuli elicit neural activity in the secondary somatosensory cortical areas (SII) bilaterally, which is assumed to be the basis for pain perception. Thalamic pain activation is also bilateral. Neurons in the anterior cingular gyrus receive input from the spinothalamic tract.[6] Some of this input comes directly from the spinothalamic tract and some is via the amygdala through input from the dorsal and medial thalamus and the hippocampus.[6] Neurons in the anterior cingular gyrus also receive input from the dorsal column system, thus are part of the classical somatosensory pathways. Most of these connections are reciprocal and offer vast possibilities of modulation of the pain input. The insular cortex to which the thalamic pain nucleus (the so-called posterior complex, PO) projects is also involved in pain.[6,19] The anterior cingulate gyrus is one of the sites where opiates act to reduce pain (see Scharein, referenced by Bromm.[7])

**Allodynia* is a situation where normally innocuous stimulations (such as stroking the skin) cause a sensation of pain.

***Hyperpathia* is a situation where pain elicits an abnormally strong reaction.

C. Function of Descending Systems

The existence of the abundant descending system in the nonclassical somatosensory system is the anatomical basis for an extensive feedback that can control the activity in the ascending somatosensory pathways. These systems are important for pain. The best known descending systems of the nonclassical somatosensory pathways are those originating in the thalamus and the PAG that provide descending input to the dorsal horn cells, providing central modulation of pain signals. These matters will be discussed in the following section on pain.

D. Pain

Pain has many forms, such as *somatic pain*, *neuropathic pain*, *acute pain*, or *chronic pain* (Fig. 4.31). One classification of pain distinguishes between somatic pain and neuropathic pain. Somatic pain is associated with *somesthesia* (bodily sensation, or the conscious awareness of the body). Somatic pain, also known as *nociceptive pain*, is caused by stimulation of nociceptors and is distinctly referred to specific body locations. Somatic pain can be caused by mechanical or chemical stimulation of nociceptors, such as from acute injury or from inflammation and other forms of tissue injuries. Heat and cold can also cause pain. Neuropathic pain refers to pain caused by insult to the nervous system or by changes in the nervous system involving neural plasticity. Such pain may persist after the original cause of pain (injury, etc.) has resolved. The classification of pain into acute or chronic pain relates to the duration of pain and not to its cause or mechanisms. Finally, clinicians often refer to *idiopathic pain*, meaning pain for which the source and cause are unknown.

There are many degrees of pain, and pain may cause many reactions (Fig. 4.31). Pain may activate the autonomic nervous system and cause sweating, and changes in blood pressure and heart rate. It may cause emotional reactions such as fear, anxiety, or depression. The reactions to pain depend on external factors such as the individual's understanding of the cause of the pain and, in particular, the prospects of relief.

The perception of pain is subjective and not related to physical stimuli in the same distinct way as with sensory systems, except perhaps regarding the threshold for stimulation of pain receptors in the skin. The sensation of pain depends on factors other than the actual physical stimulus causing the pain to a much greater degree than do sensory stimulations. Intrinsic factors, such as the level of consciousness, the circumstances leading to the pain, etc., affect the perception of the degree of pain to a great extent. Viral infections can decrease the pain threshold.

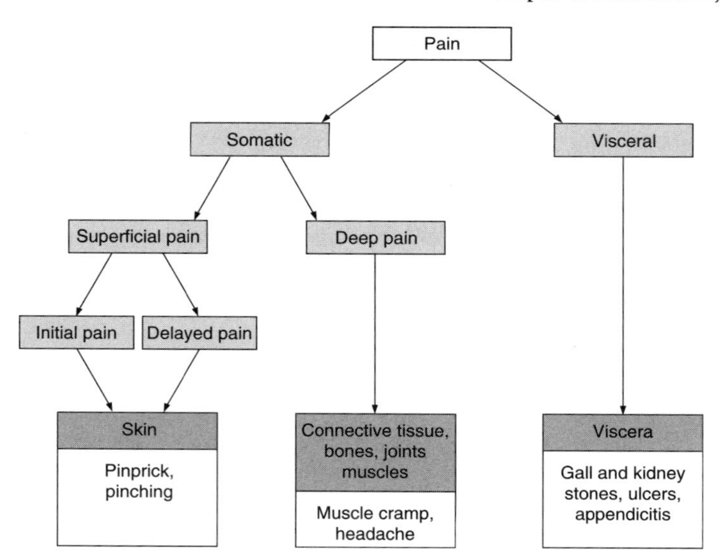

FIGURE 4.31 Schematic outline of pain. (Adapted from Schmidt, R.F., *Fundamentals of Sensory Physiology*. New York: Springer-Verlag, 1981.)

Describing the severity of pain varies widely among individuals who have been subjected to physically similar noxious stimuli, such as those experienced during dental work. This is an indication of the complexity of the sensation of pain.

Some individuals react more strongly to noxious events than others do. If the source of pain is under the person's control, it is more tolerable than pain over which the person has no control. Pain caused by unknown and possible threatening factors is perceived to be worse than pain caused by known and benign causes. The circumstances under which the pain is being caused are important in regard to how it is perceived. The reaction to pain may be termed suffering. Whether a certain form of pain is tolerable or not depends not only on the nature of the stimulation that causes the pain but also on the person's personality, as well as a large number of other factors. The saying that "the only pain that is tolerable is someone else's pain"* is valid for most individuals.

Noxious stimuli may not always elicit pain sensation, and innocuous stimuli may sometimes elicit a sensation of pain. This variability has been attributed to functional variations of cells in the dorsal horn.[23,124]

The systems that mediate pain are different in many ways from the sensory systems that respond to innocuous stimulation, the sole purpose of which is to communicate information from the environment. The message that is conveyed

* René Leriche, French surgeon, 1879–1955.

by acute pain is that of threats to our bodies either from external sources such as injuries or from internal sources such as disease. These forms of pain are essential for survival. Pain that persists after an injury has healed does not seem to have any beneficial purpose.

1. Somatic Pain

Somatic pain is pain caused by stimulation of nociceptors. Nociceptors are located in the skin, muscles, joints, tendons, and viscera. Pain from stimulation of nociceptors is known as body (somatic) pain or nociceptor pain. It is labeled acute or chronic pain depending on its duration.

Acute pain can be elicited by many different forms of stimulation of nociceptors (see Chapter 2). Some nociceptors respond to several modalities; such polymodal receptors may respond to both mechanical and chemical stimulation (see Chapter 2).

Pain receptors are activated by trauma and by chemicals associated with trauma and inflammation. Some pain receptors are specifically sensitive to certain chemicals. The best known is capsaicin, the active ingredient of red pepper.

Heat and cold are common sources of acute pain. When the increase of the skin temperature exceeds a certain value, a particular set of temperature receptors is activated, causing a painful sensation. These receptors are known as heat and cold nociceptors and are activated by extreme temperatures (>45°C or < 5°C) (Fig. 4.32).[50] The sensation of pain starts at skin temperatures below 17°C,[40,50] but the temperatures at which these nociceptors are activated can vary widely. These nociceptors are innervated by C-fibers (conduction velocity of 0.5 to 2 m/sec) similar to other nociceptors (Fig. 4.33).

Pain that is elicited by stimulation of nociceptors is not normally regarded to be a sensory modality, although pain causes conscious awareness and everybody recognizes it as a sensation. The sensation of pain caused by receptors that are located in the viscera and in somatic deep structures is different from pain caused by injuries to the skin, muscles, tendons, joints, and other peripheral structures. Often the strongest sensation of visceral pain is the associated *referred pain*, which means pain that is referred to an anatomically different location than that where the noxious stimulation has occurred.

a. Acute Pain

Acute pain may be regarded as a submodality to touch, but there are considerable differences (Fig. 4.31). Acute pain caused by a noxious stimulation of the skin such as a pinprick or burning at a small spot has two components: a short-latency pricking pain and a late-burning and more emotionally involved sensation. The short-latency component produces a distinct sensation that is

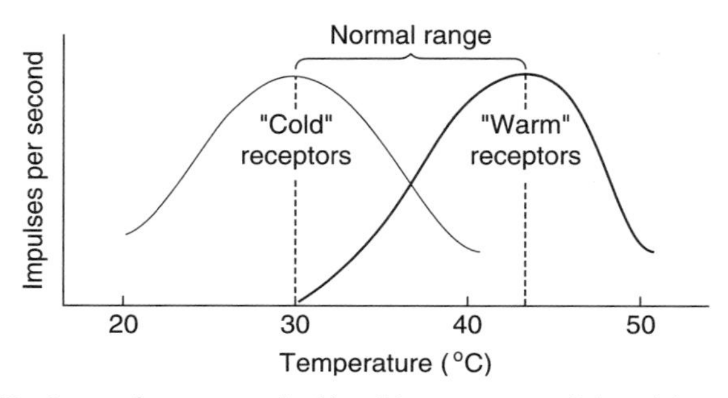

FIGURE 4.32 Range of responses of cold and heat receptors. (Adapted from Kenshalo.[50])

normally described as something that hurts. That sensation is distinctly localized to the area of the skin that is affected, thus is a spatial component ("where") that tells where on the body the pain originates. The fast phase of pain elicits the withdrawal reflex a monosynaptic spinal reflex, that activates flexor muscles.

The sensations that follow the fast component of pain sensation from stimulation of nociceptors in the skin (the slow component of pain) are often diffuse and may consist of a burning sensation of varying intensity, and that sensation may be regarded as object information ("what"). The slow component of pain often evokes emotional and unpleasant sensations that are of an urgent and primitive nature and is not as clearly localized anatomically as the short-latency component. The short-latency response of acute traumatic pain is mediated by fast-conducting nerve fibers (Aδ-fiber conducting 5 to 30 m/sec), whereas the subsequent slow and burning sensation is mediated by slow-conducting nerve fibers (C-fibers, namely unmyelinated small-diameter fibers conducting at 0.2 to 1 m/sec) (Fig. 4.33).

If the conduction in Aδ-fibers is blocked while maintaining the function in the C-fibers, such as can be done by compression of a peripheral nerve by a blood pressure cuff on the arm, the early phase of acute pain is suppressed and only a burning sensation, mediated by the C-fibers, is felt (Fig. 4.34). If early sensation to a painful stimulation is blocked, the location of such painful stimulation to the skin cannot be precisely determined. The ability to discriminate between different forms of noxious stimulation is also lost when the fast-conducting nerve fibers are blocked.

Local anesthetics (which are sodium channel blockers) affect the sensations of pain and touch in the opposite way and block the thinnest fibers first; only later and in higher dosages do they affect larger fibers. That means that analgesia (freedom from pain) can be achieved while the sensation of touch is preserved. When the C fibers are blocked, the late phase of pain is eliminated and touch is preserved (Fig. 4.34).

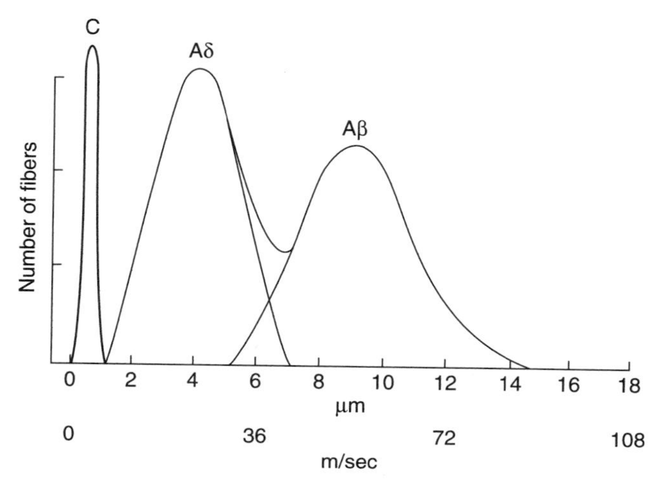

FIGURE 4.33 Distribution of conduction velocities of different kinds of nerve fibers that carry pain and sensory information. (Horizontal scale shows diameter of fibers in microns and conduction velocity in m/sec.)

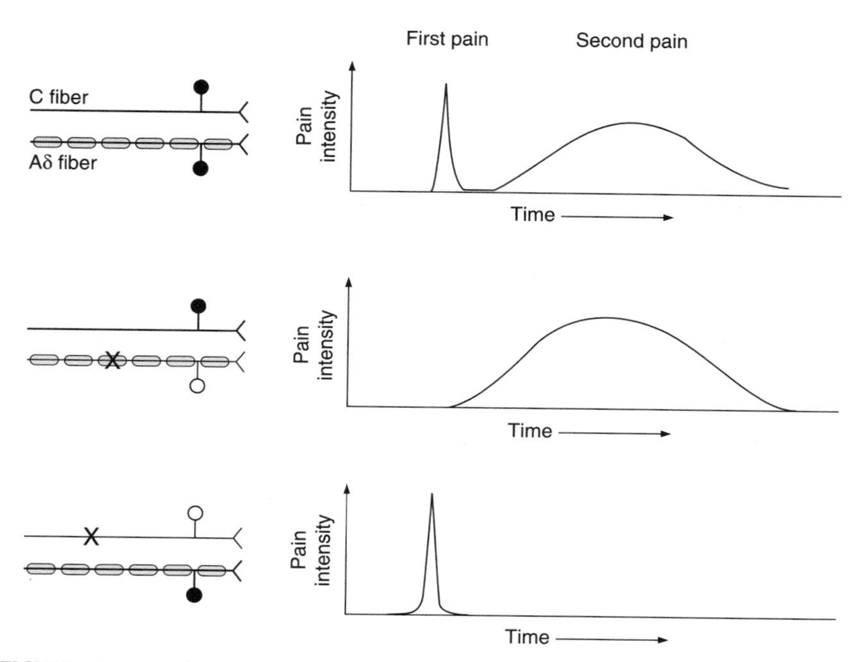

FIGURE 4.34 Two kinds of pain mediated by Aδ-and C-fibers: early pain abolished by blocking Aδ-fibers and late pain abolished after blocking of C-fibers.

The perception of somatic pain is not directly related to the physical stimulus, and the perceived intensity of pain, especially of the late component, depends on other factors. Also, the sensation of pain is not exactly related to the discharge rate of C-fibers. That observation has led to the suggestion that pain intensity is related to characteristics of the neural activity other than the discharge rate. It has been suggested that the temporal pattern of discharges of nerve fibers determines the sensations elicited.[26]

A common source of chronic pain is that of muscle spasm. Such pain can be alleviated by muscle relaxants such as the central-acting substances of the family of benzodiazepines,which reduce the excitability of α motoneurons because of reducing the excitatory influence from high CNS centers. It is also possible to alleviate such pain by physical therapy; strong massage of the tendons can often cause muscle relaxation and thereby relieve the pain. Tendon organs (Golgi organs) have an inhibitory influence on α motoneurons.

b. Chronic Pain

Chronic pain has been defined in various ways. One way would simply be to call pain that lasts a long time chronic pain.* With that definition, pain from chronic inflammation would fall in the category of chronic pain. Such pain may be caused by a continuous barrage of nerve impulses from nociceptors. Inflammatory processes such as arthritis are typical causes of chronic pain. Chronic pain often has central causes in addition to activation of nociceptors, and changes in function of the CNS may have developed over time as a result of long-lasting input from nociceptors. Chronic pain is often a mixture of nociceptor pain and neuropathic pain, although it may be believed to be elicited by activation of nociceptors.

c. Factors Affecting Sensitivity to Pain

Numerous factors can affect how a person perceives pain; sensitivity to pain is a variable quantity. For example, pain located to one part of the body is affected by pain elicited from another location. The perception of pain from one location will dominate, and pain at a certain location can elevate the threshold of pain at another location. The threshold to noxious stimulation may be lowered as a result of previous stimulation (primary hyperalgesia). The duration of the stimulation is important for the perception of its strength, which is best shown with pain elicited by heat. It has been shown that the sensation of pain from

*The U.S. National Center for Health Statistics defines a chronic condition as one of three months' duration or longer.

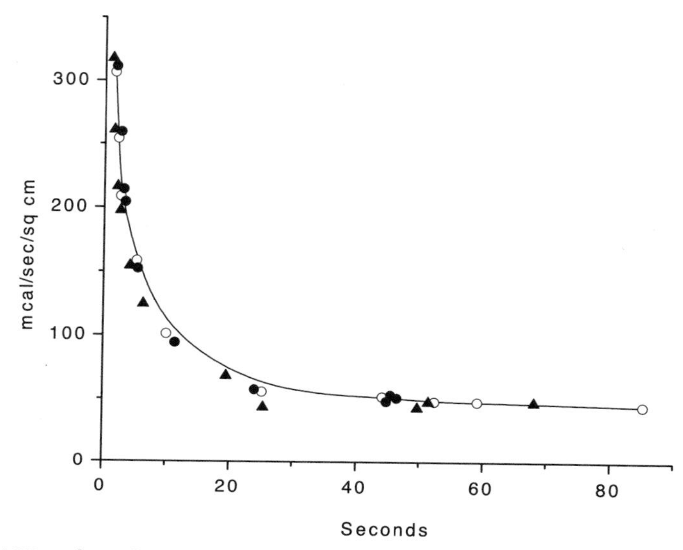

FIGURE 4.35 Relationship between the duration of stimulation with heat and the threshold of "pricking pain" in three individuals (temporal integration).[37]

heat has a considerable *temporal integration* (Fig. 4.35), so the sensation of pain is affected by the past history of stimulation.[38,115]

Many different drugs decrease the sensation of pain and are used in the treatment of pain. Understanding the effects of analgesics such as opioids has great clinical importance but the results of experimental studies are often controversial. However, opioid receptors are located in many parts of the CNS and in the spinal cord, and there is evidence that opioids can decrease the sensitivity to noxious stimulation along the entire neural axis of pain.[118]

Sensitization of Pain Receptors Two kinds of sensitization of pain receptors have been identified. One is of peripheral origin, where pain receptors can be sensitized by trauma such as burns or by chemical tissue insults, causing *hyperalgesia* (exaggerated response to painful stimuli).[15,52,91,97,108] Inflammatory processes are common causes of sensitization of pain receptors.

Sensitization of nociceptors can be caused by substances that are released in the tissue as a result of trauma or inflammation. The sensitization can spread from an area of trauma through collaterals of axons that innervate pain receptors.[91] This explains why the skin can be sensitized over a wider area than that which is directly affected by noxious stimulation such as, for instance, from tissue damage. The extension appears as a flare on the skin.[91]

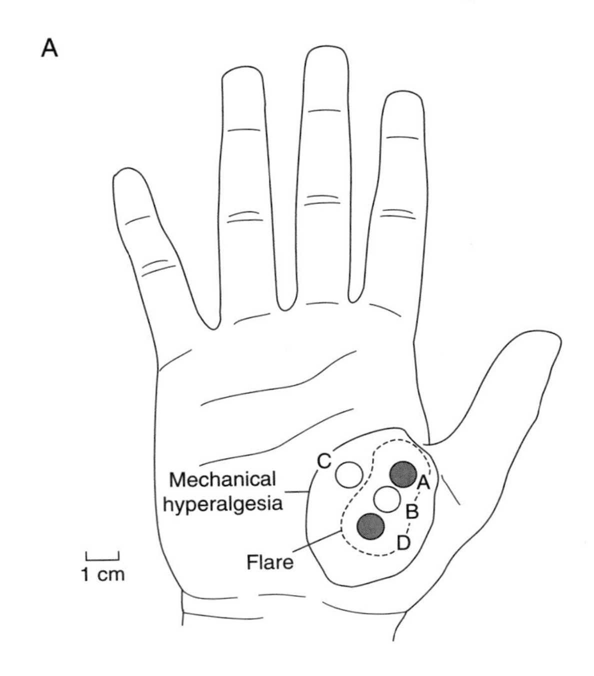

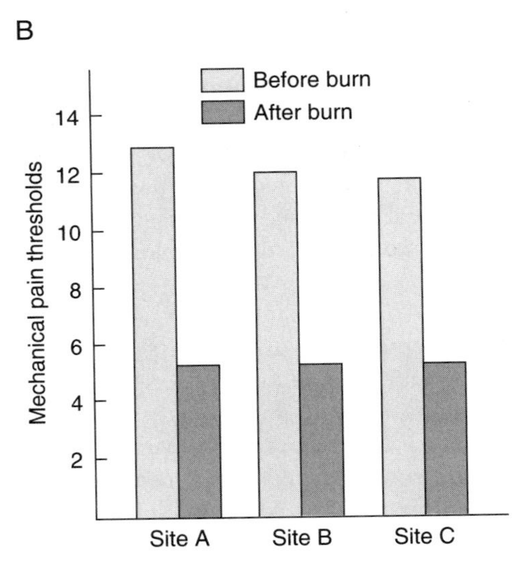

FIGURE 4.36 The effect of burns to skin (53°C for 30 sec) on hyperalgesia. (A) Mechanical hyperalgesia was recorded in sites A, B and C. (B) Threshold to pain before and after burns at these three sites. (Adapted from Raja, S.N., Campbell, J.N., and Meyer, R.A., *Brain*. 107: 1791–1188, 1984.)

The threshold of some nociceptors is normally so high that they never get activated under normal circumstances, but factors such as inflammation can sensitize such receptors so that they respond to noxious stimuli. Such receptors, named "sleeping receptors" by Schmidt,[35,99] can be activated under abnormal circumstances. Pain receptors in the teeth are examples of sleeping nociceptors.[119]

The sensitivity of nociceptors can also be increased by activation of the sympathetic nervous system. One way in which this can occur is related to the abundance of norepinephrine-secreting sympathetic nerve terminals located in close proximity to skin receptors.[94] Activation of the sympathetic nervous system can cause secretion of norepinephrine from these terminals, thus increasing the sensitivity of mechanoreceptors and pain receptors. The kind of sensitization of receptors may be so strong that receptors are activated without external stimulation.[22,83]

The involvement of the sympathetic nervous system may result in a special form of pain known as *sympathetic maintained pain*(SMP). Another classification uses the term *reflex sympathetic dystrophy* (RSD).[86] SMP is a different and more specific name for what was earlier known as *causalgia*,[94] which is defined as the burning sensation associated with nerve injuries. Attempts to clarify the classifications of these disorders have been made by the *complex regional pain syndrome* (CRPS) classification. These complex and elaborate classifications, however, may not be directly related to the clinical picture of these diseases.[80] Whatever classification is used, these pain conditions in which the sympathetic nervous system plays an important role remain poorly understood and complex.

Common for these conditions is that they produce a constant (chronic) pain because the firing of nociceptors is elicited by secretion of norepinephrine, which maintains (or increases) a high sympathetic activity, which, in turn, causes a persistent secretion of norepinephrine, a vicious cycle that maintains itself. This hypothesis is supported by the fact that sympathetic ganglion block is beneficial in treatment of these disorders.[80,87]

Central Sensitization of Pain Sensitization can also be caused by an increased responsiveness of neurons in the central nervous system (central sensitization).[120] (For a recent overviews, see Carli[8] and Devor *et al.*[22]) The central sensitization may outlast the period of stimulation of nociceptors by as much as weeks.

Sensory Input Can Modulate Pain Inhibitory input from peripheral sources to the spinal cord can modulate pain. The best known example is the inhibitory input from large (Aβ) fibers to dorsal horn cells.[64] It is known from experimental studies and from experience of patients with pain that other forms of sensory input can modulate the sensation of pain.

Gating of pain impulses at the segmental level was first described by Melzack and Wall.[64] Such gating can be activated by mechanical or electrical stimulation of the skin, and it can cause a decrease of the pain sensation. It is believed that this action occurs in the dorsal horn of the spinal cord. One theory is that neurons in the spinal cord receive input from both C-fibers (that innervate pain receptors) and A-fibers (which innervate mechanoreceptors) and that the A-fibers exert an inhibitory influence on neurons that communicate pain to the CNS.[64,117]

The fact that activity in large sensory fibers can modulate activity in neurons involved in pain is the basis for treatment of pain by *transdermal electrical nerve stimulation* (TENS).[36,117] When TENS is applied to the skin, it has been found effective for treatment of many forms of pain such as low-back pain and it is in common use. Electrical stimulation of the dorsal column of the spinal cord, another form of stimulating the somatosensory system, is also used for control of pain.[103] Such stimulation using implanted electrodes is efficient for treating neuropathic pain but acute nociceptive pain is unaffected.[71] This observation emphasizes the fundamental differences between different types of pain and the ways they can be treated by electrical stimulation.

The application of ointment containing capsaicin seems to have a effect similar to that of TENS, although the mechanisms are not understood. Capsaicin, the active ingredient in red pepper, is a stimulant of nociceptors and causes a burning sensation. It was earlier believed that its beneficial effect on pain had to do with the fact that it depletes primary small nerve endings from certain peptides (such as substance P),[25,80,96] but it has become evident more recently that its function in relieving pain is more complex and not directly related to its activation of nociceptors. Drugs are under development that would have these analgesic effects without producing the burning sensation.[127]

Some diseases such as diabetes and multiple sclerosis cause neural damage that affects large nerve fibers more than small fibers and thereby decreases sensitivity to touch while pain perception is preserved.[98] The input from large fibers (touch) has an inhibitory influence on neural conduction of pain signals in the dorsal horn of the spinal cord (see page 200) and the loss of functional large myelinated fibers is likely to be the cause of pain due to the lack of A-fiber-mediated inhibition.

Descending Spinal Activity Can Modulate Pain Activity in descending pathways can modulate ascending impulse traffic along the entire pathways of pain (see page 219). The descending pathways in the spinal cord may modulate the transmission of pain impulses at the segmental levels of the spinal cord (see Fig. 4.18).[29]

Descending activity can be elicited along the entire pain pathway.[119] For example, it has been shown that electrical stimulation of the gray matter that surrounds the third ventricle (periaquaductal gray) can reduce pain. It is believed that the stimulation of this central structure causes neural activity to descend into the spinal cord, where it reaches cells in the dorsal horn. There, it affects the transmission of pain impulses to the CNS. The descending pathways from the PAG are probably the most important modulators of pain.[119] PAG neurons are sensitive to opioids (exogenous and endogenous),[128] and it is believed that some of the analgesic effect of such substances is caused by their effect on PAG neurons.[118] These pathways, however, are very complex and incompletely understood. The involvement of the PAG in many different functions makes studies of its role in pain modulation difficult; studies of the role of supraspinal administration of morphine in decreasing the activity of dorsal horn neurons have been inconclusive.[118]

There are also other descending pathways that act on neurons on the ascending pain pathways before they reach regions of the CNS, where the awareness of pain is established. For example, recent studies have pointed to the dorsal reticular nucleus of the medulla as a possible modulator of ascending nociceptive activity.[57] These systems are believed to be the neural substrates for some of the psychological factors that can affect the perception of pain.

d. Visceral Pain and Referred Pain

Activation of visceral pain receptors does not always give rise to a sensation of pain. Irritation of the gastric mucosa and stretching of smooth muscles may be exceptions, but pain from ischemia such as of the heart does not directly produce a sensation of pain but instead a diffuse and uneasy, queasy feeling of being unwell, often accompanied by dizziness.[17] Many other disorders of visceral organs share that property. The pain often associated with many different forms of activation of pain receptors in the viscera such as ischemia of the heart muscle caused by myocardial infarction is instead often (but not always) felt as pain located in specific areas of the surface of the body. This is known as *referred pain*.[17,89,95] Referred pain may also be felt in the skeleton and in muscles, but it is more commonly felt as being located on the surface of the body (skin). Pain in other visceral organs also produces a distinct sensation of pain from specific regions of the skin, (referred pain) in addition to creating a diffuse feeling of being unwell. The location of referred pain varies from individual to individual and is more or less pronounced. It may even be totally absent.

Visceral pain felt on the surface of the body (referred pain) is probably caused by activation of cells in the dorsal horn that receive input both from receptors in the viscera and the skin (see page 200). It is also possible that convergence at more central locations can give rise to referred pain.

Knowing to which areas of the skin the pain of specific organs is referred is important for the clinical diagnosis of disorders that affect internal organs.[89] The considerable individual variation in the location and nature of referred pain is an obstacle in the diagnosis of disorders of internal organs such as the heart. For example, ischemia of the heart most commonly is felt as pain in the ulnar radiation of the left arm but also often in the neck and jaw; in many instances, it only manifests by diffuse feelings of being unwell. Ischemia of the heart muscle from myocardial infarction that does not give rise to symptoms is labeled "silent myocardial infarctions."[82]. What has been regarded as typical referred pain occurs in only a fraction of patients with myocardial infarctions (approximately half of the patients, according to some investigators). This means that a large group of patients with myocardial infarction have nonspecific symptoms, and the only symptom maybe a diffuse, uneasy feeling of ill-being or malaise. These individuals are likely to ignore their symptoms or at least delay seeking medical attention, with the result of less favorable outcomes for myocardial infarction.

It is interesting that a well-known type of pain, angina pectoris, can be relieved by TENS despite the fact that pain is assumed to originate in the heart (from ischemia).[62] Others have shown that the tenderness often felt in the chest wall by patients with angina pectoris during exercise could be relieved by applying a local anesthetic to the skin. Removing the referred pain left a true visceral pain consisting of general discomfort which the patients had never felt before.[89,90]

The prevailing hypothesis about referred pain has defined it as a convergence of the input from receptors in the skin and in viscera onto the same dorsal horn cells in the spinal cord.[17,95] It was assumed that the discharges in these dorsal horn neurons that are elicited by visceral noxious stimulation could not be distinguished from those elicited by (noxious) stimulation of the skin because the same dorsal horn neuron is shared by visceral and skin receptors. The ascending fibers from these cells therefore do not provide correct information about whether the activation is caused by visceral or cutaneous stimulation. Consequently, the CNS does not know where the activity is generated. However, other hypotheses have been proposed to explain referred pain, some of which involve facilitation, convergence at the thalamic level, neural plasticity, and axon reflexes.[17] (For reviews see Arendt-Nielsen *et al.*[1])

It is also possible that memory or neural plasticity is involved. Frequent stimulation of the skin may make an "imprint" in the cortex. Stimulation of viscera is rare, and such stimulation may therefore not make any "imprint" in the cortex.[61]

It is possible that sensory afferents in the vagus nerve are involved in producing the symptoms and nonspecific sensations such as queasy feeling or feeling unwell from stimulation of nociceptors in the viscera.

2. Neuropathic Pain

Neuropathic pain is a broad term derived from the term *neuropathy* that means disorders of the nervous system. Neuropathic pain refers to (any) pain caused by the nervous system, such as pain from damage of any kind to the nervous system or by functional changes in the morphology or function of the nervous system. Neuropathic pain can be caused by functional and perhaps anatomical changes in the CNS as a result of neural plasticity. Neuropathic pain is often chronic and not always correlated with a specific body location.[22,23,86,125]

Neuropathic pain is not caused by activation of nociceptors because there are no known nociceptors in nervous tissue, such as peripheral nerves or the CNS (although recent studies have shown evidence that the *nervi nervorum** of peripheral nerves[5,28] have nociceptors). Despite the lack of specific pain receptors, stretching or compressing of a peripheral nerve can cause pain. Evidence suggests that pain from injury to peripheral nerves is a result of neural plasticity invoked by abnormal activation of peripheral nerves or dorsal roots that cause functional changes in more central structures.[10,20,22] Ischemic or inflammatory insults to the CNS, such as strokes, are frequent causes of neuropathic pain.

a. Pathophysiologic Pain

According to Devor,[21] "pathophysiologic" pain is pain not caused by normal function or normal stimulation of nociceptors that mediate information to the CNS through Aδ- and C-fibers. Devor[21] defines three types of such pathophysiologic pain:

1. Increased sensitivity of nociceptors (peripheral sensitization)
2. Peripheral nerves acting as impulse generators
3. Increased "gain" in central pain circuits (central sensitization)

Pathophysiologic pain is thus a form of neuropathic pain. For example, the pain of a large group of pain disorders known as neuralgias (such as trigeminal neuralgia) are regarded to be forms of pathophysiologic pain. Neuralgia often affects nerve roots such as spinal dorsal roots and the trigeminal nerve root. The cause of these disorders is essentially unknown, but more recent studies point toward a central cause that may have developed because of trauma to peripheral nerves or cranial nerves.[73,75,76,98] Irritations of various kinds to these nerves can also cause pathophysiologic pain. Compression of dorsal roots of the spinal cord is a frequent cause of neuropathic and radicular** pain such as the common low-back pain, which at one time may affect as many as 80% of the adult population.[49] Such pain is often accompanied by allodynia and hyperalgesia. It is not known why certain conditions of peripheral nerve injuries in some individuals develop into chronic conditions of severe pain that seem unrelated to the primary injury, whereas similar injuries in other individuals heal and leave no permanent pain.

**Nervi nervorum* are nerves that innervate the sheaths of nerve trunks.

***Radicular* is a term related to *radix*, meaning root.

Chronic pain may be caused by injured peripheral nerves because they act as impulse generators and that activity may be interpreted as pain,[21,22] or, more likely, because the abnormal activity cause changes in the function of structures in the CNS cause the sensation of pain.

It has been claimed that pain from dorsal root compression such as occurs in the common low-back pain is a result of neural plasticity induced by stimulation of dorsal roots from the compression. This is probably the most likely cause, but it has also been shown that inflammation of nerve trunks, such as peripheral nerves, may cause pain because of stimulation of nociceptors that are located in the nervi nervorum of peripheral nerves.[5,28]

Studies in humans undergoing back operations under local anesthesia have shown that mechanical manipulation or pressure to a spinal nerve root results in pain only if the nerve root has previously been injured or stretched or is inflamed.[54] This means that at least two factors are necessary to create pain from compression or irritation of nerve roots. Similar hypotheses have been presented regarding the generation of symptoms from irritation of cranial nerve roots that can cause pain (trigeminal neuralgia) or spasm (hemifacial spasm).[73]

Recent studies in humans have shown that chronic compression of dorsal roots can result in such altered processing in the dorsal horn as increased spontaneous activity and increased receptive fields of dorsal horn cells on the affected side.[105] (For a review, see Carstens.[10]) Compression or injuries to peripheral nerves are also accompanied by changes in neural processing, as indicated by altered temporal integration as studied in humans.[74] Trauma to the body that causes acute pain can thus result in pathophysiologic pain that continues after the trauma has healed, and the location of the physiologic abnormality that generates the neural activity that is interpreted as pain has shifted from the location of the trauma to the CNS. That means that neuropathic pain can occur without the presence of a peripheral pathology. This kind of pathophysiologic pain is a source of great discomfort. This form of pain is mostly perceived as being negative, and it is difficult to see any biological advantage to the organism from such pain.

b. Sensitization of Central Pain Pathways

Central neurons in pain pathways can be sensitized by abnormal sensory stimulation. Several mechanisms for such sensitization have been proposed:[53,97,122,126]

1. Synaptic mechanisms
2. Membrane excitability
3. Phenotypical* changes, or expression of new neurotransmitters, neuromodulators, or their receptors in spinal or supraspinal neurons

*The observable constitution of an organism or the appearance of an organism resulting from the interaction of the genotype and the environment.

4. Morphological reorganization, including morphological changes in the CNS such as sprouting of afferent nerve terminals

The involvement of *N*-methyl-*D*-aspartate (NMDA) receptors has been indicated.[126] *Long-term potentiation* (LTP) of *wide-dynamic-range neurons* (WDR) in the spinal cord and brainstem are examples of plastic changes in central neurons that can cause increased excitability of neurons.[14]

That nociceptive stimulation can increase the excitability of neurons in the CNS by causing LTP of WDR neurons has been further demonstrated in animal experiments[120,122] and has been associated with pain hypersensitivity.[14] Likewise, inflammatory processes (such as arthritis) may cause changes in neural excitability that in turn can result in neuropathic pain conditions.[99]

More recently it has been shown that the sprouting of large fibers to laminae I and II of the dorsal horn invoked by nociceptive stimulation may contribute to the changes in function.[53,122] Peripheral nerve injuries may cause such sprouting of large fibers into lamina II of the dorsal column which has been hypothesized to contribute to the development of pain that often follows after nerve lesions.[123] Other effects of lesions (sectioning or crushing) to peripheral nerves (the sciatic nerve) have been observed. Thus, lesions of peripheral nerves in young animals do not cause an apparent change in the number of myelinated fibers in the dorsal column, but the number of unmyelinated fibers decreases after several weeks after the lesion.[16]

These findings illustrate the complexity of neural processing of pain signals and show that the sensation of pain depends on many factors, some of which are intrinsic central nervous system activities. For example, many individuals can be rendered analgesic by hypnosis. The fact that the perception of pain can be affected by treatment with a placebo is a further indication that the sensation of the intensity of pain is relative and depends on many factors other than the strength of the noxious stimulation or its duration.

c. *Phantom Pain*

The pain that may be experienced after amputations of limbs and referred to the amputated limb is known as *phantom pain*. It is an example of *phantom perceptions*,[43] which often include pain. Phantom pain is the most convincing indication that the anatomical location of the physiologic abnormality that causes pain can be different from the location where pain is felt. Many other forms of neuropathic pain share that property; phantom pain is just one of several phantom perceptions, including other sensations (tingling, etc.) that may be felt in an amputated limb. Tinnitus is also often regarded to be a phantom perception (see Chapter 5).

Phantom pain is assumed to be caused by changes in the function of specific CNS circuits brought about by expression of neural plasticity.[43,65] Observations about phantom pain confirm other observations indicating that the neural circuits mediating pain are far from "hard wired." This means that various exogenous or endogenous signals can alter specific connections in the brain through expression of what is generally known as neural plasticity. We have already discussed how such changes in the function can be caused by changes in synaptic efficacy or by creation of new connections (by sprouting of axons) or by elimination of connections.

It has been hypothesized that the changes in the function of the CNS that result in phantom limb symptoms are caused by stimulation that results from the trauma in connection with the amputation. This hypothesis was supported by early clinical findings that the phantom limb symptoms could be reduced if the transmission of sensory (pain) input to the CNS from the limb to be amputated is blocked by, for instance, epidural anesthesia before the amputation,[2] as suggested by Wall.[113] However, the beneficial effect of such preoperative blockage has been difficult to prove in controlled clinical studies, and the results of such studies are rather variable, some showing considerable effect while others show little or no benefit from such preemptive anesthesia. (For an overview, see Jensen and Nikolajsen[45] and Wilder-Smith.[116])

d. Signs of Neuropathic Pain

Neuropathic pain has few objective signs, and the assessment of such pain is done primarily on the basis of the patients' own assessment of the severity of their pain. One entity, temporal summation, however seems to be affected by neuropathic pain. This can be demonstrated by applying electrical stimulation to the skin consisting of repetitive impulses where the impulse frequency is varied.

Changes in the function of the somatosensory nervous system during the development of neuropathic pain can be demonstrated in studies of temporal summation of electrical stimulation of the skin or a peripheral nerve.* The threshold of pain sensation from such electrical stimulation of the skin (for instance, on the forearm) is higher than the threshold of innocuous sensation (of tingling) by an amount that depends on the repetition rate. This means that the temporal integration for pain and sensation are different (Fig. 4.36A).[74] Depending on the stimulus intensity, the sensation changes from a just noticeable tingling for low currents to a clear sensation of pain. Determination of the thresholds for the tingling sensation and for the pain sensation at different repetition rates of the impulses provides a measure of temporal integration. More specifically, the threshold of sensation is practically independent of

*Temporal integration is a form of memory that is manifest by, for example, summation of stimulations, where a stimulus that follows another stimulus produces a stronger response than the first stimulus.

the stimulus rate, indicating an absence of temporal integration. The threshold for pain decreases as the repetition rate is increased (Fig. 4.37A), a sign of temporal integration. Inspection of graphs such as those in Fig. 4.37A reveals that the threshold for pain decreases exponentially with increasing rate, indicating that neural transmission of pain activity can be modeled by a simple integrator that has a single time constant.

The threshold of sensation (of tingling), however, varies little for different rates of stimulation (Fig. 4.37A), thus displaying no noticeable degree of temporal integration of sensation. This is one example of the difference between the neural processing of innocuous and noxious stimulations.

A clear difference exists between the responses in individuals without pain and in individuals with neuropathic pain (caused by peripheral nerve injuries). In individuals with chronic pain (Fig. 4.37B), the threshold for pain is lower than normal and often only slightly higher than the threshold of sensation; the temporal integration of pain is nearly absent. This means that neural processing of pain signals is similar to the processing of neural activity that is elicited by innocuous stimuli.

e. *Abnormal Sensations in Connection with Pain*

Neuropathic pain is often accompanied by abnormal perception of acute pain stimuli and of normal somatosensory stimulation. A phenomenon known as *allodynia*[20,105] consists of perception of pain from normally innocuous stimulation of the skin. Another phenomenon that often accompanies

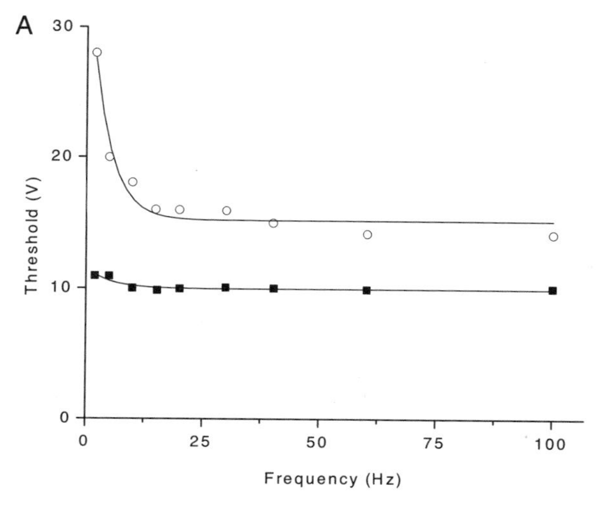

FIGURE 4.37 (A) Threshold of sensation of electrical stimulation of the skin (filled squares) and threshold of pain (open circles) shown as a function of the rate at which the stimulus impulses were presented. (B) Similar graph as in (A) obtained in an individual with chronic pain (same symbols). (Adapted from Møller, A.R. and Pinkerton, T., *Neurol. Res.*, 19: 481–488, 1997, Published by Forefront Publising.)

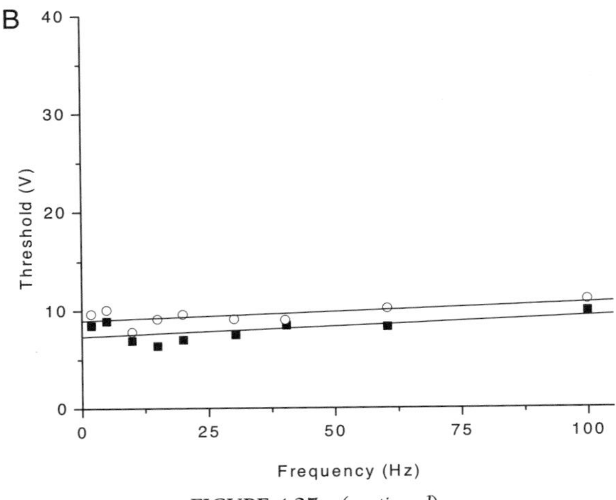

FIGURE 4.37 *(continued)*

neuropathic pain is *hyperpathia*, an exaggerated reaction to painful stimulation. Some of these phenomena can be explained by sensitization of pain receptors,[83] while others have more complex causes involving changes in central processing of pain and other somatosensory input.[14,20,86,120]

Changes in the function of WDR neurons have also been assumed to cause the *wind-up phenomenon*,[66] which consists of an increased response to stimulation that is repeated. The wind-up phenomenon is a form of an altered temporal integration and is often present in connection with pathophysiologic pain. It adds severely to the adversity of the condition. The effect of NMDA receptor activation and activation of calcium channels[126] can explain the wind-up phenomenon, where successive stimulation results in increasing response magnitude (increasing sensation of pain).

3. Neurophysiologic Basis for Neuropathic Pain

The neurophysiology of neuropathic pain has been studied extensively because of its enormous clinical importance, and we now know more about the mechanisms for neuropathic pain than that of other functions of the somatosensory system.

4. The Somatosensory Nervous System can Operate in Different Modes

One of the important keys to the understanding of pain mechanisms is that the sensory nervous system can operate in different modes, which means that stimulation that at one time is innocuous may at another time produce the sensation of pain. In that respect, the dorsal horn plays an important role, but more central structures are also involved. Woolf[23,124] has listed four different states of the dorsal horn:

1. Control state
2. Suppressed state
3. Sensitized state
4. Reorganized state

The control state is the normal situation. In this state, low-intensity somatosensory stimulation that activates low-threshold afferents will produce an innocuous sensation, and high-intensity stimulation that activates high-threshold afferents will cause a sensation of pain. In the suppressed state, a similar stimulation does not produce pain because of segmental or descending inhibition. In the third (sensitized) mode, low-intensity stimulation causes a sensation of pain due to sensitization of dorsal horn cells. This is known as (mechanical) allodynia, a painful sensation from a normally innocuous stimulation of the skin. Similar reaction to low-threshold stimulation can be caused by the sensitization of high-threshold receptors (primary hyperalgesia). High-intensity stimuli also cause pain in the sensitized state by activating high-threshold afferents because the dorsal horn cells have been sensitized. The reorganized state, as defined by Woolf,[23,124] represents a potentially irreversible situation of synaptic reorganization of the dorsal horn. In this state, both low- and high-intensity stimulation may give pain sensations. The outcome of the reorganized state is known as peripheral or central neuropathic pain, and the underlying cause is neural plasticity. This situation may occur as a result of peripheral or central morphological or functional changes.

Role of WDR Neurons Changes in pain processing in the spinal cord circuits may occur as a result of changes in function of the wide-dynamic-range (WDR) neurons, located in the dorsal horn of the spinal cord. The WDR neurons have been studied more than the other types of spinal cord neurons involved in the nonclassical somatosensory pathways. The WDR neurons normally respond to innocuous stimulation in a graded fashion over a large range of stimulus intensities (thus their name),[19,86] but they are believed to change their function as a result of abnormal circumstances (expression of neural plasticity) and then they play an important roles in processing of painful stimuli.[19,20,23,85,86,119,121] (see page 209). The fact that the WDR neurons receive input from many kinds of receptors[86] (Fig. 4.14) opens up many possibilities of interplay between different kinds of sensory input, and there is considerable evidence that the function of these neurons can change as a result of their input (neural plasticity). Such plastic changes may be responsible for neuropathic pain. Studies have indicated that changes in the function of WDR neurons may be caused either by reduced inhibitory influence (from Aβ-fibers with diameter of 6 to 12 μm,and conduction velocity of 30 to 70 m/sec) or by excessive input from pain fibers (C-fibers, see page 210).[86]

Neuropathic pain is not sensitive to treatment with medication such as opioids;[13] however, other treatments such as TENS[117] and dorsal column stimulation have been successful in controlling this kind of pain.[71]

5. Where Is the Sensation of Pain Processed?

We have already discussed the processing of pain from stimulation of nociceptors that occur in the spinal cord. The further processing that occurs in more central structures is complex and poorly understood. The connections that are known to mediate sensations of pain are many (Fig. 4.38), but the importance of many of these connections is not known with certainty.

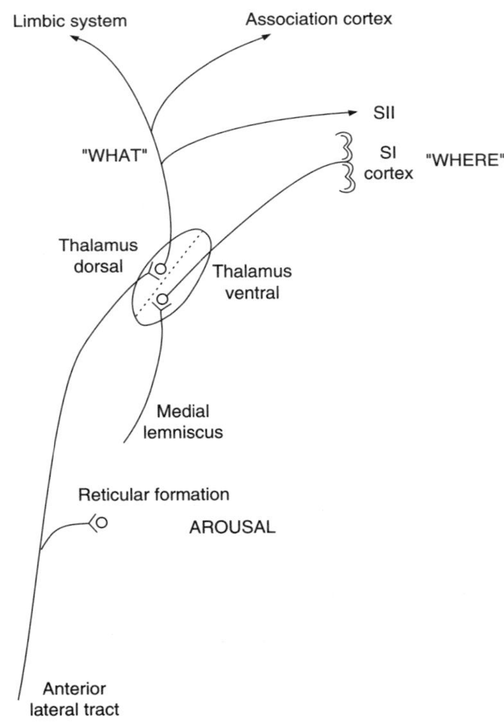

FIGURE 4.38 Schematic illustration of pathways that are involved in mediating sensation of pain.

a. *Role of the Thalamus in Pain*

Electrical stimulation of specific regions of the ventrobasal nuclei of the thalamus (VPL and VPM) is now in use for treatment for severe pain,[33,103,129,131] and many forms of deafferentation pain and pain caused by noxious stimulation are treatable by electrical stimulation of different structures. The success of such treatment confirms the involvement of these regions of the thalamus in pain.

Destructive procedures of various regions of the thalamus (thalamotomy) using stereotactic surgical methods have been used to treat severe pain, such as cancer pain, for many years.[34] The targets have been the spinoreticulothalamic tract, and that kind of treatment is aimed at nociceptive pain rather than neuropathic pain. More recently, focused gamma radiation (gamma-knife) has been used to make lesions in the nuclei of the ventrobasal thalamus to treat severe pain,[130] but newer results show that electrical stimulation seems to provide better results in alleviating intractable neuropathic pain.[34]

Studies in humans have shown that neurons in the posterior border of the ventrocaudal nucleus of the thalamus respond to painful stimuli.[61] The cells in this nucleus that respond to noxious stimulation are mainly located close to the posterior border of the nucleus, and these cells are assumed to receive input from the spinothalamic pathway.[61] This nucleus also processes innocuous stimuli (see page 202) and projects to the somatosensory cortex. Stimulation of some of these neurons produces a sensation of (noxious) heat but no other (innocuous) thermal sensations, and stimulation of neurons in the ventrocaudal nucleus of the thalamus could also produce the sensation of visceral pain, such as that of angina.[60] It is interesting that the sensation of angina induced by electrical stimulation of these neurons was observed only in a patient who had experienced angina previously which had been resolved due to opening of coronary arteries, but it was never observed in studies of 50 patients who had never experienced angina.[61] This indicates that the pain circuitry had undergone changes during the natural occurrences of angina and that such changes are a prerequisite for evoking pain sensation of that nature through electrical stimulation of neurons in that region of the thalamus.

b. *Cortical Representation of Acute Pain*

Involvement of the cerebral somatosensory cortex in pain is controversial,[119] and lesions of the primary somatosensory cortex do not relieve pain. Electrical stimulation of the SI cortex rarely produces painful sensations in awake patients,[51,84] but the SI cortex participates in the localization of pain.

These observations support other findings and have shown the two components of nociceptor pain: One determines the location on the body of the painful stimulation, and the other gives rise to the motivational, affective, and autonomic reflexes; endocrine reactions and feelings of discomfort.[88] They are processed in anatomically separate portions of the CNS. The component that determines the location of the pain may be a form of spatial ("where") information, and the component that determines the severity and nature of the pain is a form of object information ("what") (Fig. 4.38). This means that the two components of acute pain—namely the awareness of the pain and the bothersome and affective aspects of pain—are most likely mediated by different populations of neurons. Localization of the site of stimulation is assumed to be the result of projections of pain pathways to the primary sensory cortex (SI), and the perception of the degree of pain is mediated by other parts of the brain, one of which may be the secondary somatosensory cortex (SII).[51] Only the fast phase of acute pain seems to be represented in the somatosensory cortex, which is why determining the location on the body of pain ("where") is related to the fast phase of pain.

These hypotheses have been supported by recent magnetoencephalographic studies in humans[115] which have shown that the SI cortex is not activated by pain elicited by laser-induced heat. The SII of both sides were shown to be activated.

The intensity of the slow phase of pain is poorly defined which is in agreement with the hypothesis that this kind of pain is not processed in the neocortex but rather in the allocortex and subcortical structures such as the thalamus including limbic structures.

c. *Limbic System*

Activation of nociceptors cause activation of brain regions such as the limbic system,[51] which explains the many different symptoms and signs of pain such as affective, autonomic, endocrine, arousal, etc.[12] Parts of the allocortex, such as the anterior insular cortex and the cingulate gyrus, are involved in the affective aspects of pain.[51,111] The anterior cingulate gyrus, a part of the limbic system, receives input from the medial and intralaminar nuclei of the thalamus.[51]

Magnetoencephalographic studies in humans[115] have shown that limbic structures (amygdala or hippocampal formation) are activated in humans by pain elicited by laser heating of the skin. The activity in the cingulate gyrus could not be studied by the technique used in that study.

The anatomical locations that exhibit signs of increased neural activity during nociceptive stimulation may not be the structures that are responsible for the sensation of pain. Lesion studies where absence of pain sensation is related to the location of the lesion seem to provide more correct information in that respect but lesions of the pathways that carry pain signals may equally well affect perception of pain stimuli as lesions of regions that are involved in the perception of pain.

E. Itch

Itch (pruritus) is a peculiarly unpleasant sensation that causes an urge to scratch the region of the skin to which it is referred. Itch is usually referred to a distinct patch of skin, and scratching the skin relieves the sensation of itch at least for a time. The sensation of itch is assumed to be caused by a specific irritation such as that of a foreign body on the surface of the skin. Chemicals applied to the skin or in the skin such as, for instance, from mosquito bites also cause itch and an urge to scratch the area. Itch is mostly elicited from the superficial skin or mucosa and rarely referred to deeper structures. Allergic reactions often include itch and, in that form, large parts of the body surface can be affected. Itch can be elicited by a wide variety of local and systemic substances. Histamine is powerful elicitor of itch and it has been used in many studies.[24]

Kidney and liver diseases are often associated with itch, and allergic reactions often include itch. Many drugs have itch as side effects and even a psycho-physiological cause of itch is common. Likewise some psychiatric disorders are accompanied by itch.

The sensation of itch is mediated by receptors that are similar (but probably not identical) to those that mediate the sensation of pain, namely bare axons and C-fibers.[107] Painful stimuli may in fact reduce or abolish itch. Central processing of stimuli that cause the sensation of itch is different from that of painful stimuli.

Recent studies using positron emission tomography (PET) indicate that the central processing of the histamine-elicited sensation of itch is similar to that of pain but different in that itch does not seem to involve subcortical structures.[24] This conclusion was based on the observation that application of histamine increased regional cerebral blood flow (rCBF) predominantly in the contralateral somatosensory cortex and bilaterally in motor areas. Activation was also detected in prefrontal cortex and the cingulate gyrus but not in the secondary somatosensory cortex. These pathways are similar to those of pain except for the fact that no activation by itch was seen in the thalamus.

REFERENCES

1. Arendt-Nielsen, L., Laursen, R.J., and Drewes, A.M., Referred Pain as an Indicator for Neural Plasticity. In: *Nervous System Plasticity and Chronic Pain*, edited by J. Sandkühler, B. Bromm, and G.F. Gebhart. Amsterdam: Elsevier, 2000.
2. Bach, S., Noreng, M.F., and Tjellden, N.U., Phantom Limb Pain in Amputees During the First 12 Months Following Limb Amputation, after Preoperative Lumbar Epidural Blockade. *Pain*, 33: 297–301, 1988.
3. Blakemore, S.J., Wolpert, D., and Frith, C., Why Can't You Tickle Yourself? *NeuroReport*, 11: 11–16, 2000.
4. Boivie, J., Grant, G., and Silfvenius, H., A Projection from the Nucleus Z to the Ventral Nuclear Complex of the Thalamus in the Cat. *Acta Physiol. Scand.*, 80: 11A, 1970.
5. Bove, G.M. and Light, A.R., Unmyelinated Nociceptors of Rat Paraspinal Tissues. *J. Neurophysiol.*, 73: 1752, 1995.
6. Brodal, P., *The Central Nervous System*. New York: Oxford Press, 1998.

7. Bromm, B., Brain Images of Pain. *News Physiol. Sci.*, 16: 244–249, 2001.
8. Carli, G., Neuroplasticity and Clinical Pain. In: *Nervous System Plasticity and Chronic Pain*, edited by J. Sandkühler, B. Bromm, and G.F. Gebhart. Amsterdam: Elsevier, 2000.
9. Carlson, M., Characteristics of Sensory Deficits Following Lesions of Brodmann's Area 1 and 2 in the Postcentral Gyrus of Macaca mulatta. *Brain Res.*, 204: 424–430, 1980.
10. Carstens, E., Altered Spinal Processing in Animal Models of Radicular and Neuropathic Pain. In: *Nervous System Plasticity and Chronic Pain*, edited by J. Sandkühler, B. Bromm, and G.F. Gebhart. Amsterdam: Elsevier, 2000.
11. Catarina, M.J. and Julius, D., The Vanilloid Receptor: A Molecular Gateway to the Pain Pathway. *Ann. Rev. Neurosci.*, 24: 487–517, 2001.
12. Chapman, C.R., The Affective Dimension of Pain: A Model. In: *Pain and the Brain*, edited by B. Bromm and J.E. Desmedt. New York: Raven Press, 1995, pp. 283–301.
13. Cherny, N.I. and Portenoy, R.K. Practical Issues in the Management of Cancer Pain. In: *Handbook of Pain*, 4th ed., edited by P.D. Wall and R. Melzack. Edinburgh: Churchill-Livingstone, 1999, pp. 1479–1522.
14. Coderre, T.J., Katz, J., Vaccarino, A.L.,and Melzack, R., Contribution of Central Neuroplasticity to Pathological Pain: Review of Clinical and Experimental Evidence. *Pain*, 52: 259–285, 1993.
15. Coggeshall, R.E., Dougherty, P.M., Pover, C.M., and Carlton, S.M., Is Large Myelinated Fiber Loss Associated with Hyperalgesia in a Model of Experimental Peripheral Neuropathy in the Rat? *Pain*, 52: 233–242, 1993.
16. Coggeshall, R.E., Lekan, H.A., Doubell, T.P., and Woolf, C.J., Central Changes in Primary Afferent Fibers Following Peripheral Nerve Lesions. *Neuroscience*, 77: 1115–1122, 1997.
17. Cousins, M. and Power, I., Acute and Postoperative Pain. In: *Textbook of Pain*, 4th ed., edited by P.D. Wall and R. Melzack. Edinburgh: Churchill-Livingstone, 1999, pp. 447–491.
18. Craig, A.D. and Zhang, E.T., Anterior Cingulate Projection from MDVC (a Lamina I Spinothalamic Target in the Medial Thalamus of the Monkey). *Soc. Neurosci. Abstr.*, 22: 111, 1996.
19. Craig, A.D. and Dostrovsky, J.O., Medulla to Thalamus. In: *Textbook of Pain*, 4th ed., edited by P.D. Wall and R. Melzack. Edinburgh: Churchill-Livingstone, 1999, pp. 183–214.
20. Devor, M., Central Changes Mediating Neuropathic Pain. In: *Proceedings of the Fifth World Congress on Pain*, edited by R. Dubner, G. Gebhart and M. Bond. Amsterdam: Elsevier, 1988, pp. 114–128.
21. Devor, M., The Pathophysiology of Damaged Peripheral Nerves. In: *Textbook of Pain*, edited by P. D. Wall and R. Melzack. Edinburgh: Churchill-Livingstone, 1994, pp. 79–100.
22. Devor, M., and Seltzer, Z., Pathophysiology of Damaged Nerves in Relation to Chronic Pain. In: *Textbook of Pain,* 4th ed. edited by P.D. Wall and R. Melzack. Edinburgh: Churchill-Livingstone, 1999, pp. 129–164.
23. Doubell, T.P., Mannion, R.J., and Woolf, C.J.,The Dorsal Horn: State-Dependent Sensory Processing, Plasticity and the Generation of Pain. In: *Handbook of Pain*, 4th ed., edited by P.D. Wall and R. Melzack. Edinburgh: Churchill-Livingstone, 1999, pp. 165–181.
24. Drzezga, A., Darsow, U., Treede, R.D., Siebner, H., Frisch, M., Munz, F., Weilke, F., Ring, J., Schwaiger, M., and Bartenstein, P., Central Activation by Histamine-Induced Itch: Analogies to Pain Processing. A Correlational Analysis of O-15 H_2O Positron Emission Tomography Studies. *Pain*, 92: 295–305, 2001.
25. Dubner, R., Topical Capsaicin Therapy for Neuropathic Pain. *Pain*, 47: 247–248, 1991.
26. Emmers, R., *Pain: A Spike-Interval Coded Message in the Brain*. New York: Raven Press, 1981.
27. Everett, N.B., *Functional Neuroanatomy*. Philadelphia: Lea & Fibiger, 1971.
28. Fields, H.L., Baron, R., and Rowbotham, M.C., Peripheral Neuropathic Pain: An Approach to Management. In: *Textbook of Pain*, 4th ed., edited by P.D. Wall and R. Melzack. Edinburgh: Churchill-Livingstone, 1999, pp. 1523–1533.

29. Fields, H.L. and Basbaum, A.I., Central Nervous System Mechanism of Pain Modulation. In: *Textbook of Pain*, 4th ed., edited by Wall, P.D. and Melzack, R., Edinburgh: Churchill-Livingstone, 1999, pp. 309–329.
30. Fromm, G.H., Pathophysiology of Trigeminal Neuralgia. In: *Trigeminal Neuralgia*, edited by G.H. Fromme and B.J. Sessle. Boston: Butterworth-Heinemann, 1991, pp. 105–130.
31. Gordon, G., and Grant, G., Dorsolateral Spinal Afferents to Some Medullary Sensory Nuclei. An Anatomical Study in the Cat. *Exp. Brain Res.*, 46: 12–23, 1982.
32. Graybiel, A.M., The Thalamocortical Projection of the So-Called Posterior Nuclear Group: A Study with Anterograde Degeneration Methods in the Cat. *Brain Res.*, 49: 229–244, 1973.
33. Gybels, J. and Kupers, R., Subcortical Stimulation in Humans and Pain. In: *Pain and the Brain*, edited by B. Bromm and J.E. Desmedt. New York: Raven Press, 1995, p. 187–199.
34. Gybels, J.M. and Tasker, R.R., Central Neurosurgery. In: *Textbook of Pain*, 4th ed., edited by P.D. Wall and R. Melzack. Edinburgh: Churchill Livingstone, 1999, pp. 1307–1339.
35. Handwerker, H.O., Kilo, S., and Reeh, P.W., Unresponsive Nerve Fibers in the Sural Nerve of the Rat. *J. Physiol.*, 435: 229–242, 1991.
36. Hansson, P. and Lundeberg, T., Transcutaneous Electrical Nerve Stimualtion, Vibration and Acupuncture as Pain-Relieving Measures. In: *Textbook of Pain*, 4th ed., edited by P.D. Wall and R. Melzack. Edinburgh: Churchill-Livingstone, 1999, pp. 1341–1351.
37. Hardy, J.D., Wolff, H.G., and Goodell, H., *Pain Sensations and Reactions*. Baltimore: Williams Wilkins, 1952.
38. Harkins, S.W., Price, D.D., Roy, A., Itskovich, V.V., and Fei, D.-Y., Somatosensory Evoked Potentials Associated with Thermal Activation of Type IIA (Delta) Mechanoheat Nociceptive Afferents. *Intern. J. Neurosci.*, 104: 93–111, 2000.
39. Hensel, H. and Zotterman, Y., The Effect of Menthol on Thermoreceptors. *Acta Physiol. Scand.*, 24: 27–34, 1951.
40. Hensel, H., Correlations of Neural Activity and Thermal Sensation in Man. In: *Sensory Functions of the Skin in Primates*, edited by Y. Zotterman. Oxford: Pergamon Press, 1976, pp. 331–353.
41. Hyvärinen, J. and Poranen, A., Movement-Sensitive and Direction and Orientation Selective Cutaneous Receptive Fields in the Hand Area of the Post Central Gyrus in Monkeys. *J. Physiol.*, 283: 523–537, 1978.
42. Iggo, A., and Andres, K.H., Morphology of Cutaneous Receptors. *Annu. Rev. Neurosci.*, 5: 1–31, 1982.
43. Jastreboff, P.J., Phantom Auditory Perception (Tinnitus): Mechanisms of Generation and Perception. *Neurosci, Res.*, 8: 221–254, 1990.
44. Jenkins, W.M., Merzenich, M.M., Ochs, M.T., Allard, T., and Guic-Robles, E., Functional Reorganization of Primary Somatosensory Cortex in Adult Owl Monkeys after Behaviorally Controlled Tactile Stimulation. *J. Neurophysiol.*, 63: 82–104, 1990.
45. Jensen, T.S. and Nikolajsen, L., Pre-Emptive Analgesia in Postamputation Pain: An Update. In: *Nervous System Plasticity and Chronic Pain*, edited by J. Sandkühler, B. Bromm and G.F. Gebhart Amsterdam: Elsevier, 2000, pp. 493–503.
46. Johansson, R.S. and Vallbo, A.B., Tactile Sensory Coding in the Glabrous Skin of the Human Hand. *Trends Neurosci.*, 27–32, 1983.
47. Johnson, L.R., Westrum, L.E., and Henry, M.A., Anatomic Organization of the Trigeminal System and the Effects of Deafferentation. In: *Trigeminal Neuralgia*, edited by G.H. Fromm and B.J. Sessle. Boston: Butterworth-Heinemann, 1991, pp. 27–69.
48. Julius, D. and Basbaum, A.I., Molecular Mechanisms of Nociception. *Nature*, 413: 203–210, 2001.
49. Kelsey, J.A. and Whire, A.A., Epidemiology and Impact of Low-Back Pain. *Spine*, 5: 133–142, 1980.

50. Kenshalo, D.R., Correlations of Temperature Sensitivity in Man and Monkey: A First Approximation. In: *Sensory Functions of the Skin in Primates*, edited by Y. Zotterman. Oxford: Pergamon Press, 1976, pp. 305–330.
51. Kenshalo, D.R. and Douglass, D.K., The Role of the Cerebral Cortex in the Experience of Pain. In: *Pain and the Brain*, edited by B. Bromm and J.E. Desmedt. New York: Raven Press, 1995, pp. 21–34.
52. Koch, B.D., Faurot, G.F., McGuirk, J.R., Clarke, D.E., and Hunter, J.C., Modulation of Mechano-Hyperalgesia by Clinically Effective Analgesics in Rats with Peripheral Mononeuropathy. *Analgesia*, 2: 157–164, 1996.
53. Kohama, I., Ishikawa, K., and Kocsis, J.D., Synaptic Reorganization in the Substantia Gelatinosa after Peripheral Nerve Neuroma Formation: Aberrant Innervation of Lamina II Neurons by Beta Afferents. *J. Neurosci.*, 20: 1538–1549, 2000.
54. Kuslich, S.D., Ulstrom, C.L., and Micheal, C.J., The Tissue Origin of Lowback Pain and Sciatica: A Report of Pain Response to Tissue Stimulation During Operation on Lumbar Spine Using Local Anesthesia. *Orthop. Clin. N. Am.*, 22: 181–187, 1991.
55. Landgren, S. and Silfvenius, H., Nucleus Z, the Medullary Relay in the Projection Path to the Cerebral Cortex of Group I Muscle Afferents from the Cat's Hind Limb. *J. Physiol. (London)* 218: 551–571, 1971.
56. Landgren, S., Silfvenius, H., and Olsson, K.A., The Sensorimotor Integration in Area 3a of the Cat. *Exp. Brain Res.*, 75: 359–375, 1984.
57. Le Bars, D., Bouhassira, D. and Villanueva, L., Opioids and Diffuse Noxious Inhibitory Control (DNIC) in the Rat. In: *Pain and the Brain*, edited by B. Bromm and J.E. Desmedt. New York: Raven Press, 1995.
58. LeDoux, J.E., Brain Mechanisms of Emotion and Emotional Learning. *Curr. Opin. Neurobiol.*, 2: 191–197, 1992.
59. Lenz, F.A., Dostrovsky, J.O., Tasker, R.R., Yamashito, K., Kwan, H.C., and Murphy, J.T., Single-Unit Analysis of the Human Ventral Thalamic Nuclear Group: Somatosensory Responses. *J. Neurophysiol.*, 59: 299–316, 1988.
60. Lenz, F.A., Gracely, R.H., Hope, E.J., Baker, F.H., Rowland, L.H., Dougherty, P.M., and Richardson, R.T., The Sensation of Angina Can Be Evoked by Stimulation of the Human Thalamus. *Pain*, 119–125, 1994.
61. Lenz, F.A. and Dougherty, P.M., Pain Procesing in the Ventrocaudal Nucleus of the Human Thalamus. In: *Pain and the Brain*, edited by B. Bromm and J.E. Desmedt. New York: Raven Press, 1995, pp. 175–185.
62. Manheimer, C., Carlsson, C.A., Vedin, A., and Wilhelmssen, C., Transcutaneous Electrical Nerve Stimulation (TENS) in Angina Pectoris. *Pain*, 26: 291–300, 1986.
63. McKemy, D.D., Neuhausser, W.M., and Julius, D., Identification of a Cold Receptor Reveals a General Role for TRP Channels in Thermosensation. *Nature*, 416: 52–58, 2002.
64. Melzack, R. and Wall, P.D., Pain Mechanisms: A New Theory. *Science*, 150: 971–979, 1965.
65. Melzack, R., Phantom Limbs. *Sci. Am.*, 266: 120–126, 1992.
66. Mendel, L.M. and Wall, P.D., Response of Single Dorsal Cord Cells to Peripheral Cutaneous Unmyelinated Fibers. *Nature*, 206: 97–99, 1965.
67. Mense, S. and Craig, A.D.J., Spinal and Supraspinal Terminations of Primary Afferent Fibers from the Gastrocnemius-Soleus Muscle in the Cat. *Neuroscience*, 26: 1023–1035, 1988.
68. Merzenich, M.M. and Kaas, J.H., Principles of Organization of Sensory-Perceptual Systems in Mammals. In: *Progress in Psychobiology and Physiological Psychology*, edited by J.M. Sprague and A.N. Epstein. New York: Academic Press, 1980.
69. Merzenich, M.M., Kaas, J.H., Wall, J., Nelson, R.J., Sur, M., and Felleman, D., Topographic Reorganization of Somatosensory Cortical Areas 3b and 1 in Adult Monkeys Following Restricted Deafferentation. *Neuroscience*, 8: 3–55, 1983.

70. Merzenich, M.M., Nelson, R.J., Stryker, M.P., Cynader, M.S., Schoppmann, A., and Zook, J.M., Somatosensory Cortical Map Changes Following Digit Amputation in Adult Monkeys. *J. Comp. Neurol.*, 224: 591–605, 1984.
71. Meyerson, B.A. and Linderoth, B., Mechanism of Spinal Cord Stimualtion in Neuropathic Pain. *Neurol. Res.*, 22: 285–292, 2000.
72. Mitani, A. and Shimokouchi, M., Neural Connections in the Primary Auditory Cortex: An Electrophysiologic Study in the Cat. *J. Comp. Neurol.*, 235: 417–429, 1985.
73. Møller, A.R., Cranial Nerve Dysfunction Syndromes: Pathophysiology of Microvascular Compression. In: *Neurosurgical Topics*. Book 13. *Surgery of Cranial Nerves of the Posterior Fossa*, edited by D.L. Barrow Park Ridge, IL: American Association of Neurological Surgeons, 1993, pp. 105–129.
74. Møller, A.R. and Pinkerton, T., Temporal Integration of Pain from Electrical Stimulation of the Skin. *Neurol. Res.*, 19: 481–488, 1997.
75. Møller, A.R., Vascular Compression of Cranial Nerves. II. Pathophysiology. *Neurol. Res.*, 21: 439–443, 1999.
76. Møller, A.R., Symptoms and Signs Caused by Neural Plasticity. *Neurol. Res.*, 23: 565–572, 2001.
77. Mountcastle, V.B., LaMotte, R.H., and Carli, G., Detection Thresholds for Vibratory Stimuli in Humans and Monkeys: Comparison with Threshold Events in Mechanoreceptive First Order Afferent Nerve Fibers Innervating Monkey Hands. *J. Neurophysiol.*, 35: 122, 1972.
78. Mountcastle, V.B., ed., *Medical Physiology*. St. Louis: C.V. Mosby, 1974.
79. Mountcastle, V.B., ed., Neural Mechanisms in Somesthesia. In: *Medical Physiology*. St Louis: C.V. Mosby, 1974.
80. Munglani, R. and Hill, R.G., Other Drugs Including Sympathetic Blockers. In: *Textbook of Pain*, 4th ed., edited by P.D. Wall and R. Melzack. Edinburgh: Churchill-Livingstone, 1999, pp. 1233–1250.
81. Nagy, I. and Rang, H., Noxious Heat Activates All Capsaicin-Sensitive and also a Subpopulation of Capsaicin-Insensitive Dorsal Root Ganglion Neurons. *Neuroscience*, 88: 995–999, 1999.
82. Narins, C.R., Zareba, W., Moss, A.J., Goldstein, R.E., and Hall, W.J., Clinical Implications of Silent Versus Symptomatic Exercise-Induced Myocardial Ischemia in Patients with Stable Coronary Disease. *J. Am. Col. Cardiol.*, 29: 756–763, 1997.
83. Ochoa, J.L., The Newly Recognized Painful ABC Syndrome: Thermographic Aspects. *Thermology*, 2: 65–107, 1986.
84. Penfield, W., and Boldrey, E., Somatic Motor and Sensory Representation in the Cerebral Cortex of Man as Studied by Electrical Stimulation. *Brain*, 60: 389–443, 1937.
85. Price, D.D. and Dubner, R., Neurons That Subserve the Sensory-Discriminative Aspects of Pain. *Pain*, 3: 307–338, 1977.
86. Price, D.D., Long, S., and Huitt, C., Sensory Testing of Pathophysiological Mechanisms of Pain in Patients with Reflex Sympathetic Dystrophy. *Pain*, 49: 63–173, 1992.
87. Price, D.D., Long, S., Wilsey, B., and Rafii, A., Analysis of Time Course and Factors Contributing to Pain Reduction Produced by Local Anesthetics Injected into Sympathetic Ganglia. *Clin. J. Pain*, 14: 216–226, 1998.
88. Price, D.D., Psychological and Neural Mechansims of the Affective Dimension of Pain. *Science*, 288: 1769–1772, 2000.
89. Procacci, P., Zoppi, M., and Maresca, M., Heart, Vascular and Haemopathic Pain. In: *Textbook of Pain*, 4th ed., edited by P.D. Wall and R. Melzack. Hong Kong: Churchill-Livingstone, 1999, pp. 621–639.
90. Procaccio, P. and Zoppi, M., Pathophysiology and Clinical Aspects of Visceral and Referred Pain. In: *Proceedings of the Third World Congress on Pain*, edited by J.J. Bonica, U. Lindblom, and A. Iggo. New York: Raven Press, 1983, pp. 643–658.

91. Raja, S.N., Campbell, J.N., and Meyer, R.A., Evidence for Different Mechanisms of Primary and Secondary Hyperalgesia Following Heat Injury to the Glabrous Skin. *Brain*, 107: 1791–1188, 1984.
92. Randolph, M. and Semmes, G., Behavioral Consequences of Selective Subtotal Ablations in Postcentral Gyrus of Macaca mulatta. *Brain Res.*, 70: 55–70, 1974.
93. Rexed, B.A., Cytoarchitectonic Atlas of the Spinal Cord. *J. Comp. Neurol.*, 100: 297–379, 1954.
94. Roberts, W., A Hypothesis on the Physiological Basis for Causalgia and Related Pains. *Pain*, 24: 297–311, 1986.
95. Ruch, T.C., Pathophysiology of Pain. In: *The Brain and Neural Function*, edited by T.C. Ruch and H.D. Patton. Philadelphia: W.B. Saunders, 1979, pp. 272–324.
96. Salt, T.E. and Hill, R.G., Transmitter Candidates of Somatosensory Primary Afferent Fibres. *Neuroscience*, 10: 1083–1103, 1983.
97. Sandkühler, J., Benrath, J., Brechtel, C., Ryuscheweyh, R., and Heinke, B., Synaptic Mechanisms of Hyperalgesia. In: *Nervous System Plasticity and Chronic Pain*, edited by J. Sandkühler, B. Bromm, and G.F. Gebhart. Amsterdam: Elsevier, 2000, pp. 81–100.
98. Scadding, J.W. Peripheral Neuropathies. In: *Textbook of Pain*. 4th ed., edited by P.D. Wall and R. Melzack. Edinburgh: Churchill-Livingstone, 1999, pp. 815–834.
99. Schaible, H.G. and Schmidt, R.F. Time Course of Mechanosensitivity Changes in Articular Afferents During a Developing Experimental Arthritis. *J. Neurophysiol.*, 60: 2180–2195, 1988.
100. Schmidt, R.F., *Fundamentals of Sensory Physiology*. New York: Springer-Verlag, 1981.
101. Sessle, B.J. Recent Developement in Pain Research: Central Mechanism of Orofacial Pain and Its Control. *J. Endodon.*, 12: 435–444, 1986.
102. Sessle, B.J., Physiology of the Trigeminal System. In: *Trigeminal Neuralgia*, edited by G.H. Fromm and B.J. Sessle. Boston: Butterworth-Heinemann, 1991, pp. 71–104.
103. Simpson, B.A., Spinal Cord and Brain Stimulation. In: *Textbook of Pain*, 4th ed., edited by P.D. Wall and R. Melzack. Edinburgh: Churchill-Livingstone, 1999, pp. 1353–1381.
104. Srinivasan, M.A. and LaMotte, R.H., *Encoding of Shape in the Responses to Cutaneous Mechanoreceptors*. New York: Macmillan, 1991.
105. Tabo, E., Jinks, S.L., Eisle, J.H., and Carstens, E., Behavioral Manifestations of Neuropathic Pain and Mechanical Allodynia, and Changes in Spinal Dorsal Horn Neurons, Following L4-L6 Root Constriction. *Pain*, 66: 503–520, 1999.
106. Tominaga, M. and Julius, D., Capsaicin Receptor in the Pain Pathway. *Jpn. J. Pharmacol.*, 83: 20–24, 2000.
107. Torebjoerk, H.E. and Ochoa, J.L., Pain and Itch from C-Fiber Stimulation. *Soc. Neurosci. Abstr.*, 7: 228, 1981.
108. Urban, M.O. and Gebhart, G.F., Supraspinal Contribution to Hyperalgesia. *Proc. Nat. Acad. Sci. USA*, 96: 7687–7692, 1999.
109. Vallbo, A.B. and Johansson, R.S., Skin Mechanoreceptors in the Human Hand: Neural and Psychophysical Thresholds. In: *Sensory Functions of the Skin in Primates*, edited by Y. Zotterman. Oxford: Pergamon Press, 1976, pp. 185–198.
110. Vallbo, A.B., Olausson, H., and Wessberg, J., Unmyelinated Afferents Constitute a Second System Coding Tactile Stimuli of Human Hairy Skin. *J. Neurophysiol.*, 81: 2753–2763, 1999.
111. Vogt, B.A., Sikes, R.W., and Vogt, L.J., Anterior Cingulate Cortex and the Medial Pain System. In: *The Neurobiology of the Cingulate Cortex and Limbic Thalamus*, edited by B.A. Vogt and M. Gabriel Boston: Birkhauser, 1993.
112. P.D. Wall., The Presence of Ineffective Synapses and Circumstances Which Unmask Them. *Philos. Trans. Roy. Soc. (London)*, 278: 361–372, 1977.
113. Wall, P.D., The Prevention of Postoperative Pain. *Pain*, 33: 289–290, 1988.
114. Wall, P.D. and Melzack, R., *Textbook of Pain*, 4th ed. Edinburgh: Churchill-Livingstone, 1999.

115. Watanabe, S., Kakigi, R., Koyama, S., Hoshiyama, M., and Kaneoke, Y., Pain Processing Traced by Magnetoencephalography in the Human Brain. *Brain Topogr.*, 10: 1–10, 1998.
116. Wilder-Smith, O.H.G., Pre-Emptive Analgesia and Surgical Pain. In: *Nervous System Plasticity and Chronic Pain*, edited by J. Sandkühler, B. Bromm and G.F. Gebhart. Amsterdam: Elsevier, 2000, pp. 505–524.
117. Willer, J.C., Relieving Effect of TENS on Painful Muscle Contraction Produced by an Impairment of Reciprocal Innervation: An Electrophysiological Analysis. *Pain*, 32: 271–274, 1988.
118. Willer, J.C. and Le Bars, D., Electrophysiologic Studies of Morphine Analgesia in Humans. In: *Pain and the Brain*, edited by B. Bromm and J.E. Desmedt. New York: Raven Press, 1995, pp. 541–557.
119. Willis, W.D., From Nociceptor to Cortical Activity. In: *Pain and the Brain*, edited by B. Bromm and J.E. Desmedt. New York: Raven Press, 1995, pp. 1–19.
120. Woolf, C.J., Evidence of a Central Component of Postinjury Pain Hypersensitivity. *Nature*, 308: 686–688, 1983.
121. Woolf, C.J., Recent Advances in Pathophysiology of Acute Pain. *Br. J. Anaesth.*, 63: 139–146, 1989.
122. Woolf, C.J. and Thompson, S.W.N., The Induction and Maintenance of Central Sensitization Is Dependent on *N*-Methyl-D-Aspartic Acid Receptor Activation: Implications for the Treatment of Post-Injury Pain Hypersensitivity States. *Pain*, 44: 293–299, 1991.
123. Woolf, C.J., Shortland, P., and Cogershall, R.E., Peripheral Nerve Injury Triggers Central Sprouting of Myelinated Afferents. *Nature*, 355: 75–78, 1992.
124. Woolf, C.J., The Dorsal Horn: State-Dependent Sensory Processing and Generation of Pain. In: *Textbook of Pain*, 4th ed., edited by P.D. Wall and R. Melzack. Edinburgh: Churchill-Livingstone, 1994, pp. 79–100.
125. Woolf, C.J. and Mannion, R.J., Neuropathic Pain: Aetiology, Symptoms, Mechanisms, and Managements. *Lancet*, 353: 1959–1964, 1999.
126. Woolf, C.J. and Salter, M.W., Neural Plasticity: Increasing the Gain in Pain. *Science*, 288: 1765–1768, 2000.
127. Wrigglesworth, R., Walpole, C.S.J., Bevan, S., *et al.*, Analogous of Capsaicin with Agonist Activity as Novel Analgesic: Structure Activity Studies. 4. Potent, Orally Active Analgesics. *J. Med. Chem.*, 39: 4942–4951, 1996.
128. Yaksh, T.L., Central Pharmacology of Nociceptive Transmission. In: *Textbook of Pain*, 4th ed., edited by P.D. Wall and R. Melzack. Edingburgh: Churchill-Livingstone, 1999, pp. 253–308.
129. Young, R.F., Kroening, R., Fulton, W. *et al.*, Electrical Stimulation of the Brain in Treatment of Chronic Pain. Experience over 5 Years. *J. Neurosurg.*, 62: 389–396, 1985.
130. Young, R.F., Jacques, D.S., Rand, R.W., Copcutt, B.C., Vermeulen, S.S., and Posewitz, A.E., Technique of Stereotactic Medical Thalamotomy with the Leksell Gamma Knife for Treatment of Chronic Pain. *Neurol. Res.*, 17: 59–65, 1995.
131. Young. R.F., Deep Brain Stimulation for Failed Back Surgery Syndrome. In: *Textbook of Stereotactic Surgery*, edited by P.L. Gildenberg and R.R. Tasker. New York: McGraw-Hill, 1998, p. 1621.
132. Zhou, M., Silent Glutamatergic Synapses and Long-Term Facilitation in Spinal Dorsal Horn Neurons. In: *Nervous System Plasticity and Chronic Pain*, edited by J. Sandkühler, B. Bromm and G.F. Gebhart. Amsterdam: Elsevier, 2000.
133. Zigmond, M.J., Bloom, F.E., Landis, S.C., Roberts, J.L. and Squire, L.R., *Fundamental Neuroscience*. San Diego: Academic Press, 1999.
134. Zotterman, Y., *Sensory Functions of the Skin in Primates*. Oxford: Pergamon Press, 1976.

CHAPTER 5

Hearing

ABBREVIATIONS

AAF : Anterior auditory field
AES : Anterior ectosylvian areas
AI : Primary auditory field
AII : Secondary auditory cortex
ABL : Basolateral nucleus of the amygdala
ACE : Central nucleus of the amygdala
AL : Lateral nucleus of the amygdala
AVCN : Anterioventral cochlear nucleus
BIC : Brachium of the inferior colliculus
CF : Characteristic frequency
CN : Cochlear nucleus
COCB : Crossed olivocochlear bundle
DAS : Dorsal acoustic stria or stria of Monaco
DC : Dorsal cortex (of medial geniculate body)
DCN : Dorsal cochlear nucleus
DNLL : Dorsal nucleus of the lateral lemniscus
DPOAE : Distortion product otoacoustic emission
Ea : Anterior ectosylvian gyrus
Ep : Posterior ectosylvian field
FTC : Frequency threshold curves
HL : Hearing level (average threshold for normal hearing individuals)

IAS : Intermediate acoustic stria or stria of Held
IC : Inferior colliculus
ICC : Central nucleus of the inferior colliculus
ICP : Pericentral nucleus of the inferior colliculus
ICX : External nucleus of the inferior colliculus
INLL : Intermediate nucleus of the lateral lemniscus
LEA : Left ear advantage
LL : Lateral lemniscus
LSO : Lateral superior olivary nucleus
LV : Lateral ventral part of the medial geniculate body
Mc : Magnocelluar division of the medial geniculate body
MGB : Medial geniculate body
MSO : Medial superior olivary nucleus
NM : Nanometer
NTB : Nucleus of the trapezoidal body
OAE : Otoacoustic emission
OCB : Olivocochlear bundle
OV : Ovoid part of the medial geniculate body
PAF : Posterior auditory field
PMNLL : Posterior medial nucleus of the lateral lemniscus
PVCN : Posterioventral cochlear nucleus
Po : Posterior division (of thalamus)
Pom : Medial division of the posterior nucleus group
pps : Pulses per second
RE : Thalamic reticular nucleus
REA : Right ear advantage
SC : Superior colliculus
SG : Sagulum
SPL : Sound pressure level
TEOAE : Transient evoked otoacoustic emission
UOCB : Uncrossed olivocochlear bundle
VAS : Ventral acoustic stria or trapezoidal body
VNLL : Ventral nuclei of the lateral lemniscus.

ABSTRACT

Conduction of Sounds to the Receptors

1. The middle ear improves sound transmission to the cochlea.
2. The ear canal and the acoustic properties of the head increase the sound transmission to the ear drum, mostly for high frequencies.

3. The outer ear (pinna) in humans has little effect on sound transmission to the cochlea except for high frequencies.
4. The acoustic middle ear reflex (involving the stapedius muscle) is elicited by sounds above approximately 85-dB hearing levels. It decreases the sound transmission to the cochlea.
5. The basilar membrane of the cochlea separates sounds according to their frequencies (frequency analysis).
6. The frequency selectivity of the basilar membrane is nonlinear and depends on the intensity of sounds. The basilar membrane is most selective for sounds of low intensity, and the location of maximal vibration shifts along the basilar membrane as a function of the intensity of the sound.

Receptors

7. The sensory receptors for hearing are hair cells that are located along the basilar membrane and are therefore activated according to the intensity and the frequency (or spectrum) of the sounds.
8. Cochlear hair cells are transformed epithelium cells that are similar to those in the vestibular apparatus (part of the inner ear), but cochlear hair cells have no kinocilia.
9. Two types of hair cells are located along the basilar membrane—outer hair cells and inner hair cells—which are similar morphologically but have completely different functions:
   a. Inner hair cells are the auditory receptors that convert the motion of the basilar membrane into a code of nerve impulses in the fibers of the auditory nerve.
   b. Outer hair cells improve the mechanical performance of the cochlea by acting as "motors" that compensate for frictional losses in the cochlea and thereby improve the threshold of hearing by approximately 50 dB.
   c. The active role of the outer hair cells is involved in the production of sounds known as otoacoustic emissions.
10. The auditory nerve (part of the eighth cranial nerve) has two types of fibers.
    a. Type I: Many Type I auditory nerve fibers innervate each inner hair cell but each Type I nerve fiber innervates only one or two hair cells. Type I fibers connect to cells in the cochlear nucleus, which is the first relay nucleus of the ascending auditory pathways.
    b. Type II: Each fiber innervates many outer hair cells. The function of Type II auditory nerve fibers is unknown.

11. Approximately 95% of all auditory nerve fibers are Type I, and less than 5% of all auditory nerve fibers are Type II nerve fibers.

Classical Auditory Pathways

12. The cochlear nucleus has three parts: the anterior ventral, posterior ventral, and dorsal nuclei. All Type I auditory nerve fibers bifurcate twice, and each of these branches connects to cells in each of three parts of the cochlear nucleus. This is the beginning of parallel processing in the ascending auditory pathways.
13. Axons of cells in the three divisions of the cochlear nucleus cross the midline in three striae where some of the fibers or their collaterals make synaptic connections with cells in the nuclei of the superior olivary complex.
14. The fibers of these three striae join and form the lateral lemniscus, which ascends to the midbrain auditory nuclei (inferior colliculus).
15. Some fibers of the lateral lemniscus or their collaterals make synaptic contact with cells in the nuclei of the lateral lemniscus.
16. Some cells of the cochlear nucleus connect to cells in the inferior colliculus on the same side (non-crossed pathway).
17. All fibers of the lateral lemniscus are interrupted by synaptic transmission to the cells of the central nucleus of the inferior colliculus.
18. Axons from cells in the inferior colliculus reach the ventral part of the thalamic nucleus (the medial geniculate body) through the brachium of the inferior colliculus.
19. Cells of the ventral division of the medial geniculate body project to the primary auditory cortex (located in the gyrus of Heschel). Cells in the primary cortex connect to cells in the secondary auditory areas, the anterior auditory field, the posterior auditory field, and association cortices.
20. There are numerous connections between nuclei of the two sides as far peripherally as the cochlear nucleus, but not at the medial geniculate body.
21. An extensive descending system runs parallel to the ascending pathways and is mostly reciprocal to them. Descending pathways reach all the way peripherally to the cochlear hair cells.

Physiology of the Classical Auditory System

22. Each auditory nerve fiber has a receptive field with regard to the frequency and intensity of sounds. A curve that shows the threshold of the

responses as a function of frequency tonal stimuli is known as the frequency-tuning curve or the frequency threshold curve.

23. The frequency tuning of auditory nerve fibers is the result of the frequency tuning of the basilar membrane of the cochlea and is the basis for the place hypothesis of frequency discrimination.
24. Because the frequency tuning of the basilar membrane depends on the intensity of sounds, the frequency tuning of auditory nerve fibers also depends on the intensity of the sound.
25. The receptive field of an auditory nerve fiber is surrounded by inhibitory receptive fields that are the result of the nonlinear cochlear mechanics (known as two-tone suppression because it does not involve synaptic transmission).
26. The discharge patterns of auditory nerve fibers reflect the time pattern of sounds (phase-locking) and provide information about the frequency of sounds. Temporal coding of frequency is the basis for the temporal hypothesis for frequency discrimination.
27. Only a few auditory nerve fibers are able to code the intensity of sounds in the entire audible range of sound intensities.
28. Neurons in nuclei of the ascending auditory pathways have frequency tuning but the shape of frequency threshold curves (receptive fields) is different from those of auditory nerve fibers, because of convergence and interaction with inhibition.
29. Neurons in the nuclei of the auditory pathways respond better to small changes in the intensity of sounds than to steady sounds, and the receptive fields to tones with rapidly changing frequency are narrower than the receptive fields to steady tones.
30. Directional information, in the form of difference in arrival time of sounds at the two ears, is decoded in neurons of the superior olivary nuclei, which act as coincidence detectors, where axons of different length provide delay lines that make it possible to detect time differences that correspond to azimuths of −90 to +90 degrees.
31. Decoding of the temporal code of frequency into a neural code may be done by neural circuitry similar to that which decodes directional information.
32. Parallel processing that is prevalent in the auditory system provides processing of the same information in different populations of neurons. Stream segregation provides processing of different kinds of information in different populations of nerve cells.
33. Place coding of frequency (or spectrum) may be regarded as a spatial code, while the temporal code that is the basis for the temporal hypothesis for frequency discrimination may be regarded as an object code.

34. Neurons in the classical ascending auditory pathways respond only to auditory stimuli.

Nonclassical Pathways

35. The nonclassical ascending auditory pathway receives its main input from the ascending auditory pathway at the mid-brain level (inferior colliculus) but some cells in the cochlear nucleus may also be regarded as belonging to the nonclassical pathways.
36. Axons from cells in the central nucleus of the inferior colliculus relay information to the external nucleus and the cortex of the inferior colliculus.
37. Nonclassical auditory pathways project to the dorsal and medial part of the thalamic auditory nuclei.
38. Ascending pathways from the medial and dorsal nuclei of the medial geniculate body project to association cortices, structures of the limbic system such as the amygdala, and several other parts of the brain. These connections are mostly reciprocal.
39. The neurons in the cochlear nucleus, the external nucleus of the inferior colliculus, and the medial and dorsal divisions of the medial geniculate body that belongs to the nonclassical auditory pathways receive input from nonauditory sources such as the dorsal column nuclei, trigeminal sensory nuclei, vestibular nuclei, the amygdala, and the superior colliculus.
40. Pathways of the acoustic reflexes may be regarded as nonclassical pathways.

Physiology of the Nonclassical Pathways

41. Little is known about the response from neurons in the nonclassical pathways except for the findings that neurons respond less specifically to the frequency (broader tuning) and time patterns of sounds.
42. Neurons in the dorsal and medial thalamus respond preferentially to high-frequency sounds.
43. Many neurons in the auditory nonclassical pathways respond to stimulation of other sensory systems, especially the somatosensory system.

I. INTRODUCTION

Humans can discriminate between a large number of different sounds, as is evident from our ability to understand speech, recognize the voices of many different people, and discriminate among many kinds of sounds, including music. Several of the basic qualities of sounds that we can discriminate have been identified through psychoacoustic experiments using simple sounds. Some of these were described in Chapter 1.

The ability to discriminate sounds is a result of analyses that occur in the ear (the cochlea) and the auditory nervous system. The cochlea also has an intricate amplifying function that increases the ear's sensitivity by approximately 50 dB. This chapter discusses the anatomical and physiological bases for discrimination of sounds and the basis for the high sensitivity of the ear. Some of the basic functions of the auditory system were described in Chapters 2 and 3. This chapter provides a more detailed discussion of the anatomy and function of the auditory system.

II. THE EAR

The ear consists of three main parts, the outer ear (the pinna and ear canal), the middle ear, and the cochlea (inner ear) (Fig. 5.1). The outer ear and the middle ear conduct sound to the cochlea, where the sensory epithelium containing the

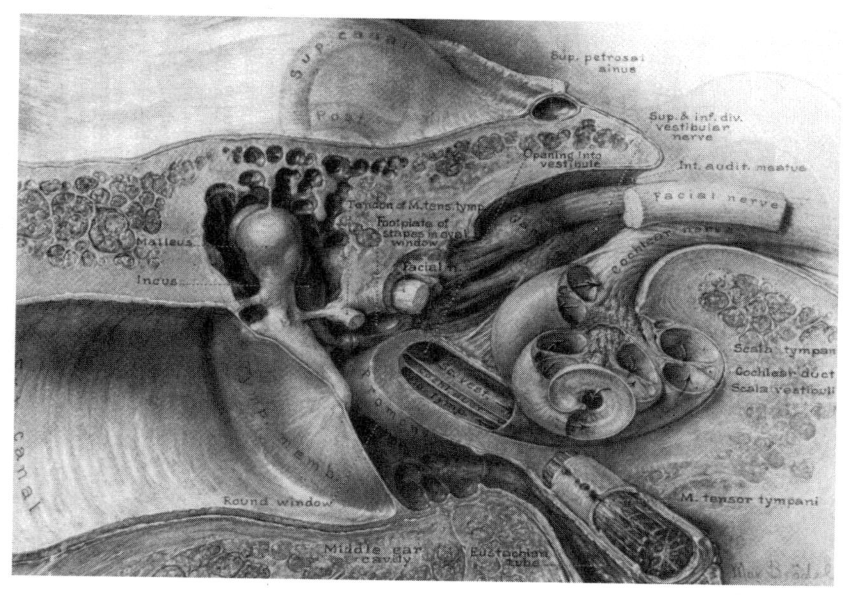

FIGURE 5.1 Cross-section of the human ear. ear showing the outer ear, the middle ear, and the inner ear. (From Brodel, M., *Three Unpublished Drawings of the Anatomy of the Human Ear*. Philadelphia: W.B. Saunders, 1946. With permission.)

sensory cells (*hair cells*) are located. The cochlea is fluid filled and enclosed by a bony capsule which has two openings: the round window that is covered by a membrane and the oval window in which the footplate of the stapes rests. Conduction of the stimulus to the receptors in the ear is more complex than in other sensory systems. The two main steps are transformation of sounds in air to vibrations of the cochlear fluid and conduction of the vibrations of the cochlear fluid to the hair cells.

A. Anatomy

1. The Outer Ear

The outer ear consists of the pinna and the ear canal. These are the only parts of the ear that can be seen with the naked eye.

2. The Middle Ear

The middle ear consists of the *tympanic membrane* and three small bones (*ossicles*): the *malleus*, *incus*, and *stapes*, which are located in the *middle ear cavity*, which is air filled. The manubrium of the malleus is embedded in the tympanic membrane. The footplate of the stapes is located in the oval window of the cochlea. These three bones are connected to each other and build a chain known as the *ossicular chain*, which leads sound to the cochlea. Two small muscles are in the middle ear: the *tensor tympani* and the *stapedius* (Fig. 5.2). The tensor tympani muscle attaches to the manubrium of the malleus and pulls the tympanic membrane inward when it contracts. The stapedius muscle pulls the stapes in a direction perpendicular to its piston-like motion in response to sound. The stapedius muscle is innervated by a branch of the *facial nerve*, and the tensor tympani is innervated by a branch of the *trigeminal nerve*.

The air pressure in the middle ear cavity must be kept close to that of the ambient pressure in order for the middle ear to function normally. The *eustachian* tube (Fig. 5.2) connects the middle ear cavity with the nasopharynx and provides pressure equalization between the environment and the middle ear cavity. This tube is normally closed but opens during swallowing. The opening of the eustachian tube is controlled by the *velum palatinum muscles* in the nasopharynx which are innervated by a branch of the trigeminal nerve.

3. The Cochlea

The cochlea is fluid filled and has three compartments: the *scala vestibuli* and the *scala tympani* and, between these, the *scala media* (Fig.5.3A).[208,215]

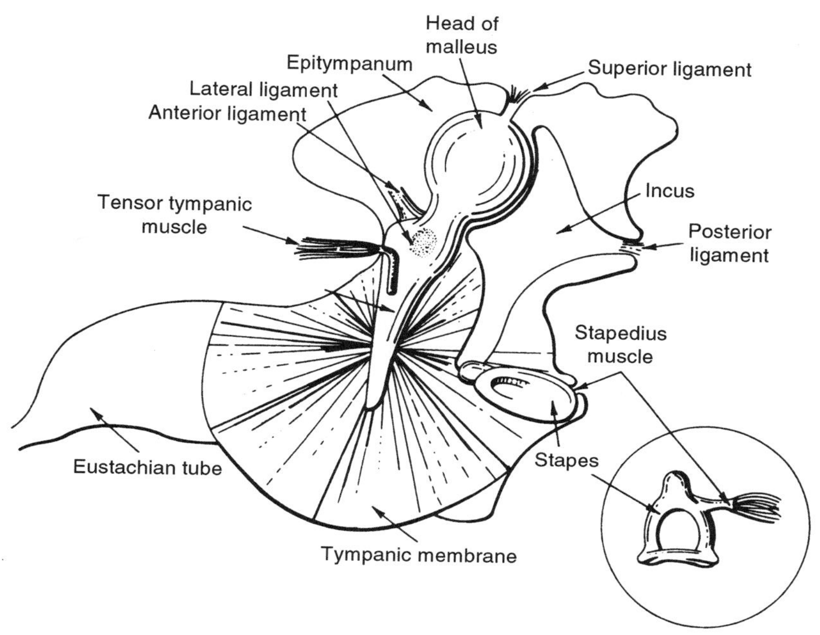

FIGURE 5.2 Schematic drawing showing the middle ear seen from the inside of the head. (From Møller, A.R., In: *Foundation of Modern Auditory Theory*, Vol. II, edited by J.V. Tobias. New York: Academic Press, 1972. pp. 133–194. With permission.)

The fluid in the scala vestibuli and scala tympani, known as *perilymph*, has a similar ionic composition as extracellular fluid (rich in sodium, poor in potassium). The fluid in the scala media (*endolymph*) is similar to intracellular fluid: rich in potassium and poor in sodium.

The *basilar membrane*, which separates the scala tympani from the scala media, is an important structure of the cochlea. The basilar membrane is set into vibration by the cochlear fluid and separates sounds according to their spectra. The sensory cells of hearing, the hair cells, are located in the *organ of Corti* (Fig. 5.3 A,B),[215] a structure located along the basilar membrane of the cochlea (Fig. 5.3B). The hair cells are modified epithelium cells that have no *kinocilium*, only *stereocilia*. The *tectorial* membrane (Fig. 5.3A) is connected to the outer hair cells by their tallest stereocilia, but it only lies in close proximity to the stereocilia of the inner hair cell. The tectorial membrane is key in the process of modification of the stimulus that reaches the inner hair cells by the outer hair cells.[268,270]

The number of cochlear hair cells given by different investigators varies only slightly. Spoendlin[217] reports that there are 15,000 (inner and outer) hair cells in

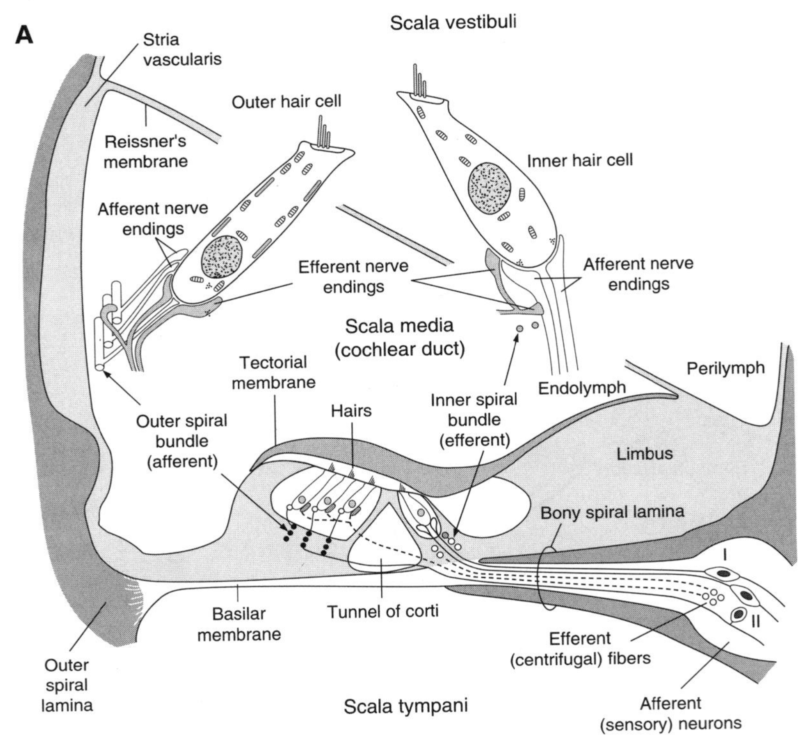

FIGURE 5.3 (A) Schematic drawing of a cross-section of the cochlea showing the organ of Corti. Outer and inner hair cells are shown as inserts. (Adapted from Shepherd, G.M., *Neurobiology*. New York: Oxford University Press, 1994.) (B) Breschet's drawings of the cochlea, spiral lamina, and cochlear nerve. The smaller sketches show the difference between the mammalian cochlea and the avian organs of hearing. (From Hawkins, J.E., In: *Physiology of the Ear*, edited by A.F. Jahn and J. Santos-Sacchi. New York: Raven Press, 1988. With permission.)

the human cochlea, while Schucknecht[200] cites the Retzius[179] estimate of 3500 inner hair cells and 12,000 outer hair cells in the human cochlea.

The normal function of the cochlea depends on the pressure in the different compartments being kept within small limits. The exact mechanisms that accomplish this are unknown, but the *cochlear aqueduct*[200] that connects the perilymphatic fluid system to the cerebrospinal system is assumed to be an important component (Fig. 5.4). The endolymphatic duct and endolymphatic sac are also assumed to be important in keeping the pressure (or rather the volume of fluid) within the different compartments of the cochlea within very tight limits.[127] The endolymphatic duct connects the endolymphatic space to the *endolymphatic sac*, a small balloon-like structure located between two layers of the *dura mater* in the *cerebellopontine angle*.[200]

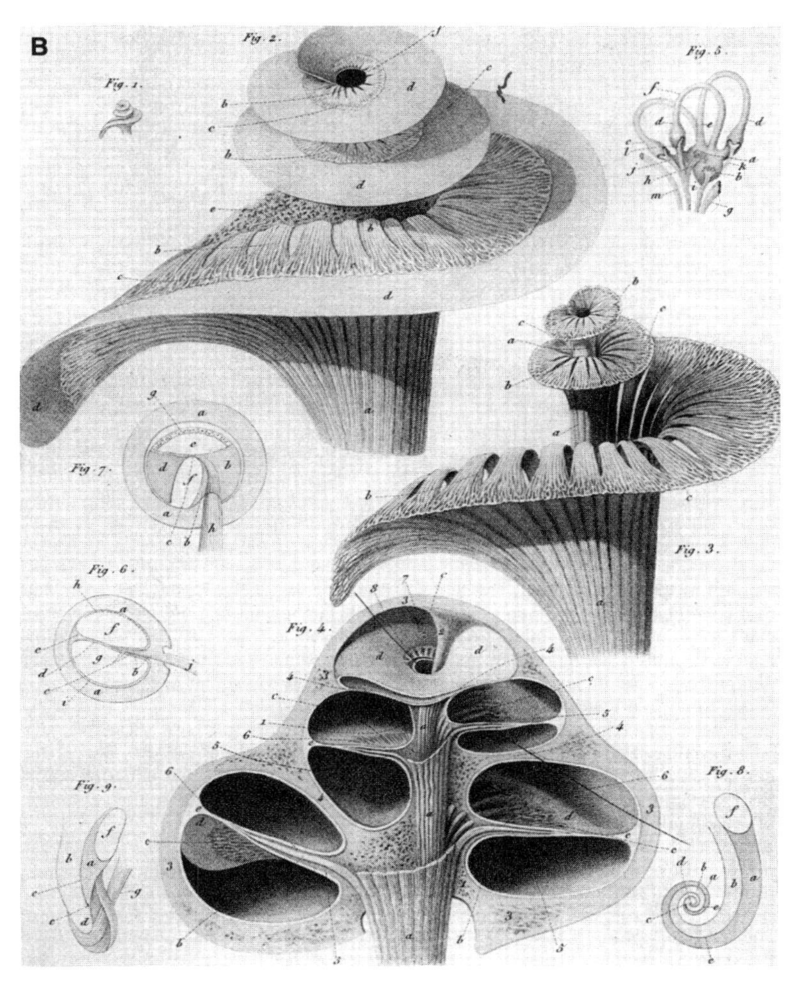

FIGURE 5.3 *(continued)*

a. Receptors in the Cochlea (Hair Cells)

The hair cells in the mammalian cochlea, the *inner hair cells* (IHCs) and *outer hair cells* (OHCs) belong to a separate group of mechanoreceptors that we call Type II receptors because the receptor cells connect to the afferent nerve fiber via a synapse-like connection (see Chapter 2). Intensive research by many scientists has resulted in accumulation of more knowledge about these receptors than about any other sensory receptors.

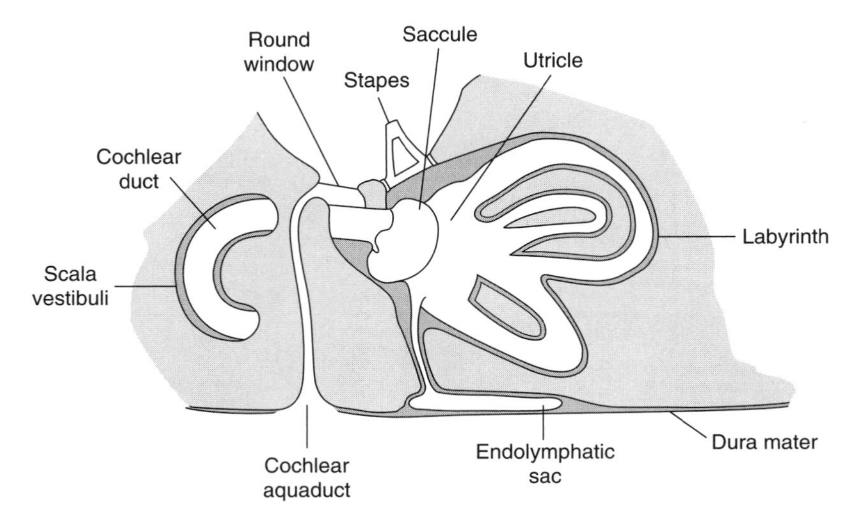

FIGURE 5.4 Schematic drawing of the fluid system of the inner ear showing communication between the perilymphatic space and the cerebrospinal space (cochlear aquaduct) and the endolymphatic sac.[127]

The hair cells in the cochlea are similar to the hair cells in the vestibular organ. The vestibular hair cells have many stereocilia and one kinocilium on the top, whereas the kinocilia of cochlear hair cells regress during ontogeny* of the organ of Corti. The hair cells of the cochlea in a mature individual, therefore, have no kinocilia, just stereocilia and a *basal body* in the place where the kinocilia had been.

Inner and outer hair cells differ only slightly in their gross morphology (inner hair cells are flask shaped whereas outer hair cells are cylindrical) (Fig. 2.7A and B),[122,196,215] but they differ dramatically in their function. Only the inner hair cells participate in neural transduction, while the outer hair cells have a mechanical function in that they act as a positive feedback at low sound intensities that serves to increase the sensitivity of the ear and sharpen the frequency tuning of the basilar membrane.[23,24]

The stereocilia are tubular protrusions of the cell and contain tightly packed, parallel *actin* filaments.[67] Actin and fibrin are proteins also found in muscles. The stereocilia protrude from a rigid circular plate (*cuticular plate*) located at one end of the cell. The basal body is located outside the cuticular plate (Fig. 2.7A and B).

*Ontogeny is the development of an individual.

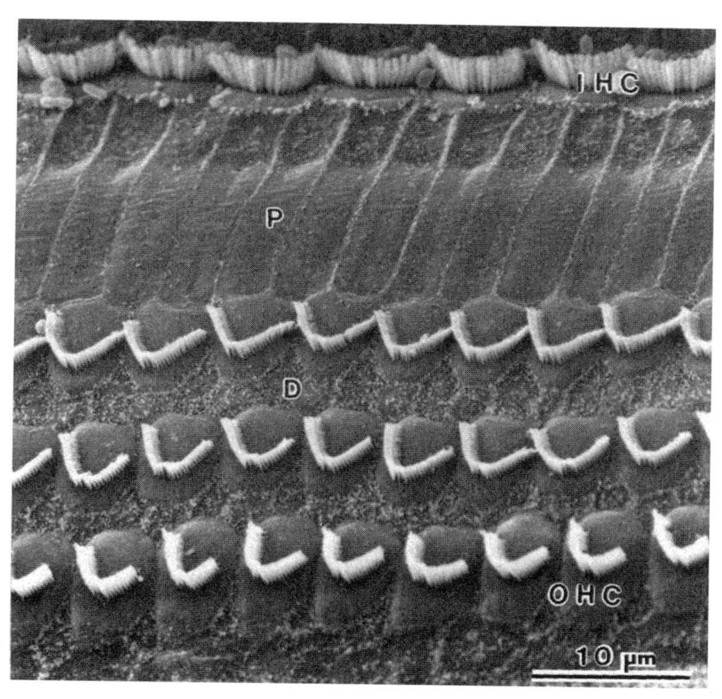

FIGURE 5.5 Scanning electron micrographs of inner hair cells (IHC) and outer hair cells (OHC) in a monkey after that the tectorial membrane has been removed. (From Harrison, R.V., and Hunter-Duvar I.M. In: *Physiology of the Ear*, edited by A.F. Jahn and J. Santos-Sacchi. New York: Raven Press, 1988, pp. 159–171. With permission.)

The diameter of the stereocilia varies between 0.1 and 0.3 µm. The diameter of the outer hair cell stereocilia is reduced where the stereocilia insert into the cuticular plate. The length of the stereocilia varies between 2 and 8 µm. Stereocilia closest to the basal body are the tallest but their length (approximately 8 µm) is only about 1/10 of the diameter of a human hair. The tips of the stereocilia are linked together (possibly by strands of *elastin*).[77,168]

The hair cells are arranged in one row of inner hair cells and three to five rows of outer hair cells along the cochlear partition (the basilar membrane) (Fig. 5.5). The stereocilia on the outer hair cells are arranged in a W-shaped manner, whereas those of inner hair cells appear as nearly straight rows when viewed perpendicularly to the surface of the basilar membrane, as in the scanning electron micrograph shown in Fig. 5.5. The stereocilia of the outer hair cells are imbedded in the *tectorial membrane* but probably not the stereocilia of the inner hair cells.

b. Innervation of Hair Cells

The cell bodies of the *auditory nerve* fibers are located in the *spinal ganglion* in the cochlear modiolus. Afferent nerve fibers connect to the bottom of the hair cells by synapse-like connections (Fig. 5.6). Hair cells are innervated by both *afferent auditory nerve* fibers of the auditory nerve (a part of CN VIII) and *efferent fibers* of the *olivocochlear bundle* (OCB). Each inner hair cell is innervated by many Type I afferent fibers (Fig. 5.7). There are approximately 20 unbranched nerve fibers per hair cell. Efferent fibers terminate on these afferent nerve fibers,[217] but some efferent fibers terminate on the cell bodies of inner hair cells (see Fig. 2.7A and B). The afferent fibers are unmyelinated close to the hair cells and may be regarded as dendrites; they become myelinated once they pass through the *hebenula perforata*. Few Type II afferent fibers terminate on outer hair cells. One afferent fiber innervates many outer hair cells,[217] and many efferent fibers terminate on the outer hair cells. Approximately 95% of auditory nerve fibers are Type I; only 5% are Type II. Type I nerve fibers communicate auditory information to higher CNS centers. The function of Type II nerve fibers is unknown.

B. Physiology of the Ear

1. Conduction of Sound to the Cochlea

The sound that reaches the cochlea is affected by the acoustic properties of the head, the ear canal, and the middle ear. The acoustic properties of the head are also the physical basis for directional hearing; the sound that reaches the two ears is different, and that difference is a function of the direction (*azimuth*) to a sound source.

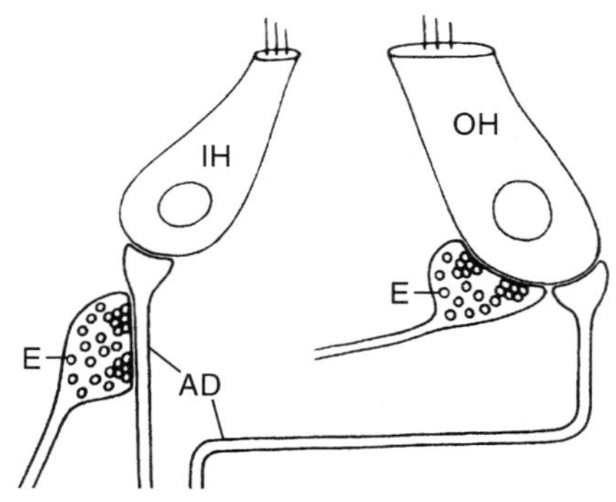

FIGURE 5.6 Schematic drawing of the innervation of hair cells. OH: Outer hair cell, IH: Inner hair cell. E: efferent synapse. AD: afferent dendrite.[217]

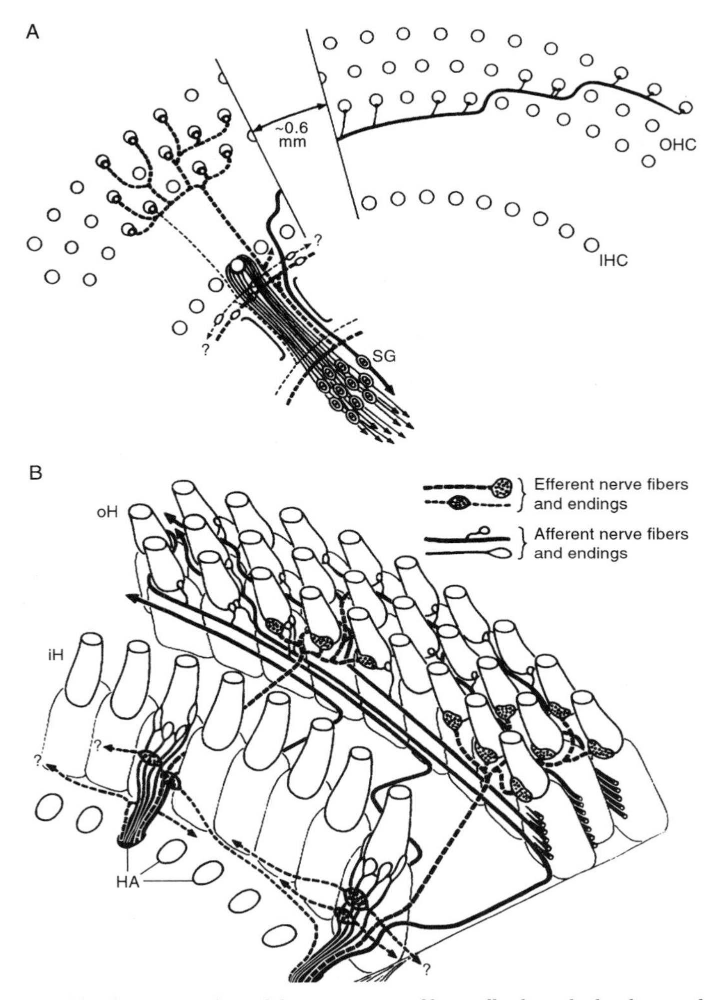

FIGURE 5.7 (A) Schematic outline of the innervation of hair cells along the basilar membrane. OH: Outer hair cells, IH: inner hair cells; SG: spiral ganglion. (B) Fiber distribution to outer hair cells (OHC) and inner hair cells (IHC). HA: Habenulae openings. (Adapted from Spoendlin, H., *Frequency Analysis and Periodicity Detection in Hearing*, edited by Plomp, R., and Smoorenburg, G.F. Leiden: A.W. Sijthoff, 1970, 2–36.)

The middle ear conducts sound from air into vibrations of the cochlear fluid. The middle ear improves sound conduction to the cochlea by matching the impedance of the two media, air and fluid of the cochlea, thereby creating a difference in the force that acts on the oval and round windows of the cochlea. This difference in force sets the cochlear fluid into motion.

a. The Head and the Outer Ear

The acoustic properties of the head affect the sound that reaches an observer from a sound source located at a distance from the person. This is because the size of the human head is not small compared to the wavelength of sound within a large part of the audible range of human hearing. The head acts as an obstacle in a sound field and distorts the sound field in a way similar to an object in a stream of water distorting the flow of water. Because of that, the sound pressure at the entrance of the ear canals of a human who is placed in a sound field is different from the sound pressure that exists when of the human subject is absent (Fig. 5.8).[206,207] The intensity of sound that reaches the ear is, in general, greater than the intensity of the sound when measured without the observer being present. This improvement of sound transmission to the cochlear fluid is dependent on the frequency (or spectrum) of the sounds. The improvement is largest in the frequency range between 2 and 5 kHz where the improvement exceeds 15 dB (Fig. 5.8).[206,207,270] The effect of the head and the ear canal has often been studied using various models of the head (Fig. 5.9).

When the sound source is not located exactly in front or exactly behind an observer, the sound that reaches the two ears will be different with regard to the

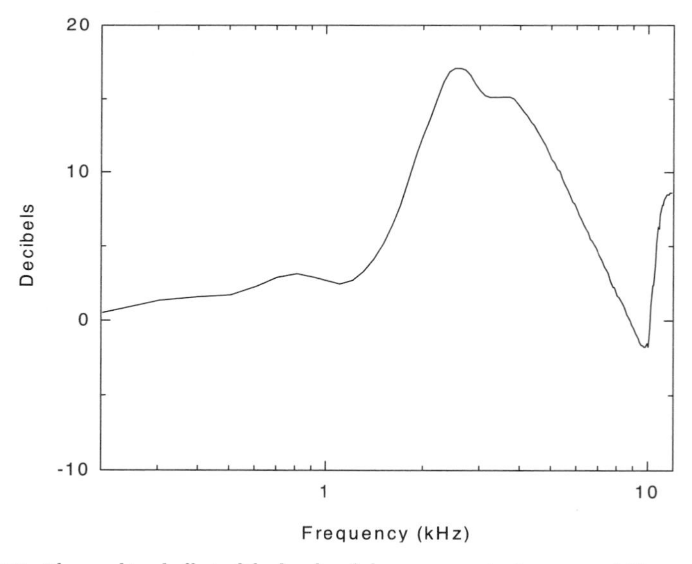

FIGURE 5.8 The combined effect of the head and the resonance in the ear canal. The graph shows the difference in the sound pressure measured close to the tympanic membrane and the sound pressure in a free field when the sound comes from a source that is located at a long distance and directly in front of the head.[207]

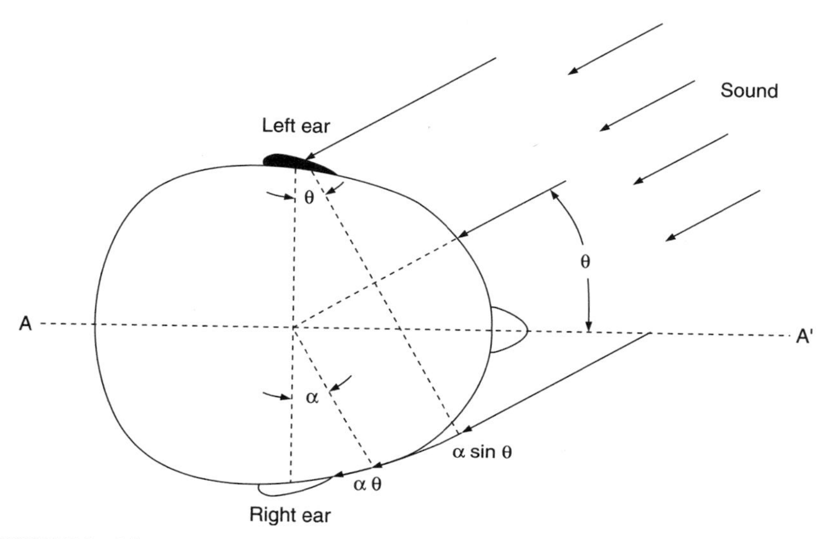

FIGURE 5.9 Schematic drawing showing a model of the head to illustrate the effect of the azimuth on sound at the ear canals in a free sound field.

time of arrival at the two ears and with regard to the intensity of the sound. The difference in time of arrival occurs because sound travels different distances to reach the two ears (Fig. 5.10).[207] Sound travels with a speed of approximately 340 m/sec, and the maximal difference in the time of arrival of sounds occurring when the head is turned 90 degrees away from the sound source is approximately 0.65 msec (Fig. 5.10). This relationship between the direction of the head to a sound source (azimuth) is one of the two physical bases for directional hearing, the ability to determine the direction to a sound source. The other factor important for directional hearing is the difference in the intensity of the sound that reaches the two ears, which is also a function of the azimuth to the sound source, but the relationship is more complex than the time difference (Fig. 5.11).[163,207] Unlike the case for difference in arrival time, the difference in sound intensity at the two ears is also dependent on the frequency of the sounds (Fig. 5.11).[163]

The intensity difference between the sounds that reach the two ears does not depend on the elevation; only the azimuth affects the intensity difference between the sounds that reach the two ears. The ability to determine the elevation of a sound source is believed to depend on differences in the spectrum of the sounds arriving at the two ears. The effect of the head on the spectrum of sounds that reach the entrance of the ear canal depends on the elevation of a sound source relative to the position of the head (Fig. 5.12).[206] It is assumed that

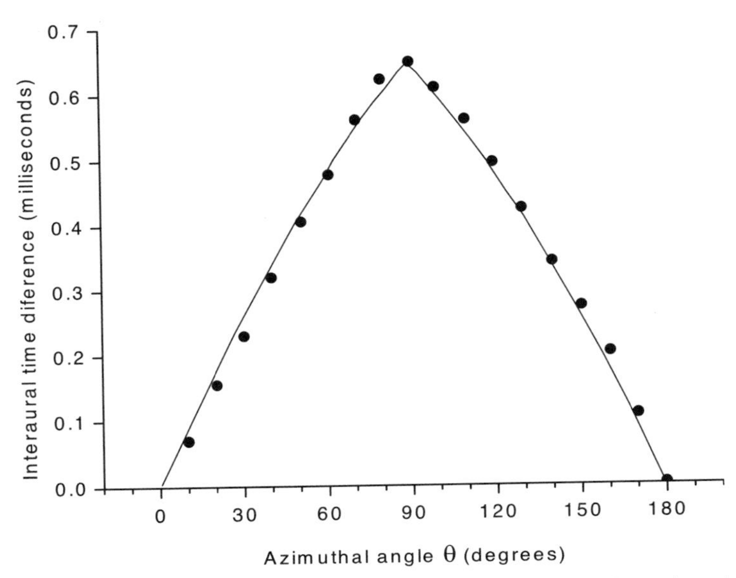

FIGURE 5.10 Calculated interaural time difference as a function of azimuth for a spherical model of the head (solid line) and measured values in a human subject (open circles).[207]

the central nervous system uses that relationship to determine the elevation of a sound source.

The neural processing that occurs in the ascending auditory pathway apparently extracts the directional information from these differences in the sound reaching the ears. The combined effect of the difference in the time of arrival and the difference in sound intensity at the two ears is the basis for creating the sense of (auditory) space.

All the considerations above refer to a person who is located in a *free sound field*, which means that sounds travel in a space where there are no obstacles. The reflections of sounds from the walls of rooms make the sound field in common rooms much more complex. Sounds that arrive at the ears of an observer in such a situation consist of the direct wave and one or more reflected waves that arrive later. The reason we are able to determine the direction to a sound source under such circumstances is assumed to be the *precedence effect*,[74,263] which means that the sound that arrives first is used by the central auditory nervous system to determine the direction to a sound source (see pages 28 and 343).

b. The Middle Ear

In order for sound to activate receptors in the cochlea, it must set the fluid that fills the cochlea into motion; however, 99.9% of the energy of sounds applied

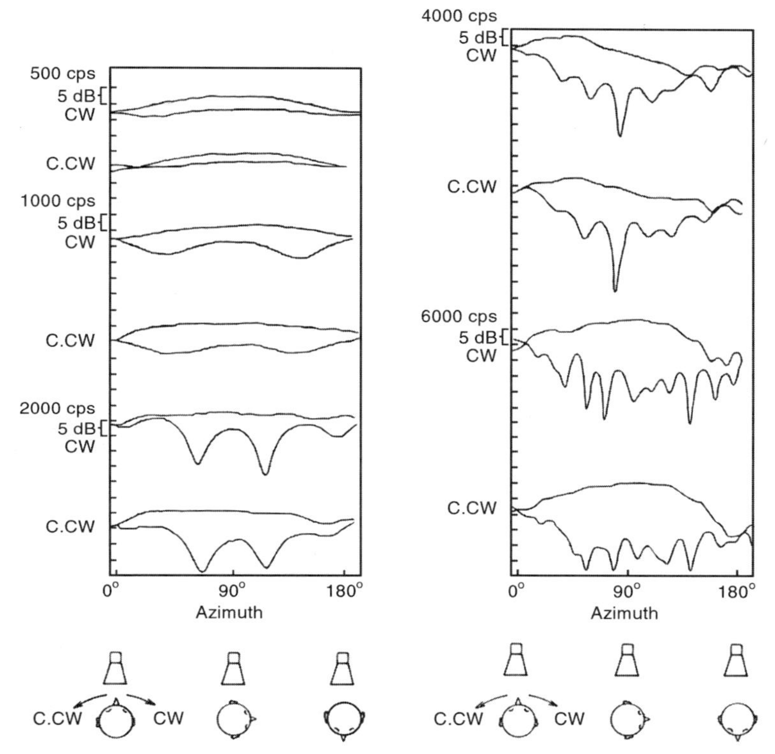

FIGURE 5.11 Difference in sound pressure at the two ears as a function of the angle to the sound source for tones of different frequencies. These graphs were obtained using a model of the head similar to the human head.[163]

directly to a fluid surface is reflected. If sound acted on one of the two windows of the cochlea directly, only 0.1% of the sound energy would be transmitted to the fluid of the cochlea. The reason that most of the sound energy is reflected is the large impedance mismatch between air and fluid. Fluid has a high impedance, and air has a low impedance. The middle ear acts as an *impedance transformer* that improves sound transmission to the cochlea by matching the high impedance of the fluid to the low impedance of air.[152] The middle ear does not provide a perfect match of the impedance of the cochlear fluid to that of air,[152,187,270] and some sound is lost to reflection from the tympanic membrane (Fig. 5.13). If the middle ear provided a perfect impedance match, the improvement of sound transmission would be approximately 36 dB compared with conducting the sound directly to the oval window (only). The actual improvement in transfer of sound energy to the cochlea provided by the middle ear is dependent on the frequency of sounds and the highest gain is

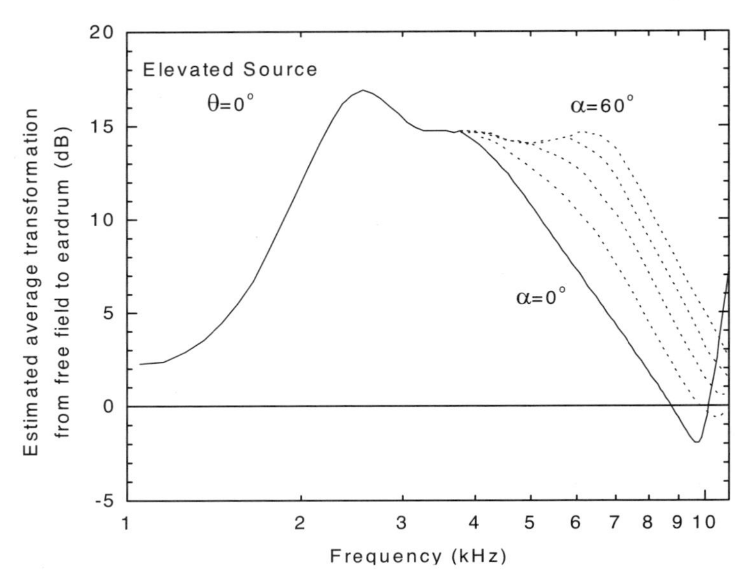

FIGURE 5.12 Estimated effect of elevation on the sound pressure at the tympanic membrane for a sound source located in front of an observer.[207]

approximately 30 dB in the ear of the cat. It is assumed that the human middle ear provides a similar improvement of sound transmission.

It is the difference between the force acting on round and the oval windows of the cochlea that sets the cochlear fluid into motion. Sound that acts directly with the same force on both windows is ineffective in setting the cochlear fluid into motion. Because the middle ear provides a much higher force on the oval window than on the round window, it establishes a large difference between the forces that act on the oval and round windows. In that way, the middle ear provides an effective transfer of sound to vibration of the fluid in the cochlea. If sound was allowed to reach both the oval and the round windows equally, the hearing loss would be much greater than what follows from loss of the transformer action of the middle ear (>60 dB).

Contraction of the middle ear muscles attenuates the sound that reaches the cochlea.[143] In humans, only the stapedius muscle contracts as an acoustic reflex. Contraction of the stapedius muscle in humans is elicited by sounds with intensities greater than approximately 85 dB above the hearing threshold. This is known as the *acoustic middle ear reflex*.[143,152] The acoustic middle ear reflex acts as an automatic gain control that is similar to the pupil of the eye. It, however, does not attenuate transient sounds because it has a long latency. It takes time to develop force in the stapedius muscle, approximately several hundred milliseconds.[135]

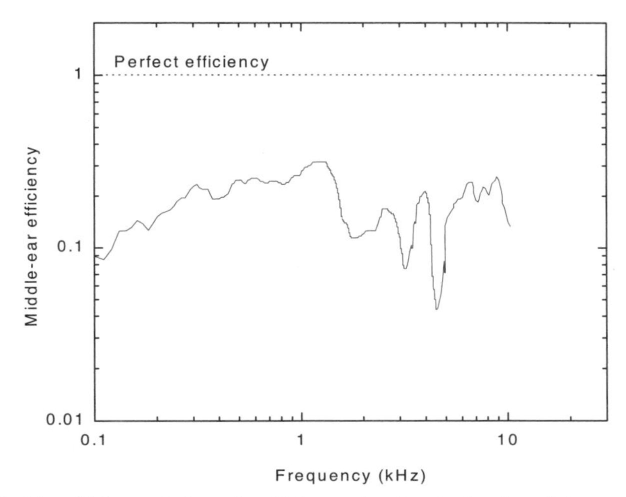

FIGURE 5.13 The efficiency of the cat's middle ear showing the fraction of the sound power that enters the middle ear and reaches the cochlea.[187]

The tensor tympani most likely does not participate in the acoustic reflex in humans, but sound elicits contraction of the tensor tympani in animals often used in auditory research such as the cat and rabbit.[20] The tensor tympani muscle contracts during swallowing, and it may play a role for exchange of air in the middle ear cavity.

2. The Cochlea

Conduction of sound to the receptors in the cochlea is more complex than in any other sensory system. Also, the processing of sounds in the cochlea is more extensive than anything occurring in other sensory systems.

The cochlea separates sounds according to their frequency (or spectrum) in such a way that the energy of sounds within specific frequency bands excites specific populations of hair cells.[14,51,270] This occurs because of the micromechanical properties of the basilar membrane and the organ of Corti and their interaction with the surrounding fluid. The micromechanical function of the cochlea is complex, and three different steps can be identified. The first step consists of the transfer of the cochlear fluid vibration to a traveling wave motion of the basilar membrane. The second step is the transfer of the basilar membrane motion into deflection of the hair cell stereocilia. The third step involves the active role of the outer hair cells, which enhances the motion of the basilar membrane.

Our understanding of the frequency selectivity of the basilar membrane has evolved during many years of research involving many investigations. The first experimental evidence for the traveling wave motion as the basis for the frequency selectivity was presented by Georg von Békésy in 1928[14] and in subsequent studies for the next two decades. (For overviews and discussion of experimental results and theoretical treatments and hypotheses, see Békésy,[14] Dallos *et al.*,[51] and Zorislocki.[266,270])

a. Conduction of Vibrations of the Cochlear Fluid to the Basilar Membrane

The stapes sets the entire volume of fluid in the cochlea into motion nearly instantaneously. The motion of the fluid in the cochlea sets the basal end of the basilar membrane in motion. That part of the basilar membrane is stiffer than other parts of the basilar membrane which facilitates transfer of energy from the fluid to the membrane. The deformation of the basal end of the basilar membrane travels toward the apex of the cochlea as a *traveling wave*. The propagation of this traveling wave is comparatively slow. The stiffness of the basilar membrane decreases gradually toward the apex; therefore, as the wave travels its amplitude increases (Fig. 5.14).[14,51,270] When the wave has traveled a certain distance on the basilar membrane, its amplitude suddenly decreases and the wave motion rapidly becomes extinct. The distance that the wave travels before becoming extinct is a direct function of the frequency of the sound. A low-frequency sound creates a wave that travels a long distance before becoming extinct, while a high-frequency sound travels only a short distance before becoming extinct. This means that the vibration of the basilar membrane reaches its maximal value for sounds of different frequency at different locations along the basilar membrane. In addition, the amplitude of the vibrations of the basilar membrane becomes not only a function of the intensity of a sound but also depends on the frequency (spectrum) of the sounds.

Because of the traveling wave motion on the basilar membrane, a pure tone causes a certain point of the basilar membrane to vibrate with higher amplitudes than any other point. The location of that point along the basilar membrane is a direct function of the frequency of the tone, and a frequency scale can be laid out along the basilar membrane with regard to the point where the wave motion has its highest amplitude. This also means that each small segment of the basilar membrane is *tuned* to a specific frequency of sounds and that the vibration of each point along the basilar membrane is a function of the frequency (or spectrum) of sounds. The frequency selectivity of the basilar membrane is a result of the traveling wave motion of the basilar membrane, which in turn is a result of the fact that the stiffness of the basilar membrane decreases along the membrane from the base toward its apex.

Because the hair cells are located along the basilar membrane, sounds of different frequencies will activate different populations of hair cells and, subsequently, different populations of auditory nerve fibers. The frequency

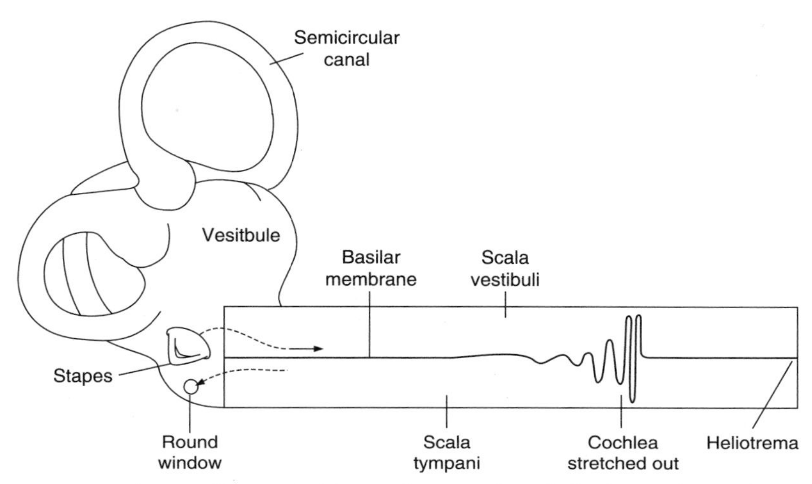

FIGURE 5.14 Schematic illustration of the traveling wave along the basilar membrane. The cochlea is shown as a straight tube in this illustration.

selectivity of the basilar membrane also results in the different hair cells and auditory nerve fibers being *tuned* to different frequencies of sounds. (see pages 293 and 318). This form of coding of the frequency (or spectrum) of sounds is a form of spatial coding that is the basis for the *place hypothesis* for frequency discrimination. This form of frequency coding can therefore be regarded as "where" type information (see Chapter 3, page 111).

Because the traveling wave progresses relatively slowly along the basilar membrane, the time it takes to reach the point of maximal deflection is longer for low-frequency sounds than for high-frequency sounds. The time it takes to reach a certain point along the basilar membrane has been estimated in animal experiments by measuring the latency of the response of auditory nerve fibers to stimulation with sounds of different frequencies.[11] These investigators found that the estimated travel time decreases with the frequency of the sounds in accordance with the fact that a frequency scale can be laid out along the basilar membrane. The travel time is approximately a logarithmic function of the frequency. The line has an equation of $\tau = a^* f^b$, where $a = 1.95$ and $b = 0.725$. At 1 kHz, the delay is approximately 2.2 msec in the cat.[11] This delay also includes the synaptic delay in the hair cells.

The description given above of the motion of the basilar membrane is simplified, and research carried out during the past 30 years has revealed that the motion of the basilar membrane is far more complex than just a traveling wave motion. Thus, in addition to the traveling wave motion of the basilar membrane, the mechanical properties of the stereocilia form a resonator, together with the tectorial membrane, that contributes to the frequency selectivity of the basilar membrane.[267,268,270]

b. Cochlear Frequency Analysis Is Nonlinear

The motion of the basilar membrane is nonlinear, which means that its properties change with sound intensity. In particular, the frequency selectivity of the basilar membrane is greater for sounds of low intensity than for sounds at high intensity, and the frequency to which a certain point of the basilar membrane is tuned is different for different sound intensities.

It was probably Honrubia and Ward[89] who first showed that the tuning of the basilar membrane is affected by sound intensity. These investigators recorded the cochlear microphonic potentials along the basilar membrane in guinea pigs and showed that the location on the basilar membrane where the cochlea microphonic potential was largest shifted toward the base of the cochlea when the sound intensity was increased.

The first direct evidence that the motion of the basilar membrane is nonlinear was published by Rhode[180] who showed that the peak of the tuning of the basilar membrane in an anesthetized monkey became more blunt when the sound intensity was increased from 70 to 90 dB sound pressure level (SPL). He also showed that the tuning became more blunt after the animal died.[181]

Rhode's finding that the tuning of the cochlea depends on the intensity of the sound[180] was unexpected and was met with skepticism. Rhode's work, however, was both confirmed and extended by many subsequent studies.[102,145,204] These studies showed that frequency selectivity (sharpness of tuning) is greatest at low sound intensities and that the tuning becomes gradually broader when the sound intensity is increased. Other investigators extended these findings when technology became available that made it possible to measure extremely small displacements of the basilar membrane (Fig. 5.15). These studies showed that the location of the maximal vibration amplitude of the basilar membrane in response to tones also shifts toward the base of the cochlea when the sound intensity is increased,[102,145,204] which causes the tuning of a specific point to shift toward lower frequencies. The reason for this nonlinearity is the action of the outer hair cells, which we will discuss later after a description of the function of the hair cells.

c. Conduction of Vibrations of the Basilar Membrane to the Receptors

The proper stimulus for hair cells is a deflection of the stereocilia. Bending of the stereocilia in the direction toward the basal body causes depolarization of the cell, which means (positive) stimulation of a hair cell.[66,191] At a first glance it seems that motion of the basilar membrane toward the scala vestibuli causes a deflection of the stereocilia toward the basal body and thus an excitatory response. However, the deflection of the stereocilia is not a direct function of the deflection of the basilar membrane because of the coupling between the

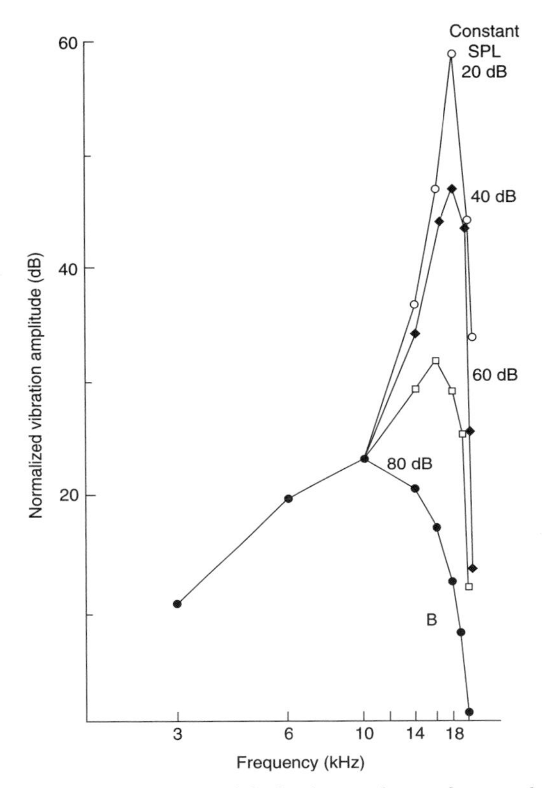

FIGURE 5.15 Amplitude of the vibration of the basilar membrane of an anesthetized guinea pig in response to pure tones at four different intensities at 20-dB intervals, shown as a function of the frequency of the tones. The curves were shifted so that they would have coincided if the cochlea had been a linear system. (Adapated from Johnstone, B.M., Patuzzi, R., and Yates, G.K., *Hear. Res.* 22: 147–153, 1986.)

membrane, and the stereocilia is complex, making the relationship between the motion of the basilar membrane and the deflection of the stereocilia on hair cells more complex and incompletely understood.

Deflection of the stereocilia of inner hair cells is different from that of outer hair cells partly because the stereocilia of inner hair cells are not imbedded in the tectorial membrane as that of outer hair cells are. Some investigators have assumed that the stereocilia of inner hair cells are deflected when the basilar membrane moves because of the flow of fluid caused by the motion of the stereocilia of the outer hair cells (Fig. 5.16).[77]

The transfer of sound to a force that deflects the stereocilia of the inner hair cells is very efficient, but it is not completely known how the motion of the basilar membrane causes the hairs of inner hair cells to become deflected.

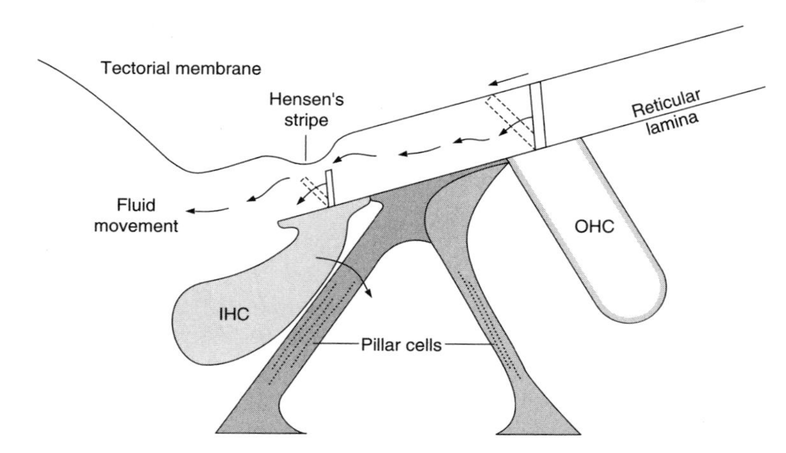

FIGURE 5.16 Illustration of a hypothesis about how flow of fluid between the tectorial membrane and the hair cells can deflect the cilia of the inner hair cells by motion of the basilar membrane. (Adapted from Geisler, C.D., *From Sound to Synapse*. New York: Oxford Press, 1998.)

The mode of motion of the basilar membrane that is most effective in deflecting the stereocilia (velocity or acceleration of the basilar membrane, velocity or acceleration of the fluid surrounding the inner hair cell stereocilia)[47,68] depends on many factors.[117,265]

The transfer of the motion of the basilar membrane to the outer hair cells is probably different from that of the inner hair cells because the tallest stereocilia of the outer hair cells are imbedded in the tectorial membrane that covers the hair cells (Fig. 5.17).[77,215] The basilar membrane moves in a direction that is nearly parallel to the long axis of the hair cells, and it has been assumed that the shearing forces caused by the motion of the basilar membrane cause deflection of the stereocilia of the outer hair cells.

d. Sensory Transduction in Cochlear Hair Cells

The hair cells are extremely sensitive mechanoreceptors, and the proper stimulus for hair cells is deflection of the stereocilia. Deflection toward the basal body (the place where the kinocilia normally is located) is excitatory, which means that it depolarizes the cell,[66,191] while deflection in the opposite direction hyperpolarizes the cells, which means that such deflection is inhibitory.

One hypothesis[79] claims that displacement of the stereocilia toward the basal body opens a cation-conducting membrane channel because of the tension in the tip link. The flow of ions through the membrane of the stereocilia is converted to intracellular

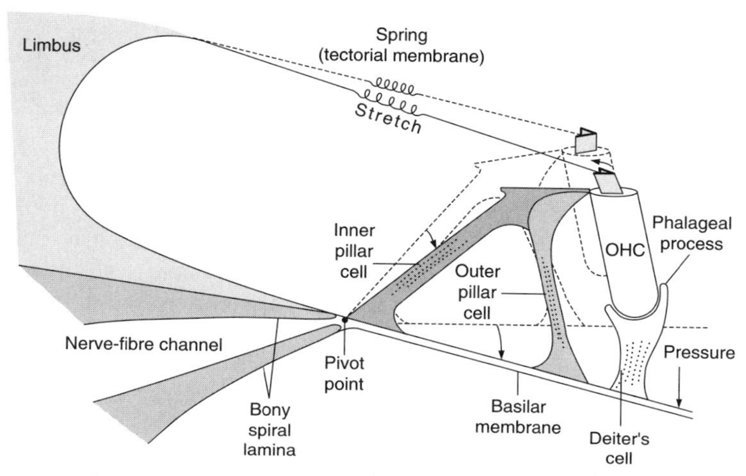

FIGURE 5.17 Illustration of a hypothesis about deflection of the cilia of the outer hair cells by motion of the basilar membrane. (Adapted from Geisher, C.D., *From Sound to Synapse*. New York: Oxford Press, 1998.)

potentials in the hair cells (Fig. 5.18).[77] When the ion channel in the stereocilia opens, positively charged ions (mostly K^+) flow into the interior of the cell. The increased concentration of K^+ ions causes a change in the membrane potential of the cell, which becomes less negative. The opening of the membrane channels causing excitation of the hair cells is very fast, with opening times of less than 50 μsec.

A popular model of the transduction process in the hair cells illustrates the process that occurs in the stereocilia by a variable resistor connected to a battery (Fig. 5.19).[77] When the resistance changes, as it does when the stereocilia are deflected, the flow of electrical current into the cell changes, and that causes the membrane potential to change. This change in the membrane potential is known as the *receptor potential* (see Chapter 2). It has been estimated that, near the threshold of hearing, displacement of the tip of the hair cell stereocilia is on the order of 0.3 nm.* Assuming a length of the cilia of 5 μm, such a displacement would correspond to an angular rotation of less than 0.003 degrees.[77,91,203] The magnitude of this displacement is smaller than the deflections caused by the random motion (Brownian motion) of the surrounding fluid molecules.[52]

The process of transforming mechanical motion into a receptor potential involves amplification as it does in other sensory cells. A deflection of the stereocilia by 10 nm produces a receptor potential of approximately 0.5 mV[53] at 100 Hz. As a comparison, the noise from thermal motion (Brownian movements) of the molecules of the surrounding fluid corresponds to deflections of the stereocilia of approximately 3.5 nm root mean square (RMS).[53] It has been calculated that the receptor potential that is the result of deflection of the stereocilia represents a 100-fold amplification of power.[77] The amplification decreases with the square of the frequency of the stimulus. The reason for the amplification is that the hair cells function as valves that control flow of

*nm = nanometer = 1/1000 μm.

electrical current from a battery (the endolymphatic potential) rather than generating the receptor potential by the mechanical energy transferred to the stereocilia. The receptor potential is thus not a result of conversion of mechanical energy to electrical energy; rather, the hair cell works as a valve that modulates a steady flow of electric current. This is one of the reasons for the high sensitivity of hair cells.

It is important to note that the stereocilia are located in the scala media (*endolymphatic space*) of the cochlea, while the bases of the hair cells are located in the scala tympani (*perilymphatic space*). The ionic composition of the fluids in these two spaces is different. The endolymphatic space holds a steady potential of approximately 80 mV (Fig. 5.19) relative to that of the vascular system.[13,77] This difference in the electrical potentials in these two spaces acts as a bias on the hair cells. Experiments where this bias has been removed show that the endolymphatic potential is necessary to maintain the normal sensitivity of hair cells.[232,233] The stria vascularis plays an important role in maintaining the unusual chemical and electrical environment in the organ of Corti.

Many factors affect the function of the hair cells. One such factor is the links that connect individual stereocilia together (tip links).[168] These tip links are important for the function of hair cells as shown by experiments where they were eliminated and the cells became nonresponsive to mechanical stimulation.[168,170]

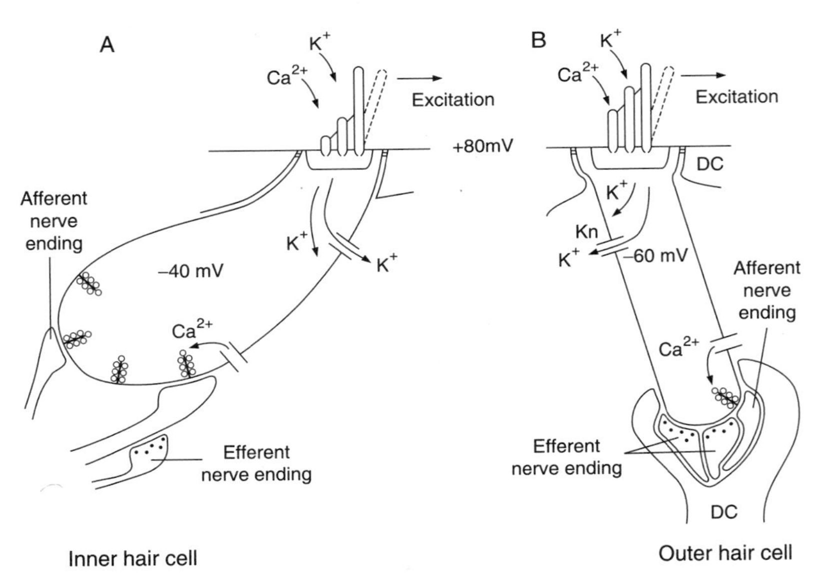

FIGURE 5.18 Schematic drawing of an inner hair cell (A) and an outer hair cell (B). Flow of ions as a result of excitation of the cells by deflection of the cilia toward and away from the modiolus. DC, Deiter's cell. Afferent nerve endings and efferent nerve endings are shown. (Adapted from Geisler, C.D., *From Sound to Synapse*. New York: Oxford Press, 1998.)

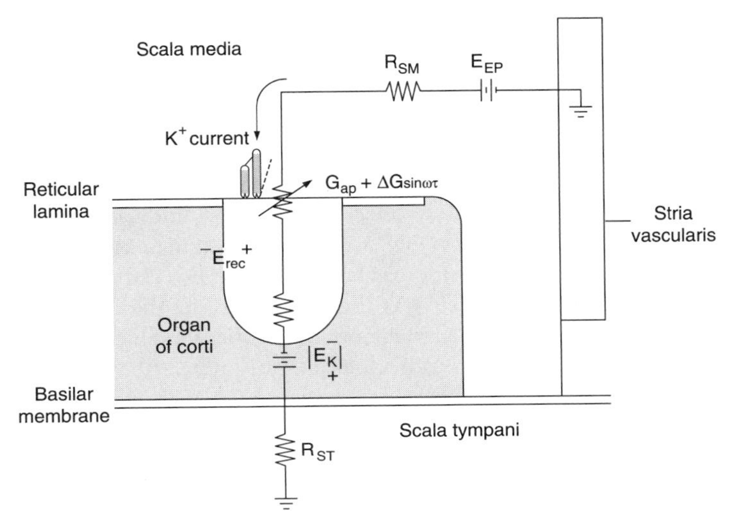

FIGURE 5.19 Illustration of how the hair cells function as variable resistors, thus acting as valves rather than converting mechanical energy into a receptor potential. Gap, resting conduction (inverse of resistance) of the apical membrane; R_{SM}, resistance of the scala media; E_{EP}, endocochlear potential; E_k, constant voltage source; sinusoidal stimulation. (Adapted from Geisler, C.D., *From Sound to Synapse*. New York: Oxford Press, 1998.)

Inner hair cells in mammals have no known specificity to mechanical stimulations,* which means that they do not contribute to the frequency selectivity of the auditory system. If an oscillatory force is applied to the stereocilia it would cause excitation of the hair cells independent of the frequency of the force over a large range of frequencies. The fact that animals such as the flying bat can hear sound above 100 kHz means that the hair cells can convert vibrations of the basilar membrane of such high frequencies. The mechanical properties of the stereocilia and their interaction with the tectorial membrane, however, add frequency-dependent properties to the transfer of the vibration of the basilar membrane to the stereocilia. (It is believed that only the stereocilia of outer hair cells are in contact with the tectorial membrane.)

The transduction of mechanical displacement of the stereocilia of hair cells into membrane potentials (receptor potentials) of inner hair cells show adaptation similar to many other mechanoreceptors.[45] Adaptation is also reflected in the discharges of single auditory nerve fibers.

*It has been shown that hair cells in some reptiles (the turtle) are tuned electrically to sound frequencies.[63] This tuning is achieved by interplay between inward and outward transports of cations through the membrane of the hair cells.

e. The Role of the Outer Hair Cells

The finding by Rhode[180] (subsequently confirmed and expanded by others[102,145,148,204]) that the tuning of the basilar membrane depended on the intensity of sounds was surprising and initially met with disbelief. The results by Evans[62] published a few years later than the work by Rhode showed that the function of the frequency selectivity of single auditory nerve fibers is dependent on metabolic energy (Fig. 5.20). This finding at that time was not associated with nonlinearity of the basilar membrane but rather was explained by a (fictive) "second filter" that was thought to modify the excitation of hair cells by the motion of the basilar membrane or by somehow modifying the initiation of neural impulses in individual auditory nerve fibers.

The nonlinearity of the cochlea was explained when Brownell[23] discovered that the outer hair cells participate actively in the motion of the basilar membrane. The discovery that the outer hair cells do not take any part in sensory transduction in the cochlea but rather have a mechanical function was a major step toward understanding some of the basic functions of the cochlea. It introduced a totally new view of the function of the basilar membrane and explained in an elegant way these earlier experimental results that showed that the tuning of the basilar membrane[180,204] and that of auditory nerve fibers[62,146,148] was intensity dependent. The action of the outer hair cells could

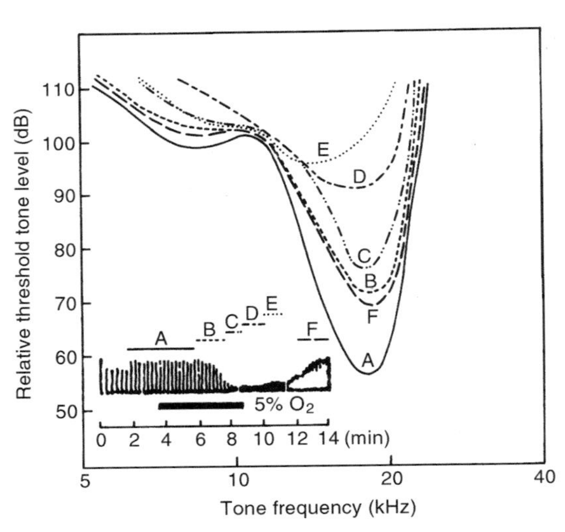

FIGURE 5.20 Frequency tuning of an auditory nerve fiber in a guinea pig before and during deprivation of oxygen. The insert shows the amplitude of the compound action potential recorded from the round window. (Adapted from Evans.[62])

explain both the intensity dependence of the frequency acuity of the tuning and why the center frequency of the tuning depended on the sound intensity.

This active function of the outer hair cells improves the sensitivity of the ear by approximately 50 dB in addition to increasing the frequency selectivity of the cochlea at low sound intensities. Outer hair cells are susceptible to injury from exposure to loud sound, *ototoxic antibiotics*, and other insults while most of these factors have much less effect on inner hair cells and commonly leave inner hair cells nearly intact.[152] The maximal hearing loss from such insults therefore seldom exceeds 50 dB, which is similar to the value of gain from the mechanical action of outer hair cells.

f. Motility of Outer Hair Cells

As mentioned above, the two types of hair cells in the cochlea, the inner and the outer hair cells, have completely different functions despite their morphological similarities. While the inner hair cells control the neural activity in the afferent axons, outer hair cells contract or expand (elongate) when the stereocilia are deflected. This mechanical motion of the outer hair cells is transferred to the basilar membrane and tectorial membrane and injects mechanical energy into the mechanical system of the organ of Corti, where it adds to the sensitivity of the ear and also is the source of the non-linearity of the cochlea.

Two kinds of motility have been identified: slow and rapid. The rapid motion can follow the vibration of sound frequencies; the upper frequency limit has not yet been established because of experimental limitations, but the motion can easily follow a 1000-Hz tone, which is probably far below the limit of the motion of the outer hair cells.[197] Externally applied electrical potentials can cause motions of these cells at frequencies as high as 24 kHz.[50] The slow motion of outer hair cells may serve the purpose of maintaining the static balance of the hair cells by adjusting their length. The slow motion occurs on a time scale of seconds and can be induced by chemical, mechanical, or osmotic stimuli.[56,198]

It is assumed that the fast motility involves the plasma membrane and not the actin–myosin system because the latter would be much too slow to explain the motion of these cells. Studies have shown that the motile elements are located along the lateral surface of the cells.[49,197]

The change in length of the outer hair cells is a function of the membrane potentials (Fig. 5.21)[197]. The maximal elongation and shortening are only a few percent of their length, but that amplitude (approximately 300 nm) is within the amplitude range of the vibrations of the basilar membrane evoked by sound stimulation. The function of the outer hair cells as active elements is controlled by the activity in the efferent fibers that terminate directly on the outer hair cells. This means that the mechanical properties of the cochlea can be altered from the central nervous system.

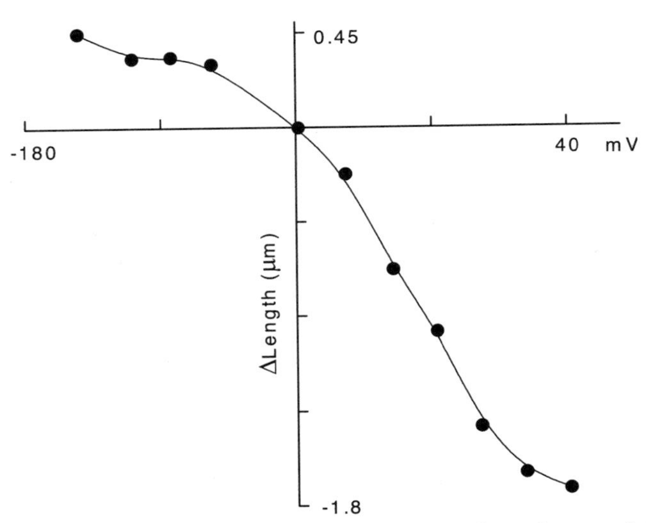

FIGURE 5.21 Changes in the length of an isolated outer hair cell as a function of changes in the membrane potential.[197]

g. *Otoacoustic Emission*

Otoacoustic emission (OAE) is a common name for sound generated in the ear. Otoacoustic emissions are complex phenomena that can be divided into several different types, suggesting the presence of more than one mechanism of generation of otoacoustic emissions. Otoacoustic emissions consist of sound generated either spontaneously, *spontaneous otoacoustic emissions* (SOAE), or in response to a sound such as a transient sound, *transient evoked otoacoustic emissions* (TEOAE) (Fig. 5.22). Otoacoustic emissions can also be elicited by continuous tones.

The motility of the outer hair cells is one of the sources of OAEs. The spontaneously generated otoacoustic emissions are most likely generated by the outer hair cells as a result of positive feedback that is so great it causes self-oscillation. TEOAE was first known as *cochlear echo*[110] because it has the properties of a sound reflected from a certain point on the basilar membrane and then transmitted backwards through the middle ear to generate a sound in the ear canal. It occurs with a latency of 5 to 15 msec, which is in agreement with the assumption that it is a reflection of the traveling wave from a certain point along the basilar membrane. The TEOAE may be caused by mechanisms similar to those for the SOAE, but it could also be the result of one or more homogeneities along the basilar membrane that would disturb the normal smooth progression of the traveling wave along the basilar membrane.[264] The result would be that some of the energy of the traveling wave would be reflected

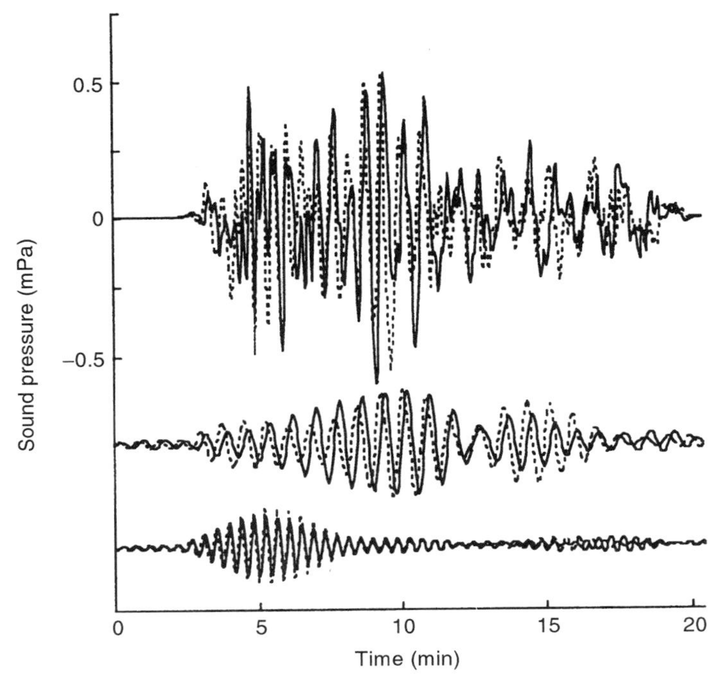

FIGURE 5.22 Click-evoked otoacoustic emission (TEOAE). The solid and the dashed lines are the responses in two different body positions. (From Büki, B., Avan, P., and Ribari, O., In: *Intracranial and Intralabyrinthine Fluids*, edited by A. Ernst, R. Marchbanks, and M. Samii, Berlin: Springer-Verlag, 1996, pp. 175–181. With permission.)

and travel in the opposite direction. The reflected wave is assumed to be amplified by the same mechanism that amplifies the normal traveling wave, namely action of the outer hair cells. The fact that the TEOAE is largest when elicited by sounds of low intensity support the hypothesis that active processes in the cochlea are involved in its generation.

Recording of the OAE is used clinically to test the function of outer hair cells.[125] The most commonly used form of the OAE is the *distortion product otoacoustic emission* (DPOAE), which is generated when two continuous tones are presented simultaneously.[125] As an example, if two tones (f_1 and f_2) with frequencies of 3.16 and 3.82 kHz respectively are used, then the difference tone, $2f_1-f_2$, will be 2.5 kHz. The difference tone can be recorded by a sensitive microphone placed in the ear canal, and the output of the microphone is filtered appropriately to suppress the stimulus tones and noise (Fig. 5.23). Unlike the TEOAE, the DPOAE is frequency specific and therefore gives information about the health of the organ of Corti along the basilar membrane.

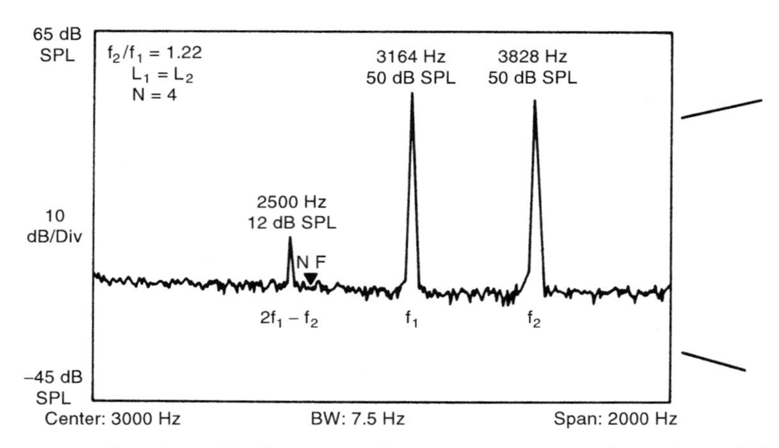

FIGURE 5.23 Illustration of the distortion product otoacoustic emission (DPOAE) recorded in the ear canal of a human subject. The $2f_1 - f_2$ component (2.5 Hz) was elicited by two tones (3.164 and 3.828 kHz) at 50 dB-sound pressure level (SPL). (Adapted from Lonsbury-Martin, B.L., and Martin, G.K., *Ear and Hearing*, 11: 144–154, 1990.)

h. Automatic Gain Control

The range of sound intensities from the threshold of hearing to the highest level where discrimination of sounds is possible is more than 100 dB, or more than 10 logarithmic units. The ratio between the energy of sound that is just detectable and the highest sound level at which normal sound discrimination is possible is 1 to 10,000,000,000. Such a large range of intensities cannot be coded directly in the discharge pattern of the auditory nerve fibers. Therefore, this enormous range of sound intensities must by reduced (compressed) before transduction into a neural code. The *amplitude compression* that occurs in the cochlea reduces that range of sound intensities so that it can be coded in the auditory nerve. The amplitude compression in the cochlea is partly a result of the active role of outer hair cells and partly a result of the transduction process in inner hair cells. At low sound intensities the outer hair cells increase the ear's sensitivity of the ear by acting as "motors" that amplify the motion of the basilar membrane. This amplification decreases with increasing sound intensities which is the same as compression of the amplitudes of the basilar membrane vibration. The transducer action of the inner hair cells provides additional compression of amplitudes, as do other mechanoreceptors (see Chapter 2). Both of these forms of amplitude compression are assumed to act nearly instantaneously, and they therefore do not affect the time pattern of the stimulation.

The acoustic middle ear reflex also exerts some amplitude compression by reducing the input to the cochlea for sounds above 85 dB hearing level (HL).[152]

The amount of gain control by the acoustic middle ear reflex is much smaller than the automatic gain control of the cochlea, and it only decreases the sound conducted to the cochlea by 15 to 20 dB.[48,152,258] The acoustic middle ear reflex is not active below approximately 85 dB HL, and it probably has little effect on sounds, the intensity of which is below approximately 90 dB HL. Studies of individuals whose acoustic middle ear reflex does not function (patients with Bell's palsy) have shown that in the absence of a functioning acoustic middle ear reflex, speech discrimination decreases for speech with sound levels above approximately 90 dB HL.[258]

The acoustic middle ear reflex acts slowly; therefore, this gain control does not affect fast changes in sound intensity and only attenuates steady sounds and sounds for which the intensity varies slowly.[135,152] This means that the acoustic middle ear reflex affects sounds in a similar way as adaptation, which attenuates slow changes and preserves fast changes.

III. THE AUDITORY NERVOUS SYSTEM

The auditory nervous system can be divided into *ascending* and *descending pathways*, similar to other sensory nervous systems. The ascending auditory pathways can be divided into the *classical* and the *nonclassical pathways*. The auditory nervous system is more complex than that of other sensory systems (for overviews, see Webster *et al.*[238]). Descriptions of the auditory pathways are normally based on results from animals such as cats and less often monkeys. Species differences regarding the anatomy and physiology of the auditory nervous system must be taken into account when results of different species are compared. In particular, the cortical regions have considerable differences between different species.

The ascending auditory pathways are organized in order from the periphery to the primary cerebral cortex which is the basis for the *hierarchical* processing of auditory information in the ascending auditory pathways. The pathways branch to form parallel pathways, where the same or different kinds of information are processed in different populations of neurons which is the basis for *parallel processing* of the same information. Some different kinds of information being processed in anatomically different parts of the CNS is known as *stream segregation*.

A. Anatomy of the Classical Ascending Auditory Pathways

The classical ascending auditory pathways perform hierarchical and parallel processing of information.[238] As the information ascends toward the cerebral

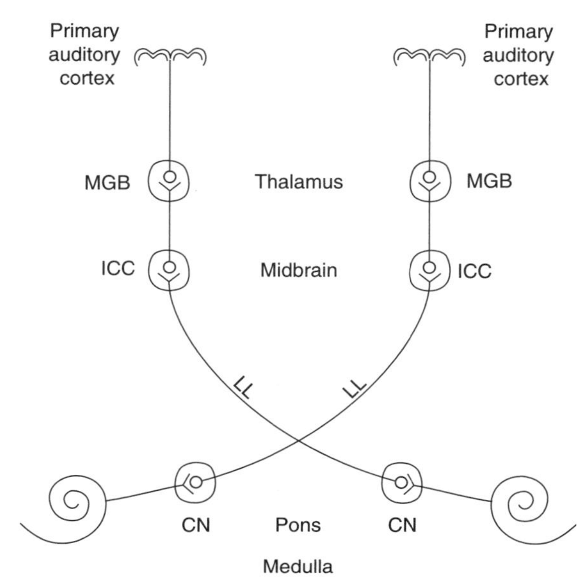

FIGURE 5.24 Simplified schematic diagram of the ascending auditory pathways.

cortex, it is transformed, restructured, and segregated for processing in different populations of neurons. The classical ascending pathway includes three nuclei in which all fibers are interrupted by synaptic transmission, namely the *cochlear nuclei*, *central nucleus* (ICC) of the *inferior colliculus* (IC), and the *medial geniculate body* (MGB) (Fig. 5.24).[159,164,237,238,245–247] The main fiber tract that connects the cochlear nuclei with the ICC is the *lateral lemniscus* (LL). The MGB projects to the *primary auditory cortex* (AI) through the *brachium of the IC* (BIC). Along the ascending pathways other nuclei interrupt some fibers (Fig. 5.25). Most notable are the nuclei of the *superior olivary complex* (SOC)[154] and the nuclei of the LL.[202,261] Most fibers from the cochlear nucleus cross the midline to form the *dorsal nucleus of the LL* (DNLL) and the *ventral nucleus of the LL* (VNLL). but there is also an ipsilateral pathway in the LL directly from the cochlear nucleus to the IC (Fig. 5.25). The ascending auditory pathways contain more nuclei than the somatosensory pathway, and there are several connections between the two sides of pathways such as the *commissure of Probst* and the *commissure of the inferior colliculus* (Fig. 5.25). Some neurons of the nuclei of the SOC receive input from both ears, and some neurons send axons ascending in the LL of both sides. These connections imply that there are connections between the two sides at the level of the SOC.

The auditory nervous system of several species of the flying bat has been studied extensively. Bats use their auditory system in a sophisticated way but that does not

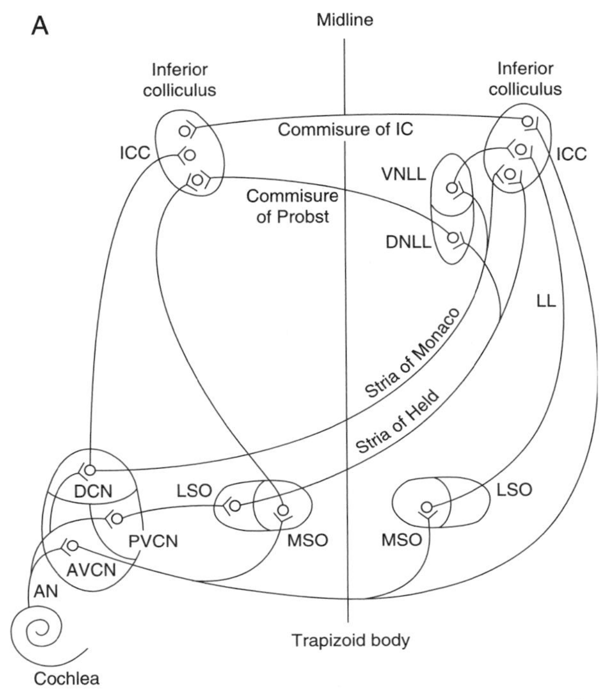

FIGURE 5.25 (A) Schematic illustration of the ascending pathways from the left cochlea to the inferior colliculus. (B) Anatomical location of the major components of the ascending auditory pathways; pathways from left auditory nerve to the medial geniculate body (MGB) are shown.

mean that their auditory system necessarily is anatomically noticeably different from that of other mammals. The auditory system serves these animals in several ways, primarily as a navigational system that enables them to fly without visual control; at the same time, bats use their auditory system for communication. This means that the bat uses hearing to handle tasks that are performed by both the visual and auditory systems in many other animals.

1. Auditory Nerve

The auditory nerve is a part of the *eighth cranial nerve* (CN VIII), which also contains the vestibular nerve. In humans, cats, and monkeys, the auditory nerve consists of bipolar nerve fibers. Studies have reported that the auditory nerve in humans has approximately 30,000 fibers; in the cat, approximately 50,000; and in the chinchilla and guinea pig, 24,000.[85] Approximately 95% of these nerve fibers are Type I auditory fibers that innervate the inner hair cells (see page 54).[193] Approximately 5% of the nerve fibers are Type II nerve fibers,

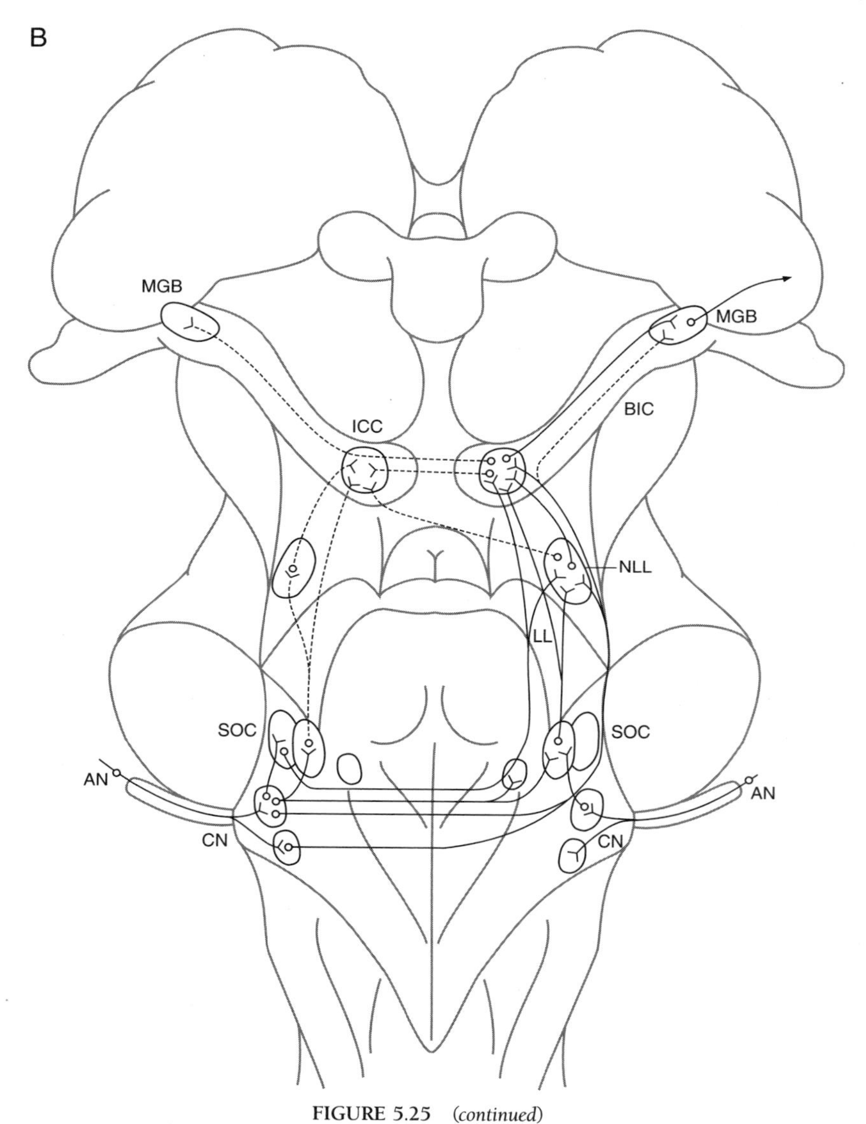

FIGURE 5.25 (*continued*)

which are smaller and innervate outer hair cells. Their function is unknown. The diameter of the Type I fibers is approximately 2.5 μm with little variation of size from one to another.[219] The average conduction velocity in the auditory nerve is approximately 20 m/sec.[150] The narrow range of fiber diameters of Type I nerve fibers means that the conduction velocities of these fibers are within a narrow range,[219] which ensures a high degree of temporal coherence of the nerve impulses that arrive at the cochlear nucleus. This is assumed to be of importance for discrimination of the frequency of sounds on the basis of the temporal principle (see page 328).

2. Cochlear Nucleus

The auditory nerve terminates in all three major divisions of the cochlear nucleus, the *anteriorventral cochlear nucleus* (AVCN), the *posteriorventral cochlear nucleus* (PVCN), and the *dorsal cochlear nucleus* (DCN).[193] Each auditory nerve fiber that arrives from the periphery bifurcates, and one branch makes synaptic contact with cells in the AVCN. The other branch bifurcates again and one of the branches terminates on cells in the PVCN, while the third branch terminates in the DCN (Fig. 5.26).

The branching of the auditory nerve to innervate different populations of neurons in the cochlear nucleus is the most peripheral anatomical basis for the parallel processing that is prominent throughout the ascending auditory system.* (Parallel processing implies that the same information can be processed in different populations of neurons whereby the same information is processed in different ways.) Branching of the auditory pathways continues along the ascending pathways toward the primary cerebral cortex, which provide the basis for further parallel processing.

The connections between the cochlear nuclei on the two sides[32,129,210] constitute the most peripheral connection between the two side's ascending auditory pathways. The connections are mainly from the AVCN of one side to the DCN of the other side, but there are also connections between the ventral cochlear nuclei on one side to the ventral cochlear nucleus on the other side.[210] Some cells in some of the nuclei of the SOC send connections to the cochlear nucleus.[216]

3. Superior Olivary Complex

Each of the three divisions of the cochlear nucleus connects to neurons in the ICC via one of the three striae. The *dorsal acoustic stria* (DAS), or *stria of Monaco*, originates in the DCN; the *intermediate acoustic stria* (IAS), or *stria of Held*, originates in the PVCN; and the *ventral acoustic stria* (VAS), or *trapezoidal body*, originates in the AVCN (Fig. 5.25).[154,202] All three striae cross the

*The fact that each inner hair cell connects to many (approximately 20) nerve fibers may be regarded as the earliest manifestation of parallel processing. However, it is not known in which way these 20 distinct neurons branch at the cochlear nucleus and whether they all arrive at the same region of the CN. It is also not known if all of these 20 fibers have the same thresholds or latency.

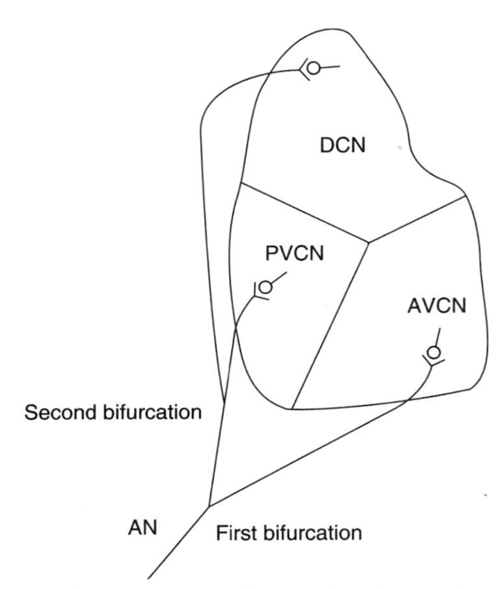

FIGURE 5.26 Schematic drawing of the cochlear nucleus showing the two bifurcations of an individual auditory nerve fiber that makes one nerve fiber innervate neurons in all three major divisions of the cochlear nucleus.[152]

midline and form the LL that terminates in the contralateral ICC, which is the midbrain relay nucleus of the ascending auditory pathways where all fibers are interrupted.

Some of the fibers of these three striae send collaterals to cells in nuclei of the SOC. Some fibers are interrupted in some of the nuclei of the SOC before they form the LL, while other connections from the cochlear nucleus pass uninterrupted to reach the (contralateral) ICC. The main nuclei of the SOC are the *medial superior olivary* (MSO) nucleus, the *lateral superior olivary* (LSO) nucleus, and the *nucleus of the trapezoidal body* (NTB).

Input from both ears reaches some of the neurons of the nuclei of the SOC, in particular those of the MSO, which receives its input from the AVCN on both sides. These connections are important for directional hearing. The length of the axons from the two sides is different for various neurons in the MSO, thus an array of delay lines is formed. The MSO neurons function as coincidence detectors and the axons from the left and right cochlear nucleus have different lengths and function as delay lines that make it possible to detect different time intervals between sounds that reach the two ears.

The axons of the MSO neurons project mainly to the ipsilateral ICC via the LL. The LSO receives its input from both AVCN and PVCN. Their axons project to both sides' ICC via the LL. The LSO neurons also receive inhibitory input from the contralateral CN via the *middle nucleus of the trapezoidal body* (MNTB).

The uncrossed connections between the cochlear nucleus and the ipsilateral ICC originate in the PVCN and form collaterals to the fibers of the trapezoidal body that originate in the AVCN. The exact extension and the importance of these uncrossed pathways are unknown.

The nuclei of the superior olivary complex show considerable species differences,[154] and because most of the anatomical research on these structures has been done in different animal species, the descriptions found in the literature and reproduced in textbooks vary considerably. Only a few anatomical studies of humans have been published.[154]

There are connections from the SOC to motor systems. Connections to the facial and the trigeminal motoneurons have been verified,[19] and these connections are the basis for the acoustic middle ear reflex (see page 352).

4. Lateral Lemniscus and Inferior Colliculus

The inferior colliculus consists of three distinct nuclei: the central nucleus (ICC), the *dorsal cortex* (DC), and the *external nucleus* (ICX) (Fig. 5.27).[7,159] The DC and ICX belong to the nonclassical auditory pathways that will be discussed later. The ICC receives all the activity carried in the LL.[165] Some fibers of the LL travel uninterrupted from the cochlear nuclei to the ICC, while other fibers of the LL are interrupted in the DNLL or the VNLL before they reach the ICC. The DNLL receives input from both ears from the AVCN and the LSO and ipsilateral input from the MSO. Axons from the DNLL form the commissure of Probst, which connects to neurons of the ICC on the opposite side (Fig. 5.25). Contralateral input to the DNLL also arrives from the opposite DNLL via the commissure of Probst.[261] The DNLL is involved in binaural hearing. The VNLL mainly receives input from the contralateral ear and projects to the ICC on the same side. In addition to the VNLL and DNLL, the *intermediate nucleus of the lateral lemniscus* (INLL) and the *posterior medial nucleus of the lateral lemniscus* (PMNLL) interrupt some fibers of the LL. These nuclei receive most of their input from the contralateral CN, and their cells send axons to the ipsilateral IC and possibly the medial division of the MGB.[262] Their neurons also receive collaterals from LL fibers.[154] The INLL is found in some animals but not in humans.[154] (See also Ehret and Romand.[59]) The LL probably gives off collaterals to the brainstem core reticular formation where it travels in the brainstem.[202] There are extensive connections between the two sides' ICC by the *commissure of the inferior colliculus* (Fig. 5.25).[6] Neurons of the ICC also connect to neurons of the ICX and DC, which belong to the nonclassical auditory pathways (see page 347). Recent evidence suggests that neurons in the ICC receive input from nonsensory parts of the CNS such as

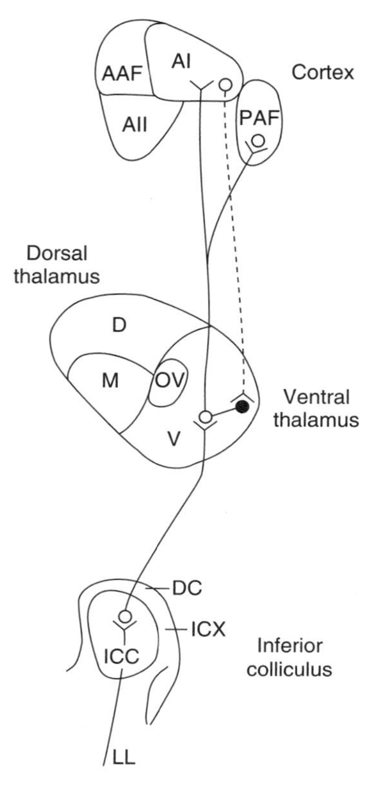

FIGURE 5.27 Schematic drawing of the ascending pathways from the central nucleus of the inferior colliculus (ICC) to the ventral portion of the thalamus and their connections to auditory cortical radiations. Most of the connections have reciprocal descending connections, only one of which is shown (between AI and the thalamus). M, medial (or magnocellular) division of thalamus; D, dorsal division; V, ventral division. OV, ovoid part of the thalamus.

the limbic system (the amygdala).[128] This has only been shown in bats so far but there are reasons to believe that similar connections may exist in mammals in general.

It has usually been regarded that all fibers of the lateral lemniscus are interrupted by synaptic transmission in the central nucleus of the ICC. This means that the ICC has been regarded as the obligatory relay nucleus for the classical ascending auditory pathways. However, recent studies[126] have shown that there are uninterrupted connections between neurons in the dorsal cochlear nucleus and the medial MGB.

These fibers may have collaterals that terminate in cells in the IC. Because these connections terminate in the medial MGB, they may be regarded as belonging to the nonclassical pathways (see page 347).

The IC has been described and often labeled as the auditory reflex center, and there is experimental evidence that some motor responses to auditory stimulation are mediated through the IC.[165] Neck muscles receive input from the IC for movement of the head toward a sound source. The IC, however, is not involved in the acoustic middle ear reflex.[19]

There are also connections from the IC to the *superior colliculus* (SC) that mediate various kinds of motor responses to sounds, especially regarding eye movements through activation of extraocular muscles. The SC may also be involved in creating a sense of space by integrating visual and auditory information.[124,253]

5. Medial Geniculate Body (MGB)

The ventral portion of the medial geniculate body belongs to the classical ascending pathways. The main output of the ICC forms the brachium of the inferior colliculus, which terminates in the ventral portion of the MGB. The ventral MGB consists of a *lateral ventral* (LV) and an *ovoid* (OV) part.[156] The dorsal MGB belongs to the nonclassical auditory pathways,[247,249] and it consists of a medial division and a dorsal division[245,246] (see page 347). The *suprageniculate thalamic nucleus* also belongs to the nonclassical part of the MGB.

The BIC consists of four parallel and separate pathways[31,73] and contains approximately 10 times more fibers than the auditory nerve (250,000 versus 25,000–30,000), which is another indication of parallel processing in the ascending auditory pathways.

The ventral MGB also receives input from the *thalamic reticular nucleus* (RE).[247] The ventral portion of the MGB receives input from the RE, which may exert control of the excitability of neurons in the MGB in general. The posterior division of the MGB (PO) receives input from the ipsilateral ICC. These regions of the MGB project to the *anterior auditory field* (AAF) (Fig. 5.28B). The fibers from the PO project to association cortices laying alongside auditory cortical areas.[82]

The ventral part of the MGB projects to the primary auditory cortex (Fig. 5.28).[247] The classical ascending pathways probably give off collaterals to the reticular formation of the brainstem but little is known about such connections. Such collaterals are probably more pronounced for the nonclassical pathways (see page 347), which are often referred to as being an important mediator of arousal. These connections to the reticular formation also mediate the *startle* response, which is a general contraction of muscles in response to impulsive and unexpected sounds.

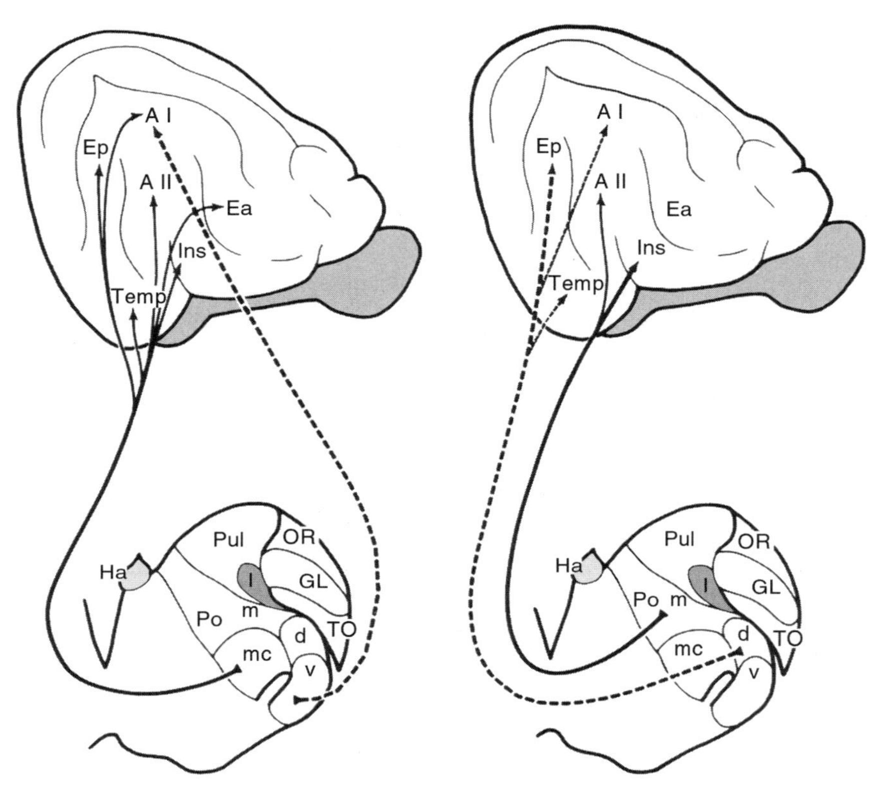

FIGURE 5.28 Projections of the different parts of the MGB to the cerebral cortex of the cat. AI, primary auditory cortex; AII, secondary auditory cortex; Ins, insular cortex; temp, temporal portion, Ep, posterior ectosylvian gyrus; Ea, anterior ectosylvian gyrus; Po, posterior nucleus group; Po m, medial division of the posterior nucleus group; mc, magnocelluar division of the MGB; d, dorsal division of the MGB; v, ventral division of the MGB. (Adapted from Diamond.[54])

6. Auditory Cortex

The primary auditory cortex in humans is located deep in the lateral fissure of the temporal lobe, in the superior temporal gyrus, Heschel's gyrus or Brodmann's area 41.[1,247] The different areas of the auditory cortex are usually defined as the areas where neurons respond to sound stimuli; the different areas are commonly labeled AI (primary auditory cortex), AII (secondary auditory cortex), Ep (posterior ectosylvian field), AAF (anterior auditory field), and PAF (posterior auditory field). The AI area receives input from the ventral MGB. The area around the AI receives input both from the MGB and the AI cortex. However, most descriptions of the anatomy and physiology of the auditory cortex refer to the auditory cortex in animals such as the cat, monkey, rat, and guinea pig.

The connections to the different layers of the AI area resemble that of the somatosensory cortex following the general rules for connections of primary sensory cortices as described in Chapter 3 (Fig. 3.10).[133,252]

There are abundant connections between primary auditory cortical areas on one side and similar areas on the other side.[40,252] The LV and the OV neurons of the MGB project mainly to layer IV of the primary auditory cortex.[133] Neurons of layer III of the primary auditory cortex project to secondary auditory cortices (AII, Ep) and to the contralateral AI.[40,247,252] Neurons of layer VI project to the MGB, and neurons of layer V project to both the MGB and IC.[55,251] The projection to both the MGB and the IC are uninterrupted. The descending projections to the IC reach mainly neurons in the ICX and DC. The descending connections from layers V and VI may be regarded as *reciprocal innervation* to the ascending connections, but they are often referred to as a separate descending auditory system.

7. Descending Pathways

Like other sensory pathways, the auditory sensory pathways include abundant descending systems. The descending pathways have often been regarded as several separate systems, but, as was discussed in Chapter 3, it may be more appropriate to regard the descending systems as reciprocal innervation that descends in parallel to the ascending pathways. There are specific characteristics of the descending pathways of the auditory system that deserve to be mentioned, however.

Two separate systems of descending pathways have been identified in connection with the classical auditory pathways. One system, the olivocochlear system originates in the superior olivary complex and reaches the hair cells of the cochlea (Fig. 5.29).[171] That system has two parts known as the olivocochlear bundles (OCB). One of these two parts, the *crossed olivocochlear bundle* (COCB) originates in the medial superior olivary nucleus. [234] The other part, the *uncrossed olivocochlear bundle* (UOCB), originates in the lateral superior olivary complex. The COCB fibers mainly terminate on outer hair cells, whereas the UOCB fibers mainly terminate on axons of inner hair cells.

The other descending system extends from the primary auditory cerebral cortex to the thalamus and the inferior colliculus.[248,250] The MGB receives abundant descending input from the same cortical area.[128,250] These connections arise from the both the AI area and from secondary cortical areas on the ipsilateral side.[55,92] Many descending fibers from the auditory cortex terminate on neurons in the ICC. Descending fibers from the cortex to the MGB are several times more

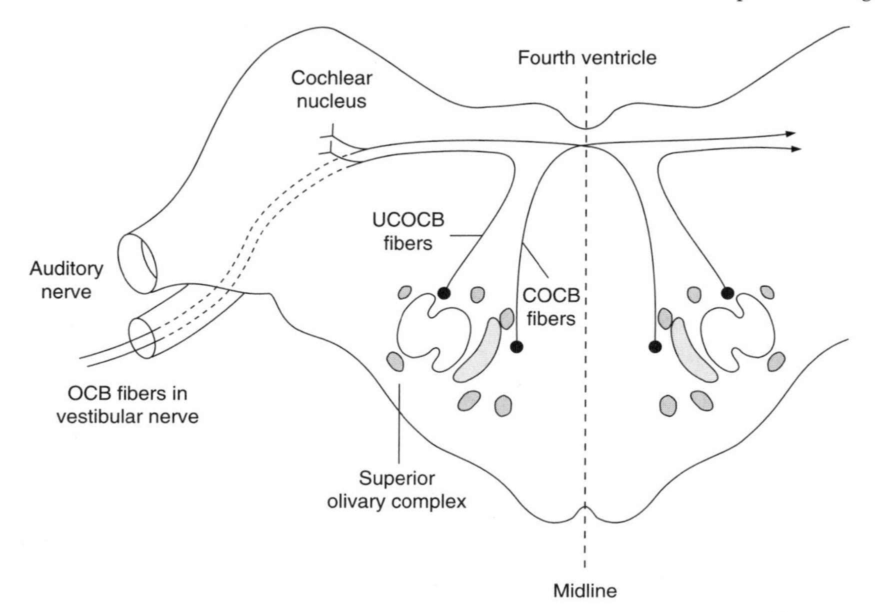

FIGURE 5.29 Olivocochlear efferent system in the cat. (Adapted from Pickles.[169])

abundant[9,10,250] than the ascending fibers between the AI and the ventral MGB. Neurons in the IC project to the DCN bilaterally.[33,42]

B. Physiology of the Classical Auditory Nervous System

The ascending pathways of the auditory system transform, separate, and restructure the information that arrives at the ear to a greater extent than is true for other sensory organs. Studies of neural processing in the classical auditory nervous system have mainly been done in animals using recordings from single auditory nerve fibers, from cells in the nuclei of the classical ascending auditory pathways, and from fibers of the fiber tracts. Such studies can reveal only part of the signal processing that takes place in the auditory nervous system, and we are far from understanding in any detail how information is transformed in the auditory pathways.

An understanding of the anatomy is the basis for understanding the physiology. It must be kept in mind, however, that anatomical connections are not always functional because the function depends on the efficacy of synapses, which can be low or ineffective (dormant synapses, see pages 81 and 164).

The efficacy of synapses can only be determined using physiologic methods, but physiologic studies are lacking regarding many of the anatomically verified connections in the auditory nervous system. The efficacy of synapses may change as a result of external or internal circumstances such as those that are related to common experimental situations where the function of specific connections is studied. Change in synaptic efficacy can lead to "functional rewiring," although anatomical connections are not altered. If that occurs as a result of experimental procedures such as, for example anesthesia, the results are not applicable to normal situations. Synaptic efficacy may also differ among animal species.

Lack of physiologic studies makes the functionality of connections that have only been verified anatomically uncertain. These problems are not unique to the auditory system (see Chapter 3), but they are perhaps of greater importance in the auditory system than in, for instance, the somatosensory system because of the complexity of the auditory system and the multitude of connections that exist in the ascending and the descending auditory pathways.

1. Information Processing in the Auditory System

The transformation of information that occurs in the ascending auditory pathways is extensive and complex. Neurophysiologic studies in animals have revealed many aspects of sensory processing. The coding and transformation of the frequency (spectrum) of sounds and the processing of the temporal pattern of sounds have been studied extensively. The dual representation of frequency of sounds as both a place representation and a temporal representation has been studied extensively, as has the processing of changes in the frequency and amplitude of sounds that are important for discrimination of natural sounds such as speech sounds.

2. Representation of Frequency (Spectrum)

The frequency or spectrum of sounds is important for discrimination of natural sounds such as speech sounds, and these features of sounds are well represented in the classical auditory pathways, both as the place representation and as a temporal representation.

The basilar membrane separates sounds according to their frequency, and the arrangement of the inner hair cells in a row along the basilar membrane of the cochlea causes the hair cells to respond according to the frequency of sounds. This is the basis for the place representation of the spectrum of sounds. The separation of sounds along the basilar membrane according to their frequency or spectrum is preserved in the anatomical organization of fibers of the auditory nerve and cells, and axons throughout the ascending auditory pathways, including the cerebral cortex, are organized according to the frequency with which they respond to the lowest threshold. The spectral separation provided by the basilar membrane of the cochlea is thus preserved throughout the auditory nervous system, including the cerebral cortex; we can perceive the basilar membrane as being projected onto populations of nerve cells in the ascending auditory pathways. This is assumed to be the basis for the place coding of frequency and may be regarded as a form of spatial ("where") information.

The frequency or spectrum of sounds is also represented in the discharge pattern of auditory nerve fibers as a temporal code of frequency. This representation is the basis for the temporal hypothesis of frequency discrimination and may be regarded as a form of object information.

a. Receptive Fields

Because each auditory nerve fiber innervates an individual inner hair cell,[217] the response of an auditory nerve fiber reflects the excitation of only that hair cell. The fact that each auditory nerve fiber terminates at an inner hair cell located at a certain position along the basilar membrane means that an individual nerve fiber will respond to sounds within limited ranges of frequencies and intensities (Fig. 5.30).[61,108,109,111] This response area is the receptive field of an individual auditory nerve fiber with regard to the frequency (spectrum) of sound and the intensity. This means that each auditory nerve fiber is tuned to a specific frequency. All neurons in the classical ascending auditory pathways have specific response areas and are tuned to one or more frequencies.

b. Receptive Fields of Auditory Nerve Fibers

The receptive fields (tuning curves) of auditory nerve fibers are typically obtained by presenting tones of different frequency and intensity to the ear of an experimental animal while recording from individual auditory nerve fibers. When the intensity of the tone is close to the threshold of hearing of the animal that is studied, a nerve fiber will respond only in a narrow range of frequencies (Fig. 5.30). A curve that follows the edge of the response area of a nerve fiber with regard to the frequency of sounds is the fiber's *frequency tuning curve* or *frequency threshold curve* (FTC) of the fiber.[61,111]

The frequency where the threshold is lowest is known as the *characteristic frequency* (CF) of the particular fiber (Fig. 5.30). When the sound intensity is increased, the frequency range over which the fiber responds becomes wider. The complete response area of an individual auditory nerve fiber is obtained by repeating that procedure while increasing the intensity of the stimulus sound until the upper level of useful hearing (Fig. 5.30). FTCs are usually obtained by determining the threshold of an auditory nerve fiber to tones of different frequencies. Different auditory nerve fibers are tuned to different frequencies and

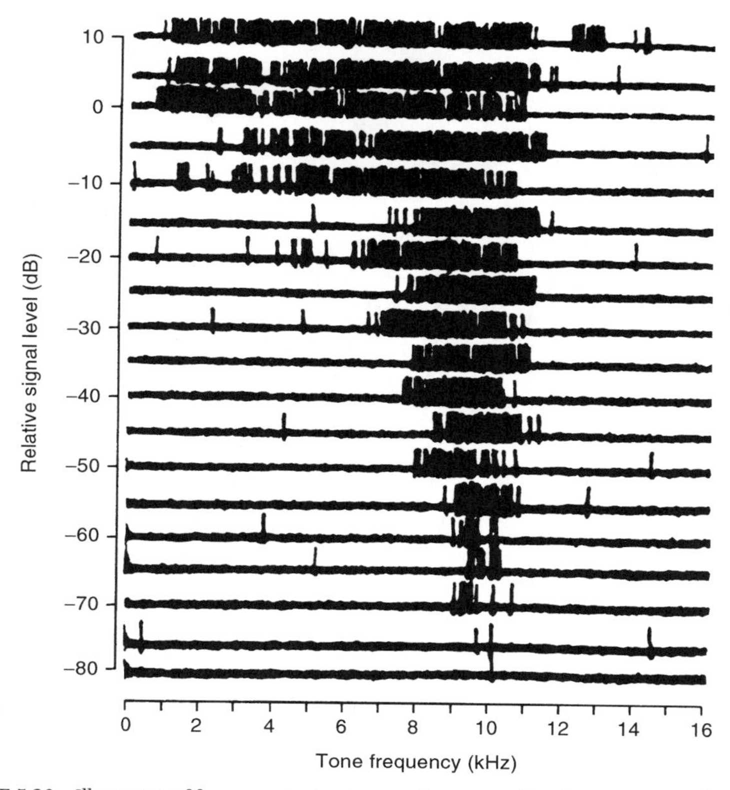

FIGURE 5.30 Illustration of frequency tuning in an auditory nerve fiber in a guinea pig, showing the area of frequency and intensity in which the nerve fiber responds by an increase in its discharge rate. (Adapted from Evans, E.F., *J. Physiol.*, 226: 263–287, 1972.)

recordings from many auditory nerve fibers yield a family of tuning curves, the CFs of which cover the entire audible frequency range of the animal from which the recordings are done (Fig. 5.31).

The frequency tuning of individual auditory nerve fibers implies that the different nerve fibers respond to sound in different parts of the audible spectrum and that representation of sound frequency may be referred to as spatial ("where") information.

Frequency Tuning Is Intensity Dependent It was mentioned earlier in this chapter that the tuning of the basilar membrane is different for different sound intensities and that the tuning of the basilar membrane becomes broader when the intensity is increased from threshold values at the same time as the

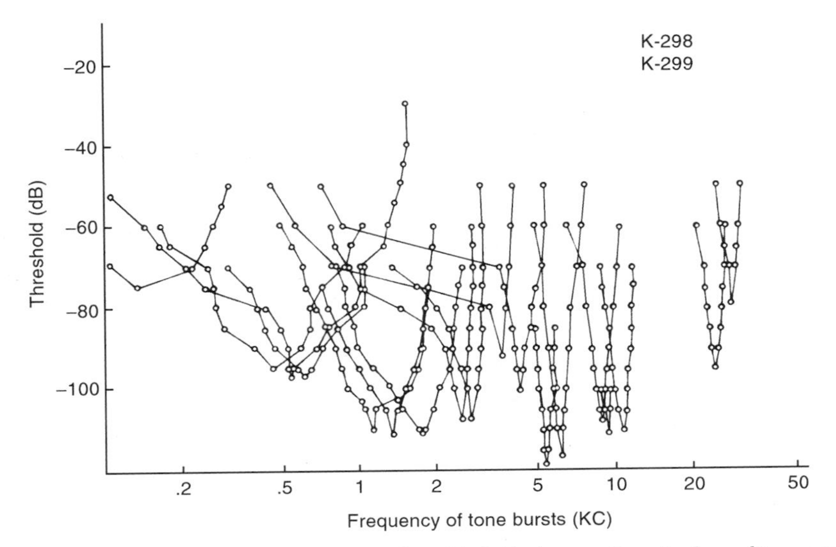

FIGURE 5.31 Tuning of representative samples of individual nerve fibers in the auditory nerve, reflecting tuning of the basilar membrane. (Adapted from Kiang *et al.*[111])

frequency to which a point is tuned shifts along the basilar membrane. Because the tuning of the basilar membrane is the basis for the tuning of auditory nerve fibers, the tuning of auditory nerve fibers also depends on the intensity of a sound and becomes broader when the sound intensity is increased.[146,148]

The nonlinearity of the basilar membrane not only causes a broadening of the tuning as the sound intensity is increased from threshold to physiologic sound levels, and the frequency to which nerve fibers are tuned changes with the sound intensity. The frequency to which a nerve fiber is tuned shifts toward lower frequencies when the sound intensity is increased from low to high sound levels (Fig. 5.32).*[,146,148,151] Also, the width of the tuning changes with the intensity of sounds in such a way that the narrowest tuning occurs at low sound intensities and the tuning gradually widens with increasing sound intensity.[148]

*The change in tuning of auditory nerve fibers when the sound intensity is changed cannot be demonstrated by obtaining the threshold of nerve fibers to tones as is done for obtaining FTCs. It can, however, be demonstrated in experiments where broadband noise is used as the stimulus and the discharge pattern is analyzed using cross-correlation analysis[146,148] or by using a special form of correlation analysis known as "reverse correlation," where the sound that precedes a neural discharge is added (averaged).[18,57]

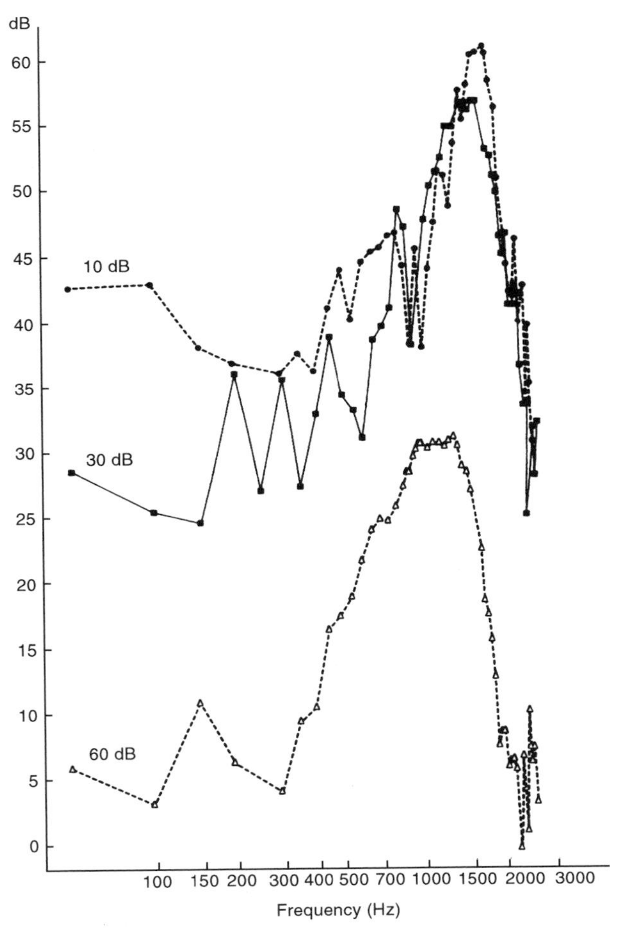

FIGURE 5.32 Illustration of the intensity-dependent tuning of an auditory nerve fiber in the rat. The curves show the tuning of an auditory nerve fiber in a rat obtained using noise as stimulus at three different sound intensities (in dB SPL). (Adapted from Møller.[151])

Suppression and Excitation in the Auditory Nerve An auditory nerve fiber's (excitatory) response area is surrounded by inhibitory areas.[194] Presentation of a tone within the inhibitory areas decreases the response elicited by a tone within the fiber's excitatory response area.[75,194] Using a technique similar to that used to determine the receptive field of an auditory nerve fiber

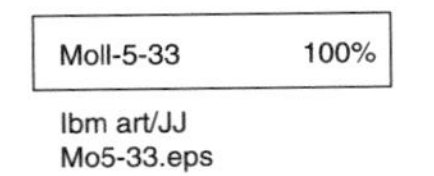

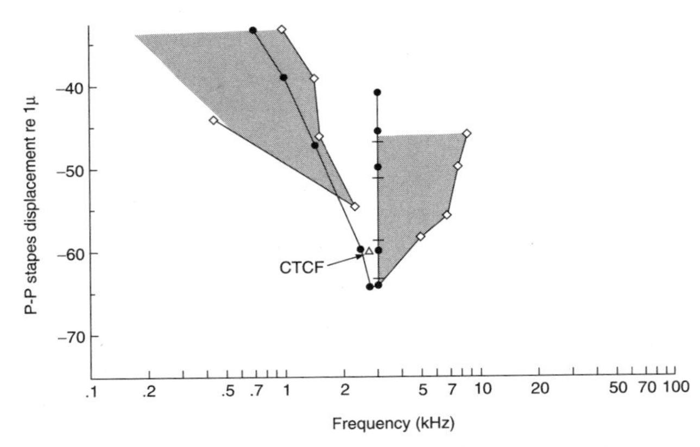

FIGURE 5.33 Schematic illustration of tuning of a nerve fiber in the auditory nerve showing an excitatory response area that is located in between two similarly shaped inhibitory areas. (Adapted from Sachs and Kiang.[194])

(but, instead of a single tone, using two tones with different frequencies), it is possible to map areas where a tone decreases the response elicited by a tone at the CF of the fiber (Fig. 5.33).

This is similar to lateral inhibition in other sensory systems. In the auditory system this inhibition is known as *two-tone suppression*. The reason for that name is that the two-tone inhibition does not involve synaptic transmission but the suppression that occurs in the auditory nerve fibers is a result of the nonlinear micromechanical properties of the cochlea, the source of which is mainly the outer hair cells. Two-tone suppression can be demonstrated in nonlinear models of the function of the cochlea such as the one proposed by Geisler and Sang,[76] which supports the assumption that it is a result of micromechanical properties of the basilar membrane. More recently, two-tone suppression has been demonstrated in recordings from hair cells, providing further evidence that two-tone suppression is established in the cochlea. Two-tone suppression therefore occurs without any delay which differs from the inhibition that occurs in the nervous system as a result of inhibitory synaptic transmission.

The phenomenon of two-tone suppression is not limited to two tones. Any number of tones or spectral components of a complex sound will interact, and different spectral components of complex sound will suppress the response evoked by other components. Because most natural sounds have energy over a broad frequency range, such sounds will activate both excitatory and suppressive areas of an auditory nerve fiber. The energy that activates the region of suppression decreases the discharge rate elicited by the sound energy which is within the excitatory response area of a nerve fiber. This causes two-tone suppression to act as a gain control.

c. *Tuning of Cells in the Ascending Auditory Pathways*

Nerve cells throughout the auditory nervous system exhibit more or less distinct receptive fields with regard to the frequency and the intensity of sounds, as can be shown by their FTCs. The shapes of the response areas of auditory nerve fibers undergo transformations as the information ascends in the ascending pathways; therefore, the shapes of the response areas of neurons in more centrally located nuclei vary and many neurons have response areas that are different from those of auditory nerve fibers.[83,94,108,120,136,155,188,223]

Some cells in the nuclei of the ascending auditory pathways have response areas that are broader than those of auditory nerve fibers; some are narrower and some have multiple peaks and are thus tuned to more than one frequency. At first glance, the tuning curves of cells in the cochlear nucleus resemble those of the auditory nerve but a few cells have tuning curves of different shapes, some with two peaks.[147] The tuning curves of cells in the nuclei of the SOC show great variations (Fig. 5.34).[83]

Generally, cells in the nuclei of the auditory pathways receive both inhibitory and excitatory input, and both these kinds of input contribute to shaping frequency selectivity and the response to time-varying sounds. The inhibitory input to most neurons comes from different frequency regions, and excitatory response areas are surrounded by inhibitory regions in most neurons.

The shapes of the FTCs of cells in the inferior colliculus vary more than in the cochlear nucleus. At least three different types of response areas with regard to the frequency of tones can be discerned in cells in the ICC: V-shaped, I-shaped, and O-shaped (Fig. 5.35).[58,94,120,155,188] The V-shaped tuning curves are similar to those of auditory nerve fibers. The I-shaped tuning curves are narrower than tuning curves of auditory nerve fibers, and the O-shaped tuning curves are closed, which means that these cells do not respond at all to sounds above a certain intensity. The inhibitory areas that are located on both sides of the excitatory response areas are more complex and more extensive than those of auditory nerve fibers or cells in the cochlear nucleus (Fig. 5.35).

Rees and co-workers[120] have shown that the ionophoretic application of bicuculline or strychnine to block GABAergic and glycinergic inhibition increases the firing rate of neurons in the IC. The shape of the frequency response areas of neurons with V-shaped tuning curves remained little changed, whereas elimination of GABAergic inhibition by ionophoretic application of bicuculline or strychnine to block glycinergic receptors changed the shapes of the frequency response areas of neurons with non-V-shaped tuning curves.

As is the case for vision and somesthesia, the receptive fields of fibers and cells in nuclei of the ascending auditory nervous system are a result of convergence and interplay between inhibition and excitation which modify the response areas of neurons. Some of these differences in the shape of the response areas of neurons in the different nuclei of the ascending auditory pathways can be explained by the complex innervation of nerve cells of the nuclei of the auditory system. The varying degrees of convergence and the abundant interplay between inhibition and excitation are responsible for much of the variations in the shapes of the tuning curves in the SOC and the ICC. Interneurons in the different nuclei and the abundant descending connections most likely also contribute to this complexity.

Cells in the ventral portion of the MGB are tonotopically organized similar to other auditory nuclei.[94,155] Tuning of cells in the ventral portion of the MGB show varying shapes.[155] Some cells have a very narrow tuning, whereas other cells are broadly tuned.

Studies of the frequency tuning of neurons in the primary auditory cortex consistently show that the tuning curves have many different shapes,[167] and

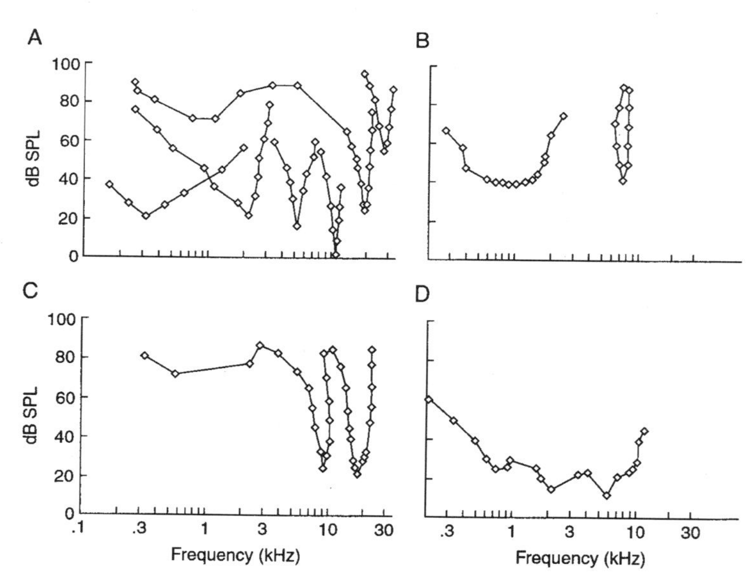

FIGURE 5.34 Examples of tuning curves of different shapes obtained in recordings from neurons in the superior olivary complex. (Adapted from Guinan, Jr. J.J., Guinan, S.S., and Norris, B.E., *Int. J. Neurosci.*, 4: 101–120, 1972. Reproduced by permission of Taylor & Francis, Inc., http://www.routledge-ny.com)

there are greater variations in the shape of the tuning curves of cortical neurons than that of neurons in nuclei at more peripheral levels. While some of the tuning curves of cortical neurons resemble tuning curves of nerve cells at more peripheral levels of the auditory system, some are very different. Some cortical tuning curves are totally circumscribed, such as the O-type FTCs in the ICC (Fig. 5.35), which means that these neurons respond within a certain intensity range, and the discharge rate as a function of sound intensity of such neurons is nonmonotonic.[38] Similar response characteristics are seen in some neurons of the MGB.[188]

Studies of the responses from neurons in the auditory forebrain and the cortex in experimental animals are hampered by the effect of anesthesia. Anesthesia alters the activity of the neurons that converge on the cell under study. For example, cortical neurons have a high and perhaps normal spontaneous activity under ketamine anesthesia, but the spontaneous activity of such neurons is very low under barbiturate anesthesia, another often used anesthesia. It can be assumed that responses to different types of stimulation depend on the kind of anesthesia used. The results of studies using various forms of anesthesia must therefore be interpreted cautiously.

d. Response to Sounds Above Threshold

Frequency threshold curves are the most common measure of frequency selectivity of fibers and cells in the ascending auditory pathways, but FTCs are threshold functions and there are reasons to assume that the frequency selectivity to sounds above threshold are different. However, the response to sounds above threshold has not been studied to any great extent, and studies of spectral processing of sounds in the auditory nervous system have been mainly based on the threshold to steady sounds while the typical natural sound appears well above threshold. Natural sounds almost always vary more or less rapidly both in intensity and with regard to their frequency or spectral composition. The response to such sounds cannot be predicted on the basis of threshold data such as those gathered from frequency threshold tuning curves. The relatively few studies that have investigated the response to sounds with varying frequencies find that such sounds produce responses that are very different from those produced by steady sounds.[138,142,214]

The few studies that have been published show that the response to sounds (for instance, tones for which the frequency is changed at different rates) is radically different from FTCs.[142] Cells in the CN and the ICC show complex response patterns to such sounds and the shape of the response areas are sharper than the threshold response to steady tones.[142] The response of single nerve cells in the cochlear nucleus neurons that show only inhibition in response to steady tones exhibit clear excitatory response areas in response to tones for which the frequency is changed at a rapid rate[138] (see page 337).

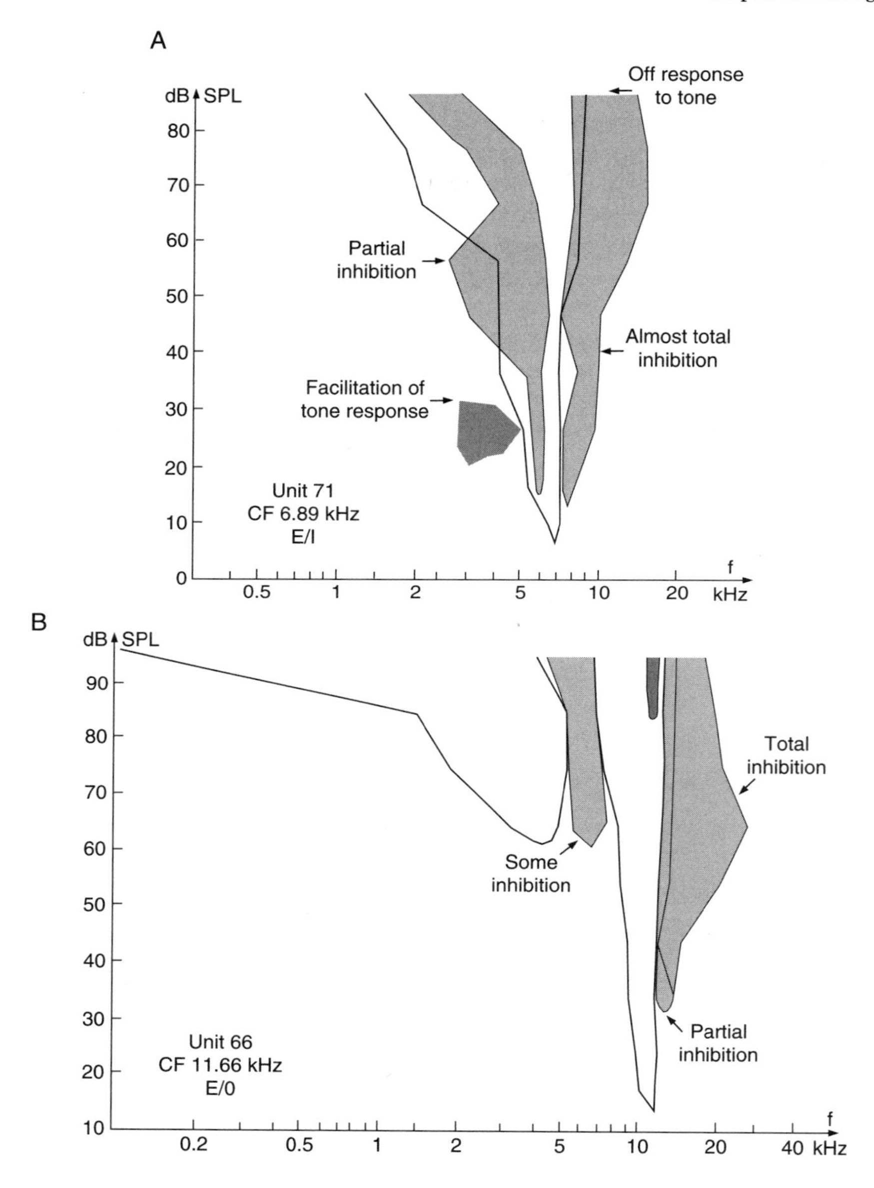

FIGURE 5.35 Four different types of tuning curves found in the inferior colliculus. (Adapted from Ehret, G. and Romand, R., *The Central Auditory Pathway*. New York: Oxford University Press, 1997.)

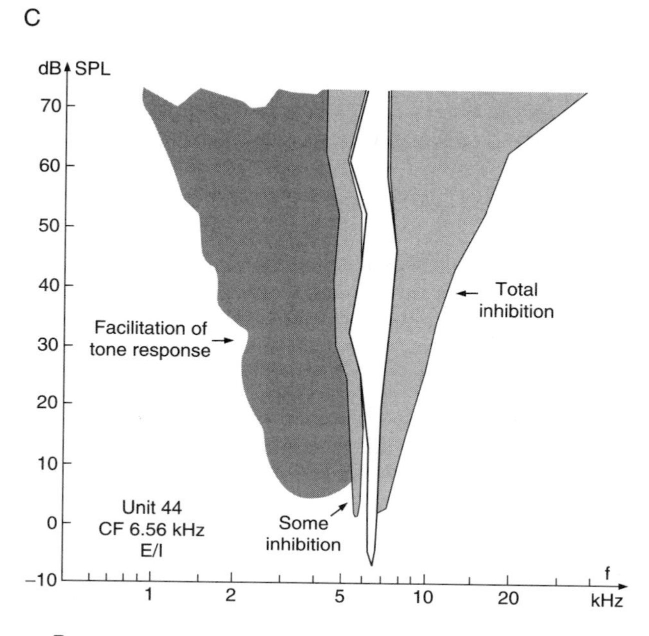

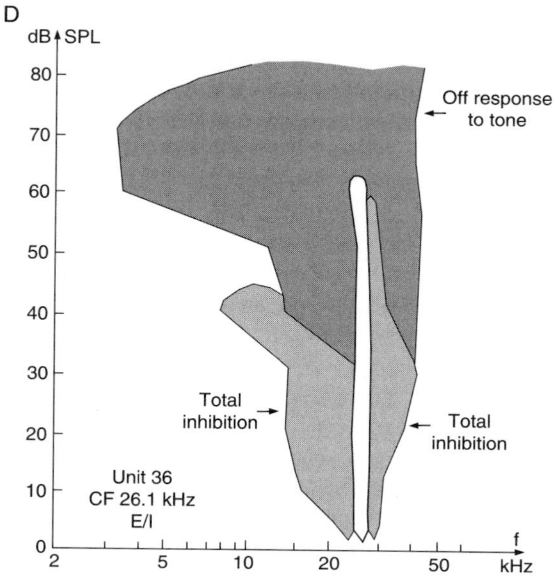

FIGURE 5.35 *(continued)*

3. Coding of the Waveform (Temporal Pattern) of Sounds in the Auditory Nerve

Auditory nerve fibers deliver a code of the time pattern of the vibration of the basilar membrane to the central nervous system which serves as the basis for the temporal hypothesis for frequency discrimination. Because each point along the basilar membrane acts as a bandpass filter that attenuates signals outside the frequency to which it is tuned, each individual nerve fiber codes the waveform of a filtered version of sounds.

The temporal representation of sound in the auditory nerve has been demonstrated in many studies[12,186] showing that the filtered version of the waveform of a sound is reproduced in the firing pattern of auditory nerve fibers. The representation of the periodicity of pure tones in the discharge pattern of auditory nerve fibers is known as *phase-locking* (Fig. 5.36). This means that the likelihood of the appearance of a discharge is greatest at a certain phase of the sound, but not all waves of the sound may elicit a nerve impulse. Such phase-locking in response to pure tones and broadband noise is most pronounced for low frequencies. Neurophysiologic studies in animals using recordings from single auditory nerve fibers have shown that phase-locking to the waveform of sounds decreases above a certain frequency. The temporal code of frequency may be regarded as an object representation of frequency compared with the place representation, which may be regarded as spatial information.

Although most studies of the temporal representation of sounds have used pure tones as stimuli, it has also been shown that the waveform of more complex sounds are reproduced in the discharge pattern of auditory nerve fibers, including combinations of two tones (Fig. 5.37),[186] noise,[146,148] and the waveform of

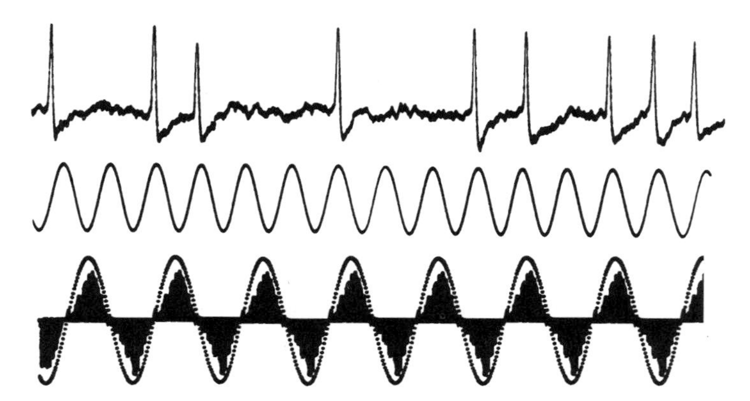

FIGURE 5.36 Illustration of phase-locking of the discharge in an auditory nerve fiber in response to a tone. (Adapted from Arthur, R.M., Pfeiffer, R.R., and Suga, N., *J. Physiol. (Lond.)*, 212: 593–609, 1971.)

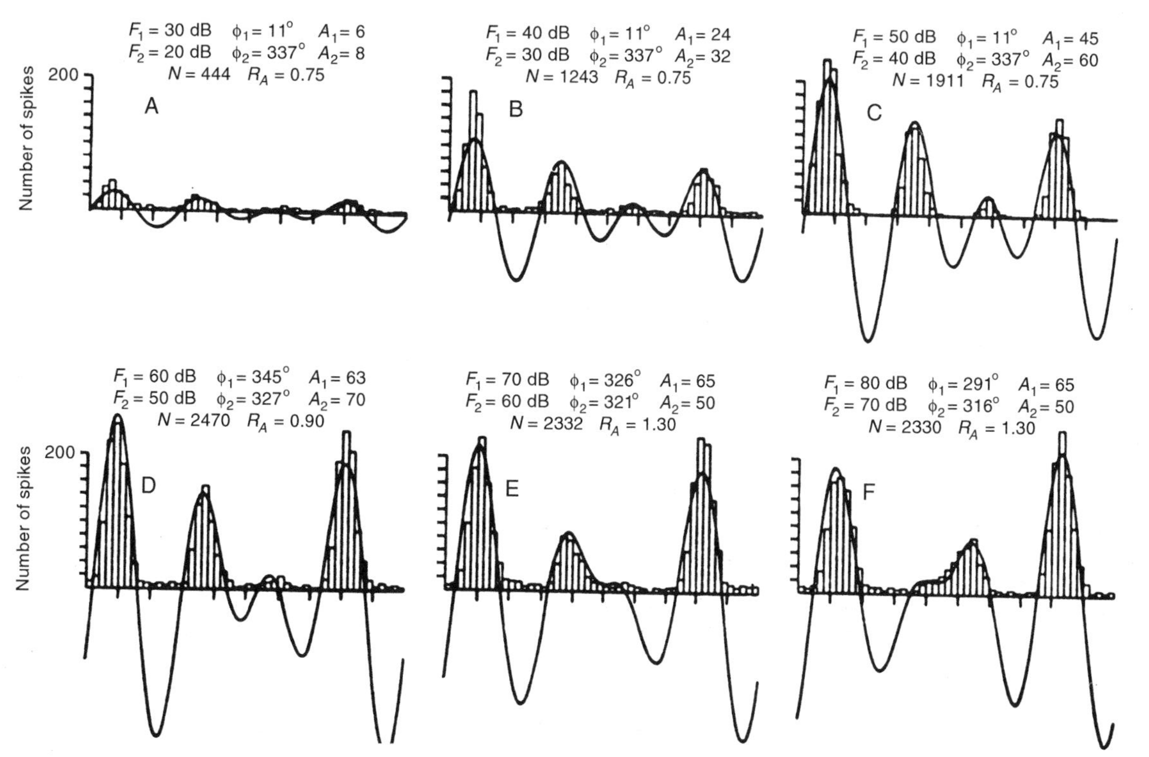

FIGURE 5.37 Illustration of phase-locking to combinations of two tones in an auditory nerve fiber of a squirrel monkey to stimulation with two tones simultaneously. (Adapted from Rose, J.E., Hind, J.E., Anderson, D.J., and Brugg, J.F., *J. Neurophysiol.* 34: 685–699, 1971.)

vowel sounds (Fig. 5.38).[256] The time patterns (waveform) of speech sounds are reproduced in the discharge pattern to an extent that makes it possible to recover information about the fundamental (voice) frequency as well as of formant frequencies from the discharge pattern of single auditory nerve fibers.

a. Preservation of the Temporal Code Through Neural Transmission

Accurate coding of the temporal pattern of sounds in the auditory nerve and preservation of the temporal code as the information ascends in the auditory pathways are essential to many basic auditory functions. It has been claimed that synaptic transmission degrades temporal coding (synaptic jitter) but it has also been shown that the convergence of many inputs to a single nerve cell, in fact, improves the precision of temporal coding[28] (see Chapter 3, page 119). Studies of binaural hearing have consistently found that the temporal pattern of sounds can be conducted with high accuracy through synaptic transmission in the nervous system. Psychoacoustic studies of the ability to determine the direction to a sound source (directional hearing) indicate that neural activity that ascends from the two ears must appear at the nerve cells in the superior olivary complex with errors (jitter) that are less than 10 μS (microseconds). This seems to be in conflict with some studies that show that phase-locking is degraded above a frequency of approximately 2.5 kHz. Such degradation would imply that preservation of timing shorter than the period length of a 2.5-kHz tone (400 μS) would be impaired. It is possible that convergence of many nerve fibers onto individual nerve cells in the cochlear nucleus may improve the accuracy of timing as a result of spatial averaging (see page 127, Chapter 3).[28,104,151] It is also possible that the way that temporal coding is determined experimentally provides a distorted picture of the representation of the temporal pattern of sounds in the nervous system.

The degradation of timing information in the auditory nerve can be represented by a low-pass filter of at least a third-order type.*[,240] While the slope of the attenuation of synchronization to pure tones is similar for several species there is a noticeable difference between the cut-off frequency for cat, guinea pig, and lizard. The cut-off frequency for cat is the highest (2.5 kHz),[101] and that of the guinea pig is 1.1 kHz.[166] The attenuation of phase-locking to tones decreases with the frequency at a rate of 100 dB/decade (30 dB/octave) above the cut-off frequency.[240] Weiss and his co-workers[112] found evidence that the attenuation of synchronization of auditory nerve impulses to stimulation with high-frequency sounds was due to three factors each responsible for 6-dB/octave (20-dB/decade) attenuation. Two of these three factors are calcium processes and the third factor is the membrane resistance and capacitance acting as a 6-dB/octave low-pass filter.[112] One of the calcium-dependant filters is an outcome of the relationship between the receptor potential and the calcium current, and the other

*A third-order, low-pass filter attenuates signals at a rate of 18 dB/octave.

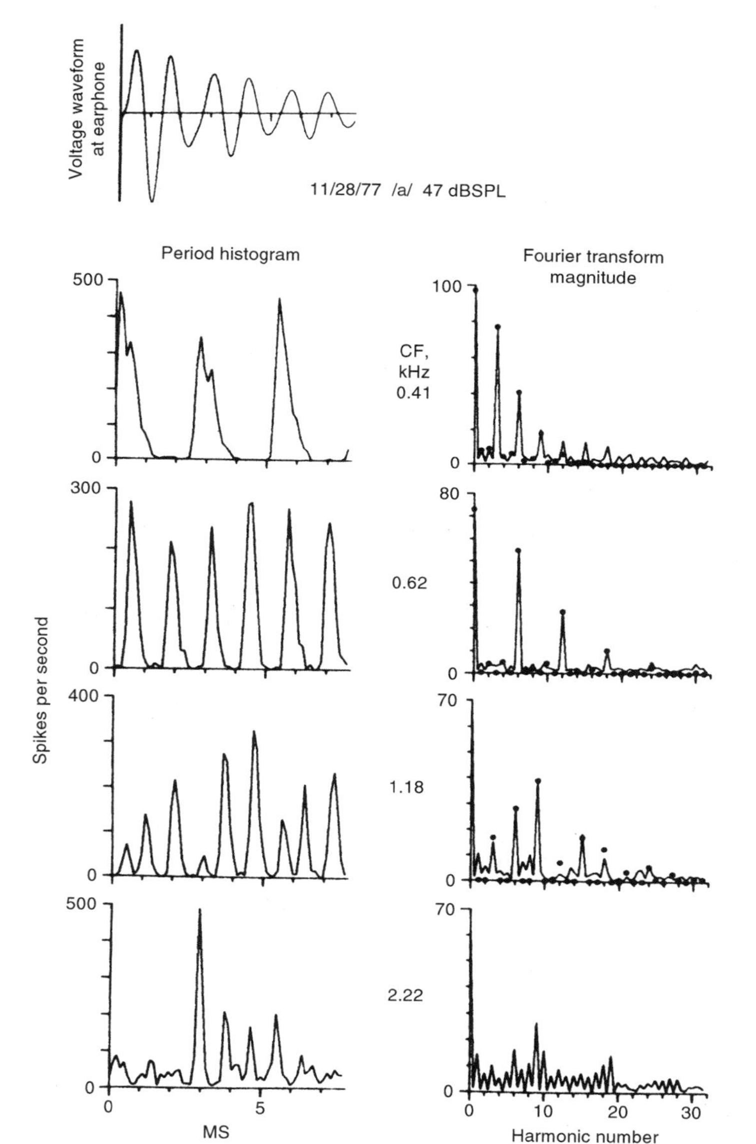

FIGURE 5.38 Illustration of phase-locking in the auditory nerve to the waveform of a synthetic vowel. The four left graphs are histograms of the response to a synthetic vowel from four different auditory nerve fibers tuned to approximately the same frequencies as the formant frequencies of the vowel. The right-hand graphs are the Fourier transforms of the histograms showing distribution of energy as a function of frequency. (Adapted from Young, E.D., and Sachs, M.B., *J. Acoust. Soc. Am.* 66: 1381–1403, 1979.)

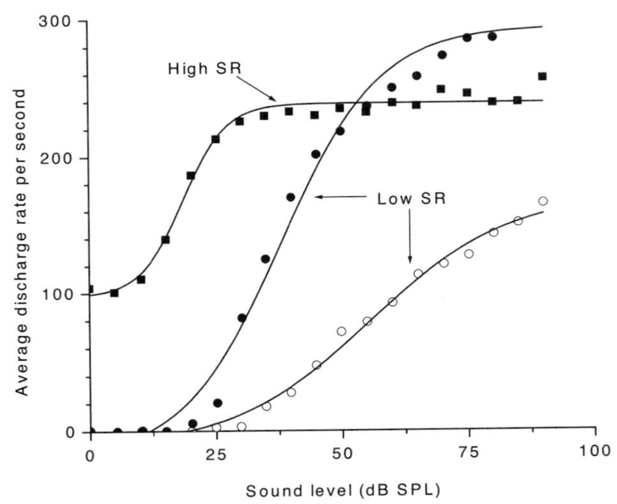

FIGURE 5.39 Stimulus response curves for three auditory nerve fibers that have different spontaneous firing rate. (Adapted from Müller *et al.*[161])

is due to the relationship between the calcium current and the calcium concentration.[112]

Determination of the effect of the frequency of stimulation on phase-locking of auditory nerve fibers is made by averaging the responses to many stimuli. That means that a variability in the timing of the response to individual stimuli will affect the results in the way it will reduce the phase-locking to high-frequency tones. The nervous system does not average the responses to many stimuli but instead averages the responses from many auditory nerve fibers, which has two important consequences: The variability in the transduction of successive stimuli does not affect the analysis performed by the nervous system, and the temporal coherence of the output of an array of many hair cells, about which little is known, will affect the analysis performed by the nervous system for extracting temporal information of sounds.

4. Coding of Sound Intensity in Auditory Nerve Fibers

The discharge rate in most single auditory nerve fibers that innervate the inner hair cells of the cochlea reach saturation at very low sound intensities, and in only a few nerve fibers does the discharge rate increase with the sound intensity over the range where an observer can discriminate the intensity of sounds (Fig. 5.39). This means that only a few auditory nerve fibers communicate the intensity of sounds over the entire audible range by their discharge rate. The dynamic range of auditory nerve fibers is related to their spontaneous firing rate. Fibers with a high spontaneous firing rate have a low threshold (high sensitivity) but small dynamic range for continuous sound stimulation. Fibers with a low spontaneous rate have higher thresholds and larger dynamic ranges. The discharge rate of some of the fibers that have a high threshold increases monotonically over the intensity range of normal hearing for stimulation with constant sounds (Fig. 5.39).[161,257] These high-threshold fibers would be able to communicate the intensity of sounds over a large range of sound intensities. It is not known what the innervation of hair cells is with respect to the organization of the fibers with a large dynamic range compared to those with a smaller dynamic range.

a. *Coding of Changes in Intensity and Frequency*

Understanding the neural coding of changes in the frequency and amplitude of sounds is important because natural sounds change both with regard to their overall intensity and with regard to the spectrum of such sounds. However, these aspects have been sparsely studied in the auditory nervous system and only a few studies have been published that have investigated the responses to sounds for which the frequency changes rapidly over large ranges. Fewer studies have concerned coding of sounds at more central parts of the ascending auditory pathways compared with those concerning the responses from more peripheral structures such as the auditory nerve and the cochlear nucleus.

Coding of Small Changes in Sound Intensity Coding of small changes in the intensity of sounds is important for understanding how the auditory system processes natural sounds. Small changes in sound intensity are generally reproduced in the discharge pattern of fibers and cells in the auditory nervous system over a large range of sound intensities. The neural coding of small changes in sound intensity can be studied by observing how the discharges of auditory nerve fibers change when the intensity of a sound, such as a tone, is increased or decreased stepwise by a small amount. However, a better method consists of studying the response to *amplitude-modulated tones* or *noise*. Such sounds are more similar to natural sounds and can be varied in controlled ways for experimentation.

Studies of the response to amplitude-modulated tones and noise use *period histograms* of the responses to show representations of the modulation waveforms in the discharge patterns. Period histograms show the average distribution of discharges over one period of the modulation. The degree of modulation of these histograms displayed as a function of the modulation frequency is known as a *modulation transfer function* (MTF). Typically, the stimuli in such experiments are tones or noises that are amplitude modulated by a sinusoidal waveform, but sounds that are modulated by noise have also been used in such experiments.

Recordings made from auditory nerve fibers in response to amplitude-modulated stimuli (Fig. 5.40) show that the modulation waveform is reproduced in the histograms of the responses to a degree that depends on the intensity of

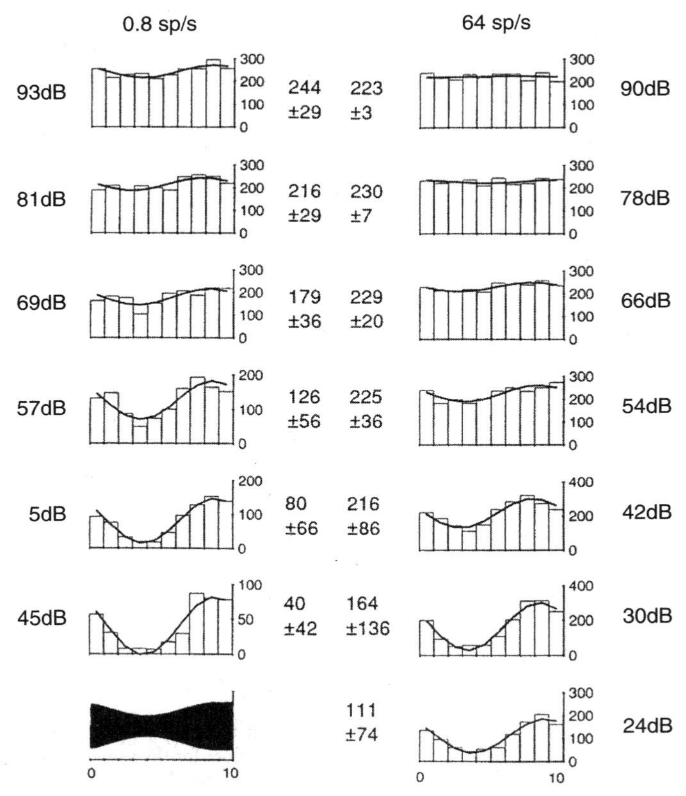

FIGURE 5.40 Period histograms of the responses from two different auditory nerve fibers in guinea pigs to amplitude modulated tones at different sound intensities. The two nerve fibers had different spontaneous activities (left column: 0.8 spikes/second; right column: 64 spikes/second). (Adapted from Cooper, N.P., Roberston, D., and Yates, G.K., *J. Neurophysiol.*, 70: 370–386, 1993.)

the sound, the frequency of the modulation, and the spontaneous rate of the fiber.[43,72,161] It is not known why nerve fibers with different rates of spontaneous activity respond differently to the modulation waveform of amplitude-modulated sounds (Fig. 5.40).[43]

The reproduction of the modulation waveform in the discharge pattern of fibers in the auditory nerve is a low-pass function (Fig. 5.41). The cut-off frequency for reproducing the modulation waveform of an amplitude-modulated tone is approximately 1 kHz in the cat (Fig. 5.41)[103] That is only about half of the cut-off frequency for phase-locking of pure tones. The reason for that difference is unknown. The reproduction of the modulation waveform in the discharge pattern of auditory nerve fibers varies little from fiber to fiber.

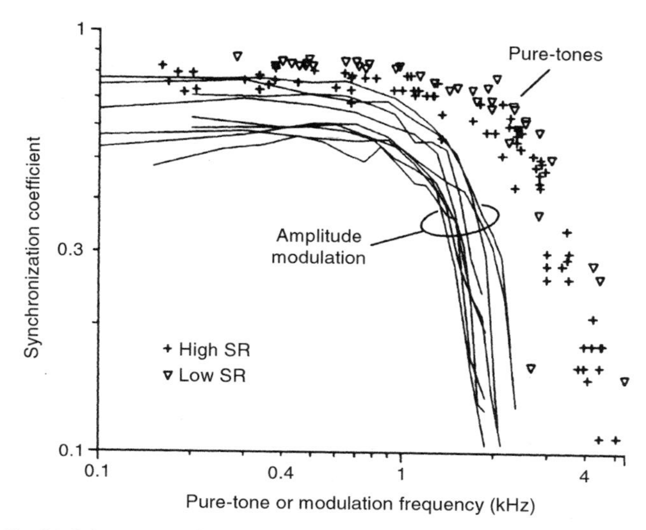

FIGURE 5.41 Modulation transfer function for auditory nerve fibers in a cat compared with phase-locking to pure tones of the auditory nerve fibers. (Adapted from Joris, P.X., and Yin, T.C.T., *J. Acoust. Soc. Am.* 91: 215–232, 1992.)

Response to Amplitude-Modulated Sounds in Nuclei of the Ascending Pathways The waveforms of amplitude-modulated tones and noise are reproduced in the discharge pattern of nerve cells of the nuclei throughout the ascending auditory pathways. In general, representation of the modulation waveform also varies between cells in different nuclei. For example, the upper limit of modulation frequencies to which neurons of the IC respond is lower than that of neurons in the cochlear nucleus. The response limit of cells in the primary cerebral cortex is even lower than that of the IC.

Cells in the cochlear nucleus respond better to the modulated waveform than do fibers of the auditory nerve.[69,70,144,260] The depth of the modulation of period histograms of the responses from cells in the cochlear nuclei may be as much as three times greater than the depth of the modulation of the histograms obtained from auditory nerve fibers in response to similar sounds. Furthermore, the range of sound intensities over which the modulation waveform is reproduced in the discharge pattern of cells in the cochlear nuclei is larger than it is for auditory nerve fibers.[69,70,139,141] While the ability to reproduce the modulation waveform of the discharge pattern of different auditory nerve fibers is similar, it varies among different types of cochlear nucleus cells.[69–71]

The differences between the response of auditory nerve fibers and cells in the cochlear nucleus may be explained by the effect of the convergence of many nerve fibers onto an individual nerve cell in the cochlear nucleus. The resulting spatial averaging, in addition to the enhancement of contrast caused by synaptic integration of excitatory and inhibitory input, may be responsible for the observed increased reproduction of small changes in the amplitudes of sounds.[28] This means that convergence of auditory nerve fibers onto cells in the cochlear nucleus increases the signal-to-noise ratio of the reproduction of amplitude modulation. This spatial averaging is similar to the signal averaging used to enhance evoked potentials with smaller amplitude than the background noise.[199] The response to the modulation waveform of tones that are presented in a background of noise is likewise enhanced in the response of cells in the cochlear nucleus.[72]

Responses from cells in the IC of urethane-anesthetized rats to amplitude-modulated tones have shown a more complex pattern than is seen in the cochlear nucleus. The modulation transfer functions were mainly bandpass type with a peak below 120 Hz,[177] thus being lower than what was found in the cochlear nucleus. This trend to respond best to lower modulation frequencies at successively higher levels of the ascending auditory pathways continues in the MGB,[46,119] and the AI cortex in anesthetized animals (cats) respond well to amplitude-modulated sounds with a low modulation frequency (Fig. 5.42).[199] It was shown many years ago that some neurons in the auditory cortex respond only to frequency-modulated sounds.[241] Later, systematic studies have shown that neurons in the AI area of the cerebral cortex respond best to tone bursts presented at a rate of 8 to 14 bursts per second.[46,199] Such sounds may be regarded as amplitude-modulated sounds where the degree of modulation is 100%. The upper

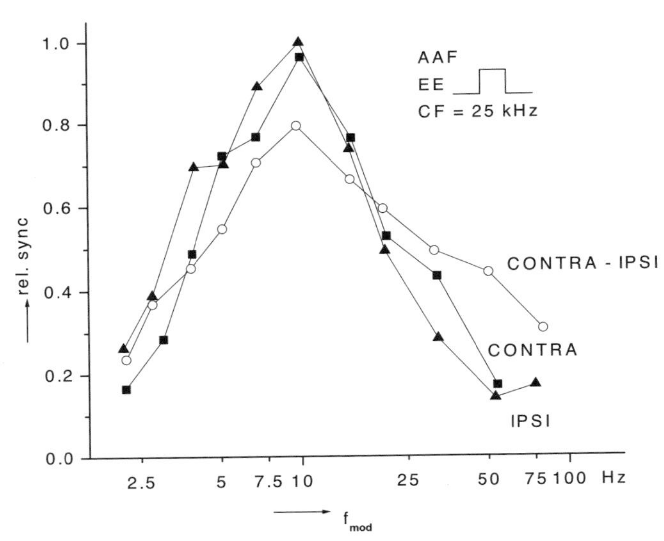

FIGURE 5.42 Modulation transfer functions obtained in the anterior auditory field (AAF) of the cerebral cortex of a cat in response to amplitude-modulated tones presented at the contralateral ear, ipsilateral ear, and bilaterally.[199]

cut-off frequency regarding reproduction of the modulation of amplitude modulated sound is lower than that of neurons in the auditory nerve and the cochlear nucleus. (For a recent review of coding in the IC, see Caseeday *et al.*[35])

Studies of coding in cortical neurons are hampered by the effects of anesthesia. Different types of anesthesia have different effects on the responses. The effect of anesthesia can change the coding of sounds from a temporally dominated pattern to a rate pattern. (For a recent review on feature detection in the auditory cortex, see Nelken.[162])

b. Response to Rapid Changes of the Frequency of Sounds in Ascending Pathways

The spectrum of natural sounds changes more or less rapidly, and these changes, together with changes in the intensity, carry most of the information that is of importance for discrimination of complex sounds. Therefore, to understand how the auditory system works, we need to know how the auditory system responds to rapid changes in sounds.

The echolocating sound of flying bats is probably the clearest example of natural sounds for which the frequency varies rapidly (frequency-modulated sounds). The auditory system of bats has been studied extensively. Bats have a sophisticated auditory system that serves these animals in several ways. The auditory system is the prime navigational system for the bat, which can fly without visual control. That means that hearing in the flying bat supports tasks that are performed by the visual system in many other animals. At the same time, bats use their auditory system for communication.

Suga[221] demonstrated that the responses to frequency-modulated sound are enhanced in the auditory nervous system.[221] This was probably the first clear demonstration of a preference of neurons in the auditory cortex for sound with rapidly changing frequency. Later, other investigators studied the response patterns of frequency-modulated sounds in neurons of the auditory nerve,[214] cochlear nucleus,[138,142,214] and IC[177,178] and found that the responses to tones with rapidly changing frequency are enhanced in the cochlear nucleus and in the IC.

When the frequency of a tone is changed slowly over a range that covers the responsive area of an auditory nerve fiber, the number of nerve impulses that are elicited varies with the frequency of the tone in accordance with the responsive area of the fiber from which the recording is made. The same is the case for cells in the cochlear nucleus and other nuclei of the ascending auditory pathways.

When the frequency of such sounds is changed rapidly, the frequency range within which a neuron in the cochlear nucleus responds becomes much narrower (Fig. 5.43).[138,142] This means that the receptive field with regard to the frequency of sounds narrows when the frequency of a tone is changed rapidly. The responses from auditory nerve fibers to similar sounds, however, show little

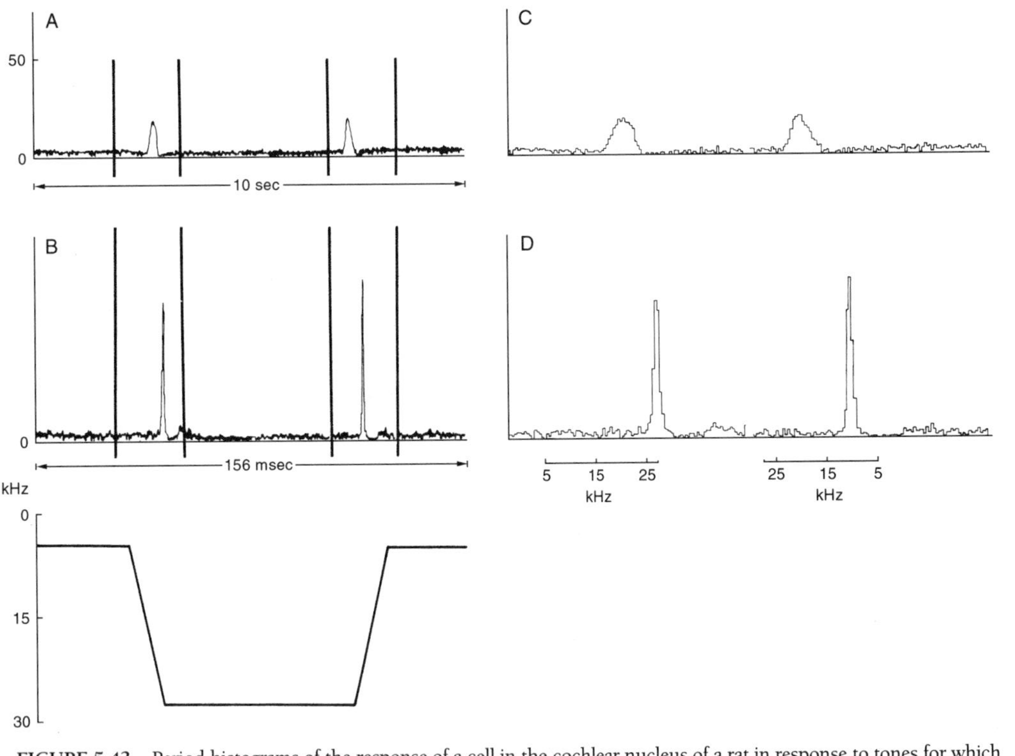

FIGURE 5.43 Period histograms of the response of a cell in the cochlear nucleus of a rat in response to tones for which the frequency was varied up and down between 5 kHz and 25 kHz at different rates. The top histograms were obtained when the frequency of the tone was varied at a slow rate (20 kHz in 1.5 sec), and the lower histogram was obtained when the frequency was changed at a high rate (20 kHz in 23 msec). (Adapted form Møller, A.R., *J. Acoust. Soc. Am.* 55: 631–640, 1974.)

change with the rate of change in frequency.[214] This means that the response to sounds with rapidly changing frequency has been transformed in the cochlear nucleus. The response to such sound by cells in the cochlear nucleus may be regarded as a form of enhancement of the response to changes in frequency of sounds, which is likely to be important for perception of natural sounds such as speech sounds.

5. The Importance of Temporal and Place Code for Speech Discrimination

Frequency discrimination is important for auditory discrimination and plays an important role for discrimination of complex sounds such as speech. The frequency or spectrum of a sound can be derived equally well from the temporal code of auditory nerve fibers (which is the basis for the temporal hypothesis of frequency discrimination) as from the spatial coding (which is the basis for the place hypothesis of frequency discrimination). As shown above, both the temporal and the spectral properties of sounds are represented in the discharge pattern of auditory nerve fibers. This means that these two properties of sound are available to the auditory nervous system. (This dual representation of frequency also means that frequency information is represented both as a spatial code and as an object code.)

However, it is not known which one of these two principles is used for frequency discrimination of complex sounds such as speech sounds or if both are important. Neurons in the ascending auditory pathways are organized according to the frequency to which they respond best (tonotopic organization), which has contributed to the credibility of the place hypothesis for frequency discrimination. Recent studies, however, have provided an increasing amount of evidence that favors the importance of the temporal hypothesis for frequency discrimination. (For more discussion about the importance of the place and temporal hypotheses for frequency discrimination, see Møller.[151,152])

Because the temporal pattern and spectrum of sounds cannot be manipulated independently, it is difficult to design experiments that can provide information about the relative importance of the spectral representation or the temporal representations of sounds for discrimination of complex sounds. Therefore, other methods have been used to study which one (or both) of these two representations is important for discrimination of frequency. For example, it has been shown that speech changed so that most of spectral characteristics are removed*,[205] loses little of its intelligibility. The results of that study emphasize

*In these experiments, the spectrum of the speech was divided into one to four frequency bands, and the energy in each band was used to modulate broadband noise. The final spectrum of the sounds was nearly flat and little affected by the speech sounds, which indicates that the spectrum of these sounds carried little information about the speech.

the importance of features other than the spectrum for discrimination of speech, thus emphasizing the temporal coding.

Other experiments have focused on the robustness of these two representations of frequency in their coding in the nervous system and it was found that in general the place representation of frequency lacks robustness with regard to being independent of the intensity of sounds.[146,148,269] This is an obstacle for use of the place principle for frequency discrimination because it is known that our perception of frequency is very little affected by the intensity of sounds.[151,220,269]

The assumed deterioration of the temporal representation for high-frequency sounds in neural transmission has been considered an obstacle in regard to the CNS using the temporal principle as a basis for frequency discrimination. However, convergence and (spatial synaptic averaging) may eliminate the obstacle for the use of the temporal preservation of coding of sounds[28] and thereby make it possible for the CNS to use temporal coding of frequency even at high frequencies.

6. Tuning to Features of Sounds Other Than Frequency

Some neurons in the inferior colliculus, in addition to being tuned to the frequency of sounds, also respond selectively to the duration of the sound.[34] This was determined in studies of the flying bat, which has been a frequent object for studies of neural coding of sounds related to echolocalization. Neurons in the bat's auditory system are tuned to the delay between two sounds. For the flying bat, the delay between two sounds would represent the delay of the echo from an object, which is a measure of the distance to the object.

7. Maps of Sound Properties

The cortical representation of sounds has often been illustrated by maps that show the anatomical location of neurons with different response properties. The most obvious spatial representation is that of the frequency to which neurons respond best (tonotopic organization). Maps that show the anatomical locations of neurons according to the frequencies of their lowest thresholds have been demonstrated for all major nuclei of the ascending pathway as well as the different cortical regions.

Suga[222] and his co-workers have demonstrated that other features of sounds that are also regarded as important for echolocation are represented in maps of the surface of the auditory cortex of the flying bat (Fig. 3.33). These investigators found that the intervals between sounds are clearly represented in cortical maps, as are the responses to the contents of harmonics in the echolocating sounds.

The bats, which often fly in large swarms, must be able to identify the echoes of their own emitted sounds in an environment of hundreds and perhaps thousands of similar sounds from other bats. Thus, the flying bat must have a problem similar to the giant "cocktail-party problem" that people experience when attempting to listen to a particular speaker in an environment where many people are talking. The harmonics of the sounds that are emitted by the flying bats are assumed to be important for identification of a bat's own echo in the environment in which many other bats are sending out sounds. The responses to these harmonics are well represented in cortical maps of responses to the bats' echolocating sounds

8. Bilateral Representation in the Classical Pathways

Many neurons in the central auditory system receive input from both ears. Bilateral input serves several functions, such as being the basis for directional hearing and facilitating discrimination of a specific sound in an environment where many other sounds are present.

Binaural interaction is prominent in more centrally located nuclei such as the ICC and MGB and in the cerebral cortices. Binaural representation of sounds in the auditory cortex has been demonstrated in studies where recordings were made from single cells in the AI cortex of cats.[26] Binaural interaction often appears in the form of alternating bands of neurons that respond with excitation from the ipsilateral ear and with inhibition from the contralateral ear (EI neurons) as well as bands where neurons respond with excitation from both ears (EE neurons).[131] Contralateral input excites almost all neurons in the (cat) ICC whereas less than 40% are excited from the ipsilateral ear.[3] Normally, approximately 23% of cells in the gerbil ICC respond with excitation to ipsilateral input.[116] Ipsilateral stimuli often evoke a late inhibitory influence on contralateral evoked responses.[7] In the ventral MGB, approximately 60% of cells respond to ipsilateral stimulation.[2]

The effect of deafening of one ear during development has been shown to elevate the threshold for binaural interaction from the affected ear by as much as 50 dB, leaving other properties of the responses little changed.[25] This indicates that a childhood hearing deficit can affect binaural hearing later, even when the hearing threshold has been restored.

Several anatomical connections may furnish bilateral representation, and it is not entirely clear where this binaural representation is established. The most peripheral binaural interaction occurs in the cochlear nucleus.[129] Some neurons in the superior olivary complex receive input from both ears and send axons along the lateral lemniscus to the IC, providing bilateral input to the ICC. The extensive connections at the midbrain level are probably responsible for a large portion of the observed binaural interaction. The commissure of the IC is a large fiber bundle that connects the two ICC, but little is known about it. The connections between the LL and the IC via the nuclei of the lateral lemniscus are mainly inhibitory for the DNLL. The commissure of Probst connects one of the nuclei of the lateral lemniscus (DNLL) with its counterpart on the other side, while some fibers connect to the contralateral ICC. Another possibility for the bilateral representation is by a direct and uncrossed connection between the cochlear nucleus and the IC (see page 307).

Excitatory input to the contralateral ICC arrives slightly earlier than input from the ipsilateral ear, indicating that ipsilateral input to the ICC is interrupted by one or more synapses between the cochlear nucleus and the ICC, most likely by nuclei in the SOC and/or the nuclei of the LL, while there are uninterrupted connections from the contralateral CN to the ICC.

There is ample evidence that sounds from one ear are represented in the cerebral cortex of both sides. Studies in humans have shown that sound in one ear gives rise to neural activity in the cerebral cortex on both sides.[36] Injuries (from tumors, strokes, etc.) of the cerebral cortex of one side in humans do not cause a noticeable difference in hearing thresholds of the two ears and speech discrimination is little affected. Only discrimination of *low-redundant speech* is noticeably reduced in the ear contralateral to the lesion.[17,118] This is despite the fact that the main auditory pathways are assumed to be dominant on one side, the side opposite the activated ear. This means that information from each ear must reach both auditory cortices but in different ways and at different degrees of processing.

The anatomical basis for the bilateral representation of sounds in the cerebral cortex may be the ample connections between the primary auditory cortices of the two sides through the *corpus callosum*. The corpus callosum connects the auditory cortices[39] as it does for the other sensory cortices (see Chapter 3). These connections have been assumed to be the more important for the bilateral cortical representation of the ears than the subcortical connections. There are no known connections between the MGBs of the two sides, but other subcortical connections between the two sides may contribute to the bilateral representation at the cortical level.

a. Directional Hearing

Directional hearing, which provides information about the direction to a sound source, is also the basis for the perception of (auditory) space. Directional information is regarded as spatial information, but it does not originate from projection of a receptor surface onto a population of neurons as a place coding of frequency such as in the visual and somatosensory systems. Rather, it is a result of manipulation and comparison of information from the two ears, which is why it is regarded as a "computational" representation of directional information. The directional information that provides a perception of (auditory) space is three dimensional and therefore requires information from the two ears to be projected onto a three-dimensional array of neurons in order to be represented in the nervous system.

Animals such as cats that are predatory animals have well-developed directional hearing. The intensity of the mid-frequency regions of a sound varies as a function of the direction to the sound source, and spectral variations are important for information about azimuth and elevation, while time differences between the arrival of a sound at the two ears determine the azimuth. There is evidence that neurons in the MSO act as coincidence detectors and, together with axons of different lengths that act as variable delay lines (see page 307) become selective regarding the interaural time interval, thus the azimuth to a sound source.[80,254] This is according to the hypothesis of Jeffress[97] for directional hearing (see also Joris *et al.*[104]). The DCN has been reported to be associated with directional hearing in cats and lesions in the DCN and dorsal stria impair directional hearing. Neurons of the DCN project to specific neurons in the ICC (type O).

While humans have the capability of directional hearing, many of the tasks that humans perform with their auditory system can be done with one ear. This is not true for many animals; for instance, binaural hearing is absolutely essential for the bat's echolocalization. The barn owl makes extensive use of directional hearing to identify prey (mice). (For a recent review of binaural hearing, see, Yin.[255])

9. The Physiologic Basis for the Precedence Effect

The precedence effect[74,231,263] is important for directional hearing in rooms with sound-reflecting walls. As described in Chapter 1, it can be demonstrated by placing two loudspeakers in front of an observer and at different azimuths (for instance 60 degrees).[15] Sound emitted by the two speakers placed in different locations (different azimuths) will be perceived as coming from the speaker that emits the sound. If sounds are presented to both speakers with a short latency (below 1 or 2 msec), the sound will be perceived as coming from a location in between the two speakers. If the interval between the two sounds is slightly longer (3–30 msec), the sounds will be perceived as coming from the speaker that emits the first sound (see Chapter 1). When the interval between the sounds emitted by the two sources is increased beyond 30 msec, the sounds are perceived as coming from each speaker, one after the other. The precedence effect is important for determining the direction to sound sources in a room with sound-reflecting walls.

It has been shown that the response to sounds from single neurons at different levels of the ascending auditory pathways is suppressed by a preceding sound.

The effect is larger at more central structures. While the time of 50% recovery of the responses to the second sound is approximately 2 msec in the auditory nerve, it is 7 msec in the inferior colliculus and at least 20 msec in cells of the auditory cortex; some neurons have taken as long as 300 msec for 50% recovery.[65]

Why is the precedence effect so important to animals living in surroundings with little sound reflection (a free sound field)? It may be an epiphenomenon that is not really used by animals in their normal environment, or perhaps we have not yet asked the appropriate questions.

10. Hemispherical Dominance

It is known that the two hemispheres process sounds in slightly different ways. Sound guided to the left ear in right-handed individuals offers a small advantage with regard to discrimination of complex sounds (left ear advantage, LEA) but not for discrimination of pure tones.[88,93] Conversely, the left hemisphere is dominant in regard to speech discrimination, and speech presented to the right ear is easier to discriminate than speech presented to the left ear (*right ear advantage*, REA).[88,93] These hemispheric dominances are related to handedness, and all right-handed individuals and more than half of left-handed individuals have the above-described hemispheric dominances. Approximately 10% of the population is left handed, which means that more than 95% of all individuals have these hemispherical dominances.

The hemispheric dominances increase with age, giving elderly individuals a greater REA for speech and greater LEA for complex sounds compared with young individuals.[99] This change in interaural asymmetry is reflected in event-related potentials.[100]

11. Stream Segregation

The term *stream segregation* was originally used to describe how objective ("what") and spatial ("where") information is processed in different populations of neurons in the visual system (see Chapter 3, page 111).[228] Only recently have similar separations of processing of information been shown to occur in the auditory system; other qualities of sensory stimuli are separated into streams that are processed by different populations of neurons.[106,175] The basis for stream segregation in the auditory system is the anatomical arrangement of the cortical projection of the ventral MGB, which in the monkey projects in parallel to two areas of the cerebral cortex, usually a rostral area on the *supertemporal plane* as well as the AI area.[173,175] The *caudomedial cortical area* only receives input indirectly from the MGB via the AI. The neurons in this area are often responding to the spatial location of a complex sound and thus the spatial properties of sounds. Neurons in the lateral surface of the superior temporal gyrus respond best to complex sounds such as species-specific communication sounds, which means that these neurons code different types of information. Experimental evidence of stream segregation of auditory information has been demonstrated in studies of monkeys,[172,175,185] which have shown that different populations of neurons respond to different kinds of sounds (complex sounds versus pure tones).[227]

Different processing of echolocation sounds and communication sounds of the bat and sounds emitted by prey are other examples of separation of the processing of different kinds of information. Fuzessery and co-workers[176] found that the same neurons in the auditory cerebral cortex of the flying bat processed both low-frequency sounds from passive listening and high-frequency sounds used in echolocalization for general orientation. These two streams of information were separated at the level of the inferior colliculus and processed in parallel up to the cortex, where the two streams joined again. Some ICC units thus represent functionally segregated pathways that play complementary roles in the processing of auditory information. The different cortical representations of features that are important for echolocation in bats are also a form of stream segregation where different types of information are processed in different populations of neurons.

The frequency representation according to the place principle may be regarded as a spatial representation of frequency, and the temporal representation of frequency may be regarded as an object representation. These two neural representations of frequency may be other examples of stream segregation, and it may be assumed that the place coding of frequency is processed in a different population of neurons than the temporal coding of frequency.

Yet another form of stream segregation would concern a separation of processing of the temporal and place coding of sounds as a basis of frequency discrimination. The temporal coding may be regarded as object ("what") information, whereas the place coding of the spectrum of sound may be regarded as spatial representation ("where"). This would be similar to the representation of the surface of the body or the representation of the visual field. The spatial coding of frequency is one dimensional, while coding of the visual fields and the body surface is two dimensional. Direct experimental evidence for that kind of stream segregation is sparse. Indications for such segregation of temporal and spectral information, however, have been presented in studies showing that temporal and spectral information is processed in separate populations of neurons in the midbrain, where some neurons in the VNLL in the bat have been found to specialize in temporal coding.[44]

These neurons are broadly tuned and have a poorly defined spectral selectivity, whereas neurons in other parts of these nuclei are sharply tuned with regard to the frequency of sounds and therefore code the spectrum of sounds according to the place principle. The former type of neurons may be regarded as processors of temporal information (object coding), whereas the latter type of neurons that code the spectrum by place represent a form of spatial coding.

Some neurons in the cochlear nucleus, which were also broadly tuned, responded with precisely timed firings, whereas other neurons fired with temporal variations similar to the fibers of the auditory nerve.[137] Other studies have demonstrated that the precision of timing of the discharges varies among the different types of neurons in the cochlear nucleus and it has been confirmed that some cells fire with less temporal dispersion than do auditory nerve fibers.[104]

12. Interpretation of Experimental Results Regarding Neural Coding of Sounds

Most of our knowledge about the coding of sounds in the auditory cortex comes from experiments in animals. Because the cortex of these animals is different anatomically from that of humans it is difficult to apply results obtained in animals to humans. Also, because most animal studies have been performed on subjects under anesthesia, the results may only be partially applicable to the awake animal. Very little physiological data are available from the human auditory cortex so it is not known in what way the anatomical differences are reflected in function.[90] The description of the anatomy and physiology of the auditory cortex above was based on data from animals, mostly cats. Keep in mind that the physiologic data are mostly obtained in anesthetized animals, which further complicates comparisons with the function of the auditory cortex in awake humans.

C. Descending Systems

As in other sensory systems, the auditory pathways include extensive descending pathways. The function of descending pathways has been studied to a lesser degree than the ascending systems, except for the olivocochlear bundle, which has been studied extensively. Considerable experimental evidence suggests that activity in the olivocochlear efferent fibers that descend from the nuclei of the superior olivary complex and terminate on hair cells in the cochlea can affect the flow of information from the inner hair cells into the auditory nerve. Experimental evidence of the functions of the more central portions of this vast descending system is scant.

Electrophysiologic studies have shown that sound can evoke activity in single fibers of the olivocochlear bundle.[64,182] There is evidence that electrical stimulation of the OCB can affect the discharge pattern in single auditory nerve fibers.[84,242] and that activity in the COCB can affect the mechanical properties of the outer hair cells and thereby alter the micromechanical properties of the cochlea. Most of these studies employed electrical stimulation of the OCB at the floor of the fourth ventricle,[160] but some studies have shown that natural activation of the COCB (from contralateral sound stimulation) can affect cochlear function through its effect on the outer hair cells[41] and the discharge pattern of auditory nerve fibers.[235,236] This means that descending neural activity can change the processing of sounds that occurs in the cochlea before the excitation of inner hair cells. The uncrossed olivocochlear bundle primarily modulates the afferent activity in the nerve fibers that innervate inner hair cells.[234]

Few studies have been published regarding the function of the more central descending pathways of the auditory nervous system.[259] For example, little is known about the function of the abundant descending tracts from the auditory cortex that terminate in the thalamus. One study has shown that (partial) abolition of input to that system from the cortex can change the frequency tuning of cells in the MGB as well as in the ICC (see Chapter 3, page 143).[259] Other studies have shown that cortical activity can influence the activity in the MGB.[192]

IV. THE NONCLASSICAL ASCENDING PATHWAYS

Much less is known about the nonclassical ascending auditory pathways compared to the classical pathways. The anatomy has been less studied, and only a few studies have concerned the function of the nonclassical pathways. Consequently, both the anatomy and physiology of the nonclassical auditory pathways are poorly understood. Like the nonclassical pathways of other sensory systems, the nonclassical auditory pathways are assumed to have evolved earlier than the classical pathways. Therefore, it may be assumed that these systems have less capability than the classical ascending system to process sounds that are the basis for the fine discrimination of sounds that higher mammals possess.

A. Anatomy of the Nonclassical Ascending Pathway

The anatomy of the nonclassical auditory pathways is different from the classical pathways in several ways. The nonclassical auditory system receives its input from the classical pathways at several levels. The *dorsal cortex* (DC) and the *external nucleus* (ICX) of the inferior colliculus belong to the nonclassical pathways, and these nuclei receive input from the ICC that belongs to the classical auditory pathways (Fig. 5.44).[5,7,81,224] There are also inputs to the nonclassical pathways from nuclei at lower levels of the classical ascending auditory pathways, such as the cochlear nucleus.[96,211]

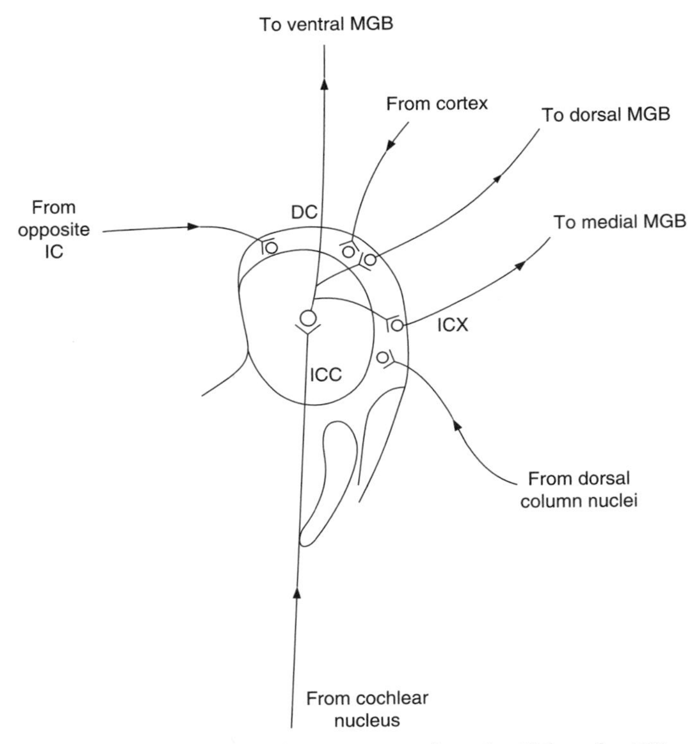

FIGURE 5.44 Schematic drawing of the connections from the ICC to the ICX and DC and connections from these nuclei to other structures. Also shown are efferent input from the cerebral cortex and connections to the DC and ICX from the somatosensory system (dorsal column nuclei) and from the opposite IC.

While the pathways of the nonclassical auditory systems from the inferior colliculus project to the ventral portion of the auditory thalamic nuclei (MGB), the neurons of the ICX and DC project to the *dorsal and medial parts* of the MGB. The neurons of the DC project to the dorsal parts of the MGB (Fig. 5.45 D) which also receive input from the neurons of the ICX, and neurons of the ICX project to the medial (or magnocellular) division of the MGB. The lateral portion of the posterior thalamus (PO) also receives auditory input (mainly from the ICC).[94] Most connections have reciprocal descending connections. There are uninterrupted descending connections from the auditory cortex to the DC and ICX (not shown in Fig. 5.45).

It has recently been shown that there are uninterrupted connections from neurons in the dorsal cochlear nucleus to the medial portions of the MGB.[126]

Neurons of the dorsal portion of the thalamus project to the AII part of the auditory cortex, and neurons of the medial part of the thalamus project to the

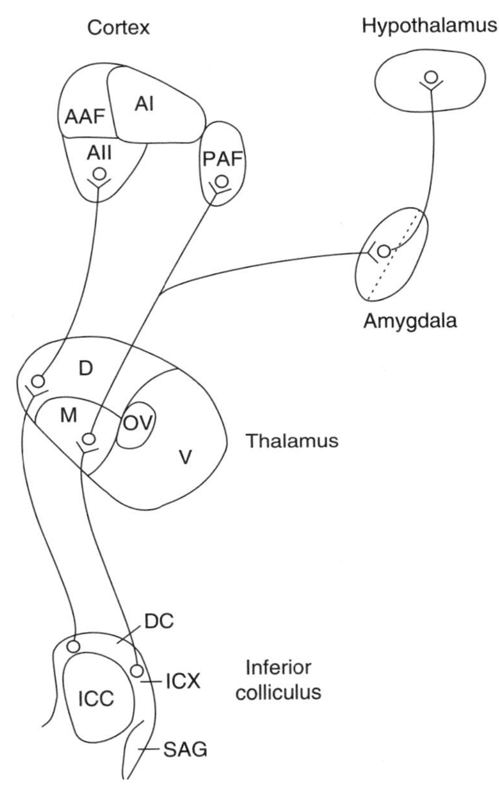

FIGURE 5.45 Schematic drawing of the ascending connections from the ICX and DC to the thalamic nuclei (MGB) and some of their cortical radiations. M. medial (or magnocellular) division of MGB; D, dorsal division; V, ventral division; OV, ovoid part of the MGB. Connections from the MGB to auditory cortical areas and to the basolateral nuclei of the amygdala are also shown.

PAF part of the cerebral cortex. Some neurons in the medial division of the MGB also project to the AI and AAF. These areas receive polysensory input (auditory, somatosensory, and visual).[37]

1. Nonclassical pathways receive input from nonauditory structures

It is an important feature of the nonclassical auditory system that it receives input from other sensory systems, most notably from the somatosensory system (Fig. 5.44).[4,5,7,96,211,239] Studies have shown anatomical connections between the nonclassical auditory pathways and the somatosensory pathways at several levels, including the cochlear nucleus and the IC. It is believed that the most import connections from the somatosensory system are those that terminate in the inferior colliculus where the ICX receives somatosensory input from the dorsal column nuclei,[4,5,7] but there are also connections from the somatosensory system (dorsal column nuclei) to the dorsal cochlear nucleus;[229,239] more recently, it has been shown that there are connections from the (sensory) trigeminal nucleus to the cochlear nucleus and the superior olivary complex.[96,211] The connections from the dorsal column nuclei to the inferior colliculus (ICX and DC) are mostly from neurons that innervate the forelegs and paws[7] (Fig. 5.47).

The medial division of the MGB also receives input from the visual and somatosensory systems and the vestibular system.[16] The *suprageniculate thalamic*

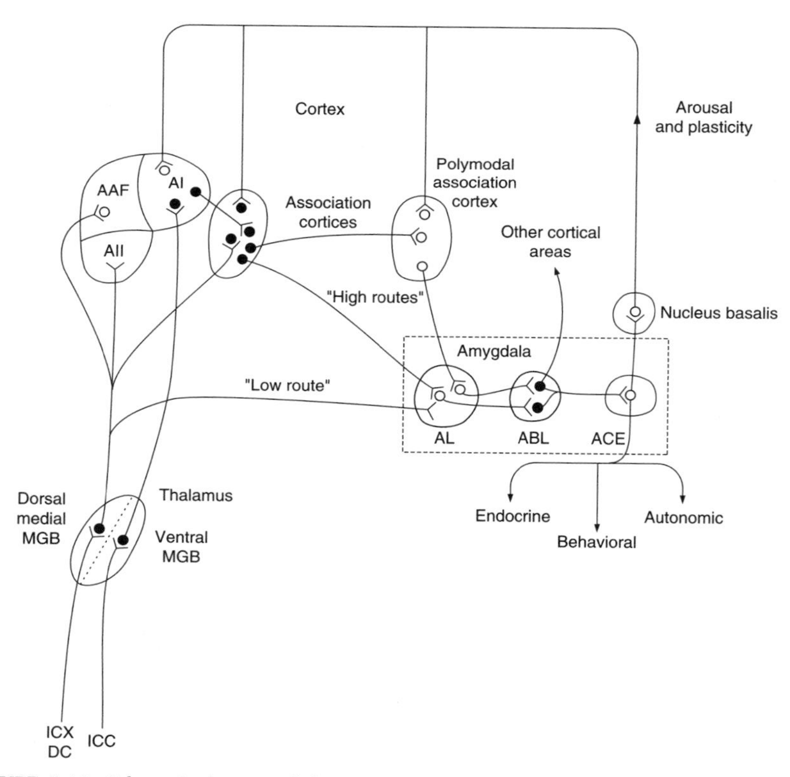

FIGURE 5.46 Schematic drawing of the connections between the classical and the nonclassical auditory systems and the amygdala showing both the "high route" and the "low route." Connections between the nuclei of the amygdala and other CNS structures are also shown. (From Aitkin, L.M., *The Auditory Midbrain, Structure and Function in the Central Auditory Pathway*. Clifton, New Jersey: Humana Press, 1986. With permission.)

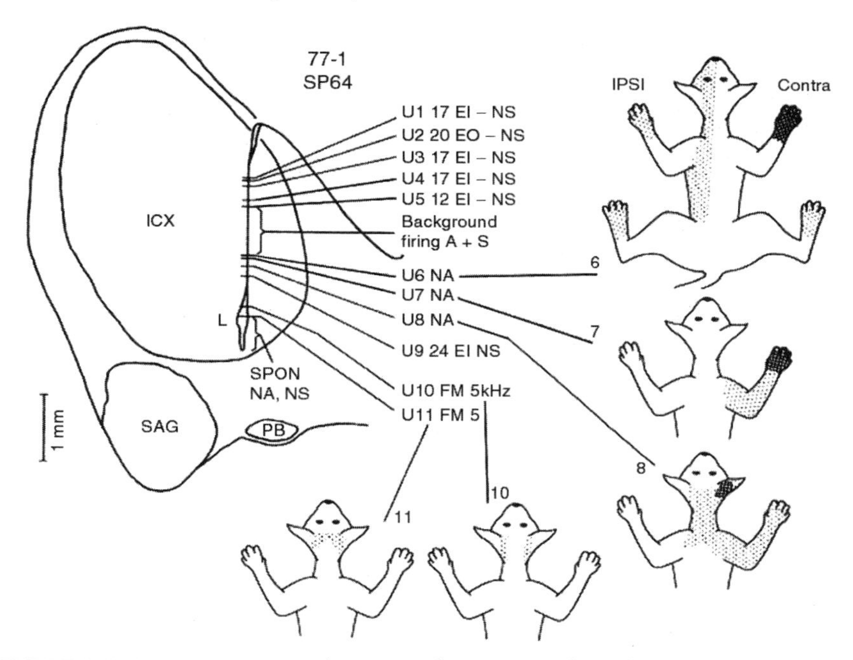

FIGURE 5.47 Response properties of neurons in the ICX to sound stimulation and to stimulation of the somatosensory system. (From Aitkin, L.M., *The Auditory Midbrain, Structure and Function in the Central Auditory Pathway*. Clifton, NJ: Humana Press, 1986. With permission.)

nucleus, which projects to the *insular cortex*, seems to receive input from the brainstem structures. The *caudodorsal division* of the MGB receives input from the *nucleus sagulum* (SG) and the *pericentral nucleus of the IC* (ICP), which also sends axons to the ventrolateral nucleus of the MGB. The *thalamic reticular nucleus*, which modulates sensory input (other thalamic regions), receives its input from the fibers that leave the LV of the MGB.[189,212]

Referring to the diversity of the nonclassical ascending auditory pathways and the fact that some parts receive input that is not only auditory but also of other modalities, some investigators[190] distinguish between two different nonclassical auditory systems, namely *diffuse* (non-tonotopic) and *polysensory*.

2. Nonclassical Auditory Systems Project to Nonauditory Regions

The fact that the nonclassical auditory pathway uses the dorsal and medial portion of the thalamus (as other nonclassical sensory pathways) is important in several ways, as has been discussed in Chapter 3. One such important feature is that connections from the dorsal and medial thalamus provide a subcortical route to the basolateral nucleus of the amygdala via its lateral nucleus (see pages 91, 141, and 156). That path (the low route[121]) supplies structures of the limbic system with information that has undergone little processing and is probably little influenced by extrinsic as well as intrinsic neural activity because it originated in a subcortical nucleus.

The recent findings of uninterrupted connections between neurons in the dorsal cochlear nucleus and the medial nucleus of the MGB[126] provide an even shorter (and probably faster) route for auditory information to reach the basolateral nucleus of the amygdala.

Auditory information can also reach limbic structures through the primary auditory cortex and association cortices (high route).[21,121] This route communicates highly processed information to the amygdala, compared with the low route, which communicates faster, carrying less possessed information to limbic structures (see also Chapter 3).

The anatomy of the nonclassical pathways and their connections to other structures are incompletely known, and different authors have described the anatomical organization of this part of the auditory nervous system in different ways. One reason for that is that much fewer investigators have studied the nonclassical pathways than have studied the classical pathways.

3. Acoustic Reflexes

Acoustic reflexes may be regarded as a part of nonclassical auditory pathways, similar to the pupil light reflex in vision. Sound can elicit involuntary muscle contractions and, aside from the general (startle) reflex that involves many muscles and which can be elicited by different sensory modalities, specific acoustic reflexes such as the acoustic middle ear reflex may be regarded as part of nonclassical auditory pathways, similar to the pupil reflex in vision.

4. Pathways for the Acoustic Middle Ear Reflexes

The acoustic middle ear reflex in humans is due to the stapedius muscle. The tensor tympani muscles also contract as an acoustic reflex in many animals. The stapedius muscle is innervated by a branch of the facial nerve that exits from the main trunk approximately 1 cm central to the emergence of the facial nerve from the stylomastoid foramen.[152,200] The pathways for the acoustic stapedius reflex involve the auditory nerve, the ventral cochlear nucleus, the superior olivary complex, and the facial motonucleus.[19] There is also a direct pathway from the ventral cochlear nucleus to the ipsilateral facial motonucleus. Connections from the ventral cochlear nucleus reach the contralateral superior olivary complex, which has connections to the facial motor nucleus on that side and mediates the reflex contraction of the contralateral stapedius muscle.[19]

The tensor tympani contracts as an acoustic reflex in many animals. That muscle is innervated by a branch of the trigeminal nerve. The neural pathways are similar to that of the stapedius reflex but there is no direct pathway from the ventral cochlear nucleus to the trigeminal motonucleus. All connections are through the MSO of the superior olivary complex.[19]

5. Descending Pathways (Reciprocal Connections)

As is the case for the classical auditory system, most ascending connections in the nonclassical pathways have descending counterparts that are mainly reciprocal connections. An extensive descending system is associated with the nonclassical ascending pathways, most of which may be regarded as reciprocal innervations thus being similar to that of the classical auditory pathways. Our knowledge about the connections of these systems is even sparser than those of the ascending nonclassical auditory pathways. However, it has been shown that there are considerable, mostly ipsilateral, connections from the primary and secondary auditory cortices to the DC and the ICX. The connections proceed to the IC uninterrupted and their targets are mostly neurons in the peripheral parts of the IC, the ICX, and DC.[55] The descending connections originate in layer V of the cerebral cortex (in the rat). The ICC receives nonsignificant connections from the auditory cortex.[55] The ICX and the DC also receive descending connections from the thalamus (posterior, intralaminar thalamic nuclei, medial MGB, suprageniculate nucleus, and the nucleus of the brachium of the BIC).[250] The dorsal MGB receives abundant descending input from extra auditory cortical areas.[247]

Similar reciprocal connections are also found in the connections from the nonclassical auditory pathways to nonauditory structures, such as the nuclei of the amygdala. It has recently been found that fibers from the amygdala (basolateral nucleus) project directly to neurons of the ICC, which belongs to the classical auditory pathway.[128] Connections from the ICC to nuclei of the nonclassical pathways (ICX and DC) that project to the amygdala complete a circle consisting of connections between the nuclei of the IC and amygdala nuclei.

B. Physiology of the Nonclassical System

A few early studies showed that the frequency tuning of neurons in the nonclassical auditory pathways was considerably broader than that of neurons in the classical system.[7,8] It was also shown that neurons in the nonclassical system responded less distinctly to auditory stimuli than neurons in the classical pathways, which is why the nonclassical pathways were named the "diffuse" pathways and the classical pathways were regarded to be the "distinct" or "specific" systems by some investigators.[59]

1. Dorsal Cortex and External Nucleus of the Inferior Colliculus

In the superficial layer of the DC, neurons are difficult to activate regularly with simple tonal stimuli, tend to habituate rapidly, are broadly tuned, and seem to prefer complex sounds over simple sounds such as tones.[3,7] Thus, vocal stimuli were found to be more efficient in activating cells in the ICX and DC than tones at CF or noise (82 and 72%, respectively, compared with 27% in cells of the ICC).[8] The cells of the ICX and DC are also reported to be broadly tuned and respond with less temporal precision than neurons in the classical pathways, where neurons normally respond with great temporal precision and are sharply tuned with regard to the frequency (spectrum) of sounds. Other studies have shown less difference between the tuning in the ICC and the ICX, and Syka and his collaborators[224] have shown that some neurons in the nonclassical pathways are narrowly tuned and respond distinctly to sound, thus responding more like neurons in the ICC.

These differences between the results obtained by different investigators may be due to species differences. They may also be related to the way the border between the ICX and the ICC is defined. The frequency specificity in neurons of the ICX and DC that are located far from their border to the ICC is less than in those located in the ICX close to its border with the ICC. Neurons of regions that are located close to the border of the ICC but regarded as belonging to the ICX seem to have characteristics more similar to those of the ICC than neurons of the ICX located near its surface and thus farther away from the ICC.[3,130,224] This means that the border between the ICC, which belongs to the classical pathways, and the ICX and DC may not be distinct and neurons for which the response characteristics were reported to have belonged to the nonclassical pathways may in fact have belonged to the classical pathways.

2. Polysensory Neurons

Some neurons in the ICX receive both auditory and somatosensory input from brainstem and cortical sources.[3,7] Some neurons respond to either acoustic or somatosensory input, and about 10% respond to both auditory and somatosensory input (Fig. 5.46).[7] Cells responding to somatosensory input have large receptive fields, often centered on distal forelimbs and paws.[5,7] Neurons in the ICX respond in different ways to stimulation of the somatosensory system. In some neurons the response evoked by auditory stimulation is inhibited by stimulation of the somatosensory system, while the response of other neurons to sounds is enhanced by somatosensory stimulation.[5,7] The response to sound can thus be either enhanced or inhibited by the input from the somatosensory system. The effect of somatosensory stimulation on the response to auditory stimulation of the ICX and DC is also evident from recordings of evoked potentials from the ICX.[225]

The fact that neurons in the nonclassical auditory pathways respond to stimulation of the somatosensory system,[5,7,225] as well as to other sensory modalities,[16] has led some investigators to call the nonclassical pathways the "polysensory" system; some investigators have gone on to distinguish between "diffuse" and "polysensory" parts of the nonclassical auditory pathways. The auditory response of neurons in the classical ascending pathways are assumed to be unaffected by somatosensory stimulation, although it is difficult to find any experimental evidence for that.

3. Dorsal and Medial Thalamus

The dorsal and medial portions of the auditory thalamus are the thalamic relay nuclei of the nonclassical ascending auditory pathways. Little is known about the information processing that occurs in that part of the MGB, but it has been shown that the response pattern of neurons in the medial and dorsal divisions is different from that of neurons in the ventral division. The neurons in the medial and dorsal portions fire in longer bursts than neurons in the ventral division, the response of which is mostly transient.[119] Many neurons in the dorsal MGB have been shown to fire repetitively (chopper type) in response to tone bursts while such neurons are rare in the ventral MGB.

4. Implications of Subcortical Auditory Input to the Amygdala

The direct (subcortical) connection that exists between the dorsal and medial auditory thalamus and the amygdala may explain why some sounds evoke emotional reactions, why tinnitus in some individuals can evoke fear and phonophobia, and why affective symptoms often accompany severe tinnitus.[123,149]

Some investigators have shown evidence that limbic structures (probably the amygdala) are involved in some forms of tinnitus.[123] The fact that somatosensory input (such as stimulation of the median nerve at the wrist) can affect the perception of tinnitus indicates that the nonclassical auditory pathways are involved in generation of tinnitus in some instances.[149] Other investigators have shown evidence of interaction between other forms of somatosensory or motor activity and tinnitus.[29,225]

The dorsal cochlear nucleus has been implicated in some forms of tinnitus by some investigators.[107] Connections between the somatosensory system to neurons in the dorsal cochlear nucleus that have been identified using morphological methods (from the dorsal column nuclei[96] and the trigeminal system[211]) may become open during pathologies such as some forms of tinnitus.

The fact that the nonclassical auditory pathways receive input not only from the ear but also from the somatosensory system has been used in studies of the involvement of the nonclassical auditory system in tinnitus.[149] In these studies, it has been shown that the perception of tinnitus in certain patients changes when the median nerve is stimulated electrically. Because the classical pathways do not receive input from the somatosensory system it was concluded that these observations indicated that the nonclassical auditory pathways were involved in the generation of tinnitus.

Studies of individuals (adults) who did not have tinnitus and were presented with a loud sound to one ear showed that the perception of the loudness of such a sound was not noticeably affected by similar electrical stimulation of the median nerve. In later studies, it was shown that children experience a noticeable change in the loudness of sounds when the median nerve is stimulated electrically.[153]

The studies showing interaction between the auditory and the somatosensory system in patients with tinnitus have later been confirmed and expanded to show that muscle contractions can also affect the perception of tinnitus in some patients with tinnitus.[29,30] Whether the latter occurs through stimulation of the somatosensory system or through activation of the motor control system is unknown.

It was shown in a recent study that somatosensory stimulation affects the perception of loudness of broadband sounds (repetitive clicks) in children but not in adults.[153] This was taken as a sign that children may use the nonclassical auditory pathways in hearing but that this ability gradually disappears during childhood,[153] and adults rarely show such signs.[149,153] The nonclassical system thus does not seem to be used normally in adult humans but is known to be activated under certain pathologic conditions.[149] This means that the connections from the classical auditory system to the nonclassical system may be variable and that they are open in young individuals, but these connections gradually close during maturation. The fact that some individuals with tinnitus show signs of the involvement of the nonclassical auditory pathways indicates that the connections between the classical and the nonclassical pathways can be activated under certain conditions. This resembles the phenomenon of unmasking of dormant synapses as described by Wall[230] in the dorsal horn of the spinal cord, thus being an expression of neural plasticity.

Considerable evidence suggests that sensory systems do not operate independently but that cross-modal interactions are common.[209] The fact that somatosensory stimulation can interfere with auditory information[149,153] may be an exception of Johannes Müller's hypothesis about the specific sensory energies (see Chapter 3).

5. The Acoustic Middle Ear Reflex

The acoustic middle ear reflex is bilateral, which means that sound presented to one ear elicits a muscle contraction in both ears. The contraction in the muscle of the contralateral ear is slightly weaker and has a slightly higher threshold than the ipsilateral response.[19,135,152]

The acoustic middle ear reflex in humans can best be elicited by tones or sounds below 2000 Hz, and the threshold elicited from the contralateral ear is approximately 85 dB HL[152]. In humans, the shortest latency of the earliest detectable response when recorded using changes in the ear's acoustic impedance is approximately 25 msec, and the latency increases with decreasing sound intensity, reaching 100 msec or more near the threshold of the reflex.[134,152]

In animals, contractions of both the tensor tympani and the stapedius muscles can be elicited as an acoustic reflex. The tensor tympani reflex has a slightly higher threshold than the stapedius reflex.

V. NEURAL PLASTICITY

There is considerable evidence of neural plasticity in the cerebral cortex. Expressions of neural plasticity have also been demonstrated in subcortical structures of ascending auditory pathways of both young and adult animals. Published studies show that deprivation of input and overstimulation are the strongest promoters of neural plasticity at the levels of the cochlear nucleus and IC as well as the cortex. Thus, studies have shown that eliminating acoustic input from one ear during development (by cochlear ablation) resulted in large changes in the responses from single cells in the ICC in the gerbil.[116] After cochlear ablation, the number of cells in the ICC ipsilateral to the unaffected ear that responded with excitation increased from approximately 28% in controls to 90% in the gerbils with one cochlea ablated. The responses of these ipsilaterally driven cells resembled that of the normal contralateral responding cells. These changes could be caused by outgrowth (sprouting) of new connections or by unmasking of normally dormant synapses (or perhaps both).

Reversible changes of input to one ear (by ear canal ligation) for 3 to 5 months caused prolongation of the latency of cells in the contralateral ICC, an increased ipsilateral suppression in the contralateral ICC, and a decreased suppression in the ipsilateral ICC.[213] Bilateral ear canal ligation has less effect on the balance between excitation and inhibition than unilateral hearing impairment (ligation of ear canals) during development.

The results of these studies thus emphasize the importance of individuals being exposed to environmental sounds during development and that such exposure should be a balanced input to the two ears, which means that there are strong indications of benefit from fitting hearing-impaired children with double hearing aids at an early age.

A. Neural Plasticity Can Alter the Threshold and Perception of Sounds

More important, perhaps, than the plasticity induced by total deprivation is the effect of reduced input that may occur in natural surroundings affecting many people. Considerable evidence has been presented that adequate sound stimulation is essential for the normal development of the young auditory system, but it is also becoming increasingly obvious that sound stimulation is essential to maintain normal hearing functions. For example, Willott *et al.*[244] have shown that an "augmented" sound environment can affect the threshold of hearing in a beneficial way and delay the development of age-related hearing loss. This means that neural plasticity can also be demonstrated in adult animals. If this is applicable to humans, it means that age-related hearing loss is more complex than previously assumed and may not be (only) an effect on outer hair cells.

Other animal experiments have shown that deprivation of sound may have considerable effects on information processing in the auditory pathways. For example, Gerken *et al.*[78] have shown that sound deprivation can cause hypersensitivity and changes in temporal integration in nuclei of the ascending auditory pathways of the cat.

Neural plasticity in the auditory system is best known from the auditory cortex. Signs of reorganization of the auditory cortex have been shown in animal experiments using electrophysiologic recordings.[95,98,105,132,183,201] Thus, plasticity of the representation of frequency in maps of cortical neurons has been demonstrated in several studies.[183] Such cortical reorganization is similar to that which has been demonstrated in other sensory modalities, studied most extensively in the somatosensory system.[174] Reorganization of the auditory cortex has been shown after conditioning (association learning). Electrical stimulation of the cortex causes similar reorganization.[195] The reorganization extends peripherally to the inferior colliculus mediated by descending systems and it can be abolished by destruction of the primary auditory cortex.[195]

It has recently been demonstrated that electrical stimulation of the nucleus basalis (an acetylcholine pathway) enhances the effect of auditory stimulation to induce plastic changes in the functional organization of the auditory cortex.[114] Thus, reorganizations of the AI and IC are augmented by stimulation of the cholinergic forebrain or by application of acetylcholine to the cortex.[114]

Also, processing of temporal information can be modified through expression of neural plasticity. Kilgard and Merzenich[113,115] showed that pairing electrical stimulation of the nucleus basalis with stimulation by tones presented at different repetition rates could change the maximal rate to which neurons could respond (from the normal maximal rate of 12 pps [pulses per second] to a significantly higher rate after training with 15-pps tone pulses paired with electrical stimulation of the nucleus basalis). After training with 7.5-pps tone bursts, the maximal rate to which neurons responded decreased.

B. Neural Plasticity from Overstimulation

Overstimulation can also induce neural plasticity, and it has been shown that noise-induced hearing loss has central components, as demonstrated in animal experiments.[157,158,243] Noise exposure is normally assumed to exert its temporary and permanent effects on the cochlea, where the loss of hair cells as a result of noise exposure has been studied extensively.[60,87,184,218] However, the finding that noise exposure also can cause permanent changes in the morphology of the auditory nervous system[157,158] means that the difficulties that people with noise-induced hearing loss experience in understanding speech may be explained by changes in the auditory nervous system rather than (only) on the effect of cochlear deficits.

Animal experiments have also shown acute changes in the function of the auditory nervous system after noise exposure. Temporal integration in neurons in the IC of the rat changed after exposure to loud sounds.[226] This form of plasticity probably involves changes in synaptic efficacy rather than morphological changes because it occurs with little delay after sound exposure.

REFERENCES

1. Aitkin, L.M., *The Auditory Cortex*. London: Chapman & Hall, 1990.
2. Aitkin, L.M. and Webster, W.R., Medial Geniculate Body of the Cat: Organization and Responses to Tonal Stimuli in Neurons in Ventral Division. *J. Neurophysiol.*, 35: 365–380, 1972.
3. Aitkin, L.M. and Moore, D.R., Inferior Colliculus: II. Development of Tuning Characteristics and Tonotopic Organization in Central Nucleus of the Inferior Colliculus of the Neonatal Cat. *J. Neurophysiol.*, 38: 1208–1216, 1975.
4. Aitkin, L.M., Dickhaus, H., Schult, W., and Zimmermann, M., External Nucleus of Inferior Colliculus: Auditory and Spinal Somatosensory Afferents and Their Interactions. *J. Neurophysiol.*, 41: 837–847, 1978.
5. Aitkin, L.M., Kenyon, C.E., and Philpott, P., The Representation of Auditory and Somatosensory Systems in the External Nucleus of the Cat Inferior Colliculus. *J. Comp. Neurol.*, 196: 25–40, 1981.
6. Aitkin, L.M. and Phillips, S.C., The Interconnections of the Inferior Colliculi through Their Commissure. *J. Comp. Neurol.*, 228: 210–216, 1984.

7. Aitkin, L.M. *The Auditory Midbrain, Structure and Function in the Central Auditory Pathway*. Clifton, NJ: Humana Press, 1986.
8. Aitkin, L.M., Tran, L., and Syka, J., The Responses of Neurons in Subdivisions of the Inferior Colliculus of Cats to Tonal, Noise and Vocal Stimuli. *Exp. Brain Res.*, 98: 53–64, 1994.
9. Andersen, P., Knight, P.L., and Merzenich, M.M., The Thalamocortical and Corticothalamic Connections of AI, AII, and the Anterior Field (AAF) in the Cat: Evidence for Two Largely Segregated Systems of Connections. *J. Comp. Neurol.*, 194: 663–701, 1980.
10. Andersen, P., Snyder, R.L., and Merzenich, M.M., The Topographic Organization of Corticocollicular Projections from Physiologically Identified Loci in the AI, AII, and Anterior Auditory Cortical Fields of the Cat. *J. Comp. Neurol.*, 191: 479–494, 1980.
11. Anderson, D.J., Hind, D.J., and Brugge, J.F., Temporal Position of Discharges in Single Auditory Nerve Fibers within the Cycle of a Sine-Wave Stimulus: Frequency and Intensity Effects. *J. Acoust. Soc. Am.*, 49: 1131–1139, 1971.
12. Arthur, R.M., Pfeiffer, R.R., and Suga, N., Properties of "Two Tone Inhibition" in Primary Auditory Neurons *J. Physiol. (London)*, 212: 593–609, 1971.
13. Békésy von, G., D-C Resting Potentials Inside the Cochlear Partition. *J. Acoust. Soc. Am.*, 24: 72–76, 1952.
14. Békésy von, G., *Experiments in Hearing*. New York: McGraw-Hill, 1960.
15. Blauert, J., *Spatial Hearing: The Psychophysics of Human Sound Localization*. Cambridge, MA: MIT Press, 1983.
16. Blum, P.S., Abraham, L.D., and Gilman, S., Vestibular, Auditory, and Somatic Input to the Posterior Thalamus of the Cat. *Exp. Brain Res.*, 34: 1–9, 1979.
17. Bocca, E., Clinical Aspects of Cortical Deafness. *Laryngoscope*, 68: 301, 1958.
18. Boer de, E., Correlation Studies Applied to the Frequency Resolution of the Cochlea. *J. Aud. Res.*, 7: 209–217, 1967.
19. Borg, E., On the Neuronal Organization of the Acoustic Middle Ear Reflex. A Physiological and Anatomical Study. *Brain Res.*, 49: 101–123, 1973.
20. Borg, E. and Møller, A.R., The Acoustic Middle Ear Reflex in Unanesthetized Rabbit. *Acta Otolaryngol.*, 65: 575–585, 1968.
21. Brodal, P., *The Central Nervous System*. New York: Oxford Press, 1998.
22. Brodel, M., *Three Unpublished Drawings of the Anatomy of the Human Ear*. Philadelphia: W.B. Saunders, 1946.
23. Brownell, W.E., Observation on the Motile Response in Isolated Hair Cells. In: *Mechanisms of Hearing*, edited by W.R. Webster and L.M. Aitkin. Melbourne: Monash University Press, 1983, pp. 5–10.
24. Brownell, W.E., Bader, C.R., Bertrand, D., and Ribaupierre de, Y., Evoked Mechanical Responses of Isolated Cochlear Hair Cells. *Science*, 227: 194–196, 1985.
25. Brugge, J.F., Orman, S.S., Coleman, J.R., Chan J.C., and Phillips, D.P., Binaural Interactions in Cortical Area AI of Cats Reared with Unilateral Atresia of the External Ear Canal. *Hear. Res.*, 20: 275–287, 1985.
26. Brugge, J.F., Reale, R.A., and Hind, J.E., The Structure of Spatial Receptive Fields of Neurons in Primary Auditory Cortex of the Cat. *J. Neurosci.*, 16: 4420–4437, 1996.
27. Büki, B., Avan, P., and Ribari, O., The Effect of Body Position on Transient Otoacoustic Emission. In: *Intracranial and Intralabyrinthine Fluids*, edited by A. Ernst, R. Marchbanks, and M. Samii. Berlin: Springer-Verlag, 1996, pp. 175–181.
28. Burkitt, A.N. and Clark, G.M., Analysis of Integrate-and-Fire Neurons: Synchronization of Synaptic Input and Spike Output. *Neural Comp.*, 11: 871–901, 1999.
29. Cacace, A.T., Lovely, T.J., McFarland, D.J., Parnes, S.M., and Winter, D.F., Anomalous Cross-Modal Plasticity Following Posterior Fossa Surgery: Some Speculations on Gaze-Evoked Tinnitus. *Hear. Res.*, 81: 22–32, 1994.

30. Cacace, A.T., Cousins, J.P., Parnes, S.M., McFarland, D.J., Semenoff, D., Holmes, T., Davenport, C., Stegbauer, K., and Lovely, T.J., Cutaneous-Evoked Tinnitus. II. Review of Neuroanatomical, Physiological and Functional Imaging Studies. *Audio. Neuro. Otol.*, 4: 258–268, 1999.
31. Calford, M.B. and Aitkin, L.M., Ascending Projections to the Medial Geniculate Body of the Cat: Evidence for Multiple, Parallel Auditory Pathways through the Thalamus. *J. Neurosci.*, 3: 365–380, 1983.
32. Cant, N.B. and Gaston, K.C., Pathways Connecting the Two Cochlear Nuclei. *J. Comp. Neurol.*, 212: 313–326, 1982.
33. Cant, N.B., *The Cochlear Nucleus: Neuronal Types and Their Synaptic Organization*. New York: Springer-Verlag, 1992.
34. Casseday, J.H., Ehrlich, D., and Covey, E., Neural Tuning for Sound Duration: Role of Inhibitory Mechanisms in the Inferior Colliculus. *Science*, 264: 847–850, 1994.
35. Casseday, J.H., Fremouw, T., and Covey, E., The Inferior Colliculus: A Hub for the Central Auditory System. In: *Integrative Functions in the Mammalian Auditory Pathway*, edited by D. Oertel, R.R. Fay, and A.N. Popper. New York: Springer, 2002, pp. 238–318.
36. Celesia, G.G., Broughton, R.J., Rasmussen, T., and Branch, C., Auditory Evoked Responses from the Exposed Human Cortex. *Electroencephalogr. Clin. Neurophysiol.*, 24: 458–466, 1968.
37. Clarey, J.C. and Irvine, D.R.F., The Anterior Ectosylvian Sulcal Auditory Field in the Cat: II. Horseradish Peroxidase Study of Its Thalamic and Cortical Connections. *J. Comp. Neurol.*, 301: 304–324, 1990.
38. Clarey, J.C., Barone, P., and Imig, T.J., Physiology of Thalamus and Cortex. In: *The Mammalian Auditory Pathway: Neurophysiology*, edited by R.R. Fay and A.N. Popper. New York: Springer-Verlag, 1992, pp. 232–334.
39. Clarke, S.F., Ribaupierre de, F., Bajo, V.M., Rouiller, E.M., and Kraftsik, R., The Auditory Pathway in Cat Corpus Callosum. *Exp. Brain Res.*, 104: 534–540, 1995.
40. Code, R.A. and Winer, J.A., Commissural Connections in Layer III of Cat Primary Auditory Cortex (AI): Pyramidal and Non-Pyramidal Cell Input. *J. Comp. Neurol.*, 242: 485–510, 1985.
41. Collet, L., Kemp, D.T., Veuillet, E., Duclaux, R., Moulin, A. and Morgon, A., Effect of Contralateral Auditory Stimuli on Active Cochlear Micro-Mechanical Properties in Human Subjects. *Hear. Res.*, 43: 251–262, 1990.
42. Conlee, J.W. and Kane, E.S., Descending Projections from the Inferior Colliculus to the Dorsal Cochlear Nucleus in the Cat: An Autoradiographic Study. *Neuroscience*, 7: 161–178, 1982.
43. Cooper, N.P., Roberston, D., and Yates, G.K., Cochlear Nerve Fiber Responses to Amplitude-Modulated Stimuli: Variations with Spontaneous Rate and Other Response Characteristics. *J. Neurophysiol.*, 70: 370–386, 1993.
44. Covey, E. and Casseday, J.H., The Monaural Nuclei of the Lateral Lemniscus in an Echolocating Bat: Parallel Pathways for Analyzing Temporal Features of Sound. *J. Neurosci.*, 11: 3456–3470, 1991.
45. Crawford, A.C., Evans, M.G., and Fettiplace, R., The Actions of Calcium on the Mechanoelectrical Transducer Current of Turtle Hair Cells. *J. Physiol. (London)*, 434: 369–398, 1991.
46. Creutzfeldt, O., Hellweg, F.C., and Schreiner, C., Thalomocortical Transformation of Responses to Complex Auditory Stimuli. *Exp. Brain Res.*, 39: 87–104, 1980.
47. Dallos, P. and Cheatham, M.A., Nonlinearities in Cochlear Receptor Potentials by Inner and Outer Hair Cells. *J. Acoust. Soc. Am.*, 86: 1790–1796, 1989.
48. Dallos, P.J., Study of the Acoustic Reflex Feedback Loop. *IEEE Trans. Bio-Med. Eng.*, 11: 1–7, 1964.
49. Dallos, P.J., Evans, B.N., and Hallworth, R., Nature of the Motor Element in Electrokinetic Shape Changes of Cochlear Outer Hair Cells. *Nature*, 350: 155–157, 1991.

50. Dallos, P.J., He, Z.Z.D., Evans, N., and Clark, B., Dynamic Characteristics of Outer Hair Cell Motility. In: *Biophysics of Hair Cell Sensory Systems*, edited by J. Duifhuis, W. Horst, van P. Dijk, and S.M. van Netten. Singapore: World Scientific, 1993, pp. 167–174.
51. Dallos, P.J., Popper, A.N., and Fay, R.R., *The Cochlea*. New York: Springer-Verlag, 1996.
52. Denk, W., Webb, W.W., and Hudspeth, A., Mechanical Properties of Sensory Hair Bundles Are Reflected in Their Brownian Motion Measured with Laser Differential Interferometer. *Proc. Nat. Acad. Sci., USA*, 86: 5371–5375, 1989.
53. Denk, W. and Webb, W.W., Forward and Reverse Transduction at the Limit of Sensitivity Studied by Correlating Electrical and Mechanical Fluctuations in Frog Saccular Hair Cells. *Hear. Res.*, 60: 89–102, 1992.
54. Diamond, I.T., The Subdivisions of Neocortex: A Proposal To Revise the Traditional View of Sensory, Motor, and Association Areas. In: *Progress in Psychobiology and Physiological Psychology*, edited by J.M. Sprague and A.N. Epstein. New York: Academic Press, 1979, pp. 2–44.
55. Druga, R., Syka, J., and Rajkowska, G., Projections of Auditory Cortex onto the Inferior Colliculus in the Rat. *Physiol. Res.*, 46: 215–222, 1997.
56. Dulon, D. and Schacht, J., Motility of Cochlear Hair Cells. *Am. J. Otol.*, 13: 108–112, 1992.
57. Eggermont, J.J., Johannesma, P.I.M., and Aertsen, A.M.H., Reverse-Correlation Methods in Auditory Research. *Q.. Rev. Biophys.* 16: 341–414, 1983.
58. Ehret, G. and Merzenich, M.M., Complex Sound Analysis (Frequency Resolution, Filtering and Spectral Integration) by Single Units of the Inferior Colliculus in the Cat. *Brain Res Rev.*, 13: 139–163, 1988.
59. Ehret, G. and Romand, R., eds., *The Central Auditory Pathway*. New York: Oxford University Press, 1997.
60. Engstrom, H., Ades, H.W., and Andersson, A., *Structural Pattern of the Organ of Corti*. Stockholm: Almquist and Wiksell, 1966.
61. Evans, E.F., The Frequency Response and Other Properties of Single Fibers in the Guinea Pig Cochlear Nerve. *J. Physiol.*, 226: 263–287, 1972.
62. Evans, E.F., Normal and Abnormal Functioning of the Cochlear Nerve. *Symp, Zool, Soc. London*, 37: 133–165, 1975.
63. Fettiplace, R., Electrical Tuning of Haircells in the Inner Ear. *Trends Neurosci.*, 10: 421–425, 1987.
64. Fex, J., Auditory Activity in Centrifugal and Centripetal Cochlear Fibers in Cat. *Acta Physiol. Scand.*, 55 (Suppl. 189) 5–68, 1962.
65. Fitzpatrick, D.C., Kuwada, S., Kim, D.O., Parham, K., and Batra, R., Responses of Neurons to Click-Pairs as Simulated Echoes: Auditory Nerve to Auditory Cortex. *J. Acoust. Soc. Am.*, 106: 3460–3472, 1999.
66. Flock, A., Transducing Mechanisms in Lateral Line Canal Organ. *Cold Spring Harbor Symp. Quant. Biol.*, 30: 133–146, 1965.
67. Flock, A. and Cheung, H.C., Actin Filaments in Sensory Hairs of Inner Ear Receptor Cells. *J. Cell Biol.*, 75: 339–343, 1977.
68. Freeman, D.M. and Weiss, T.F., Hydrodynamic Forces on Hair Bundles at Low Frequencies. *Hear. Res.*, 48: 17–30, 1990.
69. Frisina, R.D., Smith, R.L., and Chamberlain, S.C., Encoding of Amplitude Modulation in the Gerbil Cochlear Nucleus. I. A Hierarchy of Enhancement. *Hear. Res.*, 44: 99–122, 1990.
70. Frisina, R.D., Smith, R.L., and Chamberlain, S.C., Encoding of Amplitude Modulation in the Gerbil Cochlear Nucleus. II. Possible Neural Mechanisms. *Hear. Res.*, 44: 123–142, 1990.
71. Frisina, R.D., Walton, J.P., and Karcich, K.J., Dorsal Cochlear Nucleus Single Neurons Can Enhance Temporal Processing Capabilities in Background Noise. *Exp. Brain Res.*, 102: 160–164, 1994.

72. Frisina, R.D., Karcich, K.J., Tracy, T.C., Sullivan, D.M., Walton, J.P., and Colombo, J., Preservation of Amplitude Modulation Coding in the Presence of Background Noise by Chinchilla Auditory-Nerve Fibers. *J. Acoust. Soc. Am.*, 99: 475–490, 1996.
73. Galambos, R., Myers, R., and Sheatz, G., Extralemniscal Activation of Auditory Cortex in Cats. *Am. J. Physiol.*, 200: 23–28, 1961.
74. Gardner, M.B., Historical Background of the Haas or Precedence Effect. *J. Acoust. Soc. Am.*, 43: 1243–1248, 1968.
75. Geisler, C.D. and Sinex, D.G., Responses of Primary Auditory Fibers to Combined Noise and Tonal Stimuli. *Hear. Res.*, 3: 317–335, 1980.
76. Geisler, C.D. and Sang, C., A Cochlear Model Using Feed-Forward Outer-Hair-Cell Forces. *Hear. Res.*, 86: 132–146, 1995.
77. Geisler, C.D., *From Sound to Synapse*. New York: Oxford Press, 1998.
78. Gerken, G.M., Saunders, S.S., and Paul, R.E., Hypersensitivity to Electrical Stimulation of Auditory Nuclei Follows Hearing Loss in Cats. *Hear. Res.*, 13: 249–260, 1984.
79. Gillespie, P.G., Molecular Machinery of Auditory and Vestibular Transduction. *Curr. Opin. Neurbiol.*, 5: 449–455, 1995.
80. Goldberg, J.M., and Brown, P.B., B Response of Binaural Neurons of Dog Superior Olivary Complex to Dichotic Tonal Stimuli: Some Physiological Mechanisms of Sound Localization. *J. Neurophysiol.*, 32: 613–636, 1969.
81. Graybiel, A.M., Some Fiber Pathways Related to the Posterior Thalamic Region in the Cat. *Brain Behav. Evol.*, 6: 363–393, 1972.
82. Graybiel, A.M., The Thalamocortical Projection of the So-Called Posterior Nuclear Group: A Study with Anterograde Degeneration Methods in the Cat. *Brain Res.*, 49: 229–244, 1973.
83. Guinan, Jr, J.J., Guinan, S.S., and Norris, B.E., Single Auditory Units in the Superior Olivary Complex. I. Responses of Sounds and Classifications Based on Physiological Properties. *Int. J. Neurosci.*, 4: 101–120, 1972.
84. Guinan, Jr., J.J. and Gifford, M.L., Effects of Electrical Stimulation of Efferent Olivocochlear Neurons on Cat Auditory-Nerve Fibers: II. Spontaneous Rate. *Hear. Res.*, 33: 115–128, 1988.
85. Harrison, J.M. and Howe, M.E., Anatomy of the Afferent Auditory Nervous System in Mammals. In: *Handbook of Sensory Physiology*, edited by W.D. Keidel and W.D. Neff. Berlin: Springer-Verlag, 1974, pp. 283–336.
86. Harrison, R.V. and Hunter-Duvar, I.M., An Anatomical Tour of the Cochlea. In: *Physiology of the Ear*, edited by A.F. Jahn and J. Santos-Sacchi. New York: Raven Press, 1988, pp. 159–171.
87. Hawkins, J.E. Auditory Physiologic History: A Surface View. In: *Physiology of the Ear*, edited by A.F. Jahn and J. Santos-Sacchi. New York: Raven Press, 1988.
88. Hellige, J., *Hemispheric Asymmetry: What's Right and What's Left*. Cambridge, MA: Harvard University Press, 1993.
89. Honrubia, V. and Ward, P.H., Longitudinal Distribution of the Cochlear Microphonics Inside the Cochlear Duct (Guinea Pig). *J. Acoust. Soc. Am.*, 44: 951–958, 1968.
90. Howard, M.A., Volkov, I.O., Mirsky, R.A., Garell, P.C., Noh, M.D., Granner, M., Damasio, H., Steinschneider, M., Reale, R., Hind, J.E. and Brugge, J.F. Auditory Cortex on the Human Posterior Superior Temporal Gyrus. *J. Comp. Neurol.*, 416: 79–92, 2000.
91. Hudspeth, A., How the Ear's Works Work. *Nature*, 341: 97–404, 1989.
92. Huffman, R.F., and Henson, O.W., The Descending Auditory Pathway and Acousticomotor Systems: Connection with the Inferior Colliculus. *Brain Res. Rev.*, 15: 295–323, 1990.
93. Hugdahl, K., *Handbook of Dichotic Listening: Theory, Methods and Research*. New York: John Wiley & Sons, 1988.
94. Imig, T.J., and Morel, A., Tonotopic Organization in the Ventral Nucleus of the Medial Geniculate Body in the Cat. *J. Neurophysiol.*, 53: 309–340, 1985.

95. Irvine, D.R. and Rajan, R., Injury-Induced Reorganization of Frequency Maps in Adult Auditory Cortex: The Role of Unmasking of Normally-Inhibited Inputs. *Acta Otolaryngol.*, 532: 39–45, 1997.
96. Itoh, K., Kamiya, H., Mitani, A., Yasui, Y., Takada, M. and Mizuno, N., Direct Projections from Dorsal Column Nuclei and the Spinal Trigeminal Nuclei to the Cochlear Nuclei in the Cat. *Brain Res.*, 400: 145–150, 1987.
97. Jeffress, L.A., A Place Theory of Sound Localization. *J. Comp. Physiol. Psychol.*, 41: 35–39, 1948.
98. Jenkins, W.M., Merzenich, M.M., Ochs, M.T., Allard, T. and Guic-Robles, E., Functional Reorganization of Primary Somatosensory Cortex in Adult Owl Monkeys after Behaviorally Controlled Tactile Stimulation. *J. Neurophysiol.*, 63: 82–104, 1990.
99. Jerger, J., Alford, B., Lew, H., Rivera, V. and Chmiel, R., Dichotic Listening, Event-Related Potentials, and Interhemispheric Transfer in the Elderly. *Ear Hear.*, 16: 482–498, 1995.
100. Jerger, J., Moncrieff, D., Greenwald, R., Wambacq, I. and Seipel, A., Effect of Age on Interaural Asymmetry of Event-Related Potentials in Dichotic Listening Task. *J. Am. Acad. Audiol.*, 11: 383–389, 2000.
101. Johnson, D.H., The Relationship Between Spike Rate and Synchrony in Responses of Auditory-Nerve Fibers to Single Tones. *J. Acoust. Soc. Am.*, 68: 1115–1122, 1980.
102. Johnstone, B.M., Patuzzi, R. and Yates, G.K., Basilar Membrane Measurements and the Traveling Wave. *Hear. Res.*, 22: 147–153, 1986.
103. Joris, P.X. and Yin, T.C.T., Responses to Amplitude Modulated Tones in the Auditory Nerve of the Cat. *J. Acoust. Soc. Am.*, 91: 215–232, 1992.
104. Joris, P.X., Smith, P.H. and Yin, T.C.T., Coincidence Detection in the Auditory System: 50 Years after Jeffress. *Neuron*, 21: 1235–1238, 1998.
105. Kaas, J.H., Plasticity of Sensory and Motor Maps in Adult Mammals. *Annu. Rev. Neurosci.*, 14: 137–167, 1991.
106. Kaas, J.H. and Hackett, T.A., Subdivisions of Auditory Cortex and Processing Streams in Primates. *Proc. Natl., Acad. Sci. U.S.A.*, 97: 11793–11799, 2000.
107. Kaltenbach, J.A., Hyperactivity in the Dorsal Cochlear Nucleus after Intense Sound Exposure and Its Resemblance to Tone-Evoked Activity: A Physiological Model for Tinnitus. *Hear. Res.*, 140: 165–172, 2000.
108. Katsuki, Y., Sumi, T., Uchiyama, H. and Watanabe, T., Electric Responses of Auditory Neurons in Cat to Sound Stimulation. *J. Neurophysiol.*, 21: 569–588, 1958.
109. Katsuki, Y., Suga, M., and Karmo, Y., Neural Mechanisms of the Peripheral and Central Auditory System in Monkeys. *J. Acoust. Soc. Am.*, 34: 1396–1410, 1962.
110. Kemp, D.T., Stimulated Acoustic Emissions from within the Human Auditory System. *J. Acoust. Soc. Am.*, 64: 1386–1391, 1978.
111. Kiang, N.Y.S., Watanabe, T., Thomas, E.C., and Clark, L., *Discharge Patterns of Single Fibers in the Cat's Auditory Nerve*. Cambridge, MA: MIT Press, 1965.
112. Kidd, R.C. and Weiss, T.F., Mechanisms that Degrade Timing Information in the Cochlea. *Hear. Res.*, 49: 181–208, 1990.
113. Kilgard, M.P. and Merzenich, M.M., Plasticity of Temporal Information Processing in the Primary Auditory Cortex. *Nature Neurosci.*, 1: 727–731, 1998.
114. Kilgard, M.P. and Merzenich, M.M., Cortical Map Reorganization Enabled by Nucleus Basalis Activity. *Science*, 279: 1714–1718, 1998.
115. Kilgard, M.P. and Merzenich, M.M., Distributed Representation of Spectral and Temporal Information in Rat Primary Auditory Cortex. *Hear. Res.*, 134: 16–28, 1999.
116. Kitzes, L.M. and Semple, M.N., Single-Unit Responses in the Inferior Colliculus: Effects of Neonatal Unilateral Cochlear Ablation. *J. Neurophysiol.*, 53: 1483–1500, 1985.

117. Konishi, M. and Nielsen, D.W., The Temporal Relationship Between Motion of the Basilar Membrane and Initiation of Nerve Impulses in Auditory Nerve Fibers. *J. Acoust. Soc. Am.*, 53: 325, 1973.
118. Korsan-Bengtsen, (aka Møller M.B.) Distorted Speech Audiometry. *Acta Otolaryng. Stockholm*, Suppl., 310, 1973.
119. Kvasnak, E., Popelar, J., and J.S., Discharge Properties of Neurons in Subdivisions of the Medial Geniculate Body of the Guinea Pig. *Physiol. Res.*, 49: 369–378, 2000.
120. LeBeau, F.E.N., Malmierca, M.S., and Rees, A., Ionotphoresis *In Vivo* Demonstrates a Key Role of $GABA_A$ and Glycinergic Inhibition in Sharpening Frequency Response Areas in the Inferior Colliculus of Guinea Pig. *J. Neurosci.*, 21: 7303–7312, 2001.
121. LeDoux, J.E., Brain Mechanisms of Emotion and Emotional Learning. *Curr. Opin. Neurobiol.*, 2: 191–197, 1992.
122. Lim, D.J., Effects of Noise and Ototoxic Drugs at the Cellular Level in the Cochlea. *Am. J. Otolaryngol.*, 7: 73–99, 1986.
123. Lockwood, A., Salvi, R., Coad, M., Towsley, M., Wack, D., and Murphy, B., The Functional Neuroanatomy of Tinnitus. Evidence for Limbic System Links and Neural Plasticity. *Neurology*, 50: 114–120, 1998.
124. Lomber, S.G., Payne, B.R., and Cornwell, P., Role of Superior Colliculus in Analyses of Space: Superficial and Intermediate Layer Contributions to Visual Orienting, Auditory Orienting, and Visuospatial Discriminations During Unilateral and Bilateral Deactivations. *J. Comp. Neurol.*, 441: 44–57, 2001.
125. Lonsbury-Martin, B.L. and Martin,G.K. The Clinical Utility of Distortion-Product Otoacoustic Emissions. *Ear Hear.*, 11: 144–154, 1990.
126. Malmierca, M.S., Oliver, D.L., Henkel, C.K., and Merchan, M.A., A Novel Projection from Dorsal Cochlear Nucleus to the Medial Division of the Medial Geniculate Body of the Rat. *Assoc. Res. Otolaryngo. Abstr.*, 25: 176, 2002.
127. Marchbanks, R.J., Hydromechanical Interactions of the Intracranial and Intralabyrinthine Fluids. In: *Intracranial and Intralabyrinthine Fluids*, edited by A. Ernst, Marchbanks, R., and Samii, M. Berlin: Springer-Verlag, 1996.
128. Marsh, R.A., Grose, C.D., Wenstrup, J.J., and Fuzessery, Z.M., A Novel Projection from the Basolateral Nucleus of the Amygdala to the Inferior Colliculus in Bats. *Soc. Neurosci. Abstr.*, 25: 1417, 1999.
129. Mast, T.E., Binaural Interaction and Contralateral Inhibition in Dorsal Cochlear Nucleus of Chinchilla. *J. Neurophysiol.*, 62: 61–70, 1973.
130. Merzenich, M.M. and Reid, M.D., Representation of the Cochlea within the Inferior Colliculus of the Cat. *Brain Res.*, 77: 397–415, 1974.
131. Merzenich, M.M. and Kaas, J.H., Principles of Organization of Sensory-Perceptual Systems in Mammals. In: *Progress in Psychobiology and Physiological Psychology*, edited by J.M. Sprague and A.N. Epstein. New York: Academic Press, 1980.
132. Merzenich, M.M., Recanzone, G., Jenkins, W.M., Allard, T.T., and Nudo, R.J., Cortical Representational Plasticity. In: *Neurobiology of Neocortex*, edited by P. Rakic and W. Singer. New York: Wiley, 1988, pp. 41–67.
133. Mitani, A. and Shimokouchi, M., Neural Connections in the Primary Auditory Cortex: An Electrophysiologic Study in the Cat. *J. Comp. Neurol.*, 235: 417–429, 1985.
134. Møller, A.R., Intra-Aural Muscle Contraction in Man, Examined by Measuring Acoustic Impedance of the Ear. *Laryngoscope*, LXVIII: 48–62, 1958.
135. Møller, A.R., The Acoustic Reflex in Man. *J. Acoust. Soc. Am.*, 34: 1524–1534, 1962.
136. Møller, A.R., Unit Responses in the Cochlear Nucleus of the Rat to Pure Tones. *Acta Physiol. Scand.*, 75: 530–541, 1969.

137. Møller, A.R., Unit Responses in the Rat Cochlear Nucleus to Repetitive Transient Sounds. *Acta Physiol. Scand.*, 75: 542–551, 1969.
138. Møller, A.R., Unit Responses in the Rat Cochlear Nucleus to Tones of Rapidly Varying Frequency and Amplitude. *Acta Physiol. Scand.*, 81: 540–556, 1971.
139. Møller, A.R., Coding of Amplitude and Frequency Modulated Sounds in the Cochlear Nucleus of the Rat. *Acta Physiol. Scand.*, 86: 223–238, 1972.
140. Møller, A.R., The Middle Ear. In: *Foundation of Modern Auditory Theory*, edited by J.V. Tobias. New York: Academic Press, 1972, pp. 133–194.
141. Møller, A.R., Responses of Units in the Cochlear Nucleus to Sinusoidally Amplitude Modulated Tones. *Exp. Neurol.*, 45: 104–117, 1974.
142. Møller, A.R., Coding of Sounds with Rapidly Varying Spectrum in the Cochlear Nucleus. *J. Acoust. Soc. Am.*, 55: 631–640, 1974.
143. Møller, A.R., The Acoustic Middle Ear Muscle Reflex. In: *Handbook of Sensory Physiology*, edited by W.D. Keidel and W.D. Neff. New York: Springer-Verlag, 1974, pp. 519–545.
144. Møller, A.R., Dynamic Properties of Primary Auditory Fibers Compared with Cells in the Cochlear Nucleus. *Acta Physiol. Scand.*, 98: 157–167, 1976.
145. Møller, A.R., Frequency Selectivity of the Basilar Membrane Revealed from Discharges in Auditory Nerve Fibers. In: *Psychophysics and Physiology of Hearing*, edited by E.F. Evans and J.P. Wilson. London: Academic Press, 1977, pp. 197–205.
146. Møller, A.R., Frequency Selectivity of Single Auditory Nerve Fibers in Response to Broadband Noise Stimuli. *J. Acoust. Soc. Am.*, 62: 135–142, 1977.
147. Møller, A.R., *Auditory Physiology*. New York: Academic Press, 1983.
148. Møller, A.R., Frequency Selectivity of Phase-Locking of Complex Sounds in the Auditory Nerve of the Rat. *Hear. Res.*, 11: 267–284, 1983.
149. Møller, A.R., Møller, M.B., and Yokota, M., Some Forms of Tinnitus May Involve the Extralemniscal Auditory Pathway. *Laryngoscope*, 102: 1165–1171, 1992.
150. Møller, A.R., Colletti, V., and Fiorino. F.G., Neural Conduction Velocity of the Human Auditory Nerve: Bipolar Recordings from the Exposed Intracranial Portion of the Eighth Nerve During Vestibular Nerve Section. *Electroencephalogr. Clin. Neurophysiol.*, 92: 316–320, 1994.
151. Møller, A.R., Review of the Roles of Temporal and Place Coding of Frequency in Speech Discrimination. *Acta Otolaryngol.*, 119: 424–430, 1999.
152. Møller, A.R., *Hearing: Its Physiology and Pathophysiology*. San Diego: Academic Press, 2000.
153. Møller, A.R. and Rollins, P., The Non-Classical Auditory System Is Active in Children but Not in Adults. *Neurosci. Lett.*, 319: 41–44, 2002.
154. Moore, J.K., The Human Auditory Brain Stem: A Comparative View. *Hear. Res.*, 29: 1–32, 1987.
155. Morel, A., Rouiller, E.M., Ribaupierre, de Y., and Ribaupierre, de F., Tonotopic Organization in the Medial Geniculate Body (MGB) of Lightly Anesthetized Cats. *Exp. Brain Res.*, 69: 24–42, 1987.
156. Morest, D.K., The Neuronal Architecture of the Medial Geniculate Body of the Cat. *J. Anat. (London)*, 98: 611–630, 1964.
157. Morest, D.K., Ard, M.D., and Yurgelun-Todd, D., Degeneration in the Central Auditory Pathways after Acoustic Deprivation or Over-Stimulation in the Cat. *Anat. Rec.*, 193: 750, 1979.
158. Morest, D.K. and Bohne, B.A., Noise-Induced Degeneration in the Brain and Representation of Inner and Outer Hair Cells. *Hear. Res.*, 9: 145–152, 1983.
159. Morest, D.K. and Oliver, D.L. The Neuronal Architecture of the Inferior Colliculus in the Cat. *J. Comp. Neurol.*, 222: 209–236, 1984.
160. Mountain, D.C., Geisler, C.D., and Hubbard, A.E., Stimulation of Efferents Alters the Cochlear Microphonic and Sound-Induced Resistance Changes Measured in the Scala Media of the Guinea Pig. *Hear. Res.*, 3: 231–240, 1980.

161. Müller, M., Robertson, D., and Yates, G.K., Rate-Versus-Level Functions of Primary Auditory Nerve Fibres: Evidence of Square Law Behavior of All Fibre Categories in the Guinea Pig. *Hear. Res.*, 55: 50–56, 1991.
162. Nelken, I., Feature Detection by the Auditory Cortex. In: *Integrative Functions in the Mammalian Auditory Pathway*, edited by D. Oertel, R.R. Fay, and A.N. Popper. New York: Springer, 2002, pp. 358–416.
163. Nordlund. B., Physical Factors in Angular Localization. *Acta Otolaryngol. (Stockholm)*, 54: 76–93, 1962.
164. Oliver, D.L. and Morest, D.K., The Central Nucleus of the Inferior Colliculus in the Cat. *J. Comp. Neurol.*, 222: 237–264, 1984.
165. Oliver, D.L. and Huerta, M.F. Inferior and Superior Colliculi. In: *The Mammalian Auditory Pathway: Neuroanatomy*, edited by D.B. Webster, A.N. Popper, and R.R. Fay. New York: Springer-Verlag, 1992.
166. Palmer, A.R. and Russell, I.J., Phase-Locking in the Cochlear Nerve of the Guinea-Pig and Its Relation to the Receptor Potential of Inner Hair Cells. *Hear. Res.*, 24: 1–15, 1986.
167. Phillips, D.P., Orman, S.S., Musicant, A.D., and Wilson, G.F., Neurons in the Cat's Primary Cortex Distinguished by Their Response to Tones and Wide-Spectrum Noise. *Hear. Res.*, 18: 73–86, 1985.
168. Pickles, J.O., Comis, S.D., and Osborne, M.P., Cross-Links Between Stereocilia in the Guinea Pig Organ of Corti, and Their Possible Relation to Sensory Transduction. *Hear. Res.*, 15: 103–112, 1984.
169. Pickles, J.O., *An Introduction to the Physiology of Hearing*, (2nd ed). London: Academic Press, 1988.
170. Preyer, S., Hemmert, W., Zenner, H.P., and Gummer, A.W., Abolition of the Receptor Potential Response of Isolated Mammalian Outer HairCells by Hair-Bundle Treatment with Elastase: A Test of the Tip-Link Hypothesis. *Hear. Res.*, 89: 187–193, 1995.
171. Rasmussen, G.L., The Olivary Peduncle and Other Fiber Projections or the Superior Olivary Complex. *J. Comp. Neurol.*, 84: 141–219, 1946.
172. Rauschecker, J.P., Tian, B., Pons, T., and Mishkin, M., Serial and Parallel Processing in Rhesus Monkey Auditory Cortex. *J. Comp. Neurol.*, 382: 89–103, 1997.
173. Rauschecker, J.P., Parallel Processing in the Auditory Cortex of Primates. *Audiol. Neuro-Otol.*, 3: 86–103, 1998.
174. Rauschecker, J.P., Auditory Cortical Plasticity: A Comparison with Other Sensory Systems. *Trends Neurosci.*, 22: 74–80, 1999.
175. Rauschecker, J.P. and Tian, B., Mechanisms and Streams for Processing of "What" and "Where" in Auditory Cortex. *Proc. Natl. Acad. Sci. U.S.A.*, 97: 11800–11806, 2000.
176. Razak, K.A., Fuzessery, Z.M. and Lohuis, T.D., Single Cortical Neurons Serve Both Echolocation and Passive Sound Localization. *J. Neurophysiol.*, 81: 1438–1442, 1999.
177. Rees, A. and Møller, A.R., Responses of Neurons in the Inferior Colliculus of the Rat to AM and FM Tones. *Hear. Res.*, 10: 301–330, 1983.
178. Rees, A. and Møller, A.R., Stimulus Properties Influencing the Responses of Inferior Colliculus Neurons to Amplitude-Modulated Sounds. *Hear. Res.*, 27: 129–144, 1987.
179. Retzius, G., *Das Gehörorgan der Wirbeltiere. Das Gehörorgan der Reptilien, der Vogel und der Saugetire.* Stockholm: Centraltryckeriet, 1884.
180. Rhode, W.S., Observations of the Vibration of the Basilar Membrane in Squirrel Monkeys Using the Mössbauer Technique. *J. Acoust. Soc. Am.*, 49: 1218–1231, 1971.
181. Rhode, W.S., An Investigation of Post-Mortem Cochlear Mechanics Using the Mössbauer Effect. In: *Basic Mechanisms in Hearing*, edited by A.R. Møller. New York: Academic Press, 1973.
182. Robertson, D. and Gummer, M., Physiological and Morphological Characterization of Efferent Neurons in the Guinea Pig Cochlea. *Hear. Res.*, 20: 63–77, 1985.

183. Robertson, D. and Irvine, D.R., Plasticity of Frequency Organization in Auditory Cortex of Guinea Pigs with Partial Unilateral Deafness. *J. Comp. Neurol.*, 282: 456–471, 1989.
184. Robertson, D., Johnstone, B.M., and McGill, T., Effects of Loud Tones on the Inner Ear: A Combined Electrophysiological and Ultrastructural Study. *Hear. Res.*, 2: 39–53, 1990.
185. Romanski, L.M., Tian, B., Fritz, J., Mishkin, M., Goldman-Rakic, P.S., and Rauschecker, J.P., Dual Streams of Auditory Afferents Target Multiple Domains in the Primate Prefrontal Cortex. *Nature Neurosci.*, 2: 1131–1136, 1999.
186. Rose, J.E., Hind, J.E., Anderson, D.J., and Brugge, J.F., Some Effects of Stimulus Intensity on Response of Auditory Fibers in the Squirrel Monkey. *J. Neurophysiol.*, 34: 685–699, 1971.
187. Rosowski, J.J., The Effects of External- and Middle-Ear Filtering on Auditory Threshold and Noise-Induced Hearing Loss. *J. Acoust. Soc. Am.*, 90: 124–135, 1991.
188. Rouiller, E.M., Ribaupierre, de Y., Morel, A., and Ribaupierre, de F., Intensity Functions of Single Unit Responses to Tones in the Medial Geniculate Body of Cat. *Hear. Res.*, 11: 235–247, 1983.
189. Rouiller, E.M., and Ribaupierre, de F., Origin of Afferents to Physiologically Defined Regions of the Medial Geniculate Body of the Cat: Ventral and Dorsal Divisions. *Hear. Res.*, 19: 97–114, 1985.
190. Rouiller, E.M., Functional Organization of the Auditory System. In: *The Central Auditory System*, edited by G. Ehret and R. Romand. New York: Oxford University Press, 1997, pp. 3–96.
191. Russell, I.J. and Sellick, P.M., Low Frequency Characteristic of Intracellularly Recorded Receptor Potentials in Guinea-Pig Cochlear Hair Cells. *J. Physiol.*, 338: 179–206, 1983.
192. Ryugo, D.K. and Weinberger, N.M., Corticofugal Modulation of the Medial Geniculate Body. *Exp. Neurol.*, 51: 377–391, 1976.
193. Ryugo, D.K., The Auditory Nerve: Peripheral Innervation, Cell Body Morphology, and Central Projections. In: *The Mammalian Auditory Pathway: Neuroanatomy*, edited by D.B. Webster, A.N. Popper, and R.R. Fay, New York: Springer-Verlag, 1992.
194. Sachs, M.B. and Kiang, N.Y.S., Two Tone Inhibition in Auditory Nerve Fibers. *J. Acoust. Soc. Am.*, 43: 1120–1128, 1968.
195. Sakai, M.D. and Suga, N., Plasticity of the Cochleotopic (Frequency) Map in Specialized And Nonspecialized Cortices. *Proc. Natl. Acad. Sci. U.S.A.* 98: 3507–3512, 2001.
196. Santi, P., Cochlear Microanatomy and Ultrastructure. In: *Physiology of the Ear*, edited by A.F. Jahn and J. Santos-Sacchi. New York: Raven Press, 1988, pp. 173–199.
197. Santos-Sacchi, J., On the Frequency Limit and Phase of Outer Hair Cell Motility: Effects of the Membrane Filter. *J. Neurosci.*, 12: 1906–1916, 1992.
198. Schacht, J., Fessenden, J.D., and Zajic, G., Slow Motility of Outer Hair Cells. In: *Active Hearing*, edited by A. Flock, D. Ottoson, and M. Ulfendahl. Oxford: Pergamon, 1995, pp. 209–220.
199. Schreiner, C.E. and Urbas, J.V., Representation of Amplitude Modulation in the Auditory Cortex of the Cat. I. The Anterior Auditory Field (AAF). *Hear. Res.*, 21: 227–242, 1986.
200. Schucknecht, H.F., *Pathology of the Ear*. Cambridge, MA: Harvard University Press, 1974.
201. Schwaber, M.K., Neuroplasticity of the Adult Primate Auditory Cortex Following Cochlear Hearing Loss. *Am. J. Otol.*, 14: 252–258, 1993.
202. Schwartz, I.R., The Superior Olivary Complex and Lateral Lemniscal Nuclei. In: *The Mammalian Auditory Pathway: Neuroanatomy*, edited by D.B. Webster, A.N. Popper, and R.R. Fay. New York: Springer-Verlag, 1992, pp. 117–167.
203. Sellick, P.M., Patuzzi, R., and Johnstone, B.M., Modulation of Responses of Spiral Ganglion Cells in the Guinea Pig Cochlea to Low Frequency Sound. *Hear. Res.*, 7: 199–221, 1982.
204. Sellick, P.M., Patuzzi, R., and Johnstone, B.M., Measurement of Basilar Membrane Motion in the Guinea Pig Using the Mossbauer Technique. *J. Acoust. Soc. Am.*, 72: 131–141, 1982.
205. Shannon, R.V., Zeng, F,-G., Kamath. V., Wygonski, J. and Ekelid, M., Speech Recognition with Primarily Temporal Cues. *Science*, 270: 303–304, 1995.

206. Shaw, E.A.C., Transformation of Sound Pressure Level from the Free Field to the Eardrum in the Horizontal Plane. *J. Acoust. Soc. Am.*, 56: 1848–1861, 1974.
207. Shaw, E.A.C. The External Ear. In: *Handbook of Sensory Physiology*, edited by W.D. Keidel and W.D. Neff. New York: Springer-Verlag, 1974, pp. 455–490.
208. Shepherd, G.M., *Neurobiology*. New York: Oxford University Press, 1994.
209. Shinsuke, S. and Shams, L., Sensory Modalities Are Not Separate Modalities: Plasticity and Interactions. *Curr. Opin. Neurobiol.*, 11: 505–509, 2001.
210. Shore, S.E., Godfrey, D.A., Helfert, R.H., Altschuler, R.A., and Bledsoe, S.C., Connections Between the Cochlear Nuclei in Guinea Pig. *Hear., Res.*, 62: 16–26, 1992.
211. Shore, S.E., Vass, Z., Wys, N.L., and Altschuler, R.A., Trigeminal Ganglion Innervates the Auditory Brainstem. *J. Comp. Neurol.*, 419: 271–285, 2000.
212. Shosaku, A. and Sumitomo, I., Auditory Neurons in the Rat Thalamic Reticular Nucleus. *Exp. Brain Res.*, 49: 432–442, 1983.
213. Silverman, M.S. and Clopton, B.M., Plasticity of Binaural Interaction. I. Effects of Early Auditory Deprivation. *J. Neurophysiol.*, 40: 1266–1274, 1977.
214. Sinex, D.G. and Geisler, C.D., Auditory-Nerve Fiber Responses to Frequency-Modulated Tones. *Hear. Res.*, 4: 127–148, 1981.
215. Slepecky, N.B., Structure of the Mammalian Cochlea. In: *The Cochlea*, edited by P.Dallos, A.N. Popper, and R.R. Fay. New York: Springer, 1996, pp. 44–129.
216. Spangler, K.M., Cant, N.B., Henkel, C.K., Farley, G.R. and Warr, W.B. Descending Projections from the Superior Olivary Complex to the Cochlear Nucleus of the Cat. *J. Comp. Neurol.*, 259: 452–465, 1987.
217. Spoendlin, H., Structural Basis of Peripheral Frequency Analysis. In: *Frequency Analysis and Periodicity Detection in Hearing*, edited by R. Plomp and G.F. Smoorenburg. Leiden: A.W. Sijthoff, 1970, pp. 2–36.
218. Spoendlin, H., Anatomical Changes Following Noise Exposure. In: *Effects of Noise on Hearing*, edited by D. Henderson, R.P. Hamernik, D. Dosanjh, and J.H. Mills. New York: Raven Press, 1976.
219. Spoendlin, H., and Schrott, A., Analysis of the Human Auditory Nerve. *Hear. Res.*, 43: 25–38, 1989.
220. Stevens, S.S., The Relation of Pitch to Intensity. *J. Acoust. Soc. Am.*, 6: 150–154, 1935.
221. Suga, N., Responses of Cortical Auditory Neurons to Frequency Modulated Sounds in Echo-Locating Bats. *Nature*, 206: 890–891, 1965.
222. Suga, N., Auditory Neuroetology and Speech Processing: Complex-Sound Processing by Combination-Sensitive Neurons. In: *Auditory Function*, edited by G.M. Edelman, W.E. Gall, and W.M. Cowan. New York: John Wiley & Sons, 1988, pp. 679–720.
223. Suga, N., Sharpening of Frequency Tuning by Inhibition in the Central Auditory System: Tribute to Yasuji Katsuki. *Neurosci. Res.*, 21: 287–299, 1995.
224. Syka, J., Popelar, J., and Kvasnak, E., Response Properties of Neurons in the Central Nucleus and External and Dorsal Cortices of the Inferior Colliculus in Guinea Pig. *Exp. Brain Res.*, 133: 254–266, 2000.
225. Szczepaniak, W.S. and Møller, A.R. Interaction between Auditory and Somatosensory Systems: A Study of Evoked Potentials in the Inferior Colliculus. *Electroencephalogr. Clin. Neurophysiol.*, 88: 508–515, 1993.
226. Szczepaniak, W.S. and Møller, A.R., Evidence of Neuronal Plasticity within the Inferior Colliculus after Noise Exposure: A Study of Evoked Potentials in the Rat. *Electroencephalogr. Clin. Neurophysiol.*, 100: 158–164, 1996.
227. Tian, B., Reser, D., Durham, A., Kustov, A., and Rauschecker, J.P., Functional Specialization in Rhesus Monkey Auditory Cortex. *Science*, 292: 290–293, 2001.
228. Ungeleider, L.G. and Mishkin, M., Analysis of Visual Behavior. In: *Analysis of Visual Behavior*, edited by D.J. Ingle, M.A. Goodale, and R.J.W. Mansfield. Cambridge MA: MIT Press, 1982.

229. Vass, Z., Shore, S.E., Nuttall, A.L., Jabncso, G., Brechtelsbauer, P.B., and Miller, J.M., Trigeminal Ganglion Innervation of the Cochlea—A Retrograde Transport Study. *Neuroscience*, 79: 605–615, 1997.
230. Wall, P.D., The Presence of Ineffective Synapses and Circumstances Which Unmask Them. *Philos. Trans. Royal Soc. (London)* 278: 361–372, 1977.
231. Wallach, H., Newman, E.B., and Rosenzweig, M.R., The Precedence Effect in Sound Localization. *Am. J. Psychol.*, 62: 315–336, 1949.
232. Wangemann, P., Comparison of Ion Transport Mechanisms Between Vestibular Dark Cells and Strial Marginal Cells. *Hear. Res.*, 90: 149–157, 1995.
233. Wangemann, P. and Schacht, J., Homeostatic Mechanisms in the Cochlea. In: *The Cochlea*, edited by P.J. Dallos, A.N. Popper, and R.R. Fay, New York: Springer-Verlag, 1996.
234. Warr, W.B., Organization of Olivocochlear Systems in Mammals. In: *The Mammalian Auditory Pathway: Neuroanatomy*, edited by D.B. Webster, A.N. Popper, and R.R. Fay. New York: Springer-Verlag, 1992.
235. Warren, E.H. and Liberman, M.C., Effects of Contralateral Sound on Auditory-Nerve Responses. I. Contributions of Cochlear Efferents. *Hear. Res.*, 37: 89–104, 1989.
236. Warren, E.H. and Liberman, M.C., Effects of Contralateral Sound on Auditory-Nerve Responses. II. Dependence on Stimulus Variables. *Hear. Res.*, 37: 105–121, 1989.
237. Webster, D.B., An Overview of Mammalian Auditory Pathways with an Emphasis on Humans. In: *The Mammalian Auditory Pathway: Neuroanatomy*, edited by D.B. Webster, A.N. Popper and R.R. Fay. New York: Springer-Verlag, 1992, pp. 1–22.
238. Webster, D.B., Popper, A.N., and Fay, RR. Eds. *The Mammalian Auditory Pathway: Neuroanatomy*. New York: Springer-Verlag, 1992.
239. Weinberg, R.J. and Rustioni, A., A Cuneocochlear Pathway in the Rat. *Neuroscience*, 20: 209–219, 1987.
240. Weiss, T.F. and Rose, C., A Comparison of Synchronization Filters in Different Receptor Organs. *Hear. Res.*, 33: 175–180, 1988.
241. Whitfield, I.C. and Evans, E.F., Responses of Auditory Cortical Neurons to Stimuli of Changing Frequency. *J. Neurophysiol.*, 28: 655–672, 1965.
242. Wiederhold, M.L. and Kiang, N.Y.S., Effects of Electrical Stimulation of the Crossed Olivocochlear Bundle on Single Auditory-Nerve Fibers in the Cat. *J. Acoust. Soc. Am.*, 48: 950–965, 1970.
243. Willott, J.F. and Lu, S.M., Noise Induced Hearing Loss Can Alter Neural Coding and Increase Excitability in the Central Nervous System. *Science*, 16: 1331–1332, 1981.
244. Willott, J.F., Turner, J.G., and Sundin, V.S., Effects of Exposure to an Augmented Acoustic Environment on Auditory Function in Mice: Roles of Hearing Loss and Age During Treatment. *Hear. Res.*, 142: 79–88, 2000.
245. Winer, J.A. and Morest, D.K., The Medial Division of the Medial Geniculate Body of the Cat: Implications for Thalamic Organization. *J. Neurosci.*, 3: 2629–2651, 1983.
246. Winer, J.A. and Morest, D.K., The Neuronal Architecture of the Dorsal Division of the Medial Geniculate Body of the Cat. A Study with the Rapid Golgi Method. *J. Comp. Neurol.*, 221: 1–30, 1983.
247. Winer, J.A., The Functional Architecture of the Medial Geniculate Body and the Primary Auditory Cortex. In: *The Mammalian Auditory Pathway: Neuroanatomy*, edited by D.B. Webster, A.N. Popper, and R.R. Fay. New York: Springer-Verlag, 1992, pp. 222–409.
248. Winer, J.A., Larue, D.T., Diehl, J.J., and Hefti, B.J., Auditory Cortical Projections to the Cat Inferior Colliculus. *J. Comp. Neurol.*, 400: 147–174, 1998.
249. Winer, J.A., Sally, S.L., Larue, D.T., and Kelly, J.B., Origins of Medial Geniculate Body Projections to Physiologically Defined Zones of Rat Primary Auditory Cortex. *Hear. Res.*, 130: 42–61, 1999.

250. Winer, J.A., Diehl, J.J., and Larue, D.T., Projections of Auditory Cortex to the Medial Geniculate Body of the Cat. *J. Comp. Neurol.*, 430: 27–55, 2001.
251. Winer, J.A. and Prieto, J.J., Layer V in Cat Primary Auditory Cortex (AI): Cellular Architecture and Identification of Projection Neurons. *J. Comp. Neurol.*, 434: 379–412, 2001.
252. Winguth, S.D. and Winer, J.A., Corticocortical Connections of Cat Primary Auditory Cortex (AI): Laminar Organization and Identification of Supragranular Neurons Projecting to Area AII. *J. Comp. Neurol.*, 248: 36–56, 1986.
253. Wise, L.Z. and Irvine, D.R.F., Topographic Organization of Interaural Intensity Difference Sensitivity in Deep Layers of Cat Superior Colliculus: Implications for Auditory Spatial Representation. *J. Neurophysiol.*, 54: 185–211, 1985.
254. Yin, T.C.T. and Chan, J.C.K., Interaural Time Sensitivity in Medial Superior Olive of Cat. *J. Neurophysiol.*, 64: 465–488, 1990.
255. Yin, T.C.T., Neural Mechansims of Encoding Binaural Localization Cues in the Auditory Brainstem. In: *Integrative Functions in the Mammalian Auditory Pathway*, edited by D. Oertel, R.R. Fay, and A.N. Popper. New York: Springer, 2002, pp. 99–159.
256. Young, E.D. and Sachs, M.B., Representation of Steady-State Vowels in the Temporal Aspects of the Discharge Patterns of Populations of Auditory Nerve Fibers. *J. Acoust. Soc. Am.*, 66: 1381–1403, 1979.
257. Zagaeski, M., Cody, A.R., Russell, I.J., and Mountain, D.C., Transfer Characteristics of the Inner Hair Cell Synapse: Steady-State Analysis. *J. Acoust. Soc. Am.*, 95: 3430–3434, 1994.
258. Zakrisson, J.E., Borg, E., Diamant, H., and Møller, A.R., Auditory Fatigue in Patients with Stapedius Muscle Paralysis. *Acta Otolaryngol. (Stockholm)* 79: 228–232, 1975.
259. Zhang, M., Suga, N., and Yan, J., Corticofugal Modulation of Frequency Processing in Bat Auditory System. *Nature*, 387: 900–903, 1997.
260. Zhao, H. and Liang, Z., Processing of Modulation Frequency in the Dorsal Cochlear Nucleus of the Guinea Pig: Amplitude Modulated Tones. 82: 244–256, 1995.
261. Zook, J.M. and Casseday, J.H., Connections of the Nuclei of the Lateral Lemniscus in the Mustache Bat *Pteronotus parnellii*. *Neurosci. Abstr.*, 5: 34, 1979.
262. Zook, J.M. and Casseday, J.H., Origin of Ascending Projections to Inferior Colliculus in the Mustache Bat, *Pteronotus parnellii*. *J. Comp. Neurol.*, 207: 14–28, 1982.
263. Zurek, P.M.,The Precedence Effect. In: *Directional Hearing*, edited by W. Yost and G. Gourevitch. New York: Springer-Verlag, 1987.
264. Zwicker, E. and Lumer, G., Evaluating Traveling Wave Characteristics in Man by an Active Nonlinear Cochlear Pre-Processing Model. In: *Peripheral Auditory Mechanisms*, edited by J.L. Allen. Berlin: Springer, 1985, pp. 250–257.
265. Zwislocki, J.J. and Sokolich, W.G., Velocity and Displacement Responses in Auditory Nerve Fibers. *Science*, 182: 64–66, 1973.
266. Zwislocki, J.J., Five Decades of Research on Cochlear Mechanics. *J. Acoust. Soc. Am.*, 67: 1679–1685, 1980.
267. Zwislocki, J.J., Analysis of Cochlear Mechanics. *Hear. Res.*, 22: 155–169, 1986.
268. Zwislocki, J.J. and Cefaratti, L.K., Tectorial Membrane: II. Stiffness Measurements *In Vivo*. *Hear. Res.*, 42: 211–228, 1989.
269. Zwislocki, J.J., What Is the Cochlear Place Code for Pitch? *Acta Otolaryngol. (Stockholm)* 111: 256–262, 1992.
270. Zwislocki, J.J., *Auditory Sound Transmission: An Autobiographical Perspective*. Mahwah, NJ: Lawrence Erlbaum Associates, 2002.

CHAPTER 6

Vision

ABBREVIATIONS

BSC : Brachium of the superior colliculus
cGMP : Cyclic guanylic monophosphate
CN : Cranial nerve
fMRI : Functional magnetic resonance imaging
GDP : Guanylate diphosphate
GTP : Guanylate triphosphate
LGN : Lateral geniculate nucleus
M cells : Magnocellular
MT/V5 : Medial temporal (visual cortex)
nm : Nanometer = 1/1000 of a micrometer
P cells : Parvocellular
PET : Positron emission tomography
SC : Superior colliculus
SCN : Suprachiasmic nucleus
TE : Anterior inferior temporal cortex
TEO : Posterior inferior temporal cortex

ABSTRACT

The Eye

1. The eye contains the conductive apparatus, and the retina.
2. The lens projects an inverted picture onto the retina.
3. The pupil regulates the light that reaches the retina.
4. The focal distance of the lens can be changed to focus on near objects (accommodation).
5. The retina, located on the inside back wall of the eye globe, contains the photoreceptors and a neural network that connect the photoreceptors to the optic nerve.
6. The human eye has two kinds of photoreceptors, rods and cones.
   a. Rods are more sensitive to light than cones are and they are used for scotopic vision (low light).
   b. Cones are used for photopic vision (bright light). Cones contain different kinds of photopigment and provide the basis for color vision.
7. The density of cones is largest at the fovea, and there are no photoreceptors where the optic nerve leaves the retina (the blind spot).
8. The neural network in the retina consists of bipolar cells that connect the photoreceptors to ganglion cells (vertical connections). Horizontal cells and amacrine cells make connections between photoreceptors and between ganglion cells (horizontal connections).
9. Photoreceptors and bipolar, horizontal, and amacrine cells communicate by graded potentials. The earliest all-or-none potentials (discharges) occur in ganglion cells, the axons of which form the optic nerve (cranial nerve II).
10. The two principle kinds of ganglion cells are X and Y cells. X cells receive input from photoreceptors near the fovea, Y cells receive input from the periphery of the retina.
11. The receptive fields of ganglion cells consist of a combination of excitation and inhibition: "center ON" and "surround OFF" or "center OFF" and "surround ON."

Visual Nervous System

12. The optic nerves (cranial nerve II) from the two eyes merge in the optic chiasm and ascend as the optic tracts.
13. In most animals with forward-looking eyes (including humans), optic nerve fibers that originate from the temporal retina (the nasal field) continue in the optic tract on the same side while those from the

nasal retina (the temporal field) cross at the chiasm and continue in the contralateral optic tract.

Classical Visual Pathways

14. The classical visual pathway (the retinogeniculocortical pathway) uses the ventral thalamus as a relay as do other sensory systems. From the ventral thalamus the pathway projects to the primary cerebral cortex.
15. The lateral geniculate nucleus is the thalamic nucleus of the ascending visual pathways where all fibers are interrupted by synaptic connections.
16. Two kinds of cells have been identified in the lateral geniculate nucleus: M and P cells. M (magnocellular) cells have large receptive fields, and P (parvocellular) cells have small receptive fields.
17. The lateral geniculate nucleus in primates has six distinct layers: two ventral magnocellular layers and four dorsal parvocellular layers.
18. A third type of cells, konio (dust) cells, are located between the M and P layers.
19. The receptive fields of cells in the lateral geniculate nucleus are arranged as "center ON" and "surround OFF" or "center OFF" and "surround ON."
20. The primary visual cortex (the striate cortex, V1 or Area 17) projects to several other (extrastriatal) cortical regions (secondary and association cortices).
21. Two types of cortical cells have been identified: simple cells respond best to bars of light with specific orientation, and complex cells respond best to patterns that move.
22. Spatial and object information is processed separately in different regions of association cortex (stream segregation).
23. Most of the afferent connections between the cells of the lateral geniculate nucleus and those of the primary cortex are reciprocal with extensive descending connections.
24. There are extensive descending tracts. The most abundant extend from cerebral cortex to the LGN.

Nonclassical Visual Pathways

25. Two different nonclassical pathways have been identified:
    a. The superior colliculus pathway is involved in directional vision and visual reflexes.
    b. The pretectal nucleus and pulvinar pathways connect to the extrastriatal visual cortex and many other parts of the brain. These pathways mediate reflexes such as eye movements and neck movements and provide some perception of light.

26. There are extensive descending tracts, some of which may be regarded as reciprocal to the ascending tracts.

I. INTRODUCTION

The visual system has been studied extensively, and in many ways its function is better understood than any other sensory system. The visual system has been studied in great detail in the cat and Old World monkeys. It is assumed that the visual system of Old World monkeys is similar to that of humans, and the visual system of the macaque monkey seems to be most similar. There are considerable differences, however, among the visual systems of mammalian species, and the eyes of diurnal animals differ from those of nocturnal animals. Some species have color vision, while others see only black and white. In fact, few mammals other than primates can distinguish colors.* Birds, which are also diurnal animals, have color vision but use different mechanisms to achieve color vision; birds use droplets of oil to split light into its different wavelengths. Mammalian vision spans the wavelengths from 400 to 700 nm.** The range of wavelengths of human vision corresponds to the range of colors from violet to deep red. Birds can also see ultraviolet light, which mammals cannot do. Some animals (for example, some insects) can discriminate the polarization angle of polarized light, and it is believed they use that ability for navigation.

There are other species differences in the visual system in addition to the ability to discriminate color (different wavelengths of light). For example, the organization of the ascending visual pathways is different in animals with forward-directed eyes compared with animals with eyes that point laterally.

In this chapter, the unique anatomy and physiology of the visual receptor organ (the eye) and the central visual nervous system are discussed in detail, taking into consideration the variations between species. The anatomy and physiology of the visual system were compared to those of other sensory modalities in Chapters 2 and 3.

II. THE EYE

The eye contains the conductive apparatus and the *retina,* in which the sensory receptor cells are located together with a neural network that processes the information from the receptor cells. The retina is located on the back wall of the eye (see Chapter 2, page 41, Fig 2.2) where it receives light that has passed

*The bull cannot distinguish the red cloth used in a bull fight from any other color of cloth.

**The color of light is determined by its wavelength. Blue light has the shortest wavelength of visible light, and red light has the longest wavelength of what we can detect.

through the conductive apparatus of the eye. The position of the eye is controlled by six extraocular eye muscles that are innervated by three cranial nerves (CN III, CN IV, and CN VI).

A. Anatomy

1. The Conductive Apparatus

The conductive apparatus of the eye consists of the cornea, lens, and pupil, which is located in front of the lens. The lens project an inverted picture on the retina. The focal length of the lens can change (accommodation) to focus near objects on the retina, but this ability is lost with age. The pupil can regulate the amount of light that reaches the retina.

2. The Retina

The retina is located on the posterior wall of the eye globe (Fig. 6.1). It contains the sensory cells of vision, the *photoreceptors* consisting of *rods* and *cones*. The rods and cones are unevenly distributed over the retina, with cones concentrated in the *foveal* region and rods located in the periphery of the retina (Fig. 6.1). The fovea is the area of the retina where the central portion of the image is focused. That region has the greatest density of cones, and there are no blood vessels in that area. The layer of photoreceptors (rods and cones) are located behind a network of neurons connecting the sensory cells with the optic nerve (Fig. 6.2).[15,38] Light must therefore pass through that neural network to reach the photoreceptors. Light first passes a layer of *ganglion cells* and then *bipolar* and photoreceptor cells and finally reaches the light-sensitive part of the photoreceptor cells.

In some mammals, such as the cat, the backside of the eye (*tapetum*) reflects light, which therefore passes the photoreceptors twice. This lowers the visual threshold, and the animals that have a reflecting tapetum have better night vision than do humans. The eyes of such animals glow when the ambient light is low because of light reflection from the tapetum. This second pass of light through the photoreceptors may not coincide exactly with the first pass so the image may be shifted slightly and blurred. In other mammals, such as humans, the sensory cells are embedded in a dark pigment that absorbs light and prevents light from being reflected at the backside of the eye. This probably results in a better resolution than in eyes where the backside reflects light but at the expense of slightly lower sensitivity.

3. Photoreceptors

The cones and rods have characteristic morphological and functional differences (see Chapter 2, Fig. 2.9). The outer segments of cones and rods

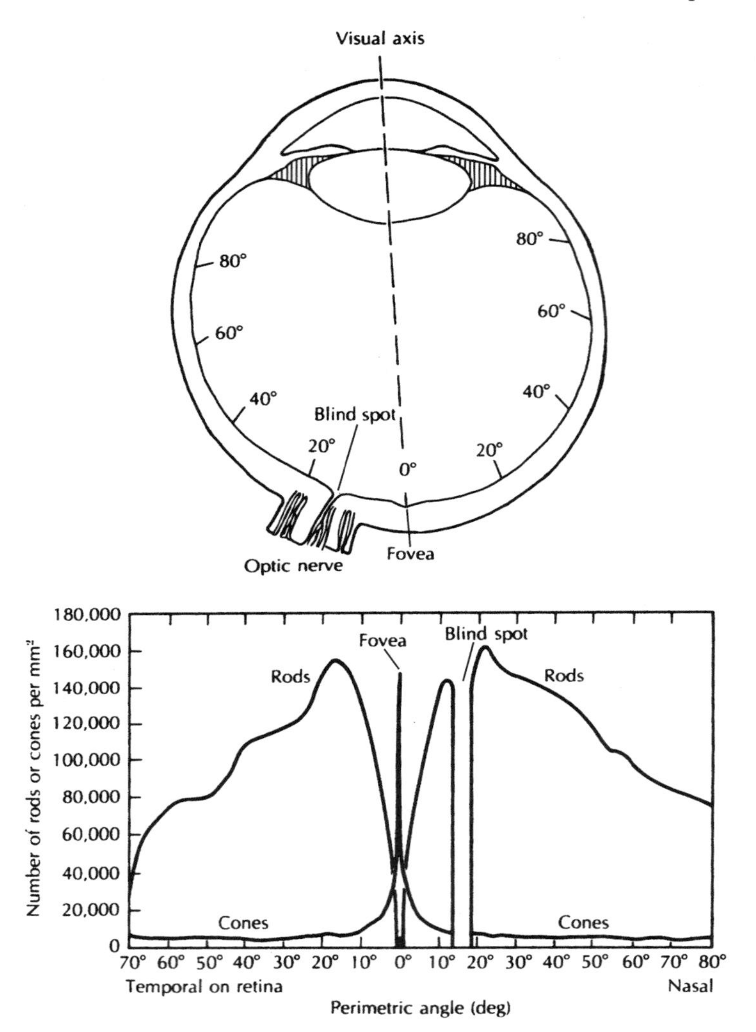

FIGURE 6.1 (Top) Cross-sectional view of the left eye. (Bottom) Density of rods as a function of the location across the retina. Notice the blind spot where the optic nerve exits the retina and where there is no photoreceptors.[9,48]

consist of modified cilia that contain *disc membranes*, which contain the *photopigment* (Fig. 2.9). The photopigment in rods is *rhodopsin*. Cones have three different kinds of photopigment, one for each of the three principle colors, blue, green, and red (see page 70). Rods account for vision in low light (*scotopic vision*). Cones have a lower sensitivity than the rods and are activated

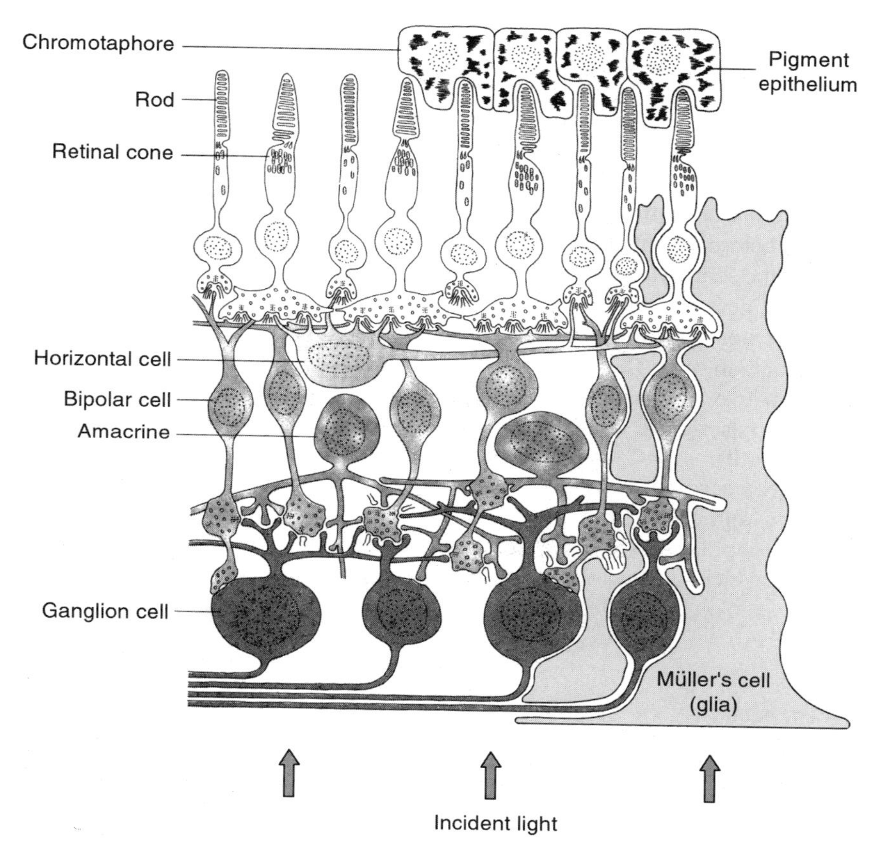

FIGURE 6.2 Schematic drawing of the organization of the primate retina. (Adapted from Schmidt, R.F. *Fundamentals of Sensory Physiology*. New York: Springer-Verlag, 1981.)

in medium and bright light (*photopic* vision). Because the cones are the basis for color vision, a certain level of light above visual threshold is necessary to see colors (Chapter 1). The human retina has approximately 20 times more rods than cones. The density of cones is highest in an area where the central visual field is projected (the fovea), and the highest density of rods is found at the location where the peripheral visual field is projected (Fig. 6.1). The rods are therefore reached by a much larger part of the visual field than that which reaches the cones. Cones provide not only color vision but also more acute vision than rods, which specialize in detecting dim light and in particular are important for detection of movement of objects. There are no sensory cells at the location where the optic nerve leaves the retina (known as the *blind* spot of the eye) (Fig. 6.1).

a. Innervation of Photoreceptors and the Retinal Neural Network

Unlike other sensory organs, the eye contains a complex network of neurons that processes the information delivered by the photoreceptors before it enters the optic nerve and is sent to the primary visual pathway (Fig. 6.2). That neural network is a part of the retina.

The information from the receptors passes at least two synapses in the retinal network before the information enters the optic nerve (Fig. 6.2). The *bipolar cells* connect the sensory cells with the ganglion cells from which the afferent fibers of the optic nerve arise. The *horizontal cells* make connections between many photoreceptor cells and similarly the *amacrine cells* make connections between many ganglion cells. The arrangement of the retinal neural network allows flow of signals in two directions perpendicular to each other: from photoreceptors to ganglion cells via the bipolar cells and laterally by the horizontal cells and the amacrine cells.

Convergence and interplay between excitation and inhibition in this network transform and restructure the coding of the image projected on the retina. Different populations of nerve cells communicate different aspects of the visual stimulus which is the beginning of parallel processing in the visual nervous system.

B. Physiology of the Eye

1. Conduction of Light to the Photoreceptors

The pupil can regulate the amount of light that reaches the retina. The lens projects an inverted image on the retina (see Chapter 2). The focal distance of the lens can be changed (accommodation) by the ciliary muscle that is innervated by the third cranial nerve and controlled by the visual nervous system. The pupils constrict in response to light (light reflex) which decreases the amount of light that reaches the retina. The pupil light reflex acts to increase the intensity range of light that the eye can process adequately. However, in humans less light attenuation is provided by the pupil than in many animals (such as the cat), and the adaptation of the receptor cells plays a more important role for regulating the sensitivity of the eye.

2. Sensory Transduction by the Receptor Cells

Because the rods are more light sensitive than cones, the sensitivity of the eye is determined by the properties of the rods. Rhodopsin, which is the photopigment of rods, has its greatest light absorption for light with a wavelength of 500 nm (nanometers), and the sensitivity of the eye is therefore greatest for light at that wavelength.

The sensitivity of the eye is dependent on prior stimulation. When the eye is not exposed to visible light its sensitivity increases gradually over many hours. When it has not been stimulated for some time the eye is said to become *dark adapted* and it reaches its highest sensitivity. Dark adaptation is a slow process that proceeds according to a two-step exponential function (Fig. 1.8). The eye is most sensitive in its dark adapted state and because that involves rods it has its greatest sensitivity to green light slightly toward the blue (approximately 500 nm) (see Chapter 1, Fig. 1.6). The sensitivity of the rods decreases gradually for light of increasingly longer or shorter wavelengths.

Excitation in photoreceptors has two steps: activation of the photopigment and activation of a second messenger, which causes a graded potential (receptor potentials) in the sensory cells. The receptor potential is the change in the membrane potential of the receptor cell, which occurs when the cell is stimulated by light. This stimulation results in a negative potential hyperpolarizing the cell membrane.[54]

In the dark, an inward (positive) membrane current that exists in the photoreceptors is known as the *dark current*. This current is caused by open cyclic guanosine monophosphate (cGMP)-gated sodium channels. This current causes the cell membrane to be polarized (approximately 40 mV). For this gate to be open, a certain amount of cGMP must be present in the cell.

In rods, light transforms rhodopsin into metarhodopsin II, which activates the membrane bond G-protein, transducin. Activation of transducin causes guanylate triphosphate (GTP) to give up a phosphate to become guanylate diphosphate (GDP). The released phosphate activates the enzyme cGMP phosphodiesterase, which causes the breakdown of available cGMP molecules to 5′GMP. This breakdown reduces the amount of cGMP in the cell, which causes the cGMP-gated channels to close. Closing of these sodium channels reduces the dark current and hyperpolarizes the cell membrane.

The fact that photoreceptors depolarize when not stimulated (in the dark) and hyperpolarize in response to light makes these receptors different from other sensory receptors (Fig. 6.3) that depolarize in response to stimulation.

3. Processing of Information in the Photoreceptors

The sensitivity of the eye depends on the properties of the photoreceptor, the medium that conducts light to the photoreceptors, and the neural network in the retina. Light adaptation and temporal summation are the result of the properties of the photoreceptors as is the role of the spectrum of light on the sensitivity of the eye. Color vision is likewise based on the properties of the photoreceptors.

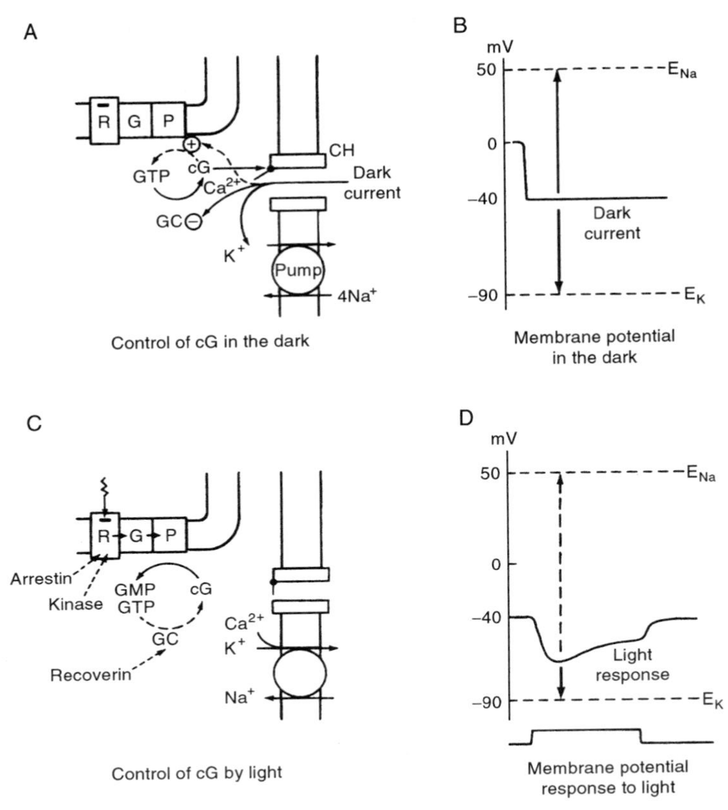

FIGURE 6.3 Schematic illustration of how light activates photoreceptors. (A) Generation of the dark current. (B) A simultaneous Na^+ current offsets the K^+ current to keep the membrane depolarized. (C) Light causes hydrolysis of cGMP that closes membrane channel. (D) Membrane potential during light stimulation that hyperpolarizes the membrane to potassium potential (Ek). (Adapted from Shepherd, G.M., *Neurobiology*, New York: Oxford University Press, 1994.)

a. Adaptation

Adaptation of the photoreceptors plays an important role for processing of information in the visual system, as it does in other sensory organs. Adaptation of the eye is a form of automatic gain control that adapts the sensitivity of the eye to the ambient illumination. The adaptation of photoreceptors provides most of the automatic gain control of the eye. The pupil also provides some automatic gain control, the range of which varies among species.

As was mentioned in Chapter 1, adaptation of the eye is often referred to as dark adaptation. Dark adaptation is the recovery of sensitivity that occurs after the eye has been exposed to bright light. After exposure to bright light the sensitivity of the eye is reduced for a long period of time (Fig. 1.8). The reduction of sensitivity is greatest for exposure to light of short wavelength (blue light), and red light is therefore less effective in reducing the sensitivity of the eye. The dark adaptation curve shows that the progressive increase in sensitivity after exposure to bright light has two steps. The initial step of the dark adaptation curve is steeper than the following segment. The first part of the dark adaptation curve represents the dark adaptation of cones and the second segment is assumed to represent rods (dashed line in Fig. 1.8).

The opposite of dark adaptation (i.e., light adaptation) occurs when the light intensity increases rapidly. Such a rapid increase in light intensity can cause a short period of impaired vision until the eye adapts to the bright light.

b. Color Vision

Color vision first appeared in reptiles and birds. It then reappeared in the line of mammals leading to primates when mammals became diurnal. Color vision present in higher orders of mammals is based on processing in the receptor cells (cones) which contain different kinds of pigment that absorb light best at specific wavelengths. Unlike these mammals, the color vision in birds and reptiles is accomplished in a different way. The photoreceptors of the retina of these animals contain droplets of oil that act as color filters. Many animals can see ultraviolet light but mammals cannot (see Chapter 1).

The eyes of primates, including humans, have three kinds of cones: one type responding best to blue light, one that responds best to green light, and one that responds best to red light (see Fig. 2.16 in Chapter 2). That is the basis for the *three-chromatic theory* for color vision. These three types of cones have photosensitive pigments that absorb light in different parts of the visible spectrum, giving the receptors their highest sensitivity at the wavelengths at which the absorption of light is maximum. The maximal sensitivity for the three types of cones occurs at light wavelengths of 420 (S pigment), 530 (M pigment), and 560 nm (L pigment).[52] Each cone probably contains only one type of pigment. The composition of these three different pigments is not known in detail.

A single type of receptor cannot code color even when it is selective with regard to the wavelength of light. This is because both a change in wavelength and a change in intensity can change the excitation of the receptor in the same way. Discrimination of color on the basis of the three-chromatic theory requires that different receptors respond differently to the wavelengths of light.

The eye in mammals that can discriminate color contains three types of photoreceptors (cones). Each type of receptor responds best to light of a specific wavelength but its sensitivity decreases gradually in response to light of a

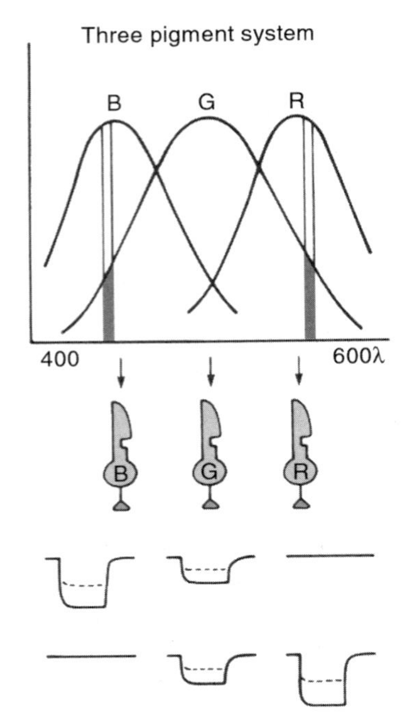

FIGURE 6.4 Illustration of how a three-pigment system can distinguish colors (wavelengths of light) independently of intensity, providing that the intensity of the light is high enough to elicit a response from at least two of the three different kinds of receptors. (Adapted from Shepherd, G.M., *Neurobiology*. New York: Oxford University Press, 1994.)

different wavelength than that at which the cones have their highest sensitivity. Therefore, light of a certain wavelength is more likely to elicit a response from one of these three types of receptor cells than from the other two receptors because the range of wavelengths to which they respond overlaps (Fig. 6.4). Because the receptive fields of the three types of cones overlap with regard to the wavelength and intensity, light of a specific wavelength may elicit a response from two or three types of receptor cells if the light is sufficiently bright. However, the strength of the response from cells of the two or three groups of cones will be different in a way that is unique for every wavelength of light within the visible spectrum. The central nervous system is able to process the information from the cones in such a way that the color of light can be determined because the response of the three different types of receptor cells provides unique information about the color of an object (Fig. 6.4). The central nervous system is able to "compute" the spectrum of light on the basis of the responses from these three different types of receptor cells, an important principle of sensory discrimination that also applies to other sensory systems that means that discrimination of a specific color does not require a specific population of photoreceptors to respond only to that color.

4. Processing of Information in the Retina

The retina has a complex network of neurons that processes visual information before it enters the optic nerve (Fig. 6.2). (The organization of the retina was recently summarized by Masland.[38]) Much of our knowledge about the processing that takes place in the neural network of the retina is based on research by Dowling published in the 1970s.[15] Dowling used the mudpuppy* for these studies and recorded the electrical potentials associated with signal processing in the retina (Fig. 6.5). The processing is characterized by interplay between excitation and inhibition.

The receptor potential in the photoreceptors is conducted via a synapse to the bipolar cells, which conduct the graded postsynaptic potential (see page 55) to the ganglion cells. The ganglion cells are the first cells in which light elicits all-or-non potentials (discharges) (Fig. 6.5).

In daylight (photopic) vision the cones are active. The cones provide visual information to the central nervous system (CNS) via the *cone pathway*. The cone pathway includes two subsets of bipolar cells, namely ON and OFF cells. Recall that the cones, like the rods, hyperpolarize when stimulated by light. The synapse that connects the photoreceptors to the ON bipolar cells is sign-inverting, which means that light causes depolarization in these cells. The connection between photoreceptors and the OFF cells is sign conserving, which means that light causes a hyperpolarization.

The rods pass information to the CNS via a slightly different pathway than the cones, known as the *rod pathway*. Rods will be excited by light of lower levels than that which can excite cones. However, a response to light can still be recorded from cones at such low light levels because each cone receives synaptic input from the rods (in the cat each cone receives input from approximately 50 rods). These synapses are electrical synapses, so the response of the cone cells cannot be isolated from the rod cells. The sensitivity to dim light benefits from the convergence of rods onto bipolar cells. This enhances the sensitivity to make it possible to create awareness of a single (or at least only a few) light quantum (photon).

The bipolar cells that receive input from rods are sign inverting but do not connect directly to ganglion cells; instead, they connect to amacrine cells that connect to ganglion cells through an excitatory synapse. These special amacrine cells connect (1) to cone bipolar cells by gap junctions (sign-conserving) that cause a depolarization (excitation) of ON center ganglion cells, and (2) to

*A large salamander (of the genus *Necturus*) that lives in water in eastern North America.

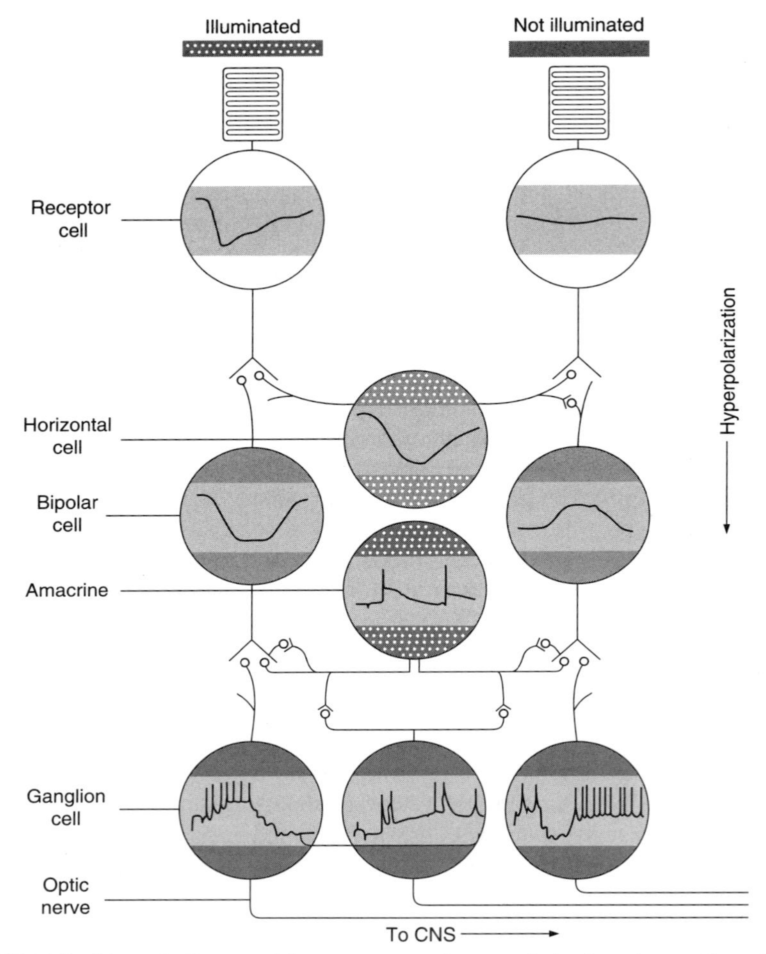

FIGURE 6.5 Schematic diagram of the arrangement of nerve cells in the retina together with the electrical potentials that are evoked by light stimulation and by darkness. (Adapted from Dowling, J.E., *Invest. Ophthalmol.* 9: 655–680, 1970. Reprinted with permission of the Association for Research in Vision and Ophthalmology.)

ganglion cells through sign-inverting chemical synapses which inhibit OFF center ganglion cells.

As mentioned earlier, the anatomical arrangement of the retina allows information to flow in two directions in the retina. One pathway connects receptor cells to ganglion cells via bipolar cells (vertical). The other pathway consisting of horizontal and amacrine cells is perpendicular to the bipolar cells. Horizontal cells interconnect receptor cells and the distal portions of bipolar cells. Amacrine cells make connections between the central portion of bipolar cells and between ganglion cells (Fig. 6.5). Both the horizontal and the vertical communications occur with graded slow potentials rather than all-or-none action potentials (discharges) as is common in neural networks. This means that the bipolar, horizontal, and the amacrine cells do not produce neural discharges but communicate by slow potentials. The earliest initiation of action potentials occurs in the ganglion cells.

Work by McCulloch's group[34] at M.I.T. showed as early as 1959 that the retinal network performs an extensive processing of visual information. They identified different kinds of retinal ganglion cells on the basis of their responses. The foundation for our understanding of the neural processing of visual information was established by the work by Kuffler[29,30] and Barlow.[5] These fundamental findings were later followed by studies by many investigators, most notable the extensive work by Hubel and Wiesel,[26,28] who described the responses of cortical cells in mammals (cats and monkey).

Kuffler[29,30] showed that the receptive fields of retinal ganglion cells consist of excitatory regions surrounded by inhibitory regions, or vice versa. The subsequent work by McCulloch's group[34] showed that responses of the frog retinal ganglion cells represent analysis and interpretation of visual information. They identified four classes of ganglion cells and showed that the ganglion cells of these classes responded in specialized fashions to (1) sharp edges, (2) curvature of the outline of objects, (3) moving edges, and (4) dimming of light intensity. They called the second group of ganglion cells "bug detectors," emphasizing the behavioral importance of these findings.

III. THE VISUAL NERVOUS SYSTEM

Like the auditory and somatic sensory systems, the visual system also has two different kinds of afferent pathways known as *classical* and *nonclassical pathways*. The classical pathway, or *retinogeniculocortical pathway,* is the main visual pathway in higher vertebrates that use the *lateral geniculate nucleus* (LGN) of the thalamus as its thalamic relay nucleus. The nonclassical pathway has at least two parts, one that projects to the *superior colliculus* (SC) and the other that involves the *pretectal nucleus* and the *pulvinar* division of the thalamus.[8,10,44] The role of the nonclassical pathways in vision includes analysis of visual space and global form perception.[35,36] It has been known for some time that the nonclassical pathways mediate various visual reflexes. The pulvinar pathway provides short and direct connections to visual cortical areas as well as to many nonvisual regions of the brain. These pathways have been identified in animals (cats and monkeys), but it is not known how prominent they are in humans.

Some early studies have identified a third pathway that provides input to the midbrain tegmentum and there is also some evidence that there are direct pathways to the *suprachiasmic nucleus* (SCN), which is involved in maintaining the circadian rhythm. The SCN projects to the hypothalamus, which, among other functions, controls endocrine functions.[8] These pathways may be regarded as a fourth nonclassical pathway. This means that, in addition to connecting to the LGN (classical pathway), the optic nerve also connects to the SC, pretectum, pulvinar of the thalamus, SCN, and hypothalamus (nonclassical pathway)

A. Anatomy of the Classical Ascending Visual Nervous System

The retinogeniculocortical pathways relay information from the retina to the primary visual cortex via the lateral geniculate nucleus (Fig. 6.6). The visual

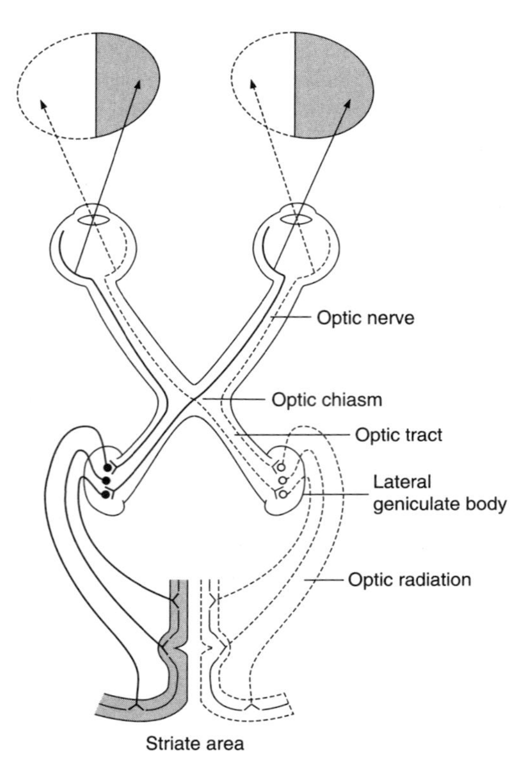

FIGURE 6.6 Schematic diagram of the main central visual pathways in human.

information is transformed, separated, and restructured as it progresses through the LGN and the primary, secondary, and association cortices. The organization of the visual nervous system is different in animals that have forward-pointing eyes compared to those whose eyes are directed laterally. Also, the central organization, such as that of the LGN, is noticeably different in various animal species.

In animals with forward-facing eyes (animals that hunt), including humans, the fibers from the lateral part of the retina generally do not cross at the optic chiasm but continue on the same side to the ipsilateral LGN. Fibers from the lateral portion of the retina corresponding to the nasal field (the lens projects an inverted picture on the retina) continues on the same side to the ipsilateral LGN. Fibers from the medial portion of the retina (corresponding to the temporal field) crosses to the opposite side to reach the contralateral LGN (Fig. 6.6). This means that information from the nasal portion of the visual field of the right eye is processed by the brain on the same side (ipsilaterally), whereas information from the temporal portion is processed in the contralateral cortex. This arrangement facilitates fusing of images from the two eyes, which is important in animals with forward-facing eyes. In animals with eyes that point sidewards, most of the fibers of the optic nerve cross at the optic chiasm and each eye is mainly represented in the opposite LGN.

1. Optic Nerve and Optic Tract

All visual information is mediated through the *optic nerve* (CN II). The fibers of the optic nerve pass through the *optic chiasm* where they reorganize and become the *optic tract* (Fig. 6.6). The optic nerve in humans has approximately 1 million fibers. The direction of the fibers from the retina is established in the optic chiasm. In higher vertebrates, most (approximately 90%) of the fibers of the optic tract belong to the classical (retinogeniculocortical) pathway, the target of which is the LGN. Only a small portion of the fibers belong to the nonclassical pathways, and these fibers reach several different nuclei such as the SC, pulvinar of the thalamus, hypothalamus (suprachiasmic nucleus), and pretectal nucleus.

The organization of the part of the optic nerve that belongs to the classical visual pathways in animals with forward-pointing eyes is best illustrated by the effect on vision from visual defects that are caused by lesions of the optic nerve and the optic tract at different locations (Fig. 6.7). If the optic nerve from one eye is severed, that eye will become totally blind (Fig. 6.7A). If the optic tract is severed on one side between the optic chiasm and the LGN, the result is *homonymous hemianopsia* (the nasal field on the same side and the temporal field of the opposite eye will be blind) (Fig. 6.7B).

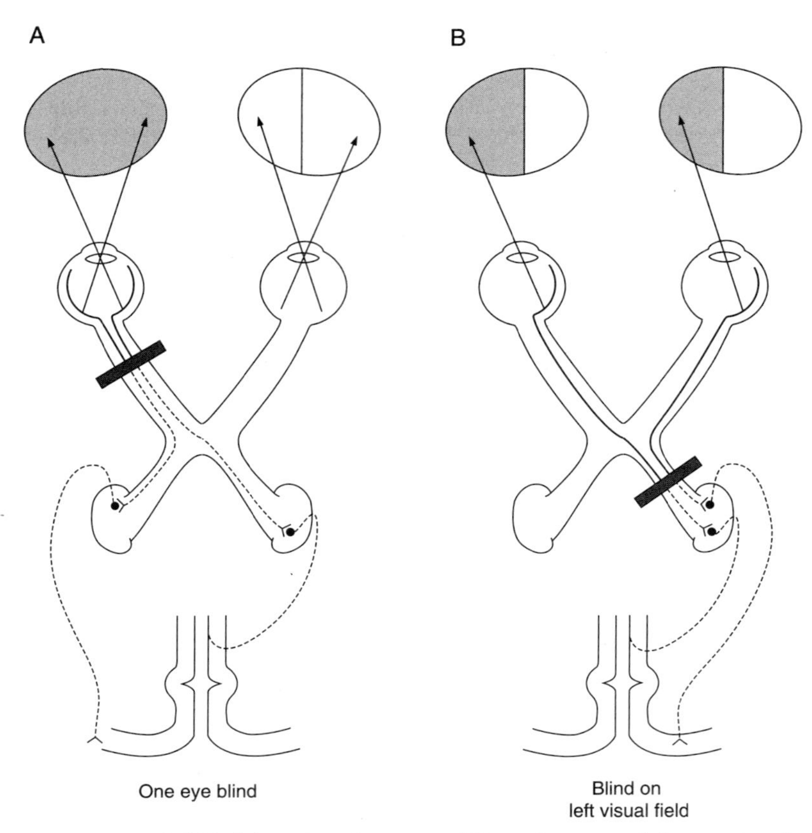

FIGURE 6.7 Visual field defects from lesions at different locations of the visual pathways. (A) Severance of one optic nerve (left). (B) Severance of one optic tract (right). (C) Severance of the optic chiasm in the midline. (D) Lesion in the right visual cortex.

Midline sectioning of the optic chiasm (Fig. 6.7C) causes *bitemporal hemianopsia* (loss of vision in the temporal field in both eyes, causing tunnel vision). Lesions at more central locations than the LGN such as the primary visual cortex will produce more complex visual defects (*scotoma*) consisting of one or more blind spots in the visual fields (Fig. 6.7D). The spots will appear in the temporal visual field opposite the side of the lesion and in the nasal field on the same side as the lesion.

2. Lateral Geniculate Nucleus

The fibers of the optic tract that belong to the classical pathway project to cells of the LGN, which is a part of the ventral thalamus. The cells of the LGN project

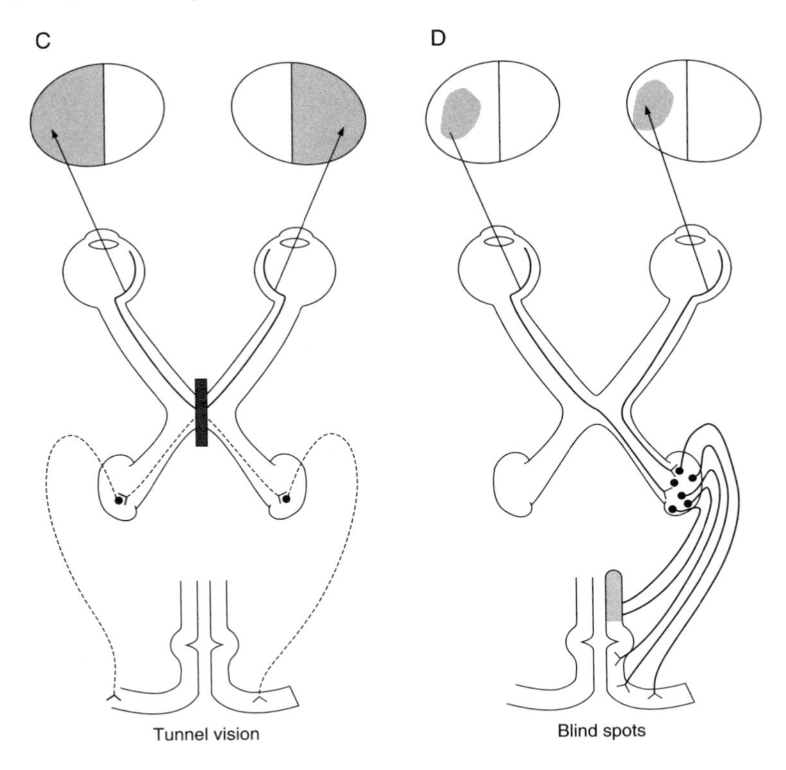

FIGURE 6.7 *(continued)*

to the primary visual (striate) cortex (Fig. 6.6). The LGN has a laminar structure that is organized in accordance with the input from different kinds of retinal ganglion cells (Fig. 6.8).[19] Each layer of the LGN receives input from one eye (monocular layers). Binocularity is only established in the cortex.

a. Classification of LGN Cells

Two classifications of retinal ganglion cells are in general use. One classification is related to the size of the cells in the LGN to which they project. At least three different types of cells have been identified in the LGN, namely the M and P cells. A third class of LGN cells, the konio cells (dust like, or tiny cells) are located between the layers of P and M cells in the LGN. This classification of retinal ganglion cells is based on the size of the cells in the LGN to which they project. The other classification is related to the way the ganglion cells respond (see page 401).

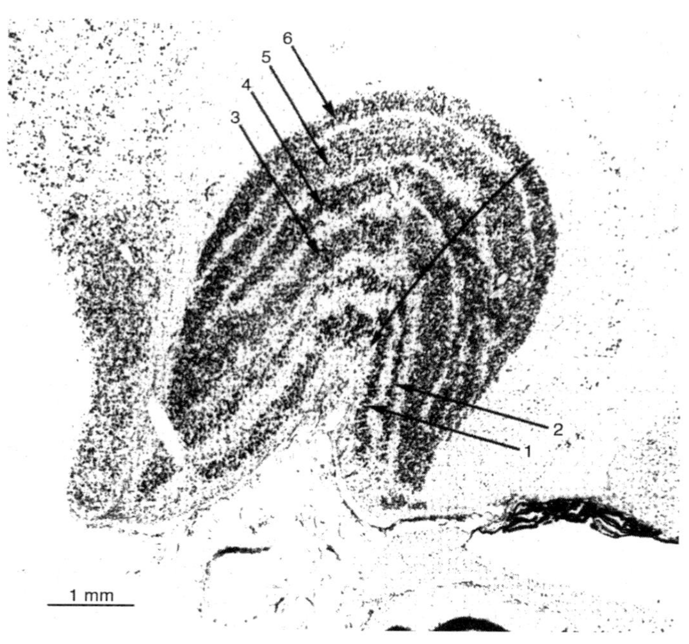

FIGURE 6.8 Six-layered laminar structure of the LGN in the macaque monkey. (Adapted from Hubel and Wiesel.[27])

b. M and P Cells

The classical part of the optic tract consists of two parallel tracts where the fibers originate from different types of ganglion cells in the retina, namely large cells (*magno*, or M cells) and small cells (*parvo*, or P cells) (Fig. 6.9). The rods and cones thus communicate with ganglion cells in parallel pathways.

The classification of M and P cells is based on the anatomical size of cells in the LGN. The M ganglion cells project to large LGN cells and P cells project to small cells in the LGN. The M and P cells comprise separate layers of the LGN known as the *magnocellular* and *parvocellular* layers. The LGN in the monkey and the cat has six layers. The P and M cells can be segregated anatomically in the LGN using the 2-deoxyglucose method[27] to show differences in metabolic activity (Fig. 6.8). These two types of cells constitute parallel processing of information because these two populations of cells process visual information of different kinds. The P cells mainly originate from ganglion cells located near the fovea, whereas the M cells are innervated by ganglion cells that represent ganglion cells in the more peripheral part of the retina. This means that the P cells serve central vision with fine discrimination and the M cells serve the peripheral visual field. The M ganglion cells have large receptive areas, and

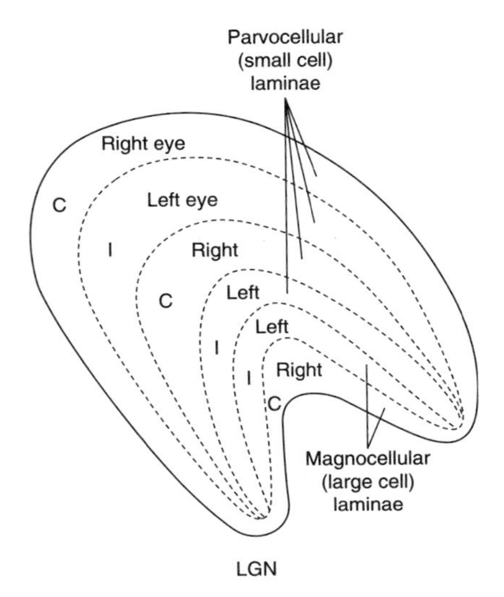

FIGURE 6.9 Organization of the M and P cells of the left LGN in a macaque monkey showing the six-layer structure. I, ipsilateral input; C, contralateral input. (Adapted from Lennie, P., and D'Zmura, M. *Crit. Rev. Neurobiol.* 3: 333–400, 1988.)

the P cells have small receptive areas. The macaque retina has 1,000,000 P cells and 100,000 M cells.

Four of the six layers in monkeys receive P cells from the two eyes in the order (from dorsal to ventral) contra–ipsi–contra–ipsi (Fig. 6.9). The two most ventral layers receive input from M cells in the order ipsi–contra. The cat has three layers with M cells (A, A1, C) and three layers with P cells (C1, C2, C3). The dorsal portion of the LGN receives input from the small ganglion cells (P cells).

c. The Konio Cells

Koniocellular cells, known as K cells in primates and W cells in the cat, occupy one third of the LGN and form three pairs of layers of cells in macaques.[25] Each pair has specific functions and projects to different structures. The middle pair mediates the output of short-wavelength cones to "blobs"* of the V1 cortex. The dorsal pair of koniocellular layers communicate low-acuity information to layer I of the V1 cortex. Neurons of the ventral pair of koniocellular layers relate to the SC, and the K layers of the LGN are the only part of the LGN that receives input from the SC.[25] Some neurons in each of the three pairs of layers of the LGN project to extrastriate cortex and these connections may explain vision in the absence of V1. There are considerable differences between the koniocellular

*"Blobs" are clusters of cells with higher metabolic activity.[37]

pathway and the M and P pathways which we discussed earlier and have been extensively studied.

3. Visual Cortex

The visual cortex that occupies the entire surface of the occipital lobe is known as Brodmann's areas 17 to 19. Area 17, the primary visual cortex, is known as the striate cortex after Gennari.* The primary (striate) cortex is also known as V1 and the several extrastriatal regions of extrastriatal cortices are known as V2, V3, V4, and MT/V5 regions. Area 17 (V1) of the visual cortex receives input from the LGN whereas Area 18 (V2, parastriate cortex) and 19 (V3, peristriate cortex) are involved in subsequent steps of visual processing. Axons from the LGN project mainly to the primary visual cortex (V1) but some relatively large cells scattered throughout the LGN project to V2 and V4.[25] Some cells in the pulvinar of the thalamus project to MT/V5 cortical regions[10,44] (see page 394).

The fact that extrastriatal cortices receive sequential input not only from the primary cortex (V1) means that visual information that activates photoreceptors at the same time may arrive at the extrastriatal cortices with different delays.

a. Organization of the Visual Cortex

The visual cortex is organized in a similar way as other sensory cortices with six layers and vertical columns (see Chapter 3, Figs. 3.7 and 3.8). These columns are organized in hyper-columns that form functional units of a region of the visual field. Each column is 30 to 100 μm wide and approximately 2 mm deep.

Just as in the LGN, neurons in the visual cortex are organized anatomically according to their origin on the retina (retinotopic organization). Approximately half of the primary cortex is devoted to the region of the fovea of the retina. Some of the clusters of cells in the V1 cortex have higher metabolic activity than others (known as "blobs") (Fig. 6.10).[37] These "blobs" are located in layer 2 of the cortex and receive input from intermediate layers of the LGN (K or W cells). The output of the LGN reaches layer 4 of the primary cortex, as is common for the primary sensory cortices (see Chapter 2). The output of M and P cells terminates in adjacent sublayers of layer 4 (Fig. 6.11).[37,53] The axons of the cells of the layers of the LGN that receive input from the (large) M ganglion cells enter layer 4C of the striate cortex (Fig. 6.11). However, only a small fraction of the synapses on these cells are from the LGN, similar to the thalamocortical connections in other sensory systems (see Chapter 3). It has been reported that only 1.3 to 1.9% of the

*Francesco Gennari was an Italian anatomist (1750–1795) who used dark staining to bind layer IV of the cortex.

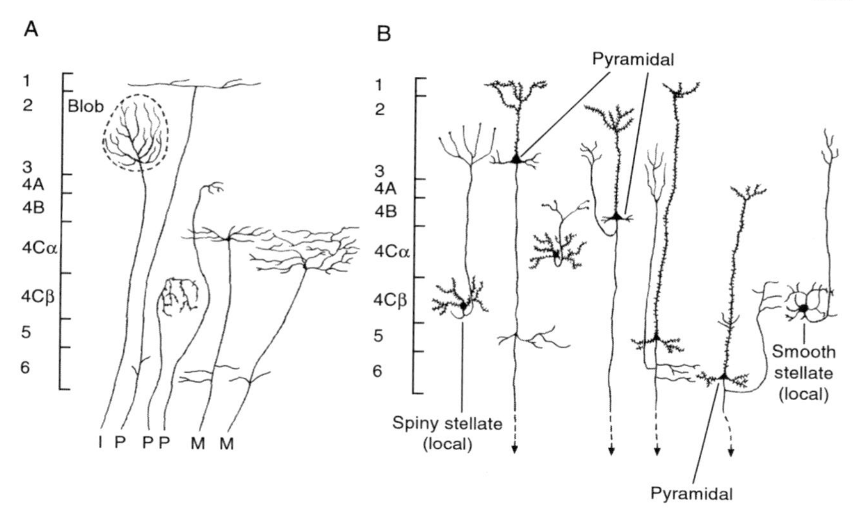

FIGURE 6.10 Several types of cells that make up the primary visual cortex. (A) "Blob" in layer II, where metabolic activity is high, which receives input from intralaminar regions of the LGN (K or W cells). (B) Distribution of various types of cells in different layers of the cortex. (Adapted from Lund, J.S., *Ann. Rev. Neurosci.* 11: 253–288, 1988 by Annual Reviews www.annualreviews.org.)

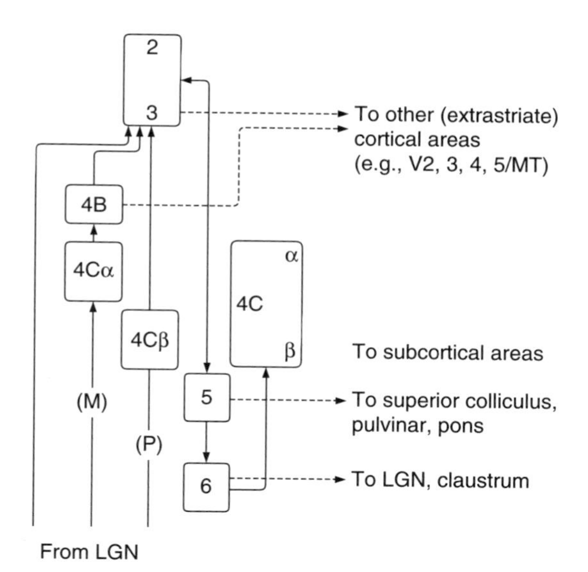

FIGURE 6.11 Connections from M and P ganglion cells to the respective layers of the LGN. (Adapted from Lund, J.S., *Ann. Rev. Neurosci.* 11: 253–288, 1988 by Annual Reviews www.annualreviews.org.)

excitatory synapses of cells in layer IVC alpha are from magnocellular layers in the LGN. The parvicellular afferents to layer IVC beta provide only 3.7 to 8.7% of the synapses on these cells.[47] The fibers from the LGN give off collateral fibers to layer VI. Cells from intralaminar regions of the LGN terminate in layers II and III. The output of the primary cortex exits from all layers except layer IVC (Fig. 6.11).[20] Connections travel horizontally in the cortex to reach other extrastriate cortical areas from layers 2 and 3.

There are connections between the modules of the primary cortex and neurons in secondary cortices. These connections originate in layers II and III of the primary cortex and reach neurons in Brodmann's area 18 (Fig. 6.11), for instance. These regions of secondary cortices specialize in different aspects of the visual input. Connections between cortical areas of one side of the brain travel in the corpus callosum to neurons in cortical areas of the opposite side.

Connections from deeper layers of the primary cortex (layers V and VI) travel to subcortical regions. Neurons of layer V project to the superior colliculus of the midbrain, the pulvinar of the posterior thalamus, and nuclei in the pons.[8] The output from layer VI reaches the LGN as a reciprocal innervation (Fig. 6.11). In fact, most of the input to the LGN originates from the cerebral cortex, indicating that the descending innervation from the cortex to the thalamus is as large in vision as it is in other sensory systems.

In summary, parallel processing in the visual system and the interplay between inhibition and excitation that was established in the retina are preserved and further developed in the LGN and primary visual cortex (Fig. 6.12). These are important factors for processing of visual information.

b. Fusion of Images

Fusion of the visual image in animals with forward-directed eyes is important for the animals' survival. The two different types of ganglion cells, M and P cells, connect to respective layers of the LGN which in turn connect to specific layers of the primary (striate) visual cortex (V1) (Fig. 6.12). The organization of the connections in the LGN and the subsequent projection of that information onto the striate cortex are involved in the fusion of the images from the two eyes (Fig. 6.13).

4. Connections from the Primary Cortex to Higher Order Visual Cortices

Considerable research efforts have been spent on understanding the role of the extrastriatal cortices and their role in processing of visual information. The results of such studies have shown that the connections between the primary striate cortex (V1) and the different parts of the higher order, extrastriatal cortices are very complex, and these regions of extrastriatal cortices are not just

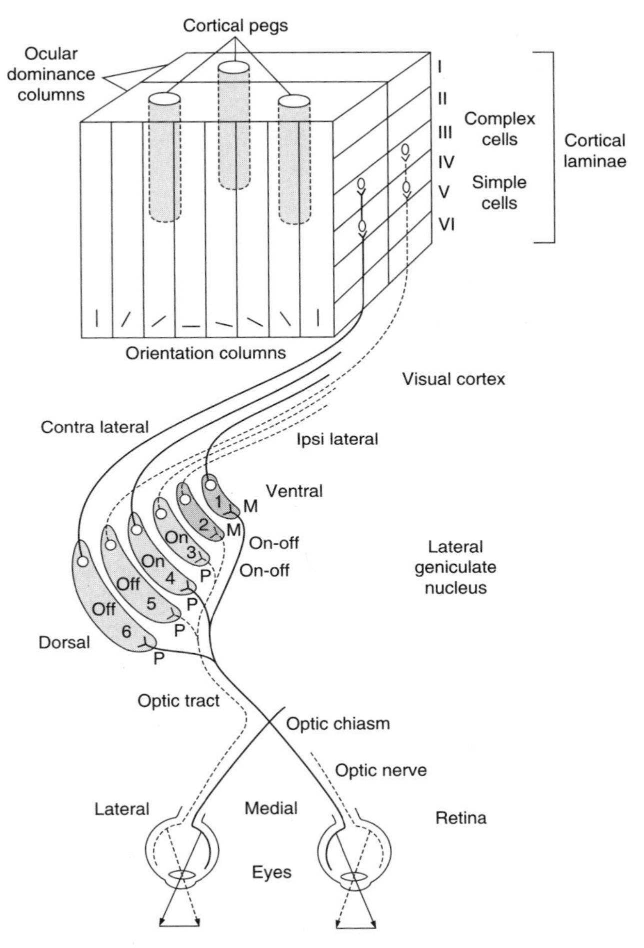

FIGURE 6.12 Summary of the organization of the classical visual pathways. (Adapted from Shepherd, G.M., *Neurobiology*. New York: Oxford University Press, 1994.)

activated in a simple sequential order (Fig. 6.14).[24] Instead, evidence has been obtained that these different parts of these extrastriate cortices process information of different kinds. In attempts to understand the intricate relationships between these cortical regions it is therefore useful to consider the concept of *stream segregation*. Stream segregation means that different kinds of information are processed in anatomically segregated populations of neurons. Solid experimental evidence supporting this concept was first obtained in studies of the visual system.[57,58] Stream segregation has now been

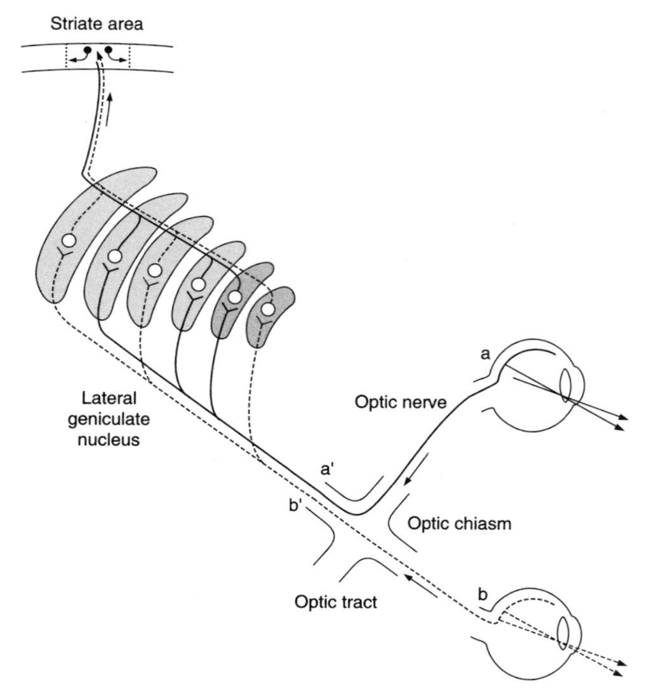

FIGURE 6.13 Connections in the visual system that are important for fusion of images. (Adapted from Brodal, P., *The Central Nervous System*. New York: Oxford Press, 1998.)

shown to be important in sensory processing in other systems such as the auditory system (discussed in Chapters 3 and 5).

5. Anatomical Basis for Stream Segregation

There is considerable evidence that (at least) two separate parallel streams of information leave the striatal cortex to reach separate regions of secondary cortices (Fig. 6.14A).[40,41,57,58] The dorsal stream located in the parietal region (MT/V5) (Fig 6.14B) specializes in processing spatial information (''where''), including motion. The other stream of information, namely the ventral stream, targets neurons in the ventral part of the extrastriatal cortex. These regions of extrastriatal cortex process object information (''what''). This organization is the anatomical basis for separate processing of spatial and object information.

It is possible that these connections demonstrated in monkeys are different in humans in view of the different anatomy of the human brain, specifically the much larger temporal lobe in humans. Considering the enormous complexity of the extrastriatal visual cortical system, most likely there are several other separate

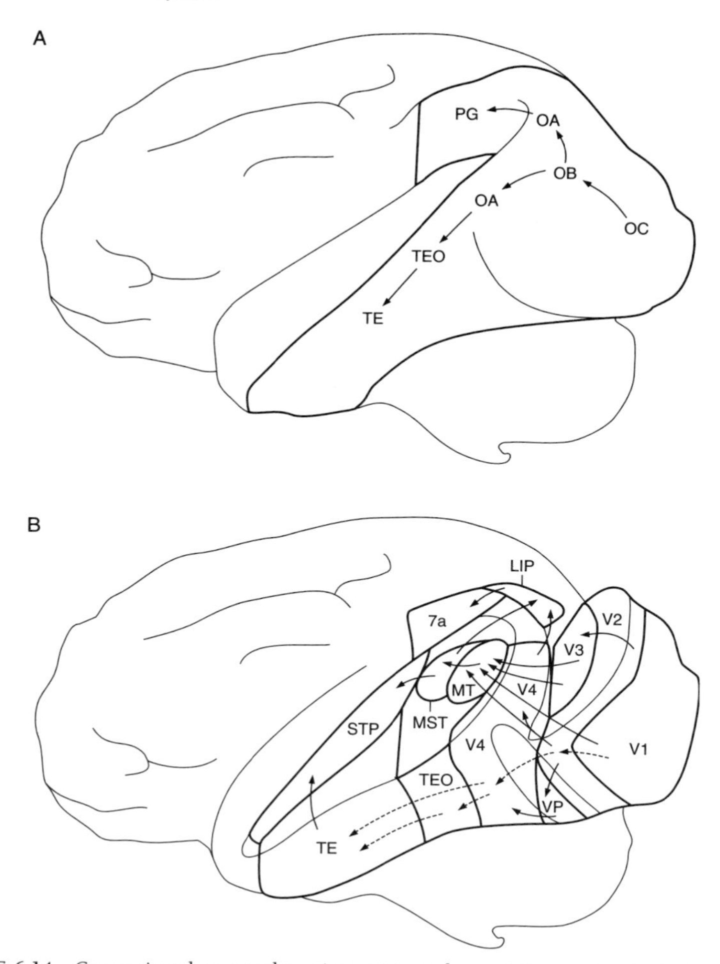

FIGURE 6.14 Connections between the striate cortex and association cortices in the brain of the monkey. (A) Anatomical separation of information into two principal streams according to Mishkin *et al.*[41] The nomenclature of cortical regions (see Fig. 6.14C) is from von Bonin and Bailey.[7] (B) More detailed descriptions of connections from the primary visual cortex (V1) to higher order visual cortices, using a lateral view of the monkey brain. (C) Some of the connections of the inferior temporal cortex to other cortical areas. The main afferent pathway is from V1 to V2 to V4 and from V4 to the posterior inferior temporal cortex and the anterior inferior temporal cortex. Neurons in the areas with faces have been found to respond selectively to faces. (Adapted from Gross *et al.*[24])

anatomical locations for processing of visual information just waiting to be discovered. The illustrations in Fig. 6.14 show only connections from lower to higher order regions (afferent) but there are extensive (mainly reciprocal) connections in the opposite direction.

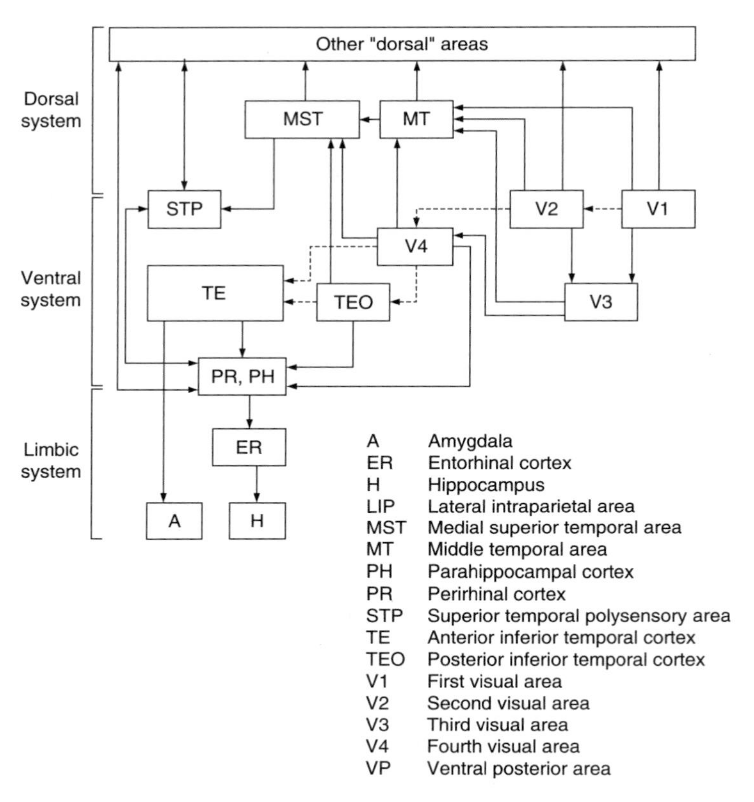

FIGURE 6.14 *(continued)*

B. Physiology of the Visual Nervous System

This section concerns the neural processing that occurs in the ascending pathways including the primary visual cortex, the extrastriatal cortices, and association cortices. The fundamental findings by Kuffler and Barlow in the early 1950s,[5,29,30] by Lettvin *et al.*,[34] and by Enroth-Cugell[18] provided the foundations for work that followed later, most notably the extensive work by Hubel and Wiesel,[26,28] which described the responses of cortical cells in mammals (cat and monkey) to stimulation with different kinds of visual patterns such as stripes with different orientations and moving stripes.

1. Information Processing in the Classical Visual Pathways

Processing of visual information in the nervous system is based on the processing that occurs in the retina. Much of the responses from neurons in the visual pathways can be traced to the responses of retinal ganglion cells, and the way neurons in the LGN and the primary visual cortex (V1) respond can be explained on the basis of modifications of the interplay between inhibition and excitation that is already present in retinal ganglion cells.

The transformation of the neural code of visual images that leaves the retina in the optic nerve continues as the information ascends in the visual pathways and in the extrastriatal cortices (V2, V3, V4, and MT/V5). Parallel processing and stream segregation contribute to the transformation and restructuring of the neural message in preparation for interpretation. Unlike the senses of hearing and touch, the sense of vision is little concerned with the time pattern of the light that reaches the eye and most of the information is contained in differences in contrast. Processing differences in contrast and changes in contrast, including movement of objects in the visual field, therefore dominates the neural processing of visual information in the ascending visual pathways.

2. Classification of Retinal Ganglion Cells

It was mentioned earlier that cells in the LGN were classified according to their size in M and P cells. Retinal ganglion cells are classified in accordance with the M and P cells that they innervate in the LGN. The M cells have large receptive fields and show rapid adaptation (respond transiently to illumination). These cells can follow rapid changes in illumination (compared with rapid adapting mechanoreceptors). The P cells that have small receptor fields respond to fine details and are assumed to be important for perception of form. The P cells are color selective and originate from the foveal area of the retina, while the M cells originate from peripheral areas of the retina. The other group of ganglion cells innervates the small cells in the LGN (konio or dust cells), known as K cells in the monkey and W cells in the cat.

Another classification of retinal ganglion cells is based on the way that ganglion cells respond to light stimulation. That classification was proposed by Enroth-Cugell and co-workers[18] and is based on the response of ganglion cells to stationary sinusoidally shaped grating patterns. It divides the ganglion cells of the retina of primates into two groups, the X and the Y ganglion cells. These investigators found that these two groups of cells responded differently to stationary sinusoidal grating patterns despite the fact that they had similar center-surround receptive fields, and they showed that X cells responded best to a certain orientation of a grating pattern (Fig. 6.15). When the pattern was reversed (180-degree shift) the X cells responded with opposite polarity of the response.

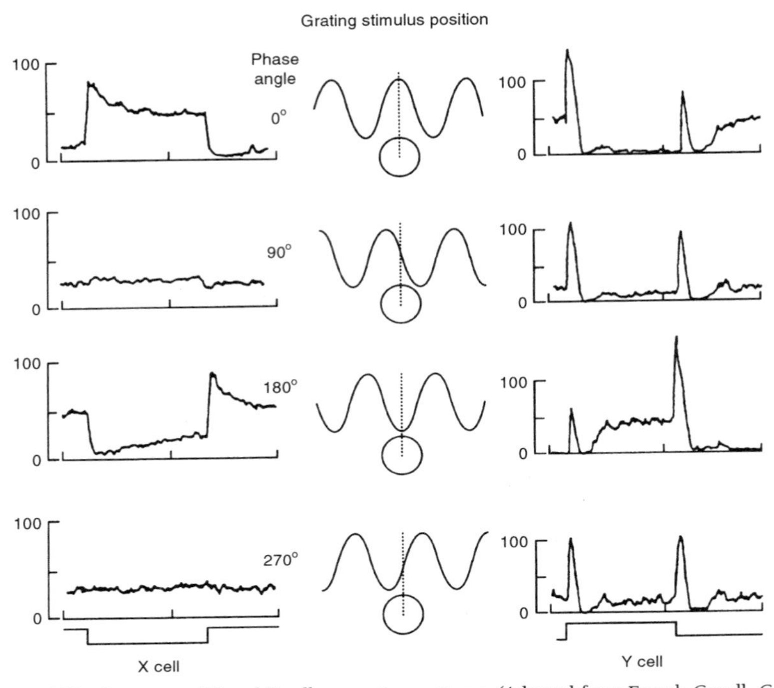

FIGURE 6.15 Response of X and Y cells to grating patterns. (Adapted from Enroth-Cugell, C., and Robson, J.G. *J. Physiol. (Lond.)*. 187: 517–552, 1966.)

In between (90-degree shift), the X cells did not respond at all. This means that X ganglion cells sum spatial stimulations in a *linear* way. The Y cells always respond positively to a grating pattern independent of its direction, and there was no null position of the grating (pattern where there was no response). This means that Y cells add input in a *nonlinear* fashion (Fig. 6.15).

The X and Y ganglion cells are different morphologically. The X ganglion cells receive input from receptors near the fovea (center of the visual field) whereas the Y ganglion cells mainly get their input from the periphery of the retina. The X cells are involved in high-resolution vision, whereas the Y cells are concerned with motion. (In the cat, the X cells are also known as β cells and the Y cells are known as α cells.)

The Y cells have a high degree of convergence, and a single ganglion cell receives input from approximately 100 bipolar cells which in turn receive input from as many as 1500 rods. This convergence enhances sensitivity and, as mentioned earlier, rods are active in dim light while the cones are inactive. The high sensitivity of the rods is helped by this convergence of many receptors upon ganglion cells. This means that Y cells are involved in night vision. The X cells are different from Y cells in many ways. In primates, the ratio of cones to bipolar cells to X cells is closer to one. In the cat, 36 cones converge onto 9 cone bipolar cells, which connect to one β ganglion cell (corresponding to X cells in primates). The cones that supply the input to the X (or β) cells are located near the fovea, and these cells are involved in fine definition vision. Because the input to the X cells comes from cones, these cells are mainly involved in daylight vision and sum input linearly, whereas the Y cells sum input nonlinearly.[18] The Y cells are concerned with detection of light and direction of movements with little emphasis on fine details, whereas the X cells communicate accurate and fine-resolution images, best served by a linear summation of input.

The conduction velocity of the axons in the optic nerve that originate in Y cells is faster than that in axons that innervated X-type ganglion cells. The X ganglion cell projects to the LGN while the Y cells project both to the LGN and to the nonclassical pathways (SC and interlaminar nuclei of the midbrain).[8]

3. Cone and Rod Pathways

The rods and cones connect to separate bipolar cells, forming a separate *cone pathway* and a *rod pathway* (Fig. 6.16). The rod and cone pathways proceed through the retina in parallel with some interactions. The rods connect to special bipolar cells where hyperpolarization of a cone leads to depolarization of a (rod) bipolar cell. This means that the connection between rods and bipolar cells is a sign-inverting synapse. These bipolar cells do not connect to ganglion cells but rather to a specific kind of amacrine cell, which connects with electrical synapses to ON center ganglion cells and by chemical synapses to (inhibit Off center) X ganglion cells.

The organization of the rod pathway favors high sensitivity and this serves to enhance the response to single photons. It is less direct than the cone pathway. At the same time, there is divergence, as in when one rod connects to two bipolar cells and from there on to two ganglion cells (Fig. 6.16).[15]

Other types of ganglion cells about which little is known convey information about ambient light levels. Some of these cells project to nonclassical pathways (SC and SCN) in addition to the LGN.

4. Receptive Fields of Ganglion Cells

The ganglion cells provide the output of the retina to the optic nerve. Receptor cells converge onto ganglion cells and the *receptive field* of ganglion cells is therefore a combination of the receptive fields of many photoreceptors. The receptive field of ganglion cells overlap, which means that a small spot of light elicits response from several ganglion cells. The receptive field of ganglion cells

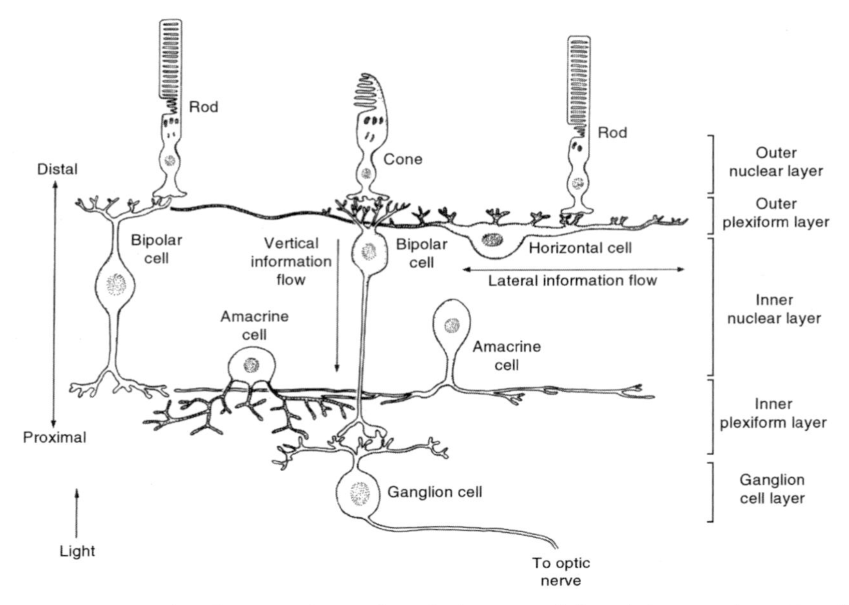

FIGURE 6.16 Rod and cone pathways through the retina. (Adapted from Dowling, J.E., *The Neurosciences: Fourth Study Program*, edited by Schmitt, F.O., and Worden, F.G. Cambridge, MA: MIT Press, 1979, 163–182.)

is roughly circular. The receptive field of most ganglion cells has two parts: the *center* and the *surround*.[29,30]

The receptive field of ganglion cells is complex as a result of specific interplay between inhibition and excitation. Our understanding of the interplay between inhibition and excitation in the retina is to a great extent based on the fundamental work by Kuffler,[30] who showed that illumination of the center of the receptive field of certain ganglion cells resulted in excitation. That response could be inhibited by shining another spot of light at locations within a region of the visual field that surrounds the area of the retina where light gives rise to excitation. Such ganglion cells are known as ON center cells (Fig. 6.17) because they are turned on by illumination of the center of their response. Other ganglion cells respond in the opposite way, namely with excitation to illumination in a ring around a center where light inhibits the response. Such cells are known as OFF center cells because they are turned off by light at the center of their receptive field.

The interplay between inhibition and excitation is the basis for *lateral* inhibition as discussed in Chapters 1 and 2. Lateral inhibition enhances contrast by suppressing even illumination. Near the border between light and dark the light areas appear lighter and the dark areas appear darker. This is explained by the interaction between inhibition and excitation as illustrated in Fig. 6.18.

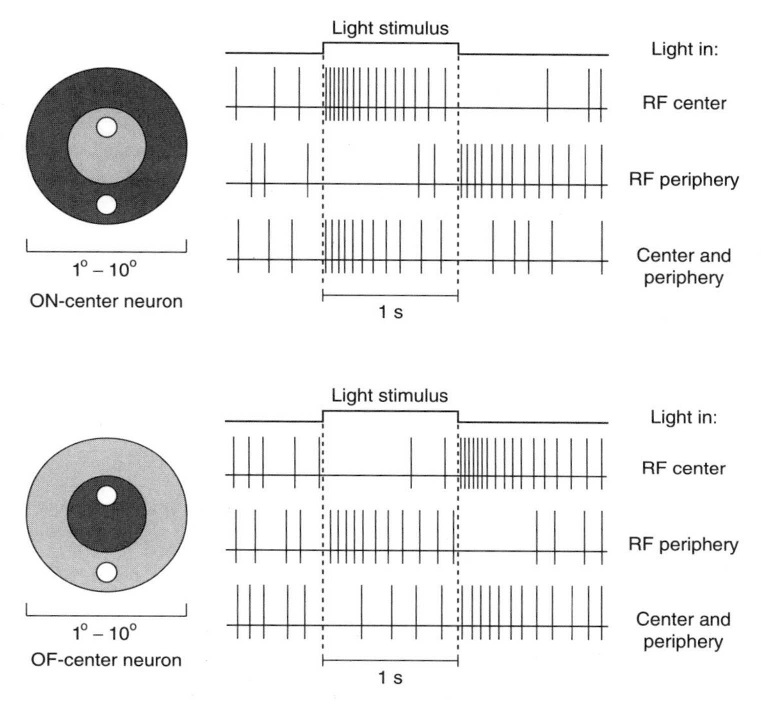

FIGURE 6.17 The response of two different types of retinal ganglion cells. RF, receptive field.

Inhibition increases when the inhibitory surround receptive field of a ganglion cell is illuminated without increasing the excitation, and the result is that the area is perceived to be darker. When the excitatory center of the receptive field of a ganglion cell is illuminated, excitation increases and the object is perceived to become gradually lighter. This means that the response of ganglion cells can explain the edge enhancement evident from psychophysical studies.

The implications of the interplay between inhibition and excitation are many, and it has a fundamental importance for processing of visual images. A diffuse illumination of ON or OFF ganglion cells produces little change in its discharge rate because it activates inhibitory and excitatory regions equally. This arrangement of excitatory and inhibitory response areas suppresses stimuli that illuminate a large portion of the visual field evenly because they excite and inhibit approximately the same number of cells (Fig. 6.18); even and constant illumination, therefore, causes little increase in the firing of ganglion cells. This is beneficial because such stimuli contain little information and the limited dynamic range of neurons is reserved for transmitting important changes in the sensory stimuli.

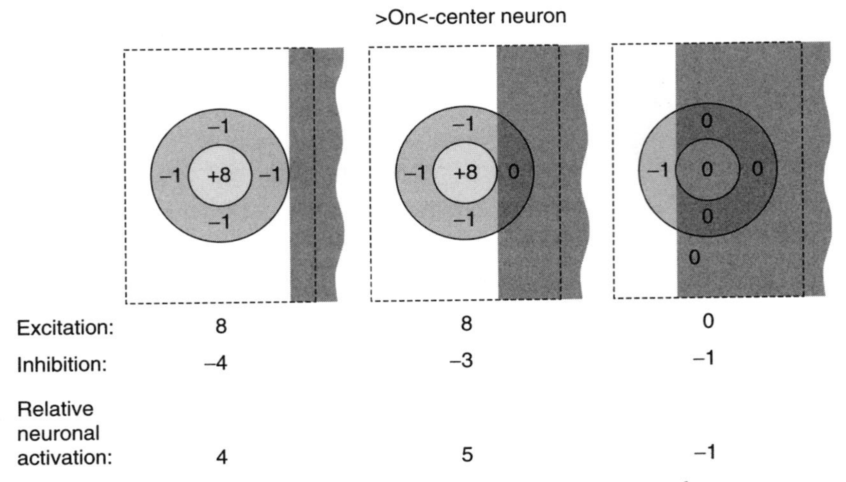

FIGURE 6.18 The effect of exposure to diffuse light that activates inhibitory and excitatory response areas of a retinal ganglion cells simultaneously. (Adapted from Schmidt, R.F., *Fundamentals of Sensory Physiology*. New York: Springer-Verlag, 1981.)

It is important to point out that the arrangements of inhibitory and excitatory fields such as those shown in the previous figures change as a result of the intensity of the illumination. Thus, the receptive field of an ON center ganglion cell in the retina decreases in size with increasing illumination (Fig 6.19). The receptive fields of ganglion cells also depend on the level of adaptation, which alters the relation between inhibition and excitation.

The dependence of the relationship between excitation and inhibition on intensity can be explained, at least to some extent, by the make up of the ganglion, horizontal, and amacrine cells in the retinal network. The excitation of ganglion cells is mediated by convergence of input from several bipolar cells onto ganglion cells; whereas, the (lateral) inhibition is mediated by horizontal and amacrine cells. This causes inhibition to have a higher threshold than excitation (Fig. 6.19).

The processing that occurs in the retina means that some of the fundamental principles of visual processing are already established in the eye (the retina). Therefore, the response of retinal ganglion cells can explain some of the psychophysical properties of vision such as the contrast enhancement lateral inhibition, as mentioned earlier (see Chapter 1).

5. Lateral Geniculate Nucleus

The LGN receives only 10 to 20% of its input from the retina, and the remaining input comes from several parts of the brain, most noticeably from the

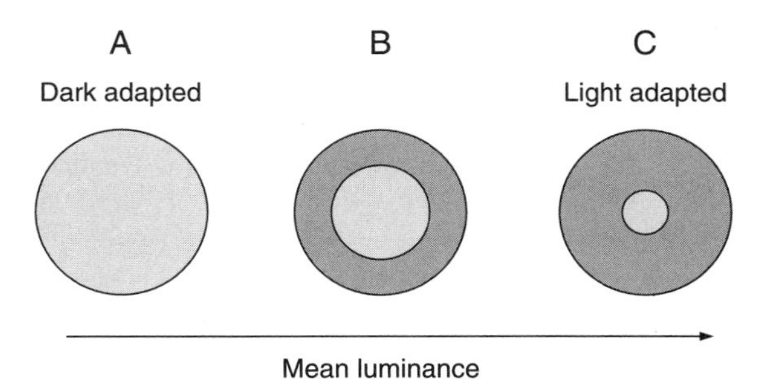

FIGURE 6.19 The effect of light exposure on the relation between inhibition and excitation in an ON center ganglion cell. (Adapted from Schmidt, R.F., *Fundamentals of Sensory Physiology*. New York: Springer-Verlag, 1981.)

cortex and the reticular formation. This is similar to the situation for hearing, where most of the input to the LGN is descending input that comes from the cortex.

The projection of the optic tract onto the LGN is retinotopic, which means that the neurons of the LGN are organized according to the location on the retina where the nerve fibers that innervate them originate, and an image that is projected on the retina activates an array of nerve cells in the LGN in accordance with the organization of stimulation of cells in the retina. The receptive fields of neurons in the LGN are similar to those of retinal ganglion cells, with circular concentric inhibitory and excitatory areas.

Signal transmission in the LGN is greatly dependent on the state of wakefulness (arousal). During sleep, little information is transferred through the LGN. This can be explained by the large input from the reticular formation to the LGN.

a. M and P Pathways in the LGN

The response characteristic of M and P cells in the LGN are much better known than those of the K (or W) cells. The response of M and P cells differ in fundamental ways. The six main differences in the way in which P and M cells in the LGN in the rhesus monkey respond are:[60]

1. P cells have smaller receptive fields than M cells.
2. The axons from M ganglion cells conduct faster than those from P cells.
3. M cells respond in a transient way to sustained stimulations (rapid adaptation).
4. P cells can respond to sustained illumination (slow adaptation).
5. M cells are more sensitive to low-contrast light than P cells are.
6. P cells are sensitive to color contrast but M cells are not.

The M cells have large receptive fields and these cells mainly respond to movement and differences in contrast, whereas the P cells have small receptive fields and provide fine visual acuity and color. M cells are not color sensitive and may play a more important role in scotopic vision than the P cells. Both cell types can either be center ON or center OFF cells.

M cells can detect 2% differences in luminescence contrast whereas the P cells require at least 10% difference in contrast to respond. M cells tend to have lower spatial resolution than P cells, which can be understood by recalling that M cells integrate input from receptor cells located over larger areas of the retina than the P cells. Why M cells have higher temporal resolution than P cells is less obvious.

The difference between M and P cells with regard to visual perception has been studied in ablation experiments in macaque monkeys (Fig. 6.20),[39] where the magnocellular (M) and parvocellular (P) regions of the LGN were lesioned (ablated) separately. Some of these experiments were aimed at the ability to distinguish the stripes of a pattern of light and dark bands. The experiments were conducted by presenting a pattern of stripes where the contrast (luminance; namely, the difference in luminance between the brightest and the darkest stripes) was varied for each separation of the stripes until the monkey could no longer discern the grating of the pattern.[39] The difference between the M and P cells is greatest for contrast gratings and less for temporal grating.

Color vision in primates depends on P cells as shown in experiments where laminae of M or P cells were ablated. Removal of P cells leads to complete color blindness. The P cells respond to color contrast regardless of brightness, whereas the response of M cells to color is dependent on brightness. The M cells do not distinguish color contrast from brightness contrast. This is just the opposite of luminance contrast, for which the M cells have superior sensitivity.

b. Konio Cell Pathways in the LGN

The konio pathway consists of three different parts, originating in the dorsal, middle, and ventral layer of the LGN. The physiologic properties of the K (or W) cells are sparsely known and the only extensive published studies were made in bush babies.[25] Generally, their properties vary widely but may be described as being somewhere between that of M and P cells.[25,43] It is interesting that the firing rate of some K cells in response to light stimuli is affected by acoustic or tactile stimulation.[42,43] This means that this pathway may resemble nonclassical polysensory pathways of other sensory systems such as the auditory system (see page 347). There are indications that the response pattern of K cells varies more among species than that of M and P cells.[43]

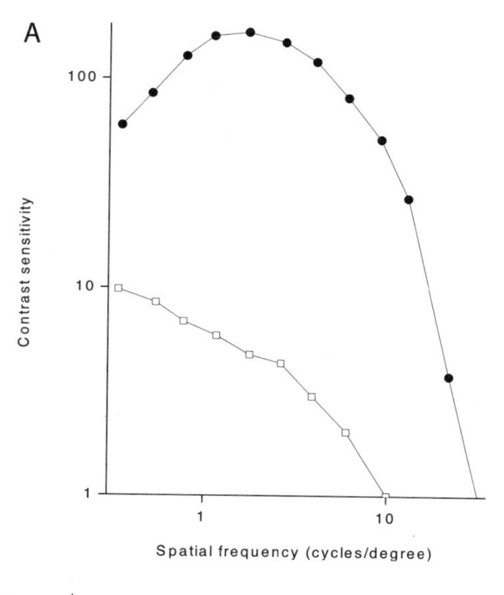

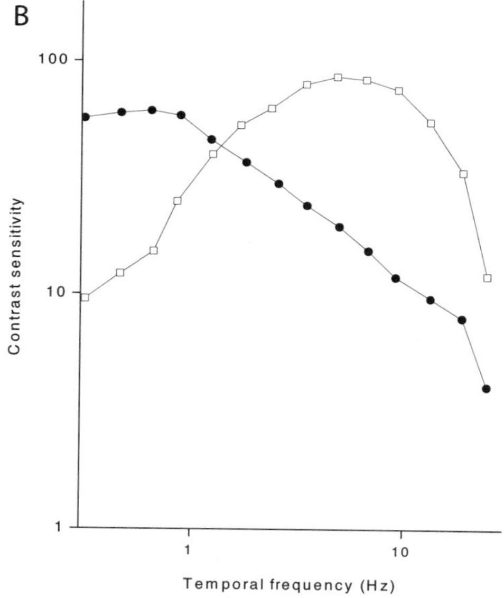

FIGURE 6.20 Differences in the response to grating stimuli of M and P cells in the LGN. The graphs show that how the contrast sensitivity is affected by selective ablation of P and M cells in the LGN of a monkey. (A) Contrast sensitivity to a grating pattern where the spatial frequency (in cycles per degree of the visual field) of dark and light stripes was varied. (B) Similar graphs as in (A) but showing the effect of changing the temporal frequency (cycles per second of dark and light stripes) of a similar grating pattern. Filled circles, P cells alone; open squares, M cells alone. (Data from Merigan and Maunsell.[39])

6. Primary (Striatal) Visual Cortex

The primary visual cortex (Brodmann's area 17) is also organized in a retinotopic fashion, which means that neurons of the columns of the cortex respond according to their innervation from the retina. Many cells in the primary visual cortex respond to stimulation of both eyes and these cells are the first cells in the visual pathways that respond to binocular stimulation.

The visual cortex is organized in columns, as are the somatosensory and auditory cortices. Each of these columns is a functional module within which processing of the input that arrives from the LGN to layer IVC occurs. The inputs from P and M cells of the LGN have slightly different targets in layer IVC (Figs. 6.10 and 6.11). The neurons in layer IVC have concentric receptive fields. Above and below that layer are simple cells that respond best to lines or bars of light of a specific orientation. Complex cells in a column receive input from simple cells. The axis of orientation of the receptive fields of neurons in one column is slightly different from that of neurons in adjacent columns. About every three quarters of one millimeter the cycle of orientation of the receptive field repeats itself. This orderly organization of the visual cortex was first demonstrated using electrophysiologic methods by Hubel and Wiesel.[26,28]

The cells of the primary visual cortex (V1) respond differently than cells in the LGN and the retina, and a spot of light is less efficient in activating cells in the visual cortex than cells in the LGN, except the input layer (layer IVC). Cells in other layers prefer stimuli with linear properties such as bars or lines. While the responses of LGN and retinal cells are dominated by their center surround ON or OFF characteristics, cells in the cortex respond best to linear contours and respond specifically to the orientation of light or dark bands. Cortical cells also respond much better to moving contours than to steady images.

This insight is a result of the research of Hubel and Wiesel,[26,27] who extended Kuffler's[30] work on the retina to higher central nervous system structures. Hubel and Wiesel identified two types of cells in the primary visual cortex, *simple cells* and *complex cells*. The response characteristics of simple cells could be explained by elongating the ON and OFF regions of cells of the retina. It was hypothesized that the complex cells acquired their characteristics by the fact that they receive their input from several simple cells.

The simple cells respond best to a bar of light with a specific orientation such as a vertical bar. A cell that responds to a vertical light bar will respond poorly or not at all to a horizontally oriented light bar (Fig. 6.21). These simple cells are arranged such that a population of cortical cells will represent all orientations of light bars. The simple cortical cells also have excitatory and inhibitory receptive fields but the size of the inhibitory and excitatory regions is larger than for LGN cells (Fig. 6.22), and the shape is not just circular, as in the case of retinal and

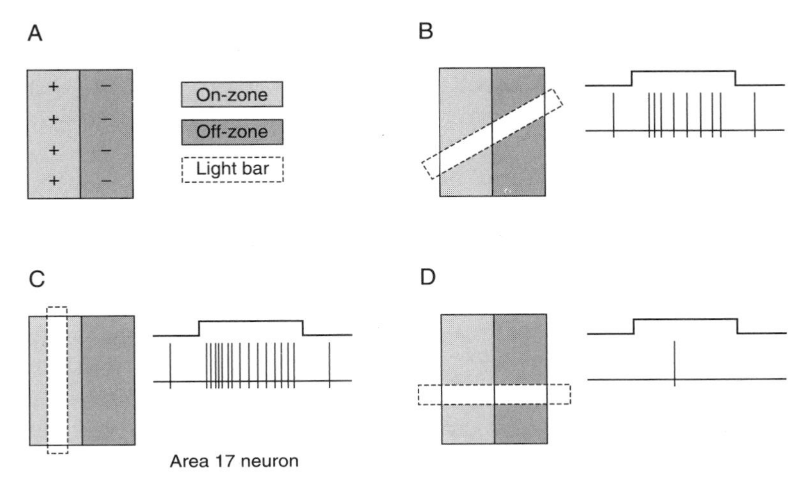

FIGURE 6.21 Responses from a simple cell in the primary visual cortex to stimulation with light bars. (Adapted from Hubel, D.H., and Wiesel, T.N., *J. Physiol. (Lond.)*. 160: 106–154, 1962.)

LGN cells. The preferential response to rectilinear pattern (bars of different orientation) could be created by combining several circular receptive fields such as could occur by many LGN neurons converging onto a single cortical neuron (Fig. 6.23).

The main difference between complex cells and simple cells is that complex cells respond vigorously to patterns that move. Complex cortical cells have larger receptive fields than simple cells but the orientation of light bars is less critical than it is for the simple cells. Light bars that move across the receptive field of complex cells are effective stimuli for many complex cells. It has been hypothesized that the input to the complex cells comes from simple cells in the cortex and not from the LGN that enters layer IVC (Fig. 6.23).

a. Fusion of Visual Images

It is important for animals with forward-pointing eyes, such as humans, to fuse the two images from the two eyes. The two eyes are represented independently in the LGN but many neurons in the striate cortex receive input from both eyes. It is therefore assumed that fusion of images occurs in the striate cortex. Fusion depends on accurate alignment of the optic axes for the two eyes, which is performed by appropriate actions of the extraocular muscles. For fusing images at a long distance the optical axes of the two eyes must be parallel, but for fusion of the image of near objects the optical axes must be at an angle to each other. This control originates from the cerebral cortex via the pathway that

A

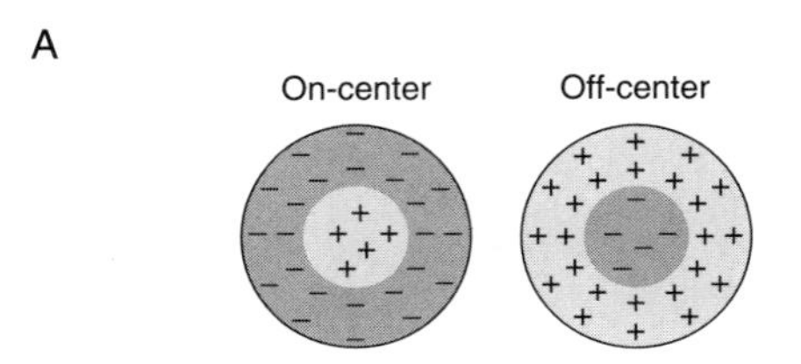

B

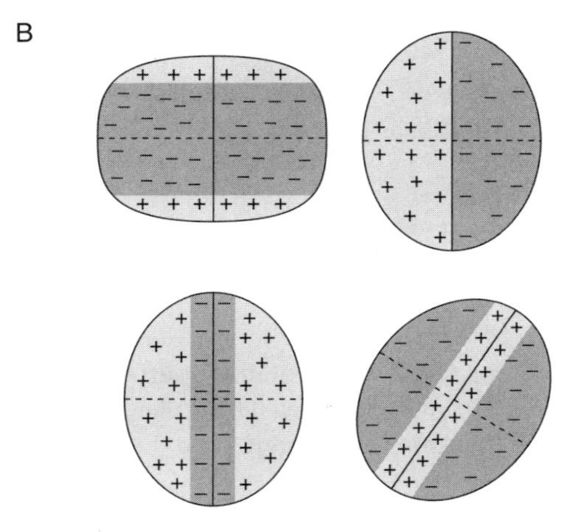

C

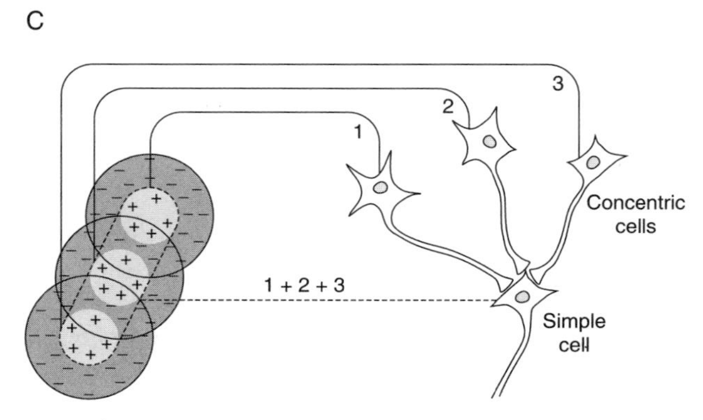

FIGURE 6.22 Illustration of the differences in response patterns of cells in the retina, the LGN, and the striate cortex.(Adapted from Hubel, D.H., and Wiesel, T.N., *J. Physiol. (Lond.)*. 160: 106–154, 1962.)

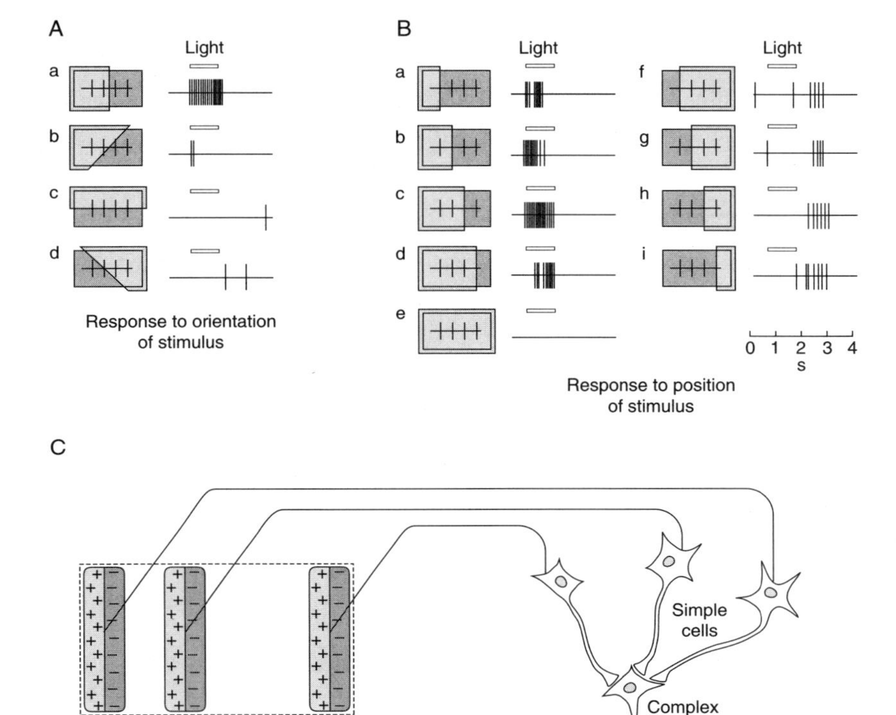

FIGURE 6.23 Responses of complex cells in the striate cortex. (Adapted from Hubel, D.H., and Wiesel, T.N., *J. Physiol. (Lond.)*. 160: 106–154, 1962.)

controls the extraocular muscles. If a perfect alignment cannot be obtained (*strabismus*) the image from one eye will be suppressed and the normal anatomical development of the striate cortex may be affected. Strabismus may lead to functional impairment of vision in the eye that is not used.

The optic nerve fibers that project to the SC mediate visual reflexes that are important for hunting and defense behavior. Most of this input originates in Y axons that receive input from the peripheral portion of the retina, which makes Y cells respond to movements in the peripheral visual field.

7. Processing in Association Cortices

Information processing in the extrastriatal cortices is very complex and poorly understood. Information reaches the different regions of the extrastriatal cortices from the primary cortex (V1) and activates neurons in the different regions (V2, V3, V4, MT/V5) in a hierarchical and parallel fashion (stream segregation). However, visual information from the periphery can bypass steps in this processing and arrive at a certain region of extrastriatal cortex ahead of information that ascends sequentially through the different regions.

a. Stream Segregation

Stream segregation was discussed in Chapter 3 and it was mentioned that the concept originated from studies of the visual system.[21,49,57-59] Ungelieder and Mishkin[57] showed that spatial visual information ("where") is processed in anatomically different regions of extrastriatal visual cortical areas located dorsally, while object information ("what") is processed in cortical regions located more ventrally (see Chapter 3, page 111). The evidence of stream segregation includes the fact that the extrastriatal cortical areas that are located ventrally in the temporal region of the brain process object type of information ("what") while the parietal and dorsal visual cortical areas, including the MT/V5 region, process spatial information. Also, the parietal and dorsal visual cortical areas are dominated by the output of M cells while the temporally located cortical regions mostly process information mediated by P cells of the retina.

This stream segregation means that different kinds of information (such as spatial and object) are processed in anatomically different cortical areas. Stream segregation is different from parallel processing, which means that the same kinds of information is processed in anatomically different regions of the CNS. Studies have presented evidence that stream segregation occurs in the visual system both in cats and in primates,[21,49,57–59] and the results have been extrapolated to humans, although such extrapolation may be questioned in view of the large anatomical differences between the relative sizes of the CNS structures involved. However, some indirect indications of stream segregation in the human visual system have been obtained through functional and behavioral studies particularly regarding face recognition and face expressions.[2,12,56]

Zeki[61] further recognized that there are several (four) parallel pathways in the visual cortex with regard to the information that is processed. The M pathway is concerned with motion and can be traced back to the M layers in the LGN, the cells of which project to layer IVB in the V1 cortex, which in turn projects to cells in MT/V5. Another distinct pathway that originates in M ganglion cells processes visual images of dynamic forms and relays information through layer IVB and from there to V3 directly via V2. The two other pathways are color pathways that originate in P cells in the LGN to layer II and III and from there to V4 directly and through V2. One of these two pathways is a color pathway and the other is a form pathway linked to color.[61]

Zeki[61] has also presented evidence that speaks against the hypothesis that there is a distinct anatomical segregation of processing of spatial and object information in vision, and he suggests that such a concept is an oversimplification of processing of visual information. Rather, there is evidence that the anatomical separation of information processing regarding different visual tasks is less distinct than what has been postulated and described as stream segregation. Zeki found that neurons in different anatomically separate regions may in fact process a wide variety of types of information rather than being as specialized as is assumed by the proponents of the stream segregation hypotheses; some neurons that perform object processing can also perform spatial tasks.[61]

Face Recognition and Interpretation of Face Expressions Recognizing individuals of ones own species is important. Many animals use scent for such identification while humans and probably higher order primates use the visual cues of the face. Nonprimate mammals use scent to establish their identity in social communication, but the sense of smell is reduced in primates, and vision has taken over as the most important sense for social identification of individuals.

There is evidence that face recognition and the interpretation of face expressions are performed in anatomically different cortical areas in the monkey.[23] There is also evidence from studies in humans that face expression and face identity are processed in different regions of the CNS,[1,11,45,46,56] whereas other investigators have not been able to find such specific regions of the CNS.[17] As was discussed in Chapter 3, face expressions are more important for communication between humans than is often assumed, and face recognition is naturally very important because it is essentially the way that we identify each other. Lesions to the amygdala eliminate emotional reactions to face expression, while the ability to recognize individuals on the basis of their face is maintained.[1] Recently, evidence has been presented that heterosexual males who view faces of the opposite sex regarded to be beautiful behave as if the view of certain faces has a reward value similar to that seen in animal experiments using cocaine.[2] These investigators used functional magnetic resonance imaging (fMRI) to show activation of the *nucleus accumbens*, which is known to be involved in addictive behaviors. Such responses are not related to the aesthetic value of the faces. The aesthetic value of faces judged as "beautiful" or "average" seems to be processed in a different way from the reactions to stimuli involving reward value.[2] These findings are in good agreement with earlier studies that show that facial expressions of emotions are processed by the amygdala, while recognition of (familiar) faces is based on neural processing in a different region of the CNS, namely the mesial occipitotemporal visual region of the cerebral cortex.[11]

Studies of the processing of face information has benefited from observations in patients with well-defined lesions in the CNS and by correlating the symptoms associated with such lesions with the anatomical locations of these lesions. Studies of individuals with *prosopagnosia*[11] (an inability to recognize familiar faces) have made it possible to determine the anatomical location of structures that are important for face recognition, and it has been shown that the damage that causes that type of deficit is located near the lingual and fusiform gyrus* ventrally and medially of the cortex that is close to the junction between the temporal and occipital cortex. This is a region of the extrastriatal cortex that has been implicated in object recognition. (It is probably necessary to have a lesion in both sides in order to have such a deficit, and more likely it was lesions of white matter (fiber tracts) that caused the symptoms[12]).

Attempts to overcome the problems of extrapolating results of locations of face recognition and interpretation of face expression from animals to human situations have been done in studies using recordings of evoked potentials and by using fMRI and positron emission tomography (PET). Electrophysiologic studies have shown that the N_{200} potential is larger in response to pictures of faces when they are oriented correctly than when shown upside down.[3,4]

IV. NONCLASSICAL VISUAL PATHWAYS

Like the auditory and the somatosensory systems, the visual system also has two major pathways, namely the classical pathway and the nonclassical pathway, through which information flows from the sensory organ to higher CNS centers. We have discussed the classical pathways, and here we will discuss the nonclassical pathways, which include visual reflexes controlling the size of the pupil and accommodation. The nonclassical pathways are less well defined and much less is known about their anatomy and physiology compared with the classical pathways.

A. Anatomy of Nonclassical Visual Pathways

A small portion of the fibers of the optic nerve (the brachium of the superior colliculus, BSC) leave the optic tract and terminate in pretectal nuclei and the superior colliculus (SC) (Fig. 6.24). It is estimated that approximately 10% of optic nerve fibers project to structures other than the LGN (10% in the monkey[8]), such as pretectal nuclei and the pulvinar of the thalamus (Fig. 6.24).[8,10,44] The nonclassical visual pathways include subcortical structures such as the contralateral superior colliculus, which controls eye movements (saccades). The pretectal nucleus of the midbrain controls the size of the pupil. The superior colliculus is concerned with reflex movements of the eye and the hand but it also integrates information from other senses, most notably from hearing, to create a sensation of space.[35] This means that there are at least two parallel nonclassical pathways for vision. Both of these pathways are phylogenetically older than the classical (retinogeniculocortical) pathway. The projections of the optic tract to the SC and the pretectal nuclei are part of the

*Fusiform gyrus is a very long gyrus that extends over the inferior aspect of the temporal and occipital lobes.

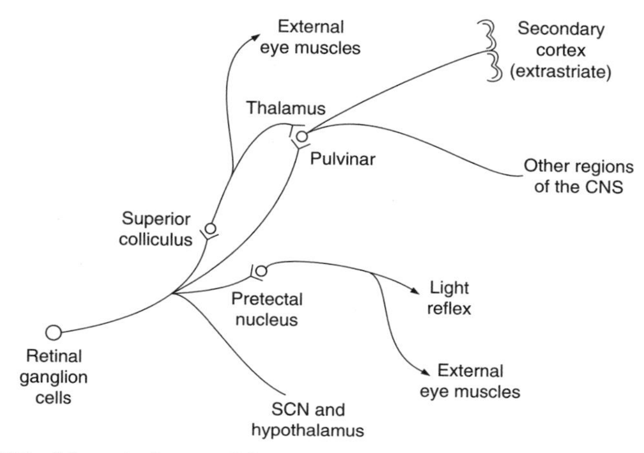

FIGURE 6.24 Schematic diagram of the main connections of the nonclassical visual pathways.

optic reflexes that control the pupil and accommodation. The connections to the pulvinar allow visual information to reach extrastriatal cortices through both short and fast pathways.

The koniogeniculate pathway may also be regarded as a nonclassical pathway because it receives input from other sensory systems (auditory and tactile)[25,42,43](see page 86).

1. The Superior Colliculus Pathway

Some optic nerve fibers project to the SC, where the projection to the superficial layer of the SC forms a map of the contralateral visual field. These SC cells control saccadic* eye movements. Cells of the SC also connect to cells in the pulvinar of the thalamus,[8,55] which in turn connects to the cerebral cortex. The deep layers of the SC receive projections from many areas of the cerebral cortex, and its cells can respond to several sensory modalities, particularly auditory and visual input.[35]

a. Visual Reflexes

The SC contributes to control of the eye position through activation of the extraocular muscles and neck muscles for control of head position. Most of the input to the pathway originates in the Y cells, which are preferentially involved in peripheral vision, and these cells emphasize motion of objects.

*Saccades are rapid eye movements that redirect (reset) the line of sight.

The size of the pupil is controlled from autonomic ganglia, where activity in a parasympathetic ganglion located in the orbit decreases the size of the pupil. Dilatation of the pupil is mediated via a sympathetic ganglion (the *superior cervical ganglion*). The sympathetic activation changes in response to the light that reaches the retina which may be regarded as a light reflex and is mediated by connections from the pretectal nuclei to the Edinger–Westphal nucleus.[8] The connections from the pretectal nucleus on one side are bilateral, which makes light that reaches only one eye control the size of the pupil of both eyes. Axons from the Edinger–Westphal nucleus reach the oculomotor motonucleus and the oculomotor nerve (CN III) to innervate the ciliary ganglion. This ganglion sends axons to the smooth muscles of the pupillary sphincter, which constricts the pupil.

The size of the pupil not only is controlled by the amount of light that reaches the retina but is also affected by the general activation of the autonomic nervous system. High sympathetic activation increases the size of the pupil as does low light. The size of the pupil is controlled by sympathetic fibers that reach the papillary radial iris muscle through the third cranial nerve. This muscle dilates the pupil. In a similar way, a strong parasympathetic activation decreases the size of the pupil. The size of the pupil is also affected by drugs such as opioids, which cause the pupil to constrict. Evaluations of the size of the pupil and its response to light, and especially the symmetry of the pupils, are important for determining brain damage.

The pretectal nucleus also receives input from the visual cortex and connects to the third cranial nerve for accommodation via the Edinger–Westphal nucleus and the pretectal nucleus. This circuit also serves to control extraocular eye muscles, which together with the vestibular–ocular reflex help keep the image on the retina steady.[8]

The SC is often referred to as the visual reflex center but it is probably more correct to refer to the pretectal nucleus located anatomically just rostral to the SC as the visual reflex center. The primary roles of these pathways are to mediate visual reflexes that control pupil size and accommodation and to control the extraocular muscles and neck muscles that help keep the image on the retina steady.

2. Pulvinar Pathway

A small part of the optic tract projects to the pulvinar division of the thalamus and the lateral posterior nucleus of the thalamus.[8] These nuclei also receive input from the SC. As other thalamic nuclei, these thalamic nuclei send axons to the cerebral cortex as well as other parts of the brain, such as structures belonging to the limbic system. The pulvinar neurons that receive visual input[10,44] project to visual cortical areas, not the primary striatal visual cortex but rather to extrastriatal visual cortical areas (MT/V5) together with other regions of the cerebral cortex (Fig. 6.24). That complex visual processing occurs in the pulvinar means that subcortical processing is important. It has mainly been processing of motion that has been shown to occur in the pulvinar,[16,22] but studies of the role of the pulvinar in visual processing is in its infancy, and later studies may reveal other kinds of processing. The fact that the pulvinar receives visual input from so many sources including the optic nerve and parts of extrastriatal cortex gives it a potential for complex processing of visual information. One of these sources is directly from the optic nerve from which the pulvinar receives raw information directly from the retina.[10,44] Because neurons in the pulvinar connect to the MT/V5 region of extrastriatal visual cortices,[55] visual information directly from the retina reaches high-order extrastriatal cortices (such as the MT/V5) with a short delay and before the information reaches more peripheral parts of the extrastriate visual cortices (V2 to V4) through their traditional routes via the LGN and primary visual cortex (V1).

It was assumed earlier that the nonclassical visual pathways mediate crude perceptions of light but recent studies elucidating the involvement of the pulvinar may indicate that at least some parts of the nonclassical pathways may mediate highly processed information of certain specific types, such as fast movements.

3. Other Nonclassical Visual Pathways

Axons of the optic tract also project to other parts of the hypothalamus, which serves to maintain circadian rhythm. It has recently been shown that there are specific sensory cells in the retina — not cones or rods — that connect to the suprachiastic nucleus (SCN) which connects to the hypothalamus, which is involved in circadian rhythm. That biologic clock must be reset regularly which is done by the above-mentioned cells in retina that connect to the SCN.[6,50] That projection may also be regarded as a part of the nonclassical ascending visual pathway. There is also evidence of connections to the pineal body, which may mediate light control of some endocrine functions (such as secretion of melatonin) that affect circadian rhythms.

B. Physiology of the Nonclassical System

Little is known about the function of the nonclassical visual pathways. Projections to the superior colliculus and pretectal nuclei mainly mediate visual reflexes such as those involving eye movements and neck movements. The projection to the extrastriatal cortex from the pulvinar neurons that receive visual information is retinotropic in a similar way as the projections from the LGN. However, although visual information reaches the cerebral cortex via this nonclassical pathway, it provides only inferior crude visual information, mainly regarding detection of light and movement of objects.[8] The control of eye muscles and accommodation, as well as light reflexes, are mediated through these pathways involving the pretectal nuclei (Fig. 6.24).

1. Visual Reflexes

Visual reflexes are mediated mainly through the SC and activate the extraocular muscles to move the eye. They also involve neck muscles, more so in animals such as the cat than in humans because cats can move their eyes much less than humans. The pretectal nucleus is also involved in light reflexes. The eye muscles also keep the optical axes of the two eyes properly aligned in parallel for far vision and so that the two images produced by the two eyes of near objects are fused. The angle between the optical axes, determined by the contractions of the extraocular muscles, are used for distance estimation of objects at short distances while estimation of the distance to objects at long distances is based on other factors such as knowledge about the size of the object being viewed.

2. Visual Activation of Limbic Structures

Face expressions provide much emotional information. This information is assumed to reach structures of the limbic system, mainly the basolateral nuclei of the amygdala, where it elicits learned emotional reactions regarding fear, pleasure, etc. Such information is assumed to be highly processed and can be modified by extrinsic and intrinsic factors. Studies of the auditory system in rats have revealed that auditory information can reach the amygdala through two different routes, namely a cortical route (cortical-associate cortices, or amygdala) and a subcortical route.[31,32] The subcortical route (low route) originates in the dorsal thalamus in the auditory system, and it may be assumed that the pulvinar route of the visual system is similar. The information that reaches the amygdala through the high route is mediated by the classical ascending pathways via the primary, secondary, and association cortices whereas the information that reaches the amygdala via the low route is mediated by the nonclassical pathway. The subcortical route is faster and probably involves little processing, whereas the cortical route to the amygdala involves a high degree of neural processing and the flow of information in that route is affected by intrinsic and extrinsic factors.

The cortical route through which visual information reaches limbic structures may be similar to what was described as the high route for the auditory system[8,13,31] (see Fig. 3.26 in Chapter 3 and Fig. 5.46 in Chapter 5). The pulvinar and the lateral posterior nucleus of the thalamus may correspond to the medial and dorsal thalamic nuclei, which in the auditory system provide a subcortical route to the amygdala known as the low route.[31] That route carries information to limbic structures that is less processed than that carried in the cortical route (high route). In addition, these pathways participate in visual reflexes and provide thalamic input to secondary visual cortices and association cortices.

REFERENCES

1. Adolphs, R.D., Tranel, H., Damasio, H., and Damasio, A.R., Fear and the Human Amygdala. *J. Neurosci.*, 15: 5879–5891, 1995.
2. Aharon, I., Etcoff, N., Ariely, D., Chabris, C.F., O'Connor, E., and Breiter, H.C., Beautiful Faces Have Variable Reward Value; fMRI and Behavioral Evidence. *Neuron*, 32: 537–551, 2001.
3. Allison, T., Ginter, H., McCarthy, C., Nobre, A., Puce, A., Luby, M., and Spencer, D.D., Face Recognition in Human Extrastriate Cortex. *J. Neurophysiol.*, 71: 821–825, 1994.
4. Allison, T., McCarthy, C., Nobre, A., Puce, A., and Belger A., Human Extrastriate Visual Cortex and the Perception of Faces, Words, Numbers, and Colors. *Cereb. Cortex*, 4: 544–554, 1994.
5. Barlow, H.B., Summation and Inhibition in the Frog's Retina. *J. Physiol. (London)*, 119: 69–88, 1953.
6. Berson, D.M., Dunn, F.A., and Takao, M., Phototransduction by Retinal Ganglion Cells That Set the Circadian Clock. *Science*, 295: 1070–1073, 2002.
7. Bonin, von, G., and Bailey, P., *The Neocortex of the Macaca mulatia*. Urbana, IL: University of Illinois Press, 1947.
8. Brodal, P., *The Central Nervous System*. New York: Oxford Press, 1998.
9. Cornsweet, T.N., *Visual Perception*. New York: Academic Press, 1970.
10. Cowey, A., Stoerig, P., and Bannister, M., Retinal Ganglion Cells Labeled from the Pulvinar Nucleus in Macaque Monkeys. *Neuroscience*, 61: 691–705, 1994.
11. Damasio, A.R., Damasio, H., and Van Hoesen, G.W., Prosopagnosia: Anatomic Basis and Behavioral Mechanisms. *Neurology*, 32: 331–341, 1982.
12. Damasio, A.R., Tranel, D., and Damasio H., Face Agnosia and the Neural Substance of Memory. *Annu. Rev. Neurosci.*, 13: 89–109, 1990.
13. Doron, N.N., and LeDoux, J.E., Cells in the Posterior Thalamus Project to Both Amygdala and Temporal Cortex: A Quantitative Retrograde Double-Labeling Study in the Rat. *J. Comp. Neurol.*, 425: 257–274, 2000.
14. Dowling, J.E., Organization of Vertebrate Retinas. *Invest. Ophthalmol.*, 9: 655–680, 1970.
15. Dowling, J.E., Information Processing by Local Circuits: The Vertebrate Retina as a Model System. In: *The Neurosciences: Fourth Study Program*, edited by F.O. Schmitt, and F.G. Worden, Cambridge, MA: MIT Press, 1979, pp. 163–182.
16. Dumbrava, D., Faubert, J., and Casanova C., Global Motion Integration in the Cat's Lateral Posterior-Pulvinar Complex. *Eur. J. Neurosci.*, 13: 2218–2226, 2001.
17. Eacott, M.J., Heywood, C.A., Gross, C.G., and Cowey, A., Visual Discrimination Impairment Following Lesions of the Superior Temporal Sulcus Are Not Specific for Facial Stimuli. *J. Neurophysiol.*, 31: 609–619, 1993.

18. Enroth-Cugell, C., and Robson, J.G., The Contrast Sensitivity of Retinal Ganglion Cells of the Cat. *J. Physiol. (London)*, 187: 517–552, 1966.
19. Fitzpatrick, D.C., Itoh, K., and Diamond, I.T., The Laminar Organization of the Lateral Geniculate Body and the Striate Cortex in the Squirrel Monkey (*Saimiri sciureuss*). *J. Neurosci.*, 3: 673–702, 1983.
20. Gilbert, C.D., and Wiesel, T.N., Intrinsic Connectivity and Receptive Field Properties in Visual Cortex. *Vision Res.*, 25: 365–374, 1985.
21. Goodale, M.A., and Milner, A.D., Separate Pathways for Perception and Action. *Trends Neurosci.*, 15: 20–25, 1992.
22. Grieve, K.L., Acuna, C., and Cudeiro, J., The Primate Pulvinar Nuclei: Vision and Action. *Trends Neurosci.*, 23: 35–39, 2000.
23. Gross, C.G., Rocha-Miranda, C.E., and Bender, D.B., Visual Properties of Neurons in Inferotemporal Cortex of the Macaque. *J. Neurophysiol.*, 35: 96–111, 1972.
24. Gross, C.G., Rodman, H.R., Gochin, P.M., and Colombo, M.W., Inferior Temporal Cortex as a Pattern Recognition Device. In: *Computational Learning and Cognition*, edited by E. Baum. Philadelphia: SIAM Press, 1993.
25. Hendry, S.H.C., and Reid, R.C., The Koniocellular Pathway. *Annu. Rev. Neurosci.*, 23: 127–153, 2000.
26. Hubel, D.H., and Wiesel, T.N., Receptive Fields, Binocular Interaction and Functional Architecture in the Cat's Visual Cortex. *J. Physiol. (London)*, 160: 106–154, 1962.
27. Hubel, D.H., and Wiesel, T.N., Ferrier Lecture: Functional Architecture of Macaque Monkey Visual Cortex. *Proc. R. Soc. London B*, 198: 1–59, 1977.
28. Hubel, D.H., *Eye, Brain, and Vision*. New York: Scientific American Library, 1988.
29. Kuffler, S.W., Neurons in the Retina: Organization, Inhibition and Excitation Problems. *Cold Spring Harbor Symp. Quant. Biol.*, 17: 281–292, 1952.
30. Kuffler, S.W., Discharge Patterns and Functional Organization of Mammalian Retina. *J. Neurophysiol.*, 16: 37–68, 1953.
31. LeDoux, J.E., Brain Mechanisms of Emotion and Emotional Learning. *Curr. Opin. Neurobiol.*, 2: 191–197, 1992.
32. LeDoux, J.E., *The Emotional Brain*. New York: Touchstone, 1996.
33. Lennie, P. and D'Zmura, M., Mechanisms of Color Vision. *Crit. Rev. Neurobiol.*, 3: 333–400, 1988.
34. Lettvin, J.Y., Maturana, H.R., McCulloch, W.S., and Pitts, W.H., What the Frog's Eye Tells the Frog's Brain. *Proc. IRE*, 47: 1940–1951, 1959.
35. Lomber, S.G., Payne, B.R., and Cornwell, P., Role of Superior Colliculus in Analyses of Space: Superficial and Intermediate Layer Contributions to Visual Orienting, Auditory Orienting, and Visuospatial Discriminations During Unilateral and Bilateral Deactivations. *J. Comp. Neurol.*, 441: 44–57, 2001.
36. Lomber, S.G., Learning To See the Trees Before the Forest: Reversible Deactivation of the Superior Colliculus During Learning of Local and Global Features. *Proc. Nat. Acad. Sci., USA*, 99: 4049–4054, 2002.
37. Lund, J.S., Anatomical Organization of the Macaque Monkey Striate Visual Cortex. *Annu. Rev. Neurosci.*, 11: 253–288, 1988.
38. Masland, R.H., The Fundamental Plan of the Retina. *Nature Neurosci.*, 4: 877–886, 2001.
39. Merigan, W.H., and Maunsell, J.H.R., Macaque Vision After Magnocellular Lateral Geniculate Lesions. *Vis. Neurosci.*, 5: 347–352, 1993.
40. Milner, A. D. and Goodale, M.A., Visual Pathways to Perception and Action. *Prog. Brain Res.*, 95: 317–337, 1993.
41. Mishkin, M., Ungeleider, L.G., and Macko, K.A., Object Vision and Spatial Vision: Two Cortical Pathways. *Trends Neurosci.*, 6: 415–417, 1983.

42. Norton, T.T., and Casagrande, V.A., Laminar Organization of Receptive-Field Properties in Lateral Geniculate Nucleus of the Bush Baby (*Galago crassicaudatus*). *J. Neurophysiol.*, 47: 715–741, 1982.
43. Norton, T.T., Casagrande, V.A., Irvin, G.E., Sesma, M.A., and Petry, H.M., Contrast Sensitivity of W-, X- and Y-Like Relay Cells in the Lateral Geniculate Nucleus of the Bush Baby, *Galago crassicaudatus*. *J. Neurophysiol.*, 59: 1639–1656, 1988.
44. O'Brien, B.J., Abel, P.L., and Olavarria, J.F., The Retinal Input to Calbindin-D28k-Defined Subdivisions in Macaque Inferior Pulvinar. *Neurosci. Lett.*, 312: 145–148, 2001.
45. Ojemann, J.G., Ojemann, G.A., and Lettich, E., Neuronal Activity Related to Faces and Matching in Human Right Nondominant Temporal Cortex. *Brain*, 115: 1–13, 1992.
46. Pallis, C.A., Impaired Identification of Faces and Places with Agnosia for Colors. *J. Neurol. Neurosurg. Psychiatry*, 18: 218–224, 1995.
47. Peters, A., Payne, B.R., and Budd, L., A Numerical Analysis of the Geniculocortical Input to Striate Cortex in the Monkey. *Cereb. Cortex*, 4: 215–229, 1994.
48. Pirenne, M.H., *Vision and the Eye*. London: Associated Book Publishers, 1967.
49. Postle, B.R., and D'Esposito, M., "What"–Then–"Where" in Visual Working Memory: An Event-Related fMRI Study. *J. Cog. Neurosci.*, 11: 585–597, 1999.
50. Rimmer, D.W., Boivin, D.B., Shanahan, T.L., Kronauer, R.E., Duffy, J.F., and Czeisler, C.A., Dynamic Resetting of the Human Circadian Pacemaker by Intermittent Bright Light. *Am. J. Physiol. Regul. Integr. Compar. Physiol.*, 279: R1574–1579, 2000.
51. Schmidt, R.F., *Fundamentals of Sensory Physiology*. New York: Springer-Verlag, 1981.
52. Schnapf, J.L., Kraft, T.W., Nunn, B.J., and Baylor, D.A., Spectral Sensitivity of Primate Photoreceptors. *Vis. Neurosci.*, 1: 255–261, 1988.
53. Shapley, R., and Perry, V.H., Cat and Monkey Retinal Ganglion Cells and Their Visual Functional Roles. *Trends Neurosci.*, 9: 229–235, 1986.
54. Shepherd, G.M., *Neurobiology*. New York: Oxford University Press, 1994.
55. Stepniewska, I., Qi, H.X., and Kaas, J.H., Projections of the Superior Colliculus to Subdivisions of the Inferior Pulvinar in New World and Old World Monkeys. *Vis. Neurosci.*, 17: 529–549, 2000.
56. Tranel, D., Damasio, A.R., and Damasio, H., Intact Recognition of Facial Expression, Gender, and Age in Patients with Impaired Recognition of Face Identity. *Neurology*, 38: 690–696, 1988.
57. Ungeleider, L.G., and Mishkin, M., Analysis of Visual Behavior. In: *Analysis of Visual Behavior*, edited by D.J. Ingle, M.A. Goodale, and R.J.W. Mansfield, Cambridge MA: MIT Press, 1982.
58. Ungeleider, L.G., and Haxby, J.V., "What" and "Where" in the Human Brain. *Curr. Opin. Neurobiol.*, 4: 157–165, 1994.
59. Wang, J., Zhou, T., Qiu, M., Du, A., Cai, K., Wang, Z., Zhou, C., Meng, M., Zhou, Y., Fan, S., and Chen, L., Relationship Between Ventral Stream for Object Vision and Dorsal Stream for Spatial Vision: An fMRI+ERP Study. *Human Brain Mapping*, 8: 170–181, 1999.
60. Wiesel, T.N., and Hubel, D.H., Spatial and Chromatic Interactions in the Lateral Geniculate Body of the Rhesus Monkey. *J. Neurophys.*, 29: 1115–1156, 1966.
61. Zeki, S. *A Vision of the Brain*. London: Blackwell Scientific, 1993.

CHAPTER 7

Chemical Senses: Olfaction and Gustation

ABBREVIATIONS

CN : Cranial nerve
IAC : Internal auditory meatus
NST : Nucleus of the solitary tract
PbN : Parabrachial nucleus
SI : Primary somatosensory cortex
VA : Visceral afferents
VPM : Ventral posterior nucleus (of the thalamus)
SI : Primary somatosensory cortex

ABSTRACT

1. The chemical senses of olfaction and gustation (taste) have served the purposes of nutrition and reproduction during the evolution of species.

TASTE

2. Taste receptors are located in taste buds found in the tongue, palate, pharynx, epiglottis, and esophagus.
   a. Taste receptors in the frontal part of the tongue are innervated by the chorda tympani, a branch of the facial nerve (CN VII). The cell bodies are located in the geniculate ganglion.
   b. Taste receptors in the posterior tongue are innervated by the lingual branch of the glossopharyngeal nerve (CN IX). The cell bodies are located in the petrosal ganglion.
   c. Taste receptors in the palate are innervated by the superficial petrosal nerve, a branch of CN VII, with cell bodies in the geniculate ganglion.
   d. Taste receptors on the epiglottis are innervated by the superior laryngeal nerve, a branch of the vagus nerve (CN X). Cell bodies are located in the nodosa ganglion.
3. Taste receptors are selective with regard to mainly four different qualities of taste—sweet, sour, bitter, and salty—and possibly a fifth, monosodium glutamate.
4. Taste information from the three cranial nerves (CN VII, CN IX, and CN X) is interrupted in the nucleus of the solitary tract.
5. The nucleus of the solitary tract projects to the ventral posterior medial nucleus of the thalamus. Taste neurons in that nucleus project to the primary somatosensory cortex and the gustatory cortex, located deep within the insula.
6. Fibers from the nucleus of the solitary tract also project to the limbic system, various motor systems (the tongue), and glands (salivation and insulin secreting cells in the pancreas).
7. The organization of the taste pathways is different in different mammalian species.

OLFACTION

8. Sensory cells of olfaction are located in the sensory epithelium deep in the back of the nose and in the vomeronasal organ.
9. Olfactory cells are selective with regard to many different odors. Many different types of sensory cells respond to the same odor, but to a different degree.
10. The sensitivity to various odors differs widely among individuals, and there are large species differences.
11. The olfactory bulb performs signal processing similar to that performed in the retina of the eye.

12. The olfactory bulb projects to allocortical regions such as the periform cortex, the uncus, the entorhinal cortex, and limbic structures such as the amygdala and the hippocampus.
13. The vomeronasal organ responds to pheromones and probably to some odorants also.
14. The vomeronasal organ projects directly and only to the amygdala, via the accessory olfactory bulb, thus a direct pathway with little neural processing. The vomeronasal pathway has similarities with the non-classical pathways of other sensory systems.

I. INTRODUCTION

Phylogenetically, *olfaction* and *gustation* have served two vital purposes in vertebrates, namely nutrition and communication for reproduction, identification of food (prey), and for identification of members of their own species. These two senses monitor intake of food and the air breathed. The sense of taste warns about poisonous food and aids in selection of food and regulation of the intake of food in general. Taste pathways have connections with parts of the brain controlling hunger and satiation. We humans mostly associate the smell and taste of food with the pleasures of eating. Enjoyment of the scent of flowers and perfumes is an obvious consequence of olfaction in humans.

Olfaction is used by many animals for identification of members of their own species as well as for identification of other species, in a similar way as humans use visual recognition of faces for identification of other humans. Olfaction, especially the *vomeronasal* system, has great importance in reproduction in many animals because of the response to *pheromones*.* It is not certain if the vomeronasal systems plays a role in human sexuality but some observations indicate that it might play a role in reproduction in humans.

While the organization of the central nervous system of taste has many similarities with that of other sensory systems the olfactory nervous system is different from that of other sensory systems in many ways. The olfactory pathways have strong and direct projections to limbic structures, indicating that smell is important in emotional, sexual, and other basic bodily functions.

The ascending neural pathways of gustation and olfaction have no distinct classical and nonclassical pathways as do hearing, somesthesia, and vision, although the sensory cells in the vomeronasal organ, also known as *Jacobson's*

*Pheromones are substances secreted by an individual and detected by another individual of the same species in whom it produces changes in sexual or social behavior. Pheromones are best known from insects, for which they play important roles in reproduction. The exact role of pheromones in higher vertebrates is not clear.

organ, project (via the *accessory olfactory bulb*) to a separate olfactory pathway that may be regarded as a nonclassical pathway similar to the nonclassical pathways in the other sensory systems. The vomeronasal pathways provide a direct route to limbic structures (especially the amygdala) like the nonclassical ascending pathways for other sensory systems (see Chapter 3). In this book, we will therefore treat the vomeronasal pathway as a separate system as we did for the other sensory systems.

While the olfactory system can differentiate among a wide variety of chemicals (*odorants*), the gustatory system can only differentiate among four different taste qualities, (bitter, salty, sour, and sweet) and most likely a fifth, (monosodium glutamate, or *umami*). It is a general impression that the flavor of food is provided by the gustatory sense, but it is in fact the olfactory sense that is the basis for discrimination of many food flavors. The mechano- and temperature-receptors in the mouth* also play important roles in providing what is regarded as the flavor of food by providing information about the texture and temperature of the food. This means that the specific flavors of food that we experience are the result of a combination of input from several senses that requires complex coordination in the central nervous system (CNS), including the association cortices, to discriminate among the flavors of foods.

Bitter substances are often toxic and may elicit a warning or even cause vomiting. Other substances, particularly sugar, elicit a sensation of pleasure. In fact, sugar is one of the few sensory stimuli, or perhaps the only one, that elicits a distinct sensation of pleasure without the sensation having been learned by experience. The reaction to most other sensory stimuli that give a sensation of pleasure is the result of a learned experience. Also, unpleasant taste or smell is mostly the result of learned experiences. One exception may be the smell of bromide, which most people regard as being unpleasant.

Some animals may have an attraction to salty substances that is related to the body's salt (NaCl) balance. Lack of salt seems to elicit a *salt hunger*, at least in certain animals, that is satisfied by stimulation of salt receptors in the tongue.

II. ANATOMY

This section describes the anatomy of the chemical sensory system, namely the receptors, the ascending and descending neural pathways, and their nuclei. The more central projections of these systems are also discussed.

*The mechanoreceptors and temperature receptors in the tongue and the palate are mainly innervated by the *trigeminal nerve*, but the glossopharyngeal nerve also innervates some mechanoreceptors in the mouth and pharynx.

A. Receptors

The receptors of the chemical senses are exclusively sensitive to chemical substances. They are Type II receptors (see Chapter 2), where the receptor cells connect to the nervous system through chemical synapses.

1. Taste Receptors

Taste cells are located in taste buds (Fig. 2.11) in the tongue, palate, pharynx, and epiglottis. In humans, taste buds are located in three different types of *papillae*: fungiform, foliate, and circumvallate (Fig. 7.1). The receptor cells of taste undergo continuous renewal, and the different types of cells that can be identified in the taste buds may be cells that are at different stages of their development.

2. Olfactory Receptors

Olfactory receptors are located in the posterior part of the nose (see Chapter 2, Fig. 2.3). The olfactory receptor cells have the form of bipolar cells with a dendrite extending distally from the cell body and an axon extending proximally (Fig. 7.2). The dendrite extends toward the surface of the epithelium and ends in a knob with cilia (or microvilli) that extend through the surface of the epithelium.* The mature cells have long dendrites with a knob on the top. Each of these knobs has several cilia that are up to 200 μm long and have diameters of 0.1 to 0.2 μm. The microvilli have specialized membranes that are sensitive to different odors. The axons of the bipolar cells pass through the cribriform plate and form the olfactory nerve (Fig. 2.10), and the axons terminate in synapses on cells in the olfactory bulb.

Like taste sensory cells, olfactory receptor cells have a short lifetime and are constantly being replaced with new cells. The olfactory epithelium in the vertebrate consists of the mature sensory cells, which have microvilli that reach the surface, the axons of which form the olfactory nerve. The olfactory epithelium also contains different kinds of supporting cells, in addition to developing sensory cells and mucous glands. The sensory cells with different morphology that can be discerned in the olfactory epithelium are probably cells at different stages of their life cycle rather than different types of cells.

> It is known that the acuity of the sense of smell decreases with age and may be explained by a decline in the number of glomeruli mitral cells. One study[14] found that of glomeruli and mitral cells in the olfactory bulb of the young adult human varied

*The olfactory receptors may be regarded as Type II receptors consisting of a cell body with cilia that are sensitive to odors and an axon with a synaptic connection to a nerve cell in the CNS.

from approximately 8000 to 40,000 and declined steadily with age at an approximate rate of 10% per decade. Individuals in their eighties and nineties had less than 30% of mitral cells and glomeruli found in young individuals.

a. The Vomeronasal Organ

The vomeronasal organ contains the receptor cells that are sensitive to pheromones. It is a fluid-filled structure located at the base of the nasal septum. The receptors terminate in the accessory olfactory bulb. While vomeronasal

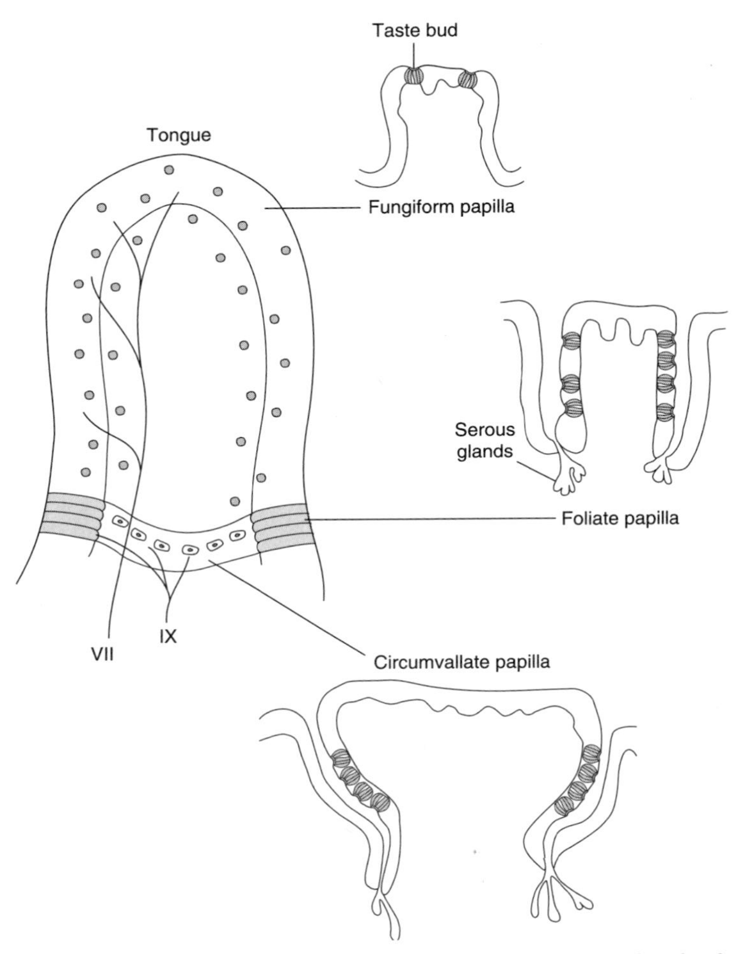

FIGURE 7.1 Location of taste buds in the three different types of papillae, and their distribution on the tongue. (Adapted from Murray.[17])

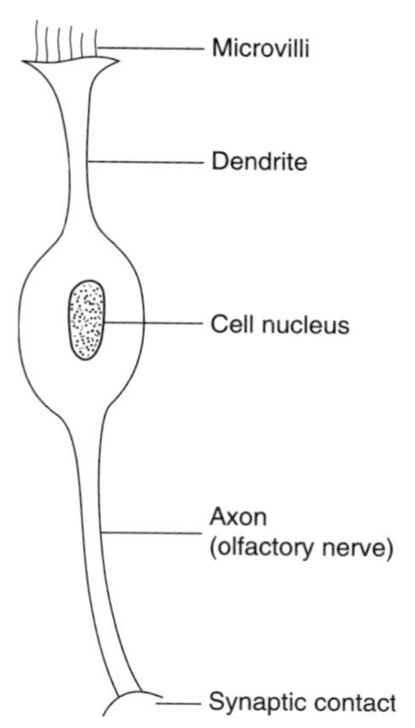

FIGURE 7.2 Schematic drawing of an olfactory sensory cell. (Adapted from Schmidt, R.F., *Fundamentals of Sensory Physiology*. New York: Springer-Verlag, 1981.)

organs have been found in many mammals in addition to reptiles,[8] it is not certain that humans have a vomeronasal organ,[4] although recent evidence for its existence in humans has been presented.[2,29] Recent evidence that humans respond to pheromones is also mounting,[7,15,31] supporting the hypothesis about the existence of a functional vomeronasal organ in humans.

B. The Media that Conduct the Stimulus to the Receptors

The media that conduct the stimulus to the receptors (the nose and the tongue) modify the stimuli to a lesser extent than is the case for hearing and the somatosensory senses.

1. The Tongue

The chemicals to which the receptors can respond must be liquid or dissolved in saliva and transported to the taste buds (Fig. 2.11), in the papillae of the tongue (Fig. 7.1) in order to reach the sensory cells that protrude through the taste pores. Substances that must be dissolved in the saliva may activate taste receptors to an extent that depends on the solubility of the substance in question, which respond with a delay depending on how long it takes to dissolve the substance in question. This results in a variation in response times in receptors. Proteins in the saliva may modulate the action of the taste substances by binding to these substances and delivering them to the taste cells, or some substances may remove or prevent taste substances from reaching the taste cells.[4]

2. The Nose

The human nose (Fig. 2.4) has three conchae around which the air sucked into the nose circulates. The olfactory epithelium is located above the superior concha. It is not known in detail how air circulates around the conchae and how it reaches the sensory epithelium. The structure of the noses of animals that have a more sensitive sense of olfaction than humans is more complex than that of a human nose and includes structures that warm and humidify the air before it reaches the sensory cells.

Because the vomeronasal organ is fluid filled, pheromones and other odors that its receptors sense must first be dissolved in fluid. That may affect the availability of different substances to the receptors differently.

C. Innervation of Receptors

The receptors of the chemical senses communicate to the CNS through synaptic contacts with the nerve cells. The olfactory receptors connect to the olfactory nerve via a complex network (Fig. 3.1B). The taste receptors, on the other hand, connect to several cranial nerves (CN VII, CN IX, CN X).

1. Innervation of Taste Receptors

The contact between the taste cells and the afferent nerve fibers has the morphological appearance of a chemical synapse (Fig. 7.3). The axons from the taste cells travel in three cranial nerves. (1) the chorda tympani and the greater superficial petrosal nerves of the facial nerve, (2) the lingual-tonsillar branch of the glossopharyngeal nerve, and (3) the superior laryngeal nerve of the vagus nerve (Fig. 7.4). Nerve fibers from the taste buds located in the frontal part of the tongue are carried in the chorda tympani of the seventh cranial nerve (the facial nerve, CN VII), while taste buds located in the posterior part of the tongue are innervated by the greater petrosal nerve, which travels in the lingual branch of the glossopharyngeal nerve (CN IX) (Figs. 7.1 and 7.4). Branches of CN IX may also innervate taste buds in the nasopharynx. The greater superficial petrosal branch of the seventh cranial nerve innervates the taste buds in the palate. The taste buds that are located in the epiglottis and the esophagus are

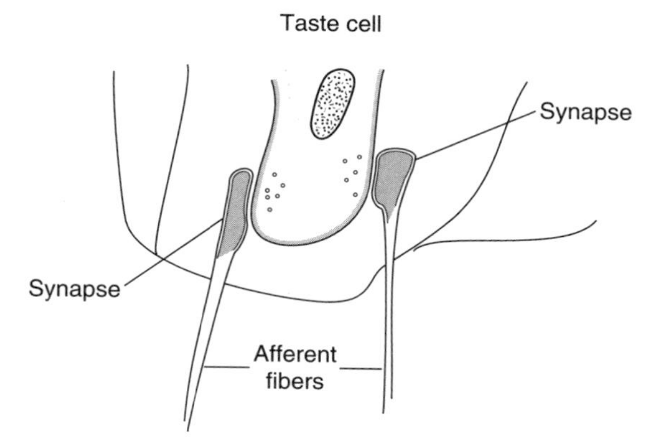

FIGURE 7.3 Synaptic contact between afferent nerve fibers and a taste receptor cell.

innervated by the superior laryngeal nerve, which is a branch of the vagus nerve (CN X).

The cell bodies of the chorda tympani (a branch of CN VII) that innervate taste receptors of the anterior part of the tongue are located in the geniculate ganglion. From there on, the nerve continues as the *intermedius* nerve and travels in the internal auditory meatus together with CN VII and CN VIII as a bundle of separate fascicules. The axons of the *lingual branch* of the glossopharyngeal nerve that innervate taste buds in the posterior third of the tongue have cell bodies in the *inferior glossopharyngeal ganglion* (petrosal ganglion). The fibers of both the chorda tympani and the greater superficial petrosal nerves have their cell bodies in the geniculate ganglion (Fig. 7.4). The cell bodies of axons of CN IX that carry taste information are located in the petrosal ganglion. The cell bodies of the branch of CN X that innervates taste receptors in the pharynx are located in the *nodosa ganglion*. That means that three different cranial nerves (CN VII, IX and X) contain taste fibers. To complicate matters further these cranial nerves are known for their anastomoses with each other.

2. Innervation of Olfactory Receptors and the Neural Network in the Olfactory Bulb

The proximal end of the olfactory receptor cells are unmyelinated axons that are regarded as part of the receptor cells (Fig. 7.2), but they are also the nerve fibers that form the olfactory nerve (CN I). These axons are collected in approximately 20 bundles (Fig. 7.5) that pass through the cribriform plate of the ethmoidal bone (Fig. 2.11).

The axons from the olfactory sensory cells make synaptic contact with cells in the glomeruli in the olfactory bulb (Fig. 7.2), located just above the sensory

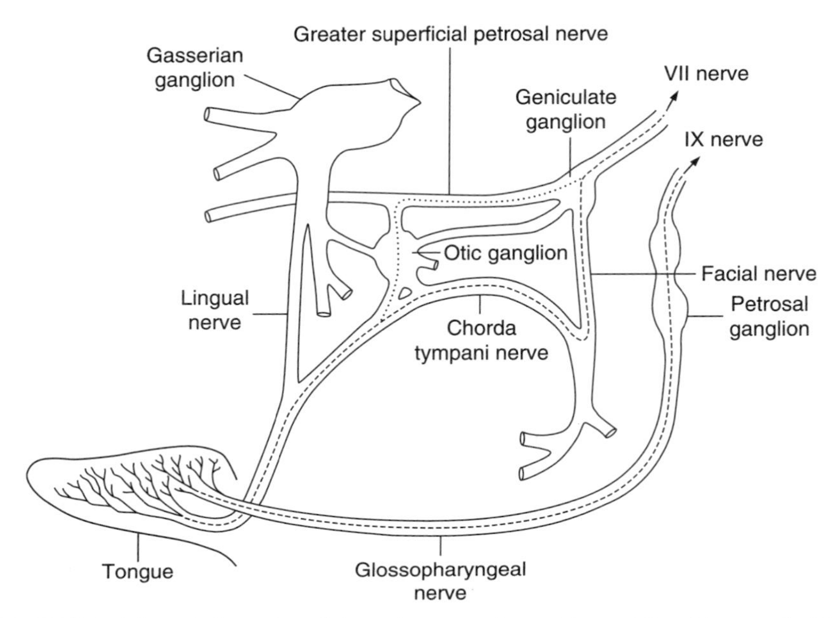

FIGURE 7.4 Schematic illustration of the major taste pathways. (Adapted from Mountcastle, V.B., ed., *Medical Physiology*. St. Louis: C.V. Mosby, 1974.)

organ. The neural network in the olfactory bulb has similarities with the neural network in the retina (Fig. 3.1).[23]

The axon of each olfactory sensory cell terminates in only one glomerulus, where it makes contact with three types of cells, namely mitral cells, tufted cells, and periglomeruli cells (Fig. 7.5).[22,25] The axons of the mitral and tufted cells project to the olfactory cortex while the periglomeruli cells are interneurons that encircle the glomeruli. Twenty to fifty glomerulus cells receive axons of several thousands of axons from sensory cells. This means an enormous convergence and a decrease of the number of neurons that transmit olfactory signals compared with the number of sensory cells.[4]

There is extensive inhibitory connections between mitral cells and granular and periglomerular cells (Fig. 7.6).[26,34] There are also efferent connections from the olfactory cortex to the olfactory bulb.

3. Innervation of Receptors in the Vomeronasal Organ

The unmyelinated axons of the bipolar receptor cells of the vomeronasal organ form the vomeronasal nerve, which penetrates the cribriform plate and terminates in the mitral cells of the accessory olfactory bulb,[8] in a similar way as the olfactory nerve.

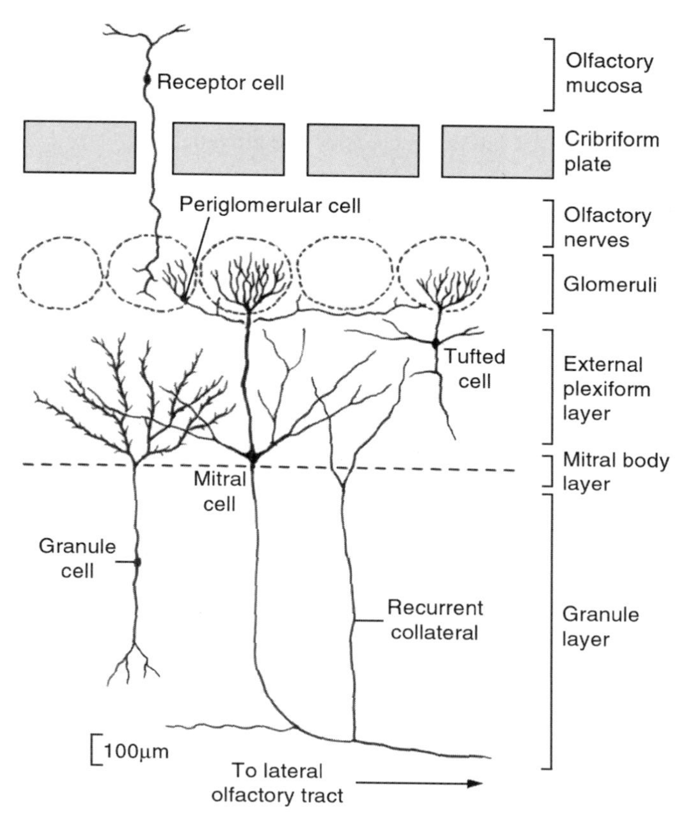

FIGURE 7.5 Connections between fibers of the olfactory nerve and three types of cells of olfactory bulb. (Adapted from Buck, L.B., *Principles of Neural Science*, 4th ed., edited by Kandel, E.R., Schwartz, J.H., and Jessell, T.M. New York: McGraw-Hill, 2000, 625–647.)

D. Gustatory and Olfactory Neural Pathways

The taste pathways have many similarities to the pathways of the other three senses but the olfactory pathways are different in many ways from other sensory pathways, both with regard to their structure and their targets. However, both these two senses (taste and olfaction) seem to have parallel pathways, one of which is the thalamocortical pathway, which mediates the conscious perception of the quality of taste and smell, and the other pathway projects directly to limbic structures, mainly the amygdala, the insula, and the hippocampus, providing emotional reactions to taste and odors. Olfaction and gustation both have short and direct connections to nuclei of the amygdala similar to the nonclassical pathways of hearing. These parallel pathways may be equivalent to the classical and the nonclassical pathways of the other three senses. Olfactory pathways primarily project to the medial nucleus of the amygdala but to some extent also to the central nuclei of the amygdala while the nonclassical auditory pathway projects to the lateral nucleus of the amygdala.

1. Taste Pathways

Taste fibers from CN VII , CN IX, and CN X merge and reach the rostral pole of the *nucleus of the solitary tract* (NST), which is the first relay nucleus for taste (Fig. 7.7).[18,26] The fibers from the petrosal ganglion terminate in cells that are

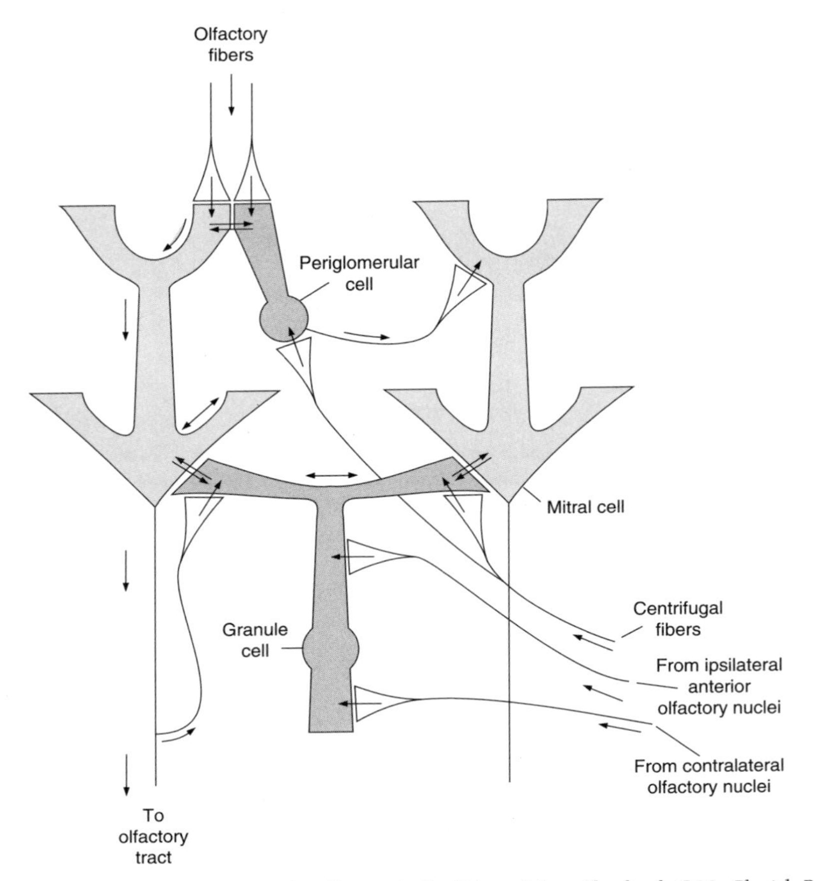

FIGURE 7.6 Connections within the olfactory bulb. (Adapted from Shepherd, G.M., *Physiol. Rev.* 52: 864–917, 1972.)

located caudal to those innervated by the facial nerve.[27] A fiber tract from the NST reaches the *ventral posterior medial* (VPM) nucleus of the thalamus, and from there fibers travel to the face region of the *primary somatosensory cortex* (SI) and to the *gustatory cortex*, which is located deep in the *insula*. These three areas are organized anatomically according to the location of the receptors on the tongue. Taste is important in feeding. Sensory pathways must therefore have connections with parts of the brain controlling hunger and satiation.

The intermedius nerve, which travels together with CN VII in the *internal auditory meatus* (IAC), is generally assumed to carry taste fibers from the chorda tympani in mammals. However, there is some doubt about whether or not that is the case in humans. Secretion from oral glands (parotid and other salivary glands) in response to taste stimuli (especially sour) is also assumed to be mediated through fibers in the nervus intermedius, which carries efferents from the superior salivatory nucleus to the lacrimal gland.[31] Nervous intermedius in humans is associated with deep ear pain that occurs in attacks of severe stabbing pain (see Chapter 4).

The taste pathways are different in different animals. In rodents, the afferent connections from the NST to cells in the VPM nucleus of the thalamus are interrupted in the *parabrachial nucleus* (PbN),[10] while in primates these afferents proceed directly from the NST to the VPM. In rodents, the neurons in the NST that receive taste fibers send axons to two main targets, namely a descending pathway that terminates in a more caudal part of the NST and an ascending pathway that reaches the medial part of the PbN in the pons.[10] In primates, neurons in the rostral NST project directly to the VPM nucleus of the thalamus, bypassing the pontine relay (Fig. 7.7).[28] Taste and olfactory information are probably integrated in the *insula* to which both systems project.[3]

The VPM nucleus belongs to the ventral thalamus associated with precise sensory information and is known as a relay for classical sensory pathways. The taste neurons in the VPM are located adjacent to, and just medial to, the thalamic nucleus that relays somatosensory information from the mouth.

In rodents, gustatory information from the thalamic nuclei projects to the *insular cortex*, and in primates (and cats) the target is the insular and *opercular cortex* (Fig. 7.7). Some studies have revealed projections from the thalamic nuclei to limbic structures such as the amygdala. There are also indications that the perirhinal cortex and insular cortex receive input directly from the gustatory nucleus in the thalamus. The insular cortex projects to structures of the limbic system and to subcortical motor areas.[28] The insular cortex sends efferent fibers to subcortical portions of the gustatory neuroaxis and to integrative cortical regions (association cortices).

The NST projects to various motor neurons that control muscles of the tongue and salivation and to the pancreas for insulin production (Fig. 7.8). Taste can elicit various reflexes through these motor connections, such as coughing, apnea, and salivation. Protective reflexes such as those elicited by bitter taste involve

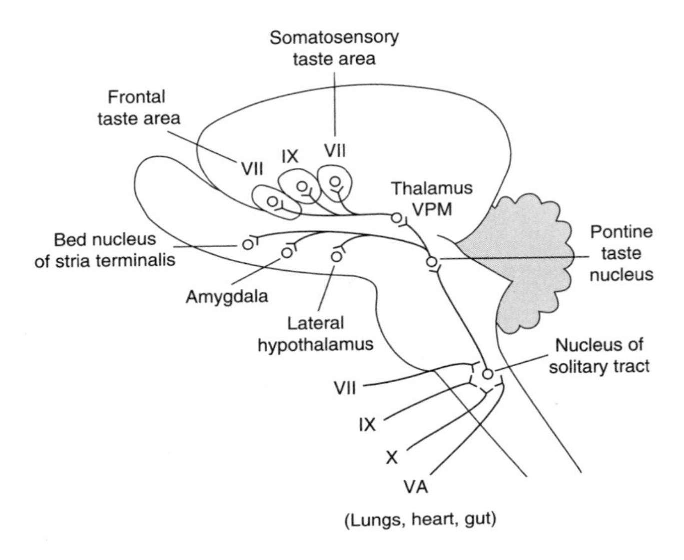

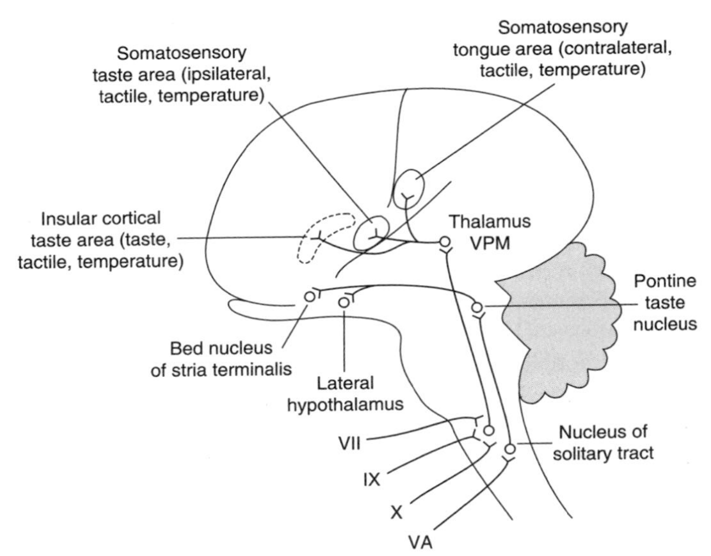

FIGURE 7.7 Schematic illustration of the major taste pathways in the rat and the cat. (Adapted from Norgren, R., *Olfaction and Taste*, edited by van der Starre, H. London: IRL Press, 1980, 288.)

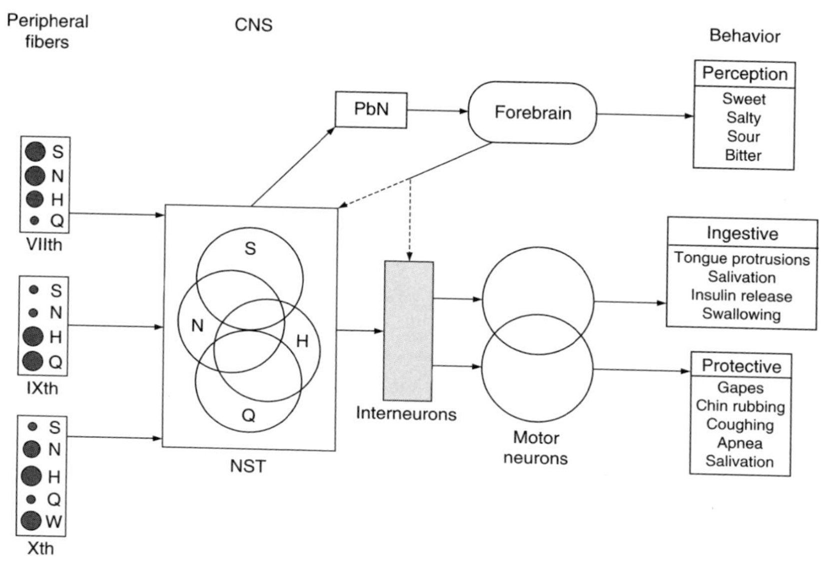

FIGURE 7.8 Schematic diagram showing the most important target of taste impulses that are mediated by the three nerves that carry taste information. (Adapted from Smith, D., and Frank, M., *Mechanisms of Taste Transduction*, edited by Simon, S.A., and Roper, S.D. Boca Raton, FL: CRC Press, 1993.)

many muscle groups and serve to prevent ingestion of poisonous food. These reflexes can include vomiting. In a few people, taste can cause general sweating (gustatory sweating*).

The taste pathways through the pontine nucleus bifurcate with one branch proceeding to the primary somatosensory cortex (SI), providing conscious perception of taste, and the other branch reaching structures of the limbic system. The latter pathway may be important for mediating the affective reactions to taste and may be regarded as an equivalent to the nonclassical pathways in the other senses.

Tactile receptors in the tongue and the mouth mediate information about the texture of food, which plays an important role in creating the flavor of food. Sensory information from the mouth including the anterior portion of the tongue is mediated by the trigeminal nerve (CN V), and sensory innervation of the posterior part of the tongue is carried in a branch of the glossopharyngeal nerve through the spinal nucleus of the trigeminal tract. Receptors that are sensitive to temperature are also important in that respect, as are the bare axons that respond

*Auriculotemporal nerve syndrome is localized flushing and sweating of the ear and cheek in response to eating.

to hot spices such as those that contain capsaicin*, which activates nociceptors (pain receptors) such as heat receptors (see Chapter 4), giving a burning sensation in the mouth. Black pepper probably activates similar receptors.

2. Central Olfaction Pathways

The organization of the ascending pathways of the olfactory system is different from that of the other primary senses, and the olfactory system is developed differently in different vertebrate species.[3,24] The olfactory bulb projects to many different regions of the CNS (Fig. 7.9) such as the *entorhinal cortex* and the *periform cortex*. The entorhinal cortex (Brodmann's area 28) is located in the parahippocampal gyrus and projects to the olfactory cortex of the *uncus*. These structures give rise to fibers to the hippocampus called the *perforant pathway*.

Only part of the ascending fibers make synaptic contacts in the thalamus,[30] and fibers from the olfactory bulb connect to cells in the *allocortex*. The allocortex connects directly to *limbic structures* such as the central nucleus of the amygdala** and from there to the medial hypothalamus. The cortical projections include the periform cortex in the tip of the temporal lobe (uncus and the entorhinal cortex) (Fig. 7.9).

The entorhinal cortex is the anatomical location most likely involved in interpretation of odors together with the *olfactory tubercle*, which is also a part of the *allocortex*. The olfactory tubercle receives input from the olfactory bulb via the *intermediate olfactory stria*[4] and connects to the hypothalamus and the medial and dorsal thalamus (Fig. 7.10).

It has been questioned whether olfaction has a noticeable thalamic connection but some investigators find evidence that the olfactory bulb indeed projects to the thalamus (the medial dorsal portion) (Fig. 7.10), and evidence has been presented that cells in the periform cortex project to the mediodorsal thalamus, the cells of which project to orbitofrontal and frontal cortical areas. The olfactory tubercle may be involved in various functions associated with the limbic system. The periform cortex connects to neocortical areas (orbitofrontal and frontal cortices) directly or via the thalamus, as does the entorhinal cortex. The entorhinal cortex also projects to the hippocampus. The thalamus to the orbitofrontal cortex is essential for discrimination of odors because individuals with injuries to the orbitofrontal cortex are unable to discriminate odors.[4]

The main targets of the mammalian olfactory bulb are the anterior olfactory nucleus, olfactory cortex, olfactory tubercle, anteriorlateral nucleus of the amygdala, and the lateral entorhinal cortex.[4,8,21,23] The olfactory bulb has direct

*Capsaicin is an ingredient of hot pepper.
**Often the amygdala nuclei are referred to as just the amygdala despite the fact that different parts of the amygdala have fundamentally different functions.

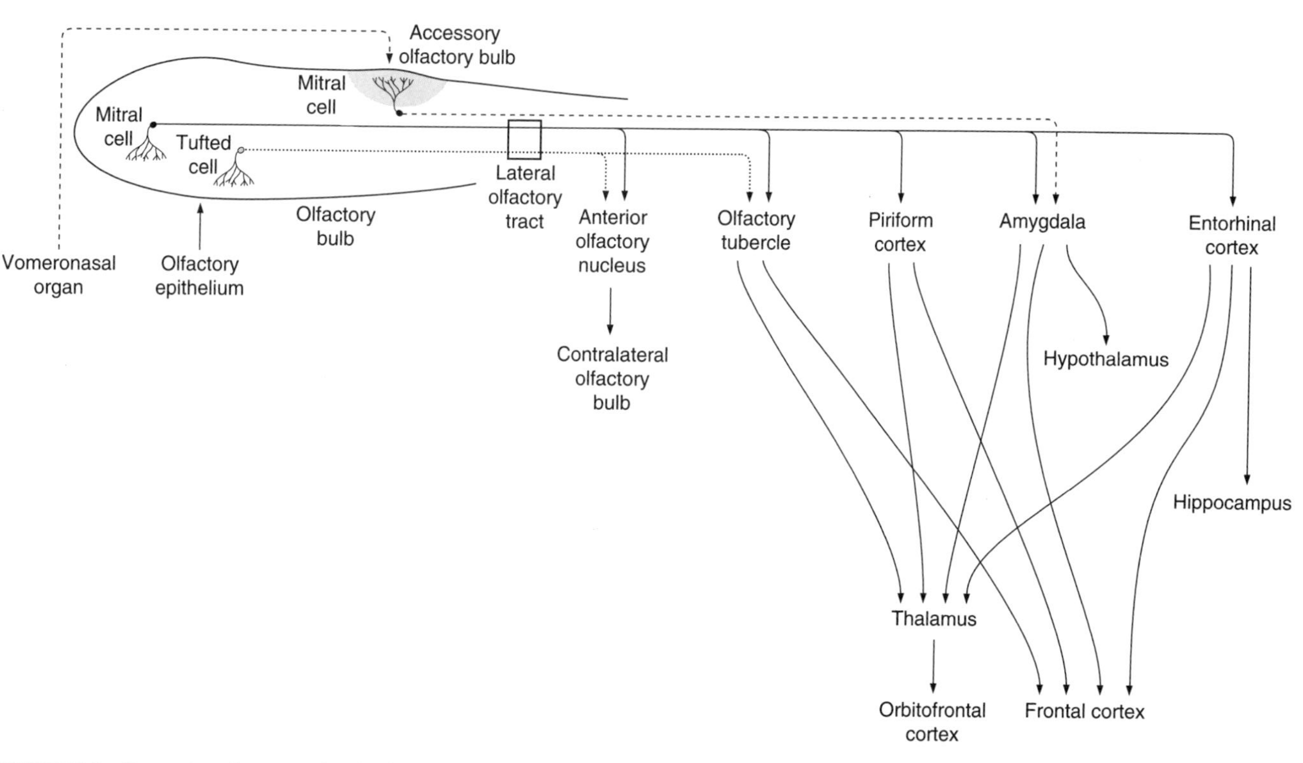

FIGURE 7.9 Connections from mitral and tufted cells in the olfactory epithelium to different regions of the brain. (Adapted from Buck, L.B., *Principles of Neural Science*, 4th ed., edited by Kandel, E.R., Schwartz, J.H., and Jessell, T.M. New York: McGraw-Hill, 2000, 625–647.)

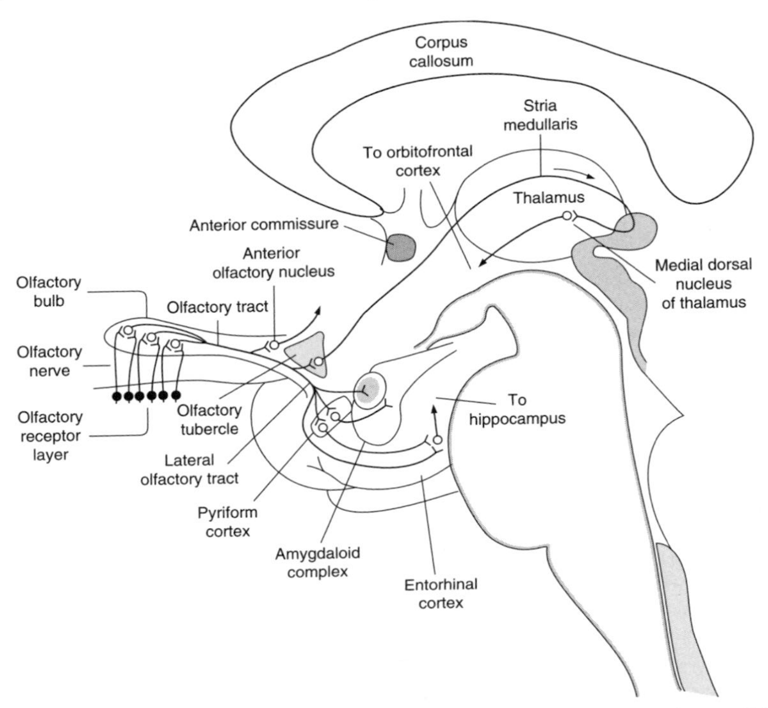

FIGURE 7.10 Anatomical location of the targets from the ascending olfactory pathways. (Adapted from Kandel, E.R., Schwartz, J.H., and Jessell, T.M., *Principles of Neural Science*, 3rd ed. Norwalk: Appleton & Lang, 1991.)

connections to the entorhinal cortex, pyriform cortex, posterior hypothalamus, and *hippocampus* (Fig. 7.10). Through the olfactory tubercle, the olfactory bulb has connections to the thalamus and the frontal cortex. The anterior olfactory nucleus projects to the contralateral olfactory bulb through the anterior commissure, which is a part of the *corpus callosum* (Fig. 7.10).[4]

The pathways through the medial dorsal portion of the thalamus that reach the orbitofrontal cortex may be responsible for perception and discrimination of odors while the pathways that reach the amygdala and hypothalamus carry emotional response and physiologic reactions to odors. These pathways may also be responsible for the role of olfaction in eating and for sexually related behavior and sexual reflexes.

a. Vomeronasal Pathway

The *vomeronasal organ* responds to pheromones (see Chapter 2) and it is assumed to be involved in (chemical) communication between individuals of the same species that are important for sexual behavior and other instinct behaviors. The sensory cells of the vomeronasal organ are innervated by the accessory olfactory bulb (Fig. 7.9). The central projections of the accessory olfactory bulb are different from those of the olfactory bulb and may be regarded as a second olfactory pathway.[8,33] The vomeronasal pathway may be equivalent to the nonclassical pathways of hearing, somesthesia, and vision.

The accessory olfactory bulb connects only to the amygdala.[23,33] The connections from the accessory olfactory bulb project to the *bed nucleus of the accessory olfactory tract*,* the *medial nucleus*, and the *posteriomedial cortical nucleus of the amygdala*.[8,13] There is no direct pathway from the vomeronasal organ to the cerebral cortex. Discrimination of pheromones and communication with pheromones may therefore occur without conscious awareness. That may be one reason why it has been difficult to establish the role of the vomeronasal system in humans.

III. PHYSIOLOGY OF THE CHEMICAL SENSES

Most of our knowledge about the physiology of the chemical senses is regarding the receptors. Olfactory and taste information reaches parts of the CNS other than those of the other three sensory systems and little is known about the processing of information in the nervous system of these senses.

A. Receptors

Both taste and olfactory receptors respond selectively to chemical substances, and the taste receptors respond to fewer types of substances than the olfactory sensory cells.

Taste cells are mainly sensitive to four qualities (bitter, sour, sweet, salty) and possibly a fifth, (monosodium glutamate, known as *umami*).[4,12] Studies of the response from single nerve fibers in the chorda tympani by Pfaffman et al.[19] showed that individual nerve fibers respond to more than one of the four chemicals to which the receptors respond but each one of the different taste substances elicits a greater response in a certain population of the afferent nerve fibers than other substances. This means that some nerve fibers respond best to sugar whereas other cells will respond better to bitter and sour, but each nerve fiber responds to several taste substances. Taste receptors have therefore been

*The bed nucleus of the stria terminalis is a basal forebrain structure that is a part of the extended amygdala.

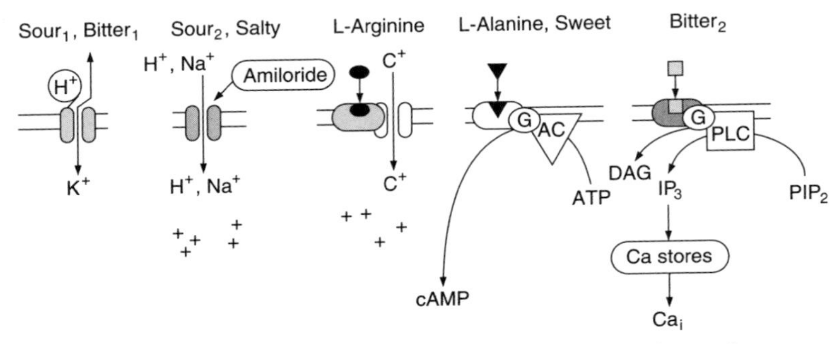

FIGURE 7.11 Illustration of the transduction process in the apical membrane of taste receptors. (Adapted from Avenet, P., Kinnamon, S.C., and Roper, S.D. *Mechanisms of Taste Transduction*, edited by Simon, S.A., and Roper, S.D. Boca Raton, FL: CRC Press, 1993.)

called *taste generalists*.[26] Taste cells that respond best to each one of these four qualities are found in all parts of the tongue.

> Substances belonging to the four families of substances we can discriminate excite taste cells through different mechanisms.[26] Salty substances cause Na^+ influx through amiloride*-sensitive Na^+ channels; sour taste is mediated by H^+ ions through the same channels or by blocking K^+ channels. Most bitter substances probably bind to a G-protein-coupled receptor. Also, some substances with a sweet taste probably bind to receptors using similar mechanisms (Fig. 7.11).

1. Olfactory Receptors

In principle, different olfactory receptor cells are sensitive to different odorants (Fig. 7.12).[5] However, sensory cells in the vertebrate olfactory system are moderately selective to odors and one odor will elicit a response from different kinds of receptor cells. That means that many different types of cells will respond to one odor but one odor will elicit the greatest response in one type of cell, a smaller response in another type of cells and no response at all in yet another population of cells. Such receptor cells are known as odor generalists. All known olfactory receptors in vertebrates belong to this category (Fig. 7.12).[5,26] This is similar to other sensory systems such as gustation and color vision, and it is assumed that the central nervous system interprets the responses from many different receptors to provide a distinct sensation of one specific odor.

*Amiloride hydrochloride, or N-amidino-3,5-diamino-6-chloropyrazinecarboxami demonohydrochloride dihydrate; a nonsteroidal compound exerting an effect similar to that of an aldosterone inhibitor (i.e., urinary sodium excretion is enhanced and potassium excretion is reduced); a potassium-sparing diuretic.

> The receptors of the olfactory sensory cells are proteins that transduce different chemicals (odorants) and activate a *second messenger*, the activation of which leads to depolarization of the receptor cells (Fig. 7.13).[26]

The adaptation of odor receptor cells may be caused by inactivation or desensitization of the receptor cells. Adaptation of the olfactory neurons in the CNS contributes to the adaptation of the olfactory system, thus similar to what occurs in other sensory systems. The physiologic responses of odor cells to constant exposure adapt at a slow rate, and adaptation of cells in the olfactory bulb is much faster than that of the receptor cells due to central processing of the stimuli (Fig. 7.14).[6,11,26]

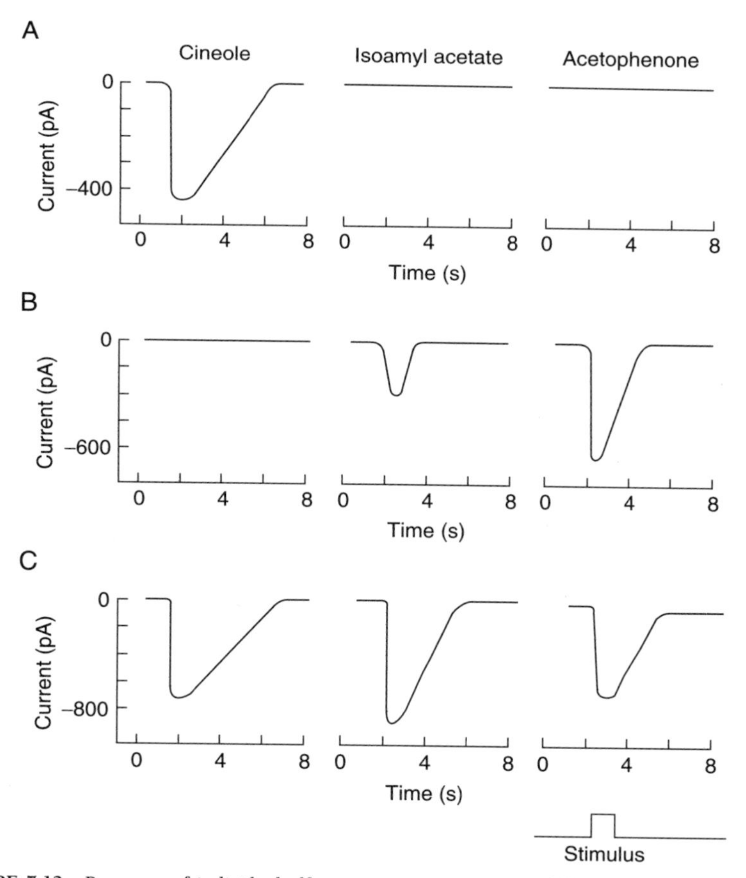

FIGURE 7.12 Response of individual olfactory sensory neurons to different odorants. (Data from Firestein *et al.*[5])

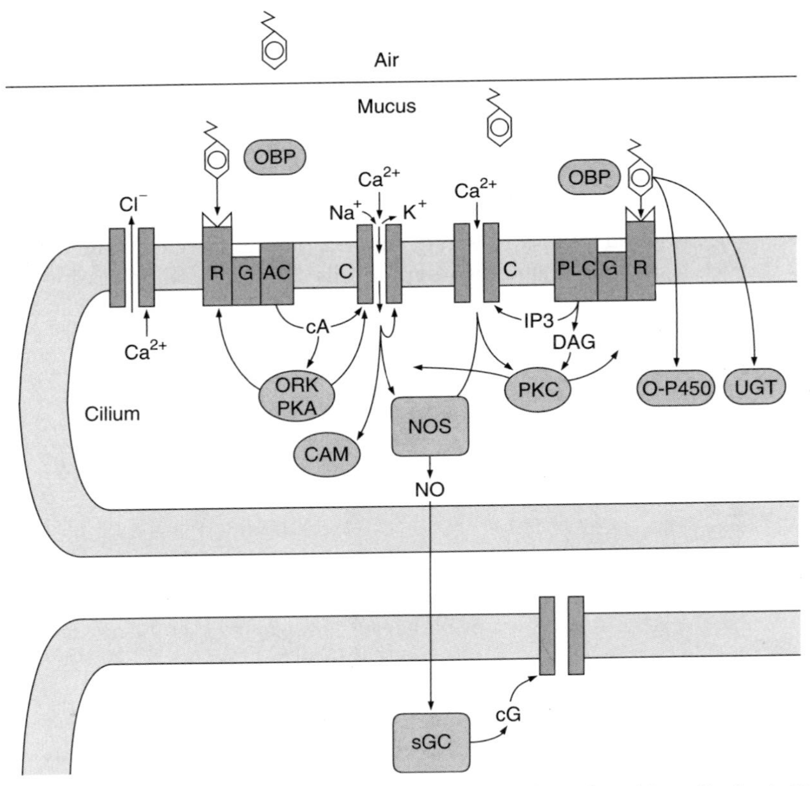

FIGURE 7.13 Sensory transduction mechanisms in olfactory cilia. (Adapted from Shepherd, G.M., *Neurobiology*, New York: Oxford University Press, 1994.)

The sensitivity of olfactory cells in humans varies considerably, and some individuals have low or no sensitivity to certain odors. The low sensitivity to specific odors, known as *specific anosmia*, has been linked to absence of the sensory cells that are sensitive to such odors. The sensitivity generally decreases with age because of loss of cells in the olfactory bulb.[14]

2. Vomeronasal Organ

The vomeronasal organ responds to special kinds of odors know as *pheromones*. Pheromones have been studied extensively in insects where each receptor cell responds only to specific pheromones and are known as odor specialists.[26] It has been suggested that the cells that respond to pheromones may also exist in mammalians, including humans, but only recently has experimental evidence for their existence been presented. Individual vomeronasal receptor cells in

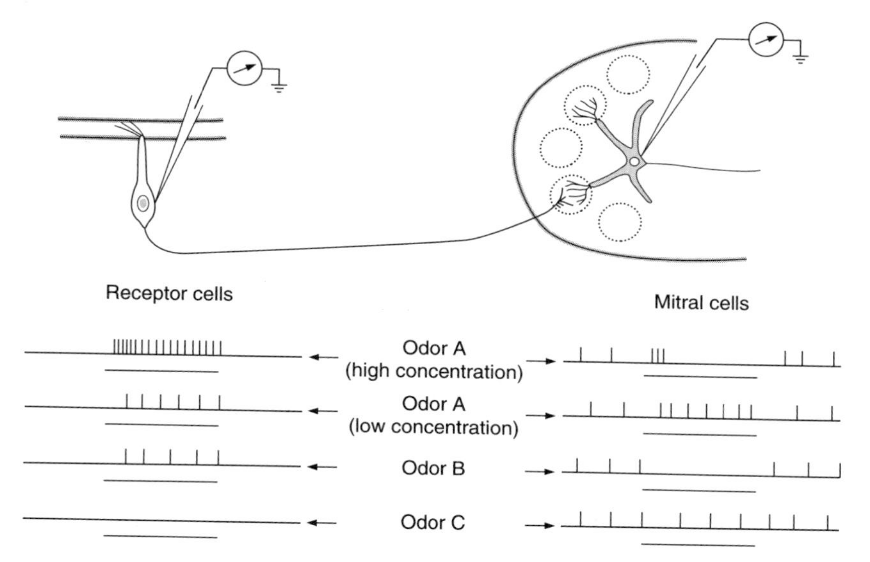

FIGURE 7.14 Illustration of the difference in adaptation of the response from two cells in the ascending olfactory pathways. (Data from Kauer[11] and Getchell.[6])

male and female mice have been found to respond selectively to only one of six pheromones.[20]

As mentioned above, the vomeronasal organ projects to the accessory olfactory bulb, the neurons of which connect directly to neurons in the amygdala. This means that the vomeronasal pathway controls endocrine functions and emotions directly and probably without conscious awareness. The vomeronasal system also provides a drive for instinctive behavior through its activation of nuclei of the amygdala. The absence of connections to the cerebral cortex and the lack of higher order processing of the information mediated by this pathway limit its functions to instinct-like reactions.

It has recently been shown that receptors in the vomeronasal organ also respond to several different odors.[20] This finding is interesting as it was earlier assumed that the vomeronasal organ and this pathway exclusively concerned responses to pheromones.[8] The concentrations of specific odors that elicited response from vomeronasal cells were lower that those that elicited response from odor cells in the olfactory epithelium, indicating a very high sensitivity of cells in the vomeronasal organ. The finding that vomeronasal neurons can also detect odorants indicates that odorants may activate the amygdala through a direct route[20] in a similar way as occur in the nonclassical pathways of other sensory systems where information can reach the amygdala directly from the dorsal and medial thalamus (see Chapter 3).

In many animals, pheromones serve as an important communication system between members of the same species. Pheromones may automatically switch on or off sexual behaviors. In some animals, stimulation of the olfactory system may even cause abortion. It is not known to what extent or exactly what role pheromones and odors play in humans for reproduction. There are, however, signs that olfaction, and probably pheromones, may have more influence on the reproductive systems in humans than what is readily observed. Thus, it is known that the menstrual cycles in women who live close together tend to synchronize[7] and is assumed to be the result of body scent.[15] (This topic was reviewed by Weller and Weller.[31]) This would indicate an unconscious influence on some basic reproductive functions. Pheromones may therefore also affect other reproductive functions. Signals that are conveyed by the vomeronasal pathway reach the amygdala directly and not the cerebral cortex. The reactions to pheromones occur without conscious awareness and without conscious control by the individual who is exposed to such substances. If pheromones are involved in the attraction between sexes it may not be associated with a specific conscious perception. This makes it difficult to detect the effect of pheromones.

B. Coding of Information in the Gustatory and Olfactory Nervous Systems

The high degree of ability to discriminate different odors and tastes, despite a low degree of selectivity of the receptors, indicates that considerable central processing occurs. The coding of chemical substances lacks emphasis on the temporal pattern, which is an obstacle in attempts to record electrical responses (evoked responses) from the nervous system to gustatory and olfactory stimulation and has hampered studies of neural processing in these two senses.

Individual nerve fibers of the nerves that innervate taste receptors innervate more than one type of receptor, which speaks against the labeled line hypothesis (see Chapter 2) for discrimination of sensory qualities. Interpretation of specific tastes must therefore be done by comparing the response from several types of receptors in a manner similar to that for other senses, such as olfaction and color vision (see Chapter 6). Also, in the olfactory system several types of receptors do respond to the same odor but to different degrees.

Some of the processing of odors occurs in the glomeruli of the olfactory bulb, which may be regarded as a part of the CNS. There is a certain spatial organization in the olfactory bulb which means that different odors are represented in anatomically different regions of the olfactory bulb. This spatial organization may be preserved in the ascending olfactory pathways and may be an important part of the coding of odors.

The olfactory nervous system seems to be hardwired (genetically programmed) to a greater extent than is the case for the other four senses.[35] Earlier studies however, have shown that the olfactory nervous system has a considerable plasticity.

The sensory cells in the vomeronasal organ that detect pheromones are odor specialists and each receptor responds only to one chemical. Discrimination between different pheromones would therefore require less central processing than what is needed as the basis for discrimination of odors.

REFERENCES

1. Avenet, P., Kinnamon, S.C., and Roper, S.D., Peripheral Transduction Mechanisms. In: *Mechanisms of Taste Transduction*, edited by S.A. Simon and S.D. Roper. Boca Raton, FL: CRC Press, 1993.
2. Bhatnagar, K.P., The Human Vomeronasal Organ. III. Postnatal Development from Infancy to the Ninth Decade. *J. Anat. (London)*, 199: 289–302, 2001.
3. Brodal, P., *The Central Nervous System*. New York: Oxford Press, 1998.
4. Buck, L.B., Smell and Taste: The Chemical Senses. In: *Principles of Neural Science*, 4th ed., edited by E.R. Kandel, J.H. Schwartz, and T.M. Jessell. New York: McGraw-Hill, 2000, pp. 625–647.
5. Firestein, S., Picco, C., and Menini, A., The Relation Between Stimulus and Response in Olfactory Receptor Cells of the Tiger Salamander. *J. Physiol. (London)* 468: 1–10, 1993.
6. Getchell, T.V. and Shepherd, G.M., Response of Olfactory Receptor Cells to Step Pulses of Odour at Different Concentrations in the Salamander. *J. Physiol. (London)*, 282: 521–540, 1974.
7. Graham, C.A. and McGrew, W.C., Menstrual Synchrony in Female Undergraduates Living on a Coeducational Campus. *Psychoneuroendocrinology*, 5: 245–252, 1980.
8. Halpern, M., The Organization and Function of the Vomeronasal System. *Annu. Rev. Neurosci.*, 10: 325–362, 1987.
9. Kandel, E.R., Schwartz, J.H., and Jessell, T.M., *Principles of Neural Science*, 3rd ed. Norwalk: Appleton & Lang, 1991.
10. Karimnamazi, H. and Travers, J.B., Differential Projections from Gustatory Responsive Regions of the Parabrachial Nucleus to the Medulla and Forebrain. *Brain Res.*, 813: 283–302, 1998.
11. Kauer, J.S., Response Patterns of Amphibian Olfactory Bulb to Odor Stimulation. *J. Physiol. (London)* 243: 675–715, 1974.
12. Matsunami, H., Montmayeur, J.-P., and Buck, L., A Family of Candidate Taste Receptors in Human and Mouse. *Nature*, 404: 601–604, 2000.
13. McDonald, A.J., Cortical Pathways to the Mammalian Amygdala. *Progr. Neurobiol.*, 55: 257–332, 1998.
14. Meisami, E., Mikhail, L., Baim, D., and Bhatnagar, K.P., Human Olfactory Bulb: Aging of Glomeruli and Mitral Cells and a Search for the Accessory Olfactory Bulb. *Ann. N. Y. Acad. Sci.*, 855: 708–715, 1998.
15. Morofushi, M., Shinohara, K., Funabashi, T., and Kimura, F., Positive Relationship Between Menstrual Synchrony and Ability To Smell 5alpha-Androst-16-En-3alpha-Ol. *Chemical Senses* 25: 407–411, 2000.
16. Mountcastle, V.B., ed., *Medical Physiology*. St. Louis: C.V. Mosby, 1974.
17. Murray, R.G., The Ultrastructure of Taste Buds. In: *The Ultrastructure of Sensory Organs*, edited by I. Friedmann. New York: Elsevier, 1973, pp. 1–81.
18. Norgren, R., Neuroanatomy of Gustatory and Visceral Afferent Systems in Rat and Monkey. In: *Olfaction and Taste*, edited by H. van der Starre. London: IRL Press, 1980, p. 288.
19. Pfaffman, C., Frank, M., Bartoshuk, L.M., and Snell, T.C., Coding Gustatory Information in the Squirrel Monkey Chorda Tympani. In: *Progress in Psychobiology and Physiological Psychology*, edited by J.M. Sprague and A.N. Epstein. New York: Academic Press, 1976, pp. 1–27.

20. Sam, M., Vora, S., Malnic, B., Ma, W., Novotny, M.V., and Buck, L.B., Odorants May Arouse Instinctive Behaviors. *Nature*, 412: 142, 2001.
21. Scalia, F. and Winans, S.S., The Differential Projections of the Olfactory Bulb and Accessory Olfactory Bulb in Mammals. *J. Comp. Neurol.*, 161: 31–56, 1975.
22. Shepherd, G.M., Synaptic Organization of the Mammalian Olfactory Bulb. *Physiol., Rev.*, 52: 864–917, 1972.
23. Shepherd, G.M., Microcircuits in the Nervous System. *Sci. Am.*, 238: 92–103, 1978.
24. Shepherd, G.M., Pedersen, P.E., and Gree, C.A., Development of Olfactory Specificity in the Albino Rat: A Model System. In: *A Psychobiological Perspective*, edited by N.A. Krasnegor, E.M. Blass, M.A. Hofer, and W.P. Smotherman. New York: Academic Press, 1987.
25. Shepherd, G.M. and Greer, C.A., Olfactory Bulb. In: *Synaptic Organization of the Brain*, 3rd ed., edited by G.M. Shepherd. New York: Oxford University Press, 1990, pp. 133–169.
26. Shepherd, G.M., *Neurobiology*, 3rd ed., New York: Oxford University Press, 1994.
27. Simon, S.A. and Roper, S.D., *Mechanisms of Taste Transduction*. Boca Raton, FL: CRC Press, 1993.
28. Smith, D. and Frank, M., Sensory Coding by Peripheral Taste Fibers. In: *Mechanisms of Taste Transduction*, edited by S.A. Simon and S. D. Roper. Boca Raton, FL: CRC Press, 1993.
29. Smith, T.D. and Bhatnagar, K.P., The Human Vomeronasal Organ. Part II: Prenatal Development. *J. Anat. (London)* 197: 421–436, 2000.
30. Stepniewska, I. and Rajkowska, G., The Sensory Projections to the Frontal Association Cortex in the Dog. *Acta Neurobiol. Exp.*, 49: 299–310, 1989.
31. Weller, L. and Weller, A., Human Menstrual Synchrony: A Critical Assessment. *Neurosci. Biobehav. Rev.*, 17: 427–439, 1993.
32. Wilson-Pauwels, A.O., Akesson, E.J., and Stewart, P.A., *Cranial Nerves*. Toronto: B.C. Decker, 1988.
33. Winans, S.S. and Scalia, F., Amygdaloid Nucleus: New Afferent Input from the Vomeronasal Organ. *Science*, 170: 330–332, 1970.
34. Yokoi, M., Mori, K., and Nakanishi, S., Refinement of Odor Molecule Tuning by Dendrodendritic Synaptic Inhibition in the Olfactory Bulb. *Proc. Natl. Acad. Sci.* USA, 92: 3371–3375, 1995.
35. Zou, Z., Horowitz, L.F., Montmayeur, J.-P., Snapper, S., and Buck, L.B., Genetic Tracing Reveals a Stereotyped Sensory Map in the Olfactory Cortex. *Nature*, 414: 173–179, 2001.

INDEX

A

AAF, *see* Anterior auditory fields
Acceleration detectors, 58, 222
Accessory olfactory bulb
  amygdala connection, 443
  bed nucleus, 156
  Jacobson's organ, 428
  mitral cells, 434
  vomeronasal organ projection, 447
Accessory olfactory tract, bed nucleus, 94, 443
Accomodation, eye, 420
Acoustic middle ear reflexes
Acoustic reflexes
  as nonclassical auditory pathways, 352
  stapedius muscle, 290
  tensor tympani participation, 291
ACTH, *see* Adrenocorticotropin
Action potentials
  ganglion cells, 387
  sensory transduction, 65
Acute pain
  characteristics, 243–246
  cortical representation, 261–262
  pain type, 241
Adaptation
  definition, 7
Adrenergic innervation, receptor sensitivity, 67–68
Adrenocorticotropin, 159
Affective reactions
  taste pathways, 439
  tinnitus, 355
Agnosia
  definition, 153
  forms, 153
AI, *see* Primary auditory cortex
AII, *see* Secondary auditory cortex
Allocortex
Allodynia, 171, 240, 257, 259
Amacrine cells
  eye innervation, 55
  photoreceptors to ganglion cells, 84
  retinal information processing, 385, 387
  retinal innervation, 55
  retinal neural network, 380
Amplitude compression, 304–305
  automatic gain control, 290
  cochlear, 304
  function, 62
  receptors, 69
Amplitude-modulated tones
  methods of study, 333
  waveform reproduction, 335
Amygdala
  accessory olfactory bulb connection, 443
  anatomy, 156–159
  basolateral nucleus, 141, 156, 217, 237
  central nucleus, 93, 141, 159, 436
  extended amygdala, 443
  function, 159–162
  ICC neural input, 312
  lateral nucleus, 141, 159, 237
  medial and central nuclei, 93
  medial nucleus, 93, 156, 436

Amygdala (*continued*)
mediodorsal nucleus relay, 101
nuclei, 440
pleasure perception, 162–163
posteriomedial cortical nucleus, 94, 443
role in arousal, 237
role in nonperceptual effects of stimulation, 237–238
species differences, 163–164
stream segregation, 113
stria terminalis, 96–97
subcortical auditory input, 355–356
thalamus projections, 217
visual activation, 420
vomeronasal organ, 447
vomeronasal pathway, 448
Angina pectoris, 252
Anteriolateral tracts, 94, 96, 216
components, 87
dorsal horn organization, 208
information processing, 145
nonclassical somatosensory system, 239–240
overview, 210–211
periaqueductal gray, 102
spinomesencephalic tract, 215
spinoreticular tract, 214–215
spinothalamic tract, 211–216
Anterior auditory fields
cat, 108
MGB projections, 313
as part of auditory cortex, 314
Anterior cingulate gyrus, 155, 216, 240
Anterior commissure, 93, 108
Anterior insular cortex, 217, 262
Anterior thalamic nucleus, 101
Anteriorventral cochlear nucleus, 92, 309
Apoptosis
neural plasticity, 171
nonclassical pathways, 87, 147
Apparent motion, 29
Arousal
amygdala, 159
ascending sensory pathways, 141–142
medial lemniscus, 202
nonclassical sensory systems, 97, 239
Association cortices
nonclassical pathways, 88
processing overview, 413–414
stream segregation, 111, 414–416
Attention
ascending sensory pathways, 142–143
Auditory cortex
frequency-modulated sound responses, 336
location and connections, 314–315
neural plasticity, 166, 358
Auditory fields
anterior fields, 108, 313–314
posterior fields, 108
Auditory nerve (CN VIII)
cochlear location, 284
cochlear nucleus comparison, 335–336
function, 307–309
receptive fields, 318–323
sound waveform coding, 328
Auditory pathways
classical ascending systems
auditory cortex, 314–315
auditory nerve, 307–309
auditory nerve fiber receptive fields, 318–323
auditory nerve fiber sound intensity coding, 332–339
auditory nerve sound waveform coding, 328–332
bilateral representation of sounds, 341–343
cell tuning, 323–325
cochlear nucleus, 309
connections to nonclassical system, 356
frequency representation, 317–318
frequency threshold curves, 325
hemispherical dominance, 344
inferior colliculus, 311–313
information processing, 317
lateral lemniscus, 311–313
medial geniculate body, 313
nervous system overview, 305
parallel processing, 309
precedence effect, 343–344
receptive fields, 318
sound property maps, 340–341
speech discrimination, 339–340
stream segregation, 344–346
superior olivary complex, 309–311
classical descending systems, 346–347
nonclassical ascending systems
acoustic middle ear reflexes, 352–353, 357
dorsal and medial thalamus, 355
dorsal cortex (of the inferior colliculus), 354

external nucleus (of the inferior colliculus), 354
nonauditory region projections, 351–352
nonauditory input, 349–351
polysensory neurons, 354–355
subcortical auditory input to amygdala, 355–356
nonclassical descending systems, 305, 353
Augmented sound environment, 358
Automatic gain control, 290
Autonomic reactions, 154–155
AVCN, *see* Anteriorventral cochlear nucleus

B

Backward masking, 11
Basilar membrane
as cochlear component, 279
cochlear fluid vibration, 292–293
cochlear frequency analysis, 294
cochlear function, 40
frequency selectivity, 59, 291–292
frequency tuning, 282
hair cell arrangement, 283
hair cell deflection, 295–296
receptor selectivity, 70–71
sound separation, 317
stimulus transmission, 57
tuning, 300
vibration conduction to receptors, 294–296
Basolateral nucleus of the amygdala
amygdala anatomy, 156
arousal, 141
nonperceptual effects of stimulation, 237
Bed nucleus, 156
of the accessory olfactory tract, 94, 443
of the stria terminalis, 94
BIC, *see* Brachium of the inferior colliculus
Binaural hearing, 27
Binaural representation of sounds, 341
Bipolar cells
ganglion cell input, 402
photoreceptors to ganglion cells, 84
retina anatomy, 377
retinal information processing, 385
retinal innervation, 55
retinal neural network, 380
Bitter substance toxicity, 428
Blind spot, 379
Blink reflex, 154
Brachium of the inferior colliculus
auditory information conduction, 92
classical ascending pathway anatomy, 306
ICC output, 313
Brainstem reticular formation
arousal, 141–142
ascending fiber tracts, 97
medial lemniscus, 202
nonperceptual effects of stimulation, 238
role in arousal, 237
spinothalamic and spinoreticular tract input, 96
Bug detectors, 387

C

Capsaicin receptor, 238
definition, 440
nociceptor sensitivity, 47
polymodal nociceptor sensitivity, 54
similarity to TENS, 250
somatic pain, 243
taste pathways, 440
Caudal-trigeminal system, 216, 239
Caudodorsal division of the medial geniculate body, 351
Caudomedial cortical area, 344
Central nucleus of the amygdala, 93, 141, 159, 436
Central nucleus of the inferior colliculus
auditory cortex descending fibers, 315
connections between sides, 311–312
descending pathways, 353
input to posterior thalamus, 348
neural plasticity, 357
Central pain pathways, 254–255
Central pain pathway sensitization, 255
wind-up phenomenon, 258
Cerebral cortex
anatomical overview, 102–104
association cortices, 107
cell types, 104–105
connections, 105–106
polysensory cortical area, 107
primary sensory cerebral cortices, 105
Cerebral cortex anatomy, 102
CF, *see* Characteristic frequency
CFF, *see* Critical fusion frequency
Characteristic frequency, 318

Chorda tympani, 93, 432
Choroids, 40
Chronic pain, 246, 254
Ciliary body, 40
Cingulate gyrus
  anterior region, 155, 216, 240
  pain processing, 262
Circadian rhythms
  hypothalamus, 419
  retina, 42
  suprachiasmic nucleus, 388, 419
Classical pathways
  auditory system
    auditory temporal coding, 328–332
    bilateral representation of sounds, 341–343
    connections to nonclassical system, 356
    descending systems, 346–347
    frequency threshold curves, 325
    hemispherical dominance, 344
    information processing, 317
    nerve fiber receptive fields, 318–323
    nerve fiber sound intensity coding, 332–339
    precedence effect, 343–344
    receptive fields, 318
    sound property maps, 340–341
    speech discrimination, 339–340
    stream segregation, 344–346
  somatosensory system
    components, 196–201
    conducting media, 192–193
    descending pathways, 217–218
    inhibition–excitation interplay, 232
    mechanoreceptor innervation, 194–196
    mechanoreceptors, 221–225
    medial lemniscus, 202
    neural plasticity, 235–236
    nonperceptual effects of stimulation, 236–238
    object shape discrimination, 233–234
    overview, 196–201, 227–228
    receptive fields, 230–232
    receptors, 190–192
    skin stimulation, 232–233
    somatosensory cortices, 202–205
    somatosensory thalamic nuclei, 202
    somatotopic organization, 228–230
    stream segregation, 235
    thermoreceptors, 225–226
  visual system
    ganglion cell responses, 401
    lateral geniculate nucleus, 390–394
    optic nerve, 389–390
    optic tract, 389–390
    primary cortex connections, 396–398
    stream segregation, 398–399
    visual cortex, 394–396
COCB, *see* Crossed olivocochlear bundle
Cochlea
  anatomy, 40, 278–284
  automatic gain control, 304–305
  basilar membrane, 292–293
  frequency analysis, 294
  frequency selectivity, 59
  hair cell innervation, 284
  hair cell sensory transduction, 296–299
  otoacoustic emission, 302–303
  outer hair cell motility, 301
  outer hair cell role, 300–301
  receptor anatomy, 47–49
  vibrations from basilar membrane to receptors, 294–296
Cochlear aqueduct, 280
Cochlear echo, *see* Otoacoustic emissions
Cochlear nucleus
  anatomy and function, 309
  anteriorventral nucleus, 92, 309
  ascending sensory pathways, 94
  classical ascending pathway anatomy, 306
  coding of modulation, 124
  coding of time pattern, 119
  discharge patterns, 121
  divisions, 92
  dorsal nucleus, 309, 316, 355
  modulated waveform response, 335
  neural plasticity, 357–358
  posteriorventral nucleus, 309
  rapidly changing frequency response, 337
  response comparison to auditory nerve fibers, 335–336
  ventral nucleus, 352
Cocktail party problem, 140
Coding of the stimulus strength, 116–118
Coding of the time pattern, 118–122
Coherent firing, 149
Coincidence detectors, 343
Cold nociceptors
  function, 191–192

innervation, 53
selectivity, 71
Color vision
cone basis, 383–385
cones, 379
photoreceptors, 381
receptor cell properties, 36
Columnar organization, cerebral cortex, 104
Commissure of Probst, 306
Commissure of the inferior colliculus, 306, 311
Common pathway, thalamus to cerebral cortex, 99
Complex regional pain syndrome, 249
Computational maps
directional information, 139
superior colliculus, 152
Cone pathway
connections, 403
retinal information processing, 385
Cones
color vision, 383
light-adapted eye, 6
light wavelength, 36
retina anatomy, 377
selectivity, 70
Cool receptors
classical somatosensory system, 226
function, 191–192
innervation, 52–53
Cornea, as eye part, 40, 377
Corpus callosum
anterior commissure, 442
bilateral representation of sounds, 342
classical ascending pathways, 91
Corticomedial amygdaloid nucleus, 156
Corticosteroids, 159
CPRS, *see* Complex regional pain syndrome
Cranial nerves
auditory nerve (CN VIII), 40, 307–309
facial nerve (CN VII), 55, 432, 436–437
glossopharyngeal nerve (CN IX), 55, 93, 194, 428, 432–433, 436
olfactory nerve (CN I), 51, 55–56, 434
optic nerve (CN II), 55, 72, 380, 385, 389–390
trigeminal nerve (CN V), 42, 54, 194, 428, 439
vagus nerve (CN X), 55, 159, 433, 436
Criteria for detection, faint stimuli, 4
Critical band
masking, 10
spatial integration, 13
Critical fusion frequency, 24
Crossed olivocochlear bundle, 313, 345

D

Dark adaptation, 379
basic state, 381
definition, 382–383
Dark current, photoreceptors, 381
DAS, *see* Dorsal acoustic stria
DC, *see* Dorsal cortex of the inferior colliculus
DCN, *see* Dorsal cochlear nucleus; Dorsal column nuclei
Dermatomes, 194–196
Determination of movement, 28–29
Difference limen, 20
Diffuse systems, 96
Directional hearing
auditory space perception, 342–343
psychoacoustic studies, 330
specialized systems in animals, 29–30
Disc membranes, photoreceptors, 378
Displacement detectors, 222
Distortion product otoacoustic emission, 303
DNLL, *see* Dorsal nucleus of the lateral lemniscus
Dormant synapses
definition, 235
neural plasticity, 166–167
unmasking, 81, 169–171, 356
Dorsal acoustic stria, 309
Dorsal cochlear nucleus
auditory nerve termination, 309
inferior colliculus projections, 316
tinnitus, 355
Dorsal column nuclei
components, 92, 198–199
dorsal column tract termination, 196
Dorsal column tract, 196, 210–211
Dorsal cortex of the inferior colliculus, 309, 347
connections from auditory cortex, 348
descending pathways, 353
Dorsal division of the thalamus
pain processing, 111
projections to cerebral cortex, 216–217
role in arousal, 237

Dorsal horn
organization and function, 208–210
role in nonclassical somatosensory system, 238–239
Dorsal nucleus of the lateral lemniscus
classical ascending pathway anatomy, 306
interruption of fibers, 311
Dorsal roots of the spinal cord
innervation, 53–54
innervation of mechanoreceptors, 194
DPOAE, *see* Distortion product otoacoustic emission

E

Ear
cochlea, *see* Cochlea
middle ear, 36–37, 39–40, 56–57, 277–278, 288–291
outer ear, 37, 277–278
Echolocation
flying bats, 150, 337
stream segregation, 345
Edinger–Westphal nucleus, 418
Efferent nerve fibers, 54
Electrotonic transmission
definition, 65
receptor sensitivity, 67
Endolymph, 279
Endolymphatic sac, 280
Endolymphatic space, 298
Entorhinal cortex, 94, 440
Ep, *see* Posterior ectosylvian field
EPSP, *see* Excitatory postsynaptic potential
Eustachian tube, 278
Excitatory postsynaptic potential, 66, 119, 121
Extended amygdala, 443
External nucleus of the inferior colliculus
anatomy, 311
connections from auditory cortex, 348
descending pathways, 353
dorsal column nuclei connections, 350
input to MGB, 348
as nonclassical pathway, 96, 146, 347
physiology, 354
polysensory neurons, 354
Extraocular muscles
eye innervation, 55
eye position control, 377
nonclassical pathways, 420
visual reflexes, 420
Extrastriatal cortex
pulvinar nuclei, 102
pulvinar pathway, 418–419
pulvinar projections, 96
stream segregation, 111, 398–399
visual information processing, 396
Eye
anatomy, 40–42, 376–377
conductive apparatus, 377
lens, 41, 377
light conduction to photoreceptors, 380
photoreceptors, 36, 41–42, 44, 50, 55, 70, 84, 377–385
retina, 377
retinal information processing, 385–387
Type II receptors, 55
Eye muscles, 55, 377, 420

F

Face identification, 113
Face recognition, 415–416
Facial dermatomes, 195–196
Facial expressions, 113, 415–416
Facial motonucleus, 352
Facial nerve (CN VII)
stapedius muscle innervation, 352
taste pathways, 436–437
taste receptors, 55, 432
Faint stimuli detection, 4
human eye light sensitivity, 6
photoreceptors, 382–383
physical stimulus transfer, 57–58
Fast motility (hair cells), 301
Fatigue, 4
Fear response, amygdala function, 160–162
Flying bat
auditory system, 115, 306–307
echolocation, 150
stream segregation, 345
Fornicate gyrus, 83
Forward masking, 10
Fovea, 377
Free sound field
directional hearing, 27–28, 288
threshold of hearing, 6
Frequency discrimination, 339
Frequency-modulated tones, flying bats, 337

Frequency selectivity
  basilar membrane, 291–292
Frequency threshold curves
  ascending auditory pathways, 325
  auditory nerve cell receptive fields, 323
  auditory nerve fibers, 70, 318
Frequency tuning curves, 70, 318
FTC, *see* Frequency tuning curves
Functional rewiring, synaptic efficacy, 317
Fusiform gyrus
  face information processing, 416

G

Ganglion cells
  bipolar cells, 84
  ON center cells, 403–406
  classification, 401–403
  eye innervation, 55
  OFF center cells, 403–406
  receptive fields, 403–406
  receptor potential, 385
  retina anatomy, 377
  retinal, classification, 391
  retinal information processing, 387
  retinal neural network, 380
  visual pathway responses, 401
  X cells, 401–403
  Y cells, 401
Generator potential, 64
Geniculocortical pathway, 112
Geniculostriate pathway, 112
Glabrous skin
  coding of stimulus strength, 117
  sensory receptors, 42
Glomeruli
  olfactory bulb, 433–434
  olfactory innervation, 55
Glossopharyngeal nerve (CN IX)
  innervation, 194
  lingual branch, 432–433
  mouth and pharynx innervation, 428
  nucleus of the solitary tract input, 93
  taste pathways, 436
  taste receptor innervation, 55
Gracilis funiculus, 197
Gracilis nucleus, 92
Grating patterns, retina, 125
Gustation
  purpose, 427
  sensory cell innervation, 55–56
  Type II receptors, 55–56
Gustatory cortex, taste pathways, 437

H

Hair cells
  basilar membrane, 59–61
  as cochlear component, 279
  cochlear receptors, 47
  function, 281–283
  inner, *see* Inner hair cells
  innervation, 284
  outer, *see* Outer hair cells
  proper stimulus, 294
  receptor sensitivity, 67
  sensory transduction, 65
Hair follicle receptors
  as rapid adapting receptors, 223
  skin location, 191
  Type I receptor types, 46
Hairy skin sensory receptors, 42
Heat receptors
  function, 191–192
  innervation, 53
  somatic pain, 243
Hebb's principle, 168
Hemifacial spasm, 254
Hemispherical dominance, auditory
    system, 344
Heschel's gyrus, 94, 314
Hierarchical processing
  ascending auditory pathways, 305
  sensory information, 82
High route
  amygdala, 156–157
  arousal, 141
  auditory information, 352
  limbic system, 156
  nonclassical somatosensory system, 239
  nonperceptual effects of stimulation, 237
Hippocampal gyrus, 155
Hippocampus
  entorhinal cortex projection, 440
  olfactory bulb connections, 442
Horizontal cells
  eye innervation, 55
  photoreceptors to ganglion cells, 84
  retinal information processing, 386–387
  retinal neural network, 380

Hyperactivity, neural plasticity, 166
Hyperacusis, 171
Hyperalgesia, 171, 247
Hyperpathia, 171, 240, 258
Hypothalamus
amygdala connections, 158–159
circadian rhythms, 419
optic tract fibers, 389
stria terminalis, 97
thalamus projections, 217

I

IAS, *see* Intermediate acoustic stria
IC, *see* Inferior colliculus
ICC, *see* Central nucleus of the inferior colliculus
ICP, *see* Pericentral nucleus of the inferior colliculus
ICX, *see* External nucleus of the inferior colliculus
Idiopathic pain, 241
IHCs, *see* Inner hair cells
Impedance transformer, middle ear, 39, 56–57, 289
Inattention, 142–143
Inferior colliculus
amplitude-modulated sounds, 336
brachium, 92, 306, 313
central nucleus, 92, 311–312, 315, 348, 353–354, 357
classical ascending pathway anatomy, 306
commissure, 306, 311
dorsal cortex, 347
external nucleus, 96, 146, 159, 311, 347–348, 350, 353–354
function, 311–313
midbrain auditory nuclei, 158
pericentral nucleus, 351
rapidly changing frequency response, 337
synaptic transmission interruption, 91
Inferior glossopharyngeal ganglion, 433
Inhibitory postsynaptic potential, 66, 114
Inhibitory response areas, 232
Inner hair cells
basilar membrane, 61
cochlear function, 40
cochlear receptors, 48
function, 85
hair cell count, 279–280
innervation, 54
receptive field, 72
receptor selectivity, 70
role in cochlear function, 281–283
stereocilia deflection, 295
tectorial membrane, 279
Insular cortex
anterior portion, 217, 262
role in pain, 240
suprageniculate thalamic nucleus projection, 351
taste pathways, 437
Integration time, 11
Intermediate acoustic stria, 309
Intermediate olfactory stria, 440
Intermedius nerve, 433, 437
Interpretation of the sensory information, 152–153
Intralaminar nuclei, 213
Intralaminar nucleus, 96
IPSP, *see* Inhibitory postsynaptic potential
Itch, 262–263

J

Jacobson's organ, *see* Vomeronasal organ

K

K cells
ganglion cell classification, 401
lateral geniculate nucleus, 407
Kindling, neural plasticity, 168–169
Kinocilium
hair cells, 279
vestibular apparatus, 47, 282
Klüver–Bucy syndrome, 162
Konio cells
classification, 391, 393–394, 401
lateral geniculate nucleus, 408
Koniogeniculate pathway, 417

L

Labeled line hypothesis
spatial information, 127–128
taste receptor innervation, 448
Lateral funiculus, 197
Lateral geniculate nucleus of the thalamus
classical pathway fibers, 100
extrastriatal cortex, 420
konio cell pathways, 408

M and P pathways, 407–408, 410
optic nerve fiber projections, 416
pulvinar nuclei proximity, 102
retinogeniculocortical pathway, 388–389
stream segregation, 414
use in retinogeniculate pathway, 387
visual image fusion, 411
Lateral inhibition
ascending somatosensory pathways, 232–233
ganglion cell receptive fields, 404–406
receptive fields, 132
spatial contrast enhancement, 136–137
spatial resolution, 25–27
Lateral lemniscus
auditory information, 92
classical ascending pathway, 306
function, 311–313
Lateral nucleus of the amygdala
arousal, 141
nonperceptual effects of stimulation, 237
thalamic nuclei connections, 159
Lateral posterior nucleus of the thalamus, 418
Lateral superior olivary nucleus, 310, 315
Leaky integrators, 69
Lens, as eye part, 41, 377
LGN, *see* Lateral geniculate nucleus of the thalamus
Light adaptation, 383
Light reflexes, 420
Limbic system
allocortex connection, 440
definition, 83
dorsomedial thalamus connections, 217
ICC neural input, 312
nonperceptual effects of stimulation, 238
pain processing, 111, 262
pleasure perception, 162
projections from olfactory pathways, 427
structures, 155–156
visual activation, 420–421
Limbic structure connection, 440
Lingual branch, glossopharyngeal nerve, 432–433
LL, *see* Lateral lemniscus
Low-redundant speech, 342
Low route
amygdala, 156, 160
arousal, 141–142
auditory information, 352
nonperceptual effects of stimulation, 237
LSO, *see* Lateral superior olivary nucleus
LV, *see* Ventral division of the thalamus

M

Mach band, 136
MAF, *see* Minimal audible field
Magnocellular division of the medial geniculate body, 98
Mammillary nucleus, 101
Manubrium of the malleus, 39
MAP, *see* Minimal audible pressure
Masking
backward masking, 11
faint stimuli detection, 4
forward masking, 10
simultaneous masking, 10
unmasking, 81, 166, 169–171, 239
M cells
classification, 391–392, 394
image fusion, 396
lateral geniculate nucleus, 407–408
retinal ganglion cell classification, 401
MD, *see* Mediodorsal division of the thalamus
Mechanoreceptors
adaptations, 58
anteriolateral pathways, 210
classical somatosensory system, 221–225
dorsal horn organization, 209
function, 7
high sensitivity, 36
information processing, adaptation effects, 68
innervation, dermatomes, 194–196
rapid adapting mechanoreceptors, 58, 117, 193, 222, 233
sensory transduction, 65
Medial division of the medial geniculate body
fear reactions, 161
ICX and DC projections, 348
reciprocal connections, 353
uninterrupted connections, 352
projections to cerebral cortex, 216–217
response patterns, 355
synaptic transmission, 100–101
Medial geniculate body of the thalamus
amplitude-modulated sounds, 336
anatomy and function, 313
caudodorsal division, 351
classical ascending pathway anatomy, 306

Medial geniculate body of the thalamus (*continued*)
classical auditory pathway, 100
descending input, 315–316
dorsal division, 160
lateral ventral part, 313
magnocellular division, 98
medial division, 161, 348, 352–353
ovoid part, 313
pulvinar nuclei, 102
tonotopical organization, 324
ventral nucleus, 313
Medial lemniscus, 92, 197
Medial nucleus of the amygdala, 93, 156, 436
Medial superior olivary nucleus
as coincidence detectors, 343
Medial temporal visual cortex
association cortices, 414
information processing, 401
perception of movement studies, 114
pulvinar pathway, 418–419
stream segregation, 398, 414
Mediodorsal division of the thalamus, 101
Medulla oblongata, 92
Meissner's corpuscles, 46–47, 191, 223
Melanopsin, retina, 42
Merkel's discs
function comparison to other receptors, 223
object shape and skin indentation, 234
receptive fields, 231
Type I receptor types, 46–47
Mesencephalic reticular formation, 101–102
Mesenphalic tegmentum, 83
Mesial occipitotemporal visual region, 113
MGB, *see* Medial geniculate body of the thalamus
Microvilli
taste receptors, 51
Type II receptor, 47
Middle ear
anatomy, 39–40
anatomy and function, 278
as ear part, 37
as impedance transformer, 56–57
cochlear sound conduction, 288–291
Middle ear cavity, 278
Minimal audible field, 5
Minimal audible pressure, 5
Mitral cells
accessory olfactory bulb, 434
olfactory bulb, 85
olfactory innervation, 55
ML, *see* Medial lemniscus
MN, *see* Mammillary nucleus
Modulation transfer function, 333
Monosodium glutamate, 428, 443
Motor system sensory reflexes, 153–154
Movement detection
directional information, 138–140
MSO, *see* Medial superior olivary nucleus
MTF, *see* Modulation transfer function
MT/V5, *see* Medial temporal visual cortex

N

Near accommodation, eye function, 41
Neocortex
ascending sensory neural pathways, 86
cell density, 104
neuron connections, 102–103
Nervi nervorum, 253–254
Neural control of sensory processing, ascending pathways
arousal, 141–142
attention, 142–143
descending pathway control, 143–145
Neural discharges
modulation, 124
signal-to-noise ratio, 126
Neural maps
characteristics, 149–152
sound property maps, 340–341
Neural plasticity
ascending somatosensory system, 235–236
auditory pathways, 357–358
cause of disease, 170–172
dormant synapses, 166–167, 169
Hebb's principle, 168
kindling, 168–169
neuropathic pain, 259
nucleus basalis, 159
from deprivation and overstimulation, 359
phantom pain, 256
somatosensory cortex, 165–166
sound threshold and perception, 358–359
wide dynamic range neurons, 259

Neuropathic pain
abnormal sensations, 257–258
central pain pathway sensitization, 254–255
neural plasticity, 259
neurophysiologic basis, 258
pathophysiologic pain, 253–254
phantom pain, 255–256
signs, 256–257
TENS treatment, 260
Nociceptive pain, 241
dorsal horn organization, 209
innervation, 207–208
polymodal, innervation, 53–54
role in nonclassical somatosensory system, 238
sensitivity, 249
somatic pain, 243
threshold, 249
Node of the Ranvier, 65
Nodosa ganglion, 433
Nonclassical pathways
auditory system
acoustic middle ear reflex, 357
descending pathways, 305, 353
dorsal and medial thalamus, 355
dorsal cortex of inferior colliculus, 354
external nucleus of the inferior colliculus, 354
polysensory neurons, 354–355
subcortical auditory input to amygdala, 355–356
somatosensory system
anteriolateral pathways, 210–216, 239–240
descending pathways, 219–220
dorsal and medial thalamus, 216–217
dorsal horn, 208–210, 238–239
receptors, 206–208, 238
trigeminal pathway, 216
Nonperceptual effects
amygdala role, 237–238
arousal, 237
overview, 236–237
NST, *see* Nucleus of solitary tract
NTB, *see* Nucleus of the trapezoidal body
Nucleus accumbens, 415
Nucleus basalis
arousal function, 141, 159
electrical stimulation, 358
neural plasticity facilitation, 160
Nucleus cuneatus, 197
Nucleus gracilis, 197
Nucleus of the solitary tract, 93, 96, 436–437
Nucleus of the trapezoidal body, 310
Nucleus raphe magnus, 98
Nucleus sagulum, 351
Nucleus Z, 199

O

OAE, *see* Otoacoustic emission
Object information
Object properties of the stimuli, 116
OCB, *see* Olivocochlear bundle
Odorants, 428
Odor generalists, 444
Odor specialists, 446, 449
OFF center ganglion cells, 403–406
OHCs, *see* Outer hair cells
Olfactory bulb
accessory bulb, 156, 428, 434, 443, 447
circuitry, 93
innervation, 55
limbic system, 155
neural connections, 84–85
neural network, 433–434
projections, 440
Olfactory cells, 70
receptor cell adaptation, 445
Olfactory cortex, 440
Olfactory epithelium
nose, 432
Olfactory nerve (CN I)
olfactory innervation, 55–56
Olfactory pathways
central olfaction pathways, 440–443
nonclassical olfactory system, 88
projection to limbic structures, 427
sensory cell innervation, 55–56
Olfactory receptors
anatomy, 51
location and lifetime, 429–430
odorant sensitivity, 444–446
selectivity, 70
vomeronasal organ, 430–431
Olfactory tubercle
entorhinal cortex, 440
olfactory bulb connections, 442

Olivocochlear bundle
electrical stimulation, 346–347
hair cell innervation, 284
origination, 315
ON center ganglion cells, 403–406
Opercular cortex, 437
Opioids, 251
Optic chiasm, 92, 389
Optic nerve (CN II)
eye innervation, 55
organization, 389–390
receptive field, 72
retinal information processing, 385
retinal neural network, 380
Optic tract, 389–390
Organ of Corti
hair cells, 279
Ossicles, middle ear, 39, 278
Otoacoustic emission, 302–303
Ototoxic antibiotics, 301
Outer ear, 37, 277–278
Outer hair cells
amplitude compression, 304
cochlear receptors, 48
function, 85
hair cell count, 279–280
innervation, 54
motility, 301
role in cochlear function, 281–283, 300–301
tectorial membrane, relation to, 279
Oval window, 39
Ovoid part of the thalamus, 313

P

Pacinian corpuscles
activation, 231
characteristics, 223
frequency selectivity, 58
Type I receptor types, 46–47
PAF, *see* Posterior auditory fields
PAG, *see* Periaqueductal gray
Pain
acute pain
anterior cingular gyrus, 240
autonomic reactions, 155
factors affecting sensitivity, 246–251
information processing, 111
limbic system, 262
modulation by sensory input, 249–250
peripheral sensitization, 249
polymodal nociceptors, 54
referred pain, 251–252
skin mechanoreceptors, 238
somatic pain, 245
visceral pain, 251–252
neuropathic pain
abnormal sensations, 257–258
allodynia, 240
central pain pathway sensitization, 254–255
neurophysiologic basis, 258
pathophysiologic pain, 251, 253–254
phantom pain, 255–256
posterior complex, 240
role of thalamus, 261
signs, 256–257
somatosensory nervous system, modes of operation, 258–260
Pain receptors, sensitization, 247–249
Paired sense organ, directional information, 138–140
Paleospinal tract, 216
Parabrachial nucleus, taste pathways, 437
Parallel processing
classical ascending pathways, 110–111, 228
ensory information, 82
Paraventricular hypothalamic nucleus, 98
Pathophysiologic pain, characteristics, 253–254
PbN, *see* Parabrachial nucleus
P cells
classification, 391–392, 394
ganglion cell classification, 401
image fusion, 396
lateral geniculate nucleus, 407–408
Periaqueductal gray
descending activity, 98, 147
fear-related pain, 159
reticular nucleus input, 102
spinomesenphalic tract termination, 215
spinothalamic tract, 211–214
Pericentral nucleus of the inferior colliculus, 351
Periform cortex, 155, 440, 442
Periglomerular cells, 85, 434
Perilymph, 40, 279
Perilymphatic space, 298
Perirhinal cortex, 437
Permanent threshold shift, 8
Petrosal ganglion, *see* Inferior glossopharyngeal

ganglion
Phantom limb symptoms, 256
Phantom perceptions, 255
Phase locking
auditory nerve fibers, 328, 332
spatial discrimination, 135
Pheromones
definition, 427
function in animals, 448
human response, 431
vomeronasal response, 84, 161, 427, 442–443, 446–447, 449
Phonophobia, 162, 171, 355–356
Photopic vision
light-adapted eye, 6
Photopigment
activation, 381
photoreceptors, 50, 378
Photoreceptors
adaptation, 382–383
color vision, 383–385
excitation steps, 381
eye anatomy, 41–42, 377–380
high sensitivity, 36
information processing, 381
innervation, 55
receptor potential, 385
selectivity, 70
Type II receptors, 47
PIN, *see* Posterior intralaminar nucleus
Pincus domes, 47, 191
Pineal body, 419
Place hypothesis, 293, 345
Pleasure perception, 162–163
PO, *see* Posterior complex of the thalamus
Polymodal nociceptors
innervation, 53–54
role in nonclassical somatosensory system, 238
somatic pain, 243
Polysensory systems
cerebral cortex, 107
neuron physiology, 354–355
nonclassical pathways, 96, 145
Population hypothesis
single nerve cells, 149
spatial information, 127–129
Postcentral gyrus, 204
Posteriomedial cortical nucleus of the amygdala, 94, 443
Posterior auditory fields, 108
Posterior complex of the thalamus
lateral portion, 348
role in pain, 240
spinothalamic fiber termination, 96, 213
Posterior ectosylvian field, 314
Posterior intralaminar nucleus, 161
Posteriorventral cochlear nucleus, 92, 309
Postsynaptic potentials, 66, 119–120, 385
Power functions, 19, 117
Precedence effect
demonstration, 28
directional hearing, 288
physiological basis, 343–344
Prefrontal cortex, 158
Pretectal nucleus
eye muscle mediation, 420
nonclassical visual pathway, 387
optic nerve fiber termination, 416
optic tract fibers, 96, 389
Primary auditory cortex
classical ascending pathway anatomy, 306
as part of auditory cortex, 314
tuning curves, 324–325
ventral MGB projections, 313
Primary auditory cortical fields, 108
Primary sensory cortex
classical ascending pathways, 82
Primary somatosensory cortex
connections, 202–203
somatotopic organization, 229–230
Primary visual cortex
information processing, 401
optic nerve and tract, 92
pulvinar pathway, 419
Propagated neural activity, 65
Pruning
anatomical connections, 87
neural plasticity, 171
Pruritus, *see* Itch
PSPs, *see* Postsynaptic potentials
PTS, *see* Permanent threshold shift
Pulvinar division of the thalamus
connections, 417
extrastriatal cortex, 102
motion processing, 419
optic tract fibers, 389, 418
visual cortex organization, 396

Pupil
adaptation, 382
dilation mediation, 418
eye anatomy, 40, 377
PVCN, *see* Posteriorventral cochlear nucleus

R

RA, *see* Rapid adapting mechanoreceptors
Radicular pain, 253
Rapid adapting mechanoreceptors
classical somatosensory system, 222
coding of stimulus strength, 117
conducting media, 193
object shape discrimination, 233
physical stimulus modification, 58
Rate hypothesis, 148
RE, *see* Thalamic reticular nucleus
Receptive fields
auditory nerve fibers, 318–323
classical ascending pathways, 230–232
color vision, 384
definition, 71–72
ganglion cells, 403–406
neural plasticity, 168
spatial information, 129
spatial resolution, 25
transformation, 129–133
Receptor information processing, 68
Receptor potentials
definition, 297
photoreceptors, 381, 385
sensory transduction, 64–65, 67
Receptors
capsaicin, 238
cold receptors, 53, 71, 191–192
cool receptors, 52–53, 191–192, 226
hair follicle receptors, 46, 191, 223
hairy skin sensory receptors, 42
heat receptors, 53, 191–192, 243
mechanoreceptors, 7, 36, 58, 65, 68, 117, 191, 193–196, 209–210, 221–225, 233
NMDA receptors, 255, 258
olfactory receptors, 51, 70, 429–431, 433, 444–446
pain receptors, 247–249
photoreceptors, 36, 41–42, 44, 50, 55, 70, 84, 377–385
polymodal receptors, 43, 238, 243
selectivity, 70–71
sensitivity, 67–68
sensory receptor innervation, 84–85
slow adapting receptors, 58, 193, 222, 233
somatosensory system receptors, 190–192
taste receptors, 42, 51
thermoreceptors, 44, 47, 71, 191–192, 225–226, 428
Type I receptors, 45–47, 52–54, 65, 67, 238
Type II receptors, 46–51, 54–56, 65–67, 84, 281–283
visceral pain receptors, 251
warmth receptors, 52–53, 71, 191–192, 226
Reciprocal connections, 82, 97
Referred pain
occurrence, 251–252
somatic pain, 243
thalamic neuron input, 213
Reflex sympathetic dystrophy, 155, 249
Reissner's membrane, 40
Reproductive systems, pheromone functions, 161, 448
Reticulospinal tract, 141
Retina
anatomy, 377
eye anatomy, 40, 42
information processing, 385–387
photoreceptors, 50
receptive field, 406
visual information processing, 401
Retinal ganglion cells, 391
Retinal neural network, 380
Retinogeniculocortical pathways, 88, 387–389
Retinotectal pathway, 112
Retinotopic organization, 394
Rhodopsin
light absorption, 380–381
metarhodopsin II, 381
photoreceptors, 50, 378
Rod pathway
connections, 403
retinal information processing, 385
Rods
dark adapted state, 381
retina anatomy, 377
rhodopsin light absorption, 380–381
Round window, 39
RSD, *see* Reflex sympathetic dystrophy
Ruffini endings, 46–47, 191, 223

S

Salt hunger, 428
SA receptors, *see* Slow adapting receptors
SC, *see* Superior colliculus
Scala media, 40, 279
Scala tympani, 40, 279
Scala vestibuli, 40, 278–279
SCN, *see* Suprachiasmic nucleus
Scotoma, 390
Scotopic vision, 8, 378
Secondary auditory cortex, 98, 108, 314
Secondary somatosensory cortex
  acute pain, 262
  connections, 204–205
  somatotopic organization, 230
Second common pathway, 103
Sensitization
  central pain pathways, 254–255
  dorsal horn cells, 259
  high-threshold receptors, 259
  pain receptors, 247–249
  somatic pain, 249
Sensory cell innervation
  Type I receptors, 52–54
  Type II receptors, 54–56
Sensory nervous system
  ascending classical pathways
    anatomy, 217–218
    developmental aspects, 146–147
    organization, 88–94
    parallel processing, 110–111
    stream segregation, 111–116
    thalamus, 100
  descending classical pathways
    control of ascending neural activity, 143–145
  nonclassical ascending pathways
    information processing, 145–147
    organization, 94–97
    thalamus, 100
  nonclassical descending pathways
    characteristics, 98
    information processing, 147
Sensory receptor innervation, 84–85
Sensory transduction
  cochlear hair cells, 296–299
  general principles, 64–67
  receptor sensitivity, 67–68
Septal area, 83
Serial processing, 203
SG, *see* Nucleus sagulum; Suprageniculate nucleus
Shape discrimination, 233–234
SI, *see* Primary somatosensory cortex
Signal detection theory, 13–15
Signal-to-noise ratio
  neural discharges, 126
  receptor sensitivity, 67
  synaptic transmission, 121
SII, *see* Secondary somatosensory cortex
Silent nociceptors, 206
Silent synapses, 235
Simultaneous masking, 10
Skin mechanoreceptors, 191
Skin stimulation, 232–233
Slow adapting receptors
  classical somatosensory system, 222
  conducting media, 193
  object shape discrimination, 233
  physical stimulus modification, 58
SMP, *see* Sympathetic maintained pain
SNR, *see* Signal-to-noise ratio
SOAE, *see* Spontaneous otoacoustic emissions
SOC, *see* Superior olivary complex
Somatic pain
  acute pain, 243–246
  causes, 243
  central sensitization, 249
  chronic pain, 246
  factors affecting sensitivity, 246–251
  modulation by sensory input, 249–250
  referred pain, 251–252
  visceral pain, 251–252
Somatosensory cortices, *see* Primary somatosensory cortex; Secondary somatosensory cortex
Somatosensory system
  classical ascending system
  descending pathways, 217–218
    components, 196–201
    inhibition–excitation interplay, 232
    medial lemniscus, 202
    neural plasticity, 235–236
    object shape discrimination, 233–234
    receptive fields, 230–232
    receptors, 190–192
    skin stimulation, 232–233
    somatosensory cortices, 202–205
    somatosensory thalamic nuclei, 202

Somatosensory system (*continued*)
somatotopic organization, 228–230
stream segregation, 235
thermoreceptors, 225–226
nonclassical descending function, 241
nonclassical somatosensory system
anteriolateral pathways, 210–216, 339–340
dorsal and medial thalamus, 216–217
dorsal horn, 208–210
receptors, 206–208
trigeminal pathway, 216
Somatosensory system receptors
capsaicin, 238
cold receptors, 53, 71, 191–192
cool receptors, 52–53, 191–192, 226
hair follicle receptors, 46, 191, 223
hairy skin sensory receptors, 42
heat receptors, 53, 191–192, 243
mechanoreceptors, 7, 36, 58, 65, 68, 117, 191, 193–196, 209–210, 221–225, 233
pain receptors, 247–249
polymodal receptors, 43, 238, 243
skin mechanoreceptors, 191
slow adapting receptors, 58, 193, 222, 233
somatosensory system receptors, 190–192
visceral pain receptors, 251
warmth receptors, 52–53, 71, 191–192, 226
Somatosensory thalamic nuclei, 202
Somatotopic organization, 228–230
Sound conduction
head and outer ear, 286–288
middle ear, 288–291
Sound intensity, auditory nerve fibers
intensity and frequency changes, 333–337
rapid frequency changes, 337–339
Sound property maps, flying bats, 340–341
Spatial averaging, 120
Spatial contrast, 136–137
Spatial discrimination, 133–135, 137
Spatial information processing
contrast enhancement, 136–137
labeled line hypothesis, 127–128
movement detection, 137–140
neural representation, 129
population hypothesis, 128–129
receptive field transformation, 129–133
spatial discrimination, 133–135
stream segregation, 111–112
Spatial integration
characteristics, 12–13
faint stimuli detection, 4
receptor sensitivity, 67
sensory information coding, 231
Speech discrimination, 339–340
Speech perception, 115
Spinal dermatomes, 194
Spinal ganglion, 284
Spinal nerves, 84
Spinal nucleus of the trigeminal tract, 439
Spinomesencephalic tract, 102, 155, 215
Spinoreticular tract, 94, 214–215
Spinothalamic pathway, 261
Spinothalamic tract, 96, 210–217
Spontaneous otoacoustic emissions, 302
Stapedius muscle, 278
acoustic middle ear reflex, 352, 357
Startle response
acoustic reflex, 352
definition, 154
mediation, 313
Stereocilia
anatomy, 282–283
cochlear hair cell sensory conduction, 296–297
cochlear receptors, 47–48
proper hair cell stimulus, 294
sensory transduction, 65
Type II receptor region, 47
vestibular hair cells, 282
Stereophonic sound, 140
Stereoscopic vision, 27, 139
Stimulus strength
coding, 116–118
perception, 19
Stream segregation
anatomical basis in visual system, 398–399
ascending auditory pathways, 305
association cortices, 414–416
face identification, 113
fear reactions, 161
primary cortex connections, 397
spatial and temporal components, 113–114
"what" and "where" information, 112
Stria of Monaco, 309
Striate cortex, 92, 94, 394, 398
Stria terminalis, 94, 96, 159
Substance P, 250
Substantia gelatinosa, 198, 208

Superior cervical ganglion, 418
Superior colliculus
computational maps, 152
inferior colliculus connections, 313
movement detection, 140
nonclassical visual pathway, 88, 387
optic nerve fiber termination, 416
optic tract fibers, 389
pulvinar nuclei, 102
visual reflexes, 413, 417–418, 420
Superior laryngeal nerve, 433
Superior olivary complex
acoustic stapedius reflex, 352
anatomy and function, 309–311
classical ascending pathway anatomy, 306
directional hearing information, 119
Superior temporal gyrus, 114
Supertemporal plane, 344
Suprachiasmic nucleus
circadian rhythms, 388, 419
optic tract fibers, 389
Suprageniculate nucleus, 161, 350–351
Surround inhibition, 132
Sylvian fissure, 204
Sympathetic maintained pain, 155, 249
Sympathetic nervous system, pain role, 249
Synapse-like structure, Type II receptor cell sensory transduction, 66
Synaptic efficacy, neural plasticity, 166, 168
Synaptic jitter, 119, 330

T

Taste
delay of response, 63
taste buds
taste receptors, 51
Taste generalists, 444
Taste pathways, 88, 96, 436–440
tongue receptors, 428
Type I receptor innervation, 54
Taste receptors
Tectal pathway, 88
Tectorial membrane
cochlear receptors, 48
embedded stereocilia, 283
inner hair cells, 61
outer hair cells, 279
Temperature receptors, *see* Thermoreceptors
Temporal code of frequency, 318
Temporal hypothesis
auditory nerve fiber discharge, 318
auditory nerve sound waveform coding, 328
frequency discrimination, 339
Temporal integration
characteristics, 11–12
faint stimuli detection, 4
pain, 247
temporal summation, 166
Temporal resolution, 20, 22, 24
receptor information processing, 68–69
Temporary threshold shift, 8
TENS, *see* Transdermal electrical nerve stimulation
Tensor tympani
acoustic reflex, 291, 357
middle ear function, 39
TEOAE, *see* Transient evoked otoacoustic emissions
Thalamic nucleus
anterior region, 101
medial geniculate body, 105
somatosensory components, 202
Thalamic reticular nucleus
arousal, 141
sensory input modulation, 351
thalamus, 101
ventral MGB input, 313
Thalamotomy, 261
Thalamus
cerebral cortex pathway, 99–100
descending pathways, 143
dorsal division, 100–101, 111, 216–217, 237, 355
lateral geniculate nucleus, 92, 100, 102, 387–389, 391, 394, 396, 406–408, 410–411, 414, 416, 420
lateral posterior nucleus, 418
medial division, 100–101, 216–217, 355
medial geniculate body, 98, 100, 102, 105, 143
nonclassical visual system, 88
posterior thalamus, 348
projections to hypothalamus and amygdala, 217
pulvinar division, 88, 96, 102, 387, 389, 396, 417–419
reticular nuclei, 101–102
role in pain, 261
somatotopic organization, 228

Thalamus (*continued*)
spinothalamic tract, 212–213
ventral division, 100, 102, 211
ventral posterior lateral nucleus, 100
ventral posterior medial nucleus, 92–93, 100, 261, 437
ventrobasal nucleus, 92, 202, 261
ventrocaudal nucleus, 261
ventroposterior lateral nucleus, 202
ventroposterior medial nucleus, 202
Thermoreceptors
classical somatosensory system, 225–226
food flavor, 428
selectivity, 71
Type I receptor types, 47
Three-chromatic theory, 383
Threshold of hearing, 4–5, 17–18
Time pattern, physical stimuli, 118–122
Tinnitus
affective symptoms, 355
dorsal cochlear nucleus, 355
tinnitus retraining, 172
TN, *see* Trigeminal nucleus
Tongue, 431–432
Transdermal electrical neural stimulation
angina pectoris relief, 252
neuropathic pain treatment, 260
pain treatment, 250
Transient evoked otoacoustic emissions, 302
Transmitter substance, Type II receptor cells, 66
Trapezoidal body, 309
Traveling wave, cochlea, 292
Trigeminal nerve (CN V)
sensory portion, 216
innervation, 194
Trigeminal neuralgia, 254
Trigeminal nucleus, 92, 197, 216
Trigeminal pathway anatomy, 216
Trigeminal sensory nucleus, 200
TTS, *see* Temporary threshold shift
Two-point discrimination
spatial discrimination, 133
spatial resolution, 24
Two-tone inhibition, 132, 322
Tympanic membrane, 39, 278
Type I receptors
mechanoreceptors, 36, 68
olfactory receptors, 51
Type II receptors
cochlear receptors, 47–49
photoreceptors, 50
role in cochlear function, 281–283
sensory transduction, 65–66
taste receptors, 51

U

Umami, *see* Monosodium glutamate
Uncrossed olivocochlear bundle, 315, 347
Uncus, 93, 155, 440
Unmasking
dormant synapses, 81, 166, 169–171
synaptic neural plasticity, 239
UOCB, *see* Uncrossed olivocochlear bundle

V

V1, *see* Primary visual cortex
Vagus nerve (CN X)
amygdala connection target, 159
superior laryngeal nerve, 433
taste pathways, 436
VAS, *see* Ventral acoustic stria
Vascular layer, eye anatomy, 40
VB, *see* Ventrobasal nucleus of the thalamus
Velocity detectors, 58, 222
Ventral acoustic stria, 309
Ventral cochlear nucleus, 352
Ventral division of the thalamus, 102, 211
Ventral nucleus of the lateral lemniscus, 306, 311
Ventral nucleus of the medial geniculate body, 313
Ventral posterior lateral nucleus of the thalamus, 100
Ventral posterior medial nucleus of the thalamus
body senses, 92
nucleus of the solitary tract, 93
role in pain, 261
taste pathways, 100, 437
Ventrobasal nucleus of the thalamus, 92, 202, 261
Ventrocaudal nucleus of the thalamus, 261
Ventroposterior lateral nucleus of the thalamus, 202
Ventroposterior medial nucleus of the thalamus, 202
Vestibular apparatus, 47
Visceral brain, 83
Visceral pain
occurrence, 251–252
receptors, 251
role of thalamus, 261

Visual cortex
organization, 394
Visual pathways
classical ascending
ganglion cell responses, 401
lateral geniculate nucleus, 390–394
optic nerve, 389–390
optic tract, 389–390
overview, 388–389
primary cortex connections, 396–398
stream segregation, 398–399
visual cortex, 394–396
nonclassical
optic tract fibers, 389
pulvinar pathway, 418
visual reflexes, 417–418
Vitrous body, 41
VNLL, *see* Ventral nucleus of the lateral lemniscus
Vomeronasal system
amygdala, 448
in humans, 431
odor response, 446–448
pathways
pheromone detection, 449
pheromone response, 84, 161, 442–443, 427
projections, 427–428
receptor innervation, 434
vomeronasal nerve, 434
vomeronasal organ
VPL, *see* Ventral posterior lateral nucleus of the thalamus
VPM, *see* Ventral posterior medial nucleus of the thalamus

W

Warmth receptors
classical somatosensory system, 226
function, 191–192
innervation, 52–53
selectivity, 71
WDR neurons, *see* Wide dynamic range neurons
"What" information
acute pain, 244, 261
auditory system, 114, 344–345
flying bat, 115–116
stream segregation, 235, 414
visual system, 112, 114
"Where" information
acute pain, 244, 261–262
association cortices stream segregation, 414
auditory nerve fibers, 319
auditory system, 114, 344–345
cochlear fluid vibration conduction, 293
flying bat, 115–116
stream segregation, 235
visual system, 112, 114
Wide dynamic range neurons
dorsal horn organization, 209–210
input to thalamus, 214
long-term potentiation, 255
nonclassical pathways, 145
role in pain, 259–260
spinothalamic tract, 211–212
Wind-up phenomenon, 258

X

X ganglion cells
classification, 401
conduction velocity, 403
input from receptors, 402

Y

Y ganglion cells
classification, 401
projections, 403

DATE DUE
